All These Evil Things

A Canadian Story of Gun Control & Government Abuse

" 'All these evil things come from
 within. It is *these* things that
 make a man unclean.' "

-- Matthew 15: 19-20

by Yogi Shan

Copyright © 2017 by Yogi Shan
All Rights Reserved

ISBN-13: 978-1-97965-304-6
ISBN-10: 1-97965-304-6

Comments, inquiries, insults, and where to find this book, to: yogishan2000@yahoo.ca

> Dedicated to my children, Oz and "A.C.", who will perhaps now understand that their father was not some crazed, wild-eyed gun-nut who loved caressing and playing with his unloaded Ak-47 at home, and babbled endlessly about freedom, corruption, revolution, and other strange things during his frequent rants.
>
> (Begrudgingly, I acknowledge AC's mother, who I dubbed with the moniker, "The Bordeaux Boomerang". And to whom I ask, "What should I feel about your faithless kiss?").
>
> To my cherished children, I have one wish: may you live, learn, and love with less pain than I did. You are the only two things of eternal value that I ever produced. And it was only by accident – it was only their mothers that saw the value in me that I was too beaten down to even dream of.
>
> This book is also fondly dedicated to former Liberal Party, Canadian Attorney-General, Allan Rock, who set the standard for "government jackass" for decades to come.

Acknowledgments
===============

I hereby acknowledge the invaluable help of the St. Denis Center, the Cornwall and Ottawa Public Libraries, and especially the four wonderful girls in the OPL Inter-Library Loans (ILL) Department, and their assistance in tirelessly filling all my strange, and bizarre ILL book and magazine article requests over the years.

I thank Lorna, for fucking me for the first and last time, the night in August 2007, mere hours before I was busted and imprisoned for seven months. The easiest excuse I've ever had for dumping a chick after a one-night stand!

From the bottom of my heart, I would like to thank the Quebec Provincial Court system, which throughout my progressing from arrest processing, to bail, to pretrial, and trial, treated me with total politeness, respect, dignity, mercy, and humanity. God bless them for being true to Justice. "_Vive le Quèbec, libre!_" ["Long live a free Quebec!"]

As opposed to the judicial system of the Government of the Province of Ontario, who treats their victims with constant abuse, disrespect, and routine humiliation, making a mockery of every precept of fairness and justice known to civilized man, to our country's eternal dishonor and disgrace.

To hell with you, Ontario! To hell with you, Canada!

And most of all, I thank my poor, long-suffering mother, JCS, for everything I have put her through in my star-crossed life. And I'm also sorry, though I always did what I felt was right. I love you...

And finally, I have to thank JLS, my ex-wife, and mother of my beloved son, for accepting my collect calls from jail, and unquestioningly finding the answers to the most arcane and bizarre questions, partial quote fragments, and technically complicated data requests, as I walked her through Internet Web sites to find the answers I needed for the factual portions of this book.

Prison Diary "-1":   A Note to the Reader
================

> "Did they get you to trade, Your heroes for ghosts?
> Hot ashes for trees? Hot air for a cool breeze?
> Cold comfort for change?
>
> Did you exchange, A walk-on part in a war,
> For a lead role in a cage?"
>
> -- "Wish You Were Here" (1975)
> Pink Floyd

> "[H]istory was a fable agreed upon..."
>
> -- "Transition" (1927)
> Will Durant

> "Alone.
>
> Listless.
>
> Breakfast table in an otherwise empty room."

Though this began as an actual prison diary while I was a "guest of the state", it should become fairly obvious that I have substantially edited my rough, pencil "chicken-scratchings", added to them, and polished them as perfectly as possible, when I got out and got access to my personal library, the Internet, used book and thrift stores, the public library, and other information and Truth sources.

New chapters about my "ancient years", long before 2007, the echoes of which still reverberate in my memory and personality, most deep within, most shallow and near the surface of my psyche, which began quaking soon after I happily quit alcohol on 19 January 2006.

They were severe, extremely disturbing, unexpected, and probably well-deserved. It was no doubt a debt that had not been forgotten and was still due.

New jokes arose from the almost constant stream of humor that emerged from my heavily medicated, post-traumatic stress disordered (PTSD) mental state. Many quotes -- of which I am quite proud -- were discovered and uncovered, and provide a scaffolding for the ideas enunciated in each chapter. Hopefully they don't distract and irritate too much, because they are highly relevant, if perhaps sometimes arcane.

New chapters arose from my readings on PTSD, and a synthesis of my former political beliefs and belief system with a new – or, actually, revived – Christian belief system.

In the words of Henry David Thoreau -- writer, pacifist, civil disobedient and fellow jail-bird -- "I trust that all good men will conspire."

* * * *

"Revolutionary movements do not spread
by contamination but by *resonance*."

-- "The Coming Insurrection" (2007)
The Invisible Committee

"The world is changed.

I feel it in the water. I feel it in the earth. I smell it
in the air.

Much that once was, is lost. For [there are] none who
now live, who remember it."

-- opening narration
"The Lord of the Rings:
The Fellowship of the Rings" (2001)

Table of Contents

"I've seen things you people **wouldn't** believe:
Attack ships on fire, off the shoulder of Orion.
I watched C-beams glitter in the dark,
Near the Tannhauser Gate.

"All those moments will be lost in time,
Like tears in rain."

-- Roy Batty, android
"Blade Runner" (1982)
Hampton Francher, David Peoples

"Just a small town girl,
Living in a lonely world,
She took the Midnight train,
Going anywhere...

"Just a city boy,
Born and raised in South Detroit.
He took the Midnight train,
Going anywhere..."

-- "Don't Stop Believin' " (1981)
Journey

Prison & Legal System Events

Beginnings

0. Context
1. A Reality Show for Real
2. Round-Up the Unusual Suspects
3. A Cheap -- I mean -- Free Press
4. And Behind Door Number Two
5. A Bullet With My Name On It
6. The Things I Left Behind
18. "The Investigation of a Citizen Above Suspicion"
19. "Bad Moon Rising"

The Ideology of Some Really Bad Days

22.	Torture of an Innocent by His Own Government
32.	The Way the Wind was Blowing
33.	Pantomime Justice
34.	The Banality of Bureaucracy
36.	"Getting Medieval" on My Ass
37.	A Witch-Hunt and a Cover-Up
47.	The Cult of the Empty Gesture

In for the Long Haul

20.	"One Day at a Time"
21.	"In the Belly of the Beast"
24.	Writer's Retreat
28.	Ode to the Sacred Bean
35.	The Way of the Fist
45.	Jail-House Triptych
46.	The Range Is **HOT**!

Wind-Up

73.	"Remember My Chains"
74.	Catching a Break, Breaking a Catch
75.	Who Mourns for Odysseus?
76.	"Out Where the Buses Don't Run"
77.	Rock on, Dude! Liberation Day
78.	Fighting the Good Fight
85.	"It's Not Over 'Till the Fat Lady Sings."
86.	Home Is Where the Hurt Is
87.	"The Pursuit of Happiness", and All That
88.	Endgame: The "White Rose" Lives!
89.	Epilogue

"Like Tears in Rain": Humorous Interludes in a Dark Time
===

9.	Why Are These Men Smiling?
25.	Prisoner of the Heart
26.	"The Hearts Filthy Lesson"
27.	No Laughing Matter
31.	The Lawyer Game
51.	Urban Guerrilla Bedtime Stories

Guns and Me, Me and My Guns
=======================

7.	A Prisoner of Politics
10.	Small Arms and the Man
11.	I Love the Smell of Cordite in the Morning

12. "The Prophet Armed"
13. "Isn't It Ironic?"
14. Is It Possible That **We're** the "Bad Guys"?
15. Armed Right vs. Armed Left

Jesus and Me
===========

23. A Friend in High Places
30. The "Commie from Mars" says _Via con Dios_
61. Damn Papists

Me, Myself, and Why
================

8. Confessions of a Fatalist
16. "Pig!"
17. Witness
29. The World According to Yogi
48. The Cult of the Lie
38. "If I Had a Rocket Launcher"
49. Huh? Eh? Duh? Er -- What?
50. A Po-Mo Paradigm
52. My Pirate Heart
53. "It's a Kinda Magic!"
56. Le Chatelier's Warning
57. The Ghost in the Machine
58. No Warning: Catastrophic Disassembly
59. Pol Pot & Me
60. The Real Ronald Reagan
62. Capitalism Deconstructed
63. Lying Liars, and their Damned Lies
64. "Unusually Harsh"
72. Nothing but a Dreamer
79. Heroes of the First World -- Disciples of Satan
81. "It's Not Easy Being Green"
82. White – I Mean Might – Makes Right
83. "You hypocrites!": The Real Canada
84. Conspiracy of the Individual

Echoes of an Earlier "Dark Time"
========================

39. Howl of a Lone(ly) Wolf
40. "Je Me Souviens" -- I Remember
41. A Knight of the Long Knives
42. An Enemy of the State
43. I, Ronin
44. You Bet Your Life -- Tales of a Gambling Man

You're What? An "Expert" in...9mm Blow Back, Fully Automatic Small Arms Design?
==

54. "Rules of the Game"
55. Servants of the Game
65. "Death's Twilight Kingdom"
66. "The Beauty of Our Weapons"
67. The Beauty of **Their** Weapons
68. "Things Fall Apart"
69. The Zen of Napalm
70. Confessions of a Weapons Expert
71. Illegal!
80. "A Terrible Beauty"

Appendix A: Chronology of Legal Events
Appendix B: The Canon
Appendix C: "Lies, Damned Lies, and Statistics"
Appendix D: Selected Bibliography

"Now they're saying I shouldn't think stuff like this. They're saying something is wrong with me -- that I should be ashamed.

Well, I'm sick of being ashamed. Aren't you? I don't mind being dejected and rejected, but I'm not gonna be ashamed about it.

At least the pain is real. You look around and you see nothing is real – but the pain is real."

-- Hard Harry Hardon
"Pump Up the Volume" (1990)

"[Secrecy] nourishes the worst excesses of power."

-- "Prying Eyes" (2012)
"New Yorker" magazine
Trevor Paglen

Prison Diary 0:					Context
						======

> "To see! To see! – this is the craving...of the rest of blind humanity. To have his path made clear for him is the aspiration of every human being in our beclouded and tempestuous existence."
>
> -- "The Mirror of the Sea" (1906)
> Joseph Conrad

> "Since then, at an uncertain hour,
> That agony returns:
> And till my ghastly tale is told,
> This heart within me burns."
>
> -- "The Rime of the Ancient Mariner" (1798)
> Samuel Taylor Coleridge

I am a scientist and an engineer, by descent, by upbringing, by education, by profession, by aptitude, and most of all, by instinct.

We are in a time of darkness.

I read the phrase "dark time" in the title of a book by the great philosopher, Hannah Arendt (who is most well-known for coining the phrase about the Nazis, "the banality of evil"). Let's hear it, so eloquently said, from Hannah, herself:

> "[T]he words "dark time" [come] from [Berthold] Brecht's famous poem 'To Prosperity', which mentions the disorder and the hunger, the massacres and the slaughterers, the outrage over injustice and the despair ' when there was only wrong and no outrage,' the legitimate hatred that makes you ugly nevertheless, the well-founded wrath that makes the voice grow hoarse.
>
> All this was real enough as it took place in public; for until the very moment when catastrophe overtook everything and everybody, it was covered up not by realities but by...official representatives who...explained away unpleasant facts and justified concerns."
>
> -- "Men in Dark Times" (1955)
> Hannah Arendt

* * * *

Jesus Christ, two millennia ago, also talked about the dark times around us. Satan was called the "Prince of Darkness". JC was, amongst other things, called the "Prince of Peace", but for good reasons he identified Satan as the "Prince of Lies". (And whether you believe in God, Christ, religion – organized or not – or Satan, one certainly cannot deny the existence of Good and Evil in this human world of ours.)

Enter the Prince of Pain, a certain Mr. Yogi Shan.

He was Canadian-born, 48, half-white/half-Tamil Sri Lankan, light brown-skinned, and the second son of immigrant parents. He was, like his parents, upper middle class, well-educated – a computer engineer with 13 years experience working at Nortel Networks, an international telecommunications computer hardware and software giant – and had no criminal record.

Yogi was considered highly intelligent and well-spoken, but a little -- how should I characterize it -- "eccentric". He read unusual books on exotic subjects, he collected "specialized" guns as a hobby, he had strange ideas, and radical political opinions. And he had a number of unusual – to say the least – "intellectual" hobbies.

I say "intellectual" hobbies, because these were areas that he didn't practice physically, but rather read about on **everything** he could get his hands on – public and university library books, purchased books, periodicals, patents, government reports (even including filing FOIA (Freedom of Information Act) requests to the American government) – on the subject matter of his current interest. Collecting, and storing away the information he found and thought worth keeping.

Explosives, organophosphorus nerve agents ("nerve gas"), nuclear weapon design, clandestine, illicit drug manufacturing and upper-level drug trafficking, and extremist politics and the left-wing "armed struggle", were five such examples of these bizarre "hobbies" of his.

His outrageous sense of humor was well known, if not always well appreciated, as he struggled with the usual vicissitudes of life, with a few extra, in August of 2007 in his hometown of Ottawa, Canada.

Jesus, an unschooled Jew born in Judea – an uninspiring provincial back-water on the out-skirts of the Roman Empire – had a large following. But he also had many skeptics, detractors, and outright mortal enemies.

After three years of controversy, and being chased around the Galilee by an assortment of mobs and bearers of ill-will, Jesus was finally betrayed and arrested by the Jewish religious authorities. They convicted Jesus of heresy, and he was passed to the Romans and executed, ostensibly as an Enemy of the State, though really as a political expedient to placate the local Jewish theological hierarchy that kept a lid, mostly, on the somewhat volatile local populace. The year was about 33 A.D. or so.

Jesus had a number of aliases, including "the Prince of Peace". It is in the Old Testament chapter, Isaiah (a prophet who lived around 600 B.C.), that Jesus Christ's coming was prophesied, and he is identified as the "Prince of Peace".

Prophet, Messiah, Savior, Son of God, King, Lord? I thought that "Prince of Peace" was the coolest title of all (with "Lamb of God" coming in a distant second).

General, President, Prime Minister, Field Marshal, Commander, Maximum Leader, Supreme Leader are some of **our** titles of important people.

Fuck that noise. "Prince of Peace" is the best moniker of all. (Actually "Supreme Leader" sounds kinda cool, too!)

[Before stumbling into it in Isaiah, the last Old Testament chapter, all these years I had thought "Prince of Peace" was from the lyrics of Beethoven's Ninth Symphony, the "Ode to Joy", a magical tune that I loved as a pre-adolescent.]

The Devil has a number of aliases too – including Satan, Lucifer, Beelzebub, and the "Prince of Darkness". But it was Christ, as reported in one of the four Gospels of the Christian New Testament, who gave the Devil a more specific name – the Prince of Lies – which identified **exactly** how the Devil influenced, manipulated, and directed men, as well as keeping evil men in power.

And this is not widely known, or appreciated.

Lies were, and **are**, the way evil attains and maintains its power. Without lies, and people willing to spout them, Jesus told us clearly that evil cannot win, survive, or even take root. It is **not** money that is the root of all evil. Jesus clearly states that **lies** are the root of all evil. Money is just the fertilizer...

By pronouncing Satan as the Prince of Lies, Jesus was clearly trying to send a message that would be remembered.

But unfortunately, it wasn't.

Jesus wanted people to know Satan's _modus operandi_, and thus identify and be ready to resist and fight the fount of all evil, the root of all evil, and the cause of all evil within the human race: the lie.

The second most important theme of Jesus' teachings (other than "Love thy neighbor as thyself") is hypocrisy. The lie expressed as a belief or action, that provides a personal advantage or increased prosperity. And that becomes a lifestyle, to the point of one even lying to oneself – usually subconsciously – believing your own lies to be the truth.

Hypocrisy is the final nail in the coffin of personal morality.

And what has been conveniently downplayed, if not forgotten, by our so-called Christian leaders and their religious hierarchy, and the religious and the pious, is that those in modern day power (including their officials, lawyers, and bureaucrats who administer it, large corporations, and the rich), are analogous to the Pharisees, the ancient Jewish theocracy that in Christ's time rigidly administered local government, public morality and mores, and who were condemned by Jesus – repeatedly, publicly, and loudly – as the biggest hypocrites of all.

Two millennia later, organized religion has fallen into such false piety, hypocrisy, and ill-repute, that most of the Western World has lost complete faith in organized religion, and even coherent religious belief of any sort. To be replaced by "nihilo-religiosity" (a word I made up).

This included the former me.

When people would ask me "If I believed in God", I would joke deadpan that "Yes, I believed he was doing a bad job..." No one ever laughed – but then again – I didn't think it was very funny, **either**.

Once, a long time ago, when I was asked why I didn't believe in religion – to the point of open contempt at its mere mention – I answered coldly, with a discussion-ending finality, "Because religion is about **faith**." It was a deliberately minimalist answer, that came with a look on my face that revealed a man with a block of solid ice for a heart.

But the way Evil operates, and the hypocrisy that is the foundation of **our** Canadian government and society is the same as Jesus Christ pointed out two thousand years ago in a not-so-quiet backwater of the Roman Empire.

And perhaps the most shocking reason for people's blindness to evil, hypocrisy, and the other many forms of surrender to evil – is the most ludicrous of all. The belief that it's always better to be on the "winning team", as it were. The need to avoid the discomfort that comes from "rocking the boat", and the suffering that "doing the right thing" will frequently entail.

That hard decisions are called "hard" precisely because of the pain you will endure for making them, even though they are the right thing to do. And, finally, that "fighting the good fight" usually means losing.

* * * *

> "The story is disconcerting. It deals with time, madness, and a perception of what the world is or isn't. It is a study of madness and dreams, of death and rebirth, set in a world coming apart."
>
> -- "12 Monkeys" DVD production notes (1998)
> Terry Gilliam

I live in Ottawa, the capital of Canada, a medium-sized city in the Province of Ontario. It is about 100 miles from Montreal, Quebec, and Upper New York State, and about 200 miles from Toronto, Canada's largest city.

When, at the age of 48, through no fault of my own, my quiet, upper middle-class life in a quiet west-end suburb of Ottawa was suddenly turned upside down, I was caught completely unawares.

On Wednesday August 22, 2007, as the end of summer was beginning to be visible, my Life abruptly and unexpectedly took a dramatic turn for the worse, when I was arrested by the police. At about 3:00 in the afternoon about a week before, after having been invited to stay at the house in Gatineau, Quebec of my 14 year old daughter and her mother, her mother literally and figuratively went nuts on me. She soon called out to my daughter to "Call '911' !", to my complete shock and confusion. (For instance, why didn't she call 911 herself?)

Relatively minor and completely false criminal charges (simple assault, and uttering threats) filed by an ex-girlfriend, snow-balled into front page news, after an illegal warrant-less search of my house caused my false arrest on extremely serious firearms charges.

Thereafter, following a Friday night press conference by the police, there were radio news reports, TV news coverage, a front page article in the "Ottawa Sun", a half page City section "Ottawa Citizen" article, and hundreds of items posted all over the Internet. My bail was denied, resulting in **seven** months of pretrial detention, before finally **all** weapons charges were dropped.

Finally, I was acquitted at trial – defending myself without a lawyer – of the original domestic assault and threat charges, ending the legal matter completely.

Whatever a sense of relief and victory that was, it was not the end of the matter for me. The entire ordeal had left me a wounded and bleeding, almost completely destroyed, mental wreck. I hate to admit it, but they had broken me.

* * * *

> "Our worst enemies here are not the ignorant and simple, however cruel, our worst enemies are the intelligent and corrupt."
>
> -- "The Human Factor" (1978)
> Graham Greene

What was just another one of many uncounted, unhappy, personal tragedies in this disposable world of our modern society, suddenly had became a serious threat to my freedom, happiness, lifestyle, health, well-being, security, status, and future.

In an instant, my life had changed forever. But it was not a cliché.

Soon, as I experienced it in stunned shock and horror, I found that my entire belief system had crumbled to dust, right before my eyes. Though I put a brave face on it, I was truly shaken to my very core. And it was the last thing I was expecting, or had anticipated.

Doubts, fears, uncertainty, and despair began to solidify, as days, then weeks, and finally months passed while I sat in jail. Hope was on life support, and I clung desperately to patience.

But in my private misery – discomfort, confusion, unhappiness, and disbelief – somehow I was able to muster the energy to begin to use the time to reflect, to question, to reconsider, to re-evaluate and re-analyze, and otherwise seek answers, explanations and options, where I initially saw none.

"Judge not, lest ye be judged," said Christ.

I was judged and exonerated. Now it's my turn. And this book – written in my usual irreverent and frequently humorous style – is my judgment of those who judged me. I have tried – as Christ told us to – to forgive, but forgiveness of my tormentors has eluded me so far, other than forgiving the woman whose mental illness created the delusion of me assaulting her. I cannot hold it against a woman whose mental disorder caused her to become so unhinged she lashed out at the father of her daughter.

However, I cannot bring myself to forgive the government officials – local police, criminal Crown prosecutors, magistrates, and judges – of Canada, who used her false allegations to destroy my life for their own malevolent or careless purposes in a shocking display of their illegal, incompetent, atrocious, and outrageous administration.

It showed their reckless indifference to allow an individual to utilize the full power of the state to accomplish petty and delusional revenge, and use it as leverage to incompetently and with malice aforethought jail me for seven months awaiting trial, ruin my reputation, and leave me with severe diagnosed mental trauma that left me 100% disabled and completely unable to work.

It cost my family almost $30,000 in legal fees, and caused my complete loss of faith in the country of my birth.

I suppose it's a bit of forgiveness when I accepted Christ's arguments and renounced revenge through violence or immoral acts. There was a time that I would have taken the liberty of issuing appointments to see God to explain themselves, to people I considered had a lot to explain. (Actually, we all have that appointment.)

* * * *

"Say not 'This is the Truth', but 'So it seems to
me to be, as I now see the things I think I see.' "

-- inscription above entrance,
German Naval Officer's School
Kiel, Germany

"[H]ard days, and many, have I seen and suffered." So says Odysseus, warrior, commander, and hero, about his 10-year long, return voyage home, victorious from the Trojan War, in Homer's Ancient Greek epic, "The Odyssey".

That and that alone I offer, to ask the reader for some moments of his or her attention. What do I have to say? What do I have to recount, teach, or explain?

Listen and I'll tell you.

"All the history of every people is symbolic. This is
to say: history and its events and its protagonists
allude to another concealed history, are the
visible manifestation of a hidden reality."

-- "Posdata" (1970)
Octavio Paz

"Just as Newton had laid bare the laws of the natural
world, men could [similarly] create a 'science of
society'. The **point** was that every human institution
could and **should** be brought to the bar of reason
for judgment."

-- "Tom Paine's Republic" (1976)
In "The American Revolution"
Eric Foner

Prison Diary 1:   A Reality Show for Real
                              ===================

> "Someone must have been telling lies about
> Josef K., for without having done anything
> wrong, he was arrested one morning."
>
> -- "The Trial" (1925)
> Franz Kafka

> "Both the _fin-de-siècle_ [end of the Century]
> atmosphere of a social order starting to decay at
> the top, and the hardening of bureaucratic error
> into institutionalized delusion, are evident at the
> very beginning of the affair."
>
> -- "Why We Cannot Forget Dreyfus"
> in "Horizon" (Spring, 1973)
> Edmond Taylor

One of the most famous opening lines in fiction, the first quote above -- "Joseph K." -- becomes even more meaningful when you know the background and context of Kafka's most famous work. Or when thrust upon you, it becomes **your** reality, as it did to me, unexpectedly descending out of the blue upon my life, one dark and disturbing evening in late August of 2007.

Because on the late evening of Wednesday August 22, 2007, what should normally have been an otherwise uneventful tail end of one of the dog days of summer, suddenly became for me, a text-book example of a **really** bad day.

Disruption, trouble, and discord were upon me. But it was just the beginning of my personal Kafkaesque journey -- but I just didn't realize it yet.

*   *   *   *

> "Keep me away from the wisdom which does not cry,
> the philosophy which does not laugh, and the
> greatness which does not bow before children."
>
> -- "Handful of Beach" (1914)
> Kahlil Gibran

Franz Kafka was a scrawny, dysfunctional, repressed, and tormented Jew, employed as a minor Czech insurance company bureaucrat, and who wrote very strange and bizarre fiction books on the side -- "The Trial", "Metamorphosis", "Amerika", "The Castle" -- as he lived in Prague's Jewish ghetto during the inter-war years -- the years between World Wars One and Two, 1919-1938 (which included the American "Roaring Twenties" followed by the "Great Depression").

In Germany, next door to Czechoslovakia, the Nazis were rising inexorably – relentlessly – out of the rubble and chaos of a defeated and bitter post-World War One Germany. When "The Trial" was finally published in 1925, 1933 was eight years away, the year Hitler finally achieved his goal of absolute power over the German nation, and one step closer to his ultimate – and rather ambitious – goal of fascist dictator of all of Europe, and then the **entire** world.

Timing is everything, and post-war Germany was fertile ground for Hitler's National Socialist Party -- commonly known as the Nazis. The end of World War One in 1918 – the "Great War" – had left the continent of Europe -- the economic, political, industrial, scientific, and cultural leader of the civilized world – like a deck of reshuffled cards.

Only one – the British – of the great European empires had survived – though terribly weakened. The remaining **four** empires had tottered, and then fallen – the German, Austro-Hungarian, Ottoman, and Russian empires – the latter under completely new management. Talk about a shake-up in European politics!

With the triumph in Russia of Lenin's Bolshevik [Majority] Party during _Krasny Oktober_ -- "Red October" -- of 1917, the Tsar was captured and executed, his dynasty permanently extinguished, private property was abolished, and the abject cruelty of the feudal Russian Empire was obliterated.

It was reborn as the first, radically novel, and unprecedented workers' state, christened "Soviet" Russia -- after the Russian word for the Worker and Peasant "Committees" that now were in charge. They were symbolized by the hammer and sickle -- the factory worker and the farmer, respectively.

The surprising triumph of Marxist class war – first enunciated by Karl Marx and Frederick Engels in the "Communist Manifesto" in 1848 (a year of revolutions all over Europe), only seven decades previously – that resulted in the birth of an actual, functioning, communist state was as radical a development as the American Revolution had been almost 150 years previously, the latter with its advanced ideas of human liberty and freedom centuries ahead of their time.

Absolute power's old names -- monarchism and empire – was a system of government, rule, power, and – most important to make a political regime viable – a relatively stable transfer of power, based on family inheritance, and a right to rule based on the charming idea that it was bequeathed from God.

But the First World War -- a war so expensive, ruinous, and costly in death and destruction, that it was optimistically called the "War to End all Wars" – had dissolved or weakened such antiquated notions of authority, having harnessed to its fullest extent the powerful advancements of the Industrial Revolution, in the mid- to late 1800's, which allowed mass production factories to equip huge armies with rifles and artillery, along with a concurrent advancements and improvements in old weaponry, along with new weaponry of unprecedented lethality.

But the principal advance -- a development of American inventor Hiram Maxim – was the introduction of the heavy machine gun – a firearm that was fed and then fired a continuous stream of ammunition from a cloth belt strung with rifle cartridges – which led to mass slaughter and stalemate, in what became known – concealing its real horror – as trench warfare.

Along with U-boats, chemical warfare, heavy artillery, and the beginnings of aerial bombings of civilians, and just barbed wire, what came to be known as "Total War" had devolved into a war of military and economic attrition between alliances of nations. And its end was five year later, with 15 million -- almost all young men, and mostly just soldiers – dead.

An entire generation of European youth wiped out.

The times they were a-changing.

The absolute power of monarchism had fractured into two opposing and contrary halves. It was now to have two names, and be based not on religion and inherited power, but on dogmatic ideas, though enforced with brutality just as before, and only slightly more rational in nature.

Dictatorship was now based upon ideology, and the new names were known as fascism and communism. With democracy threatened by both, a second World War was just a matter of time.

* * * *

"It is magnificent, but it is not war. It is madness."

-- Field Marshal Pierre Bosquet (1854)
on the "Charge of the Light Brigade"

Kafka was certainly deeply troubled by many things in his tiny, unimportant world. But he may have had a sense, or realized that he was witnessing a mighty sunset, that rare and disquieting (to its witnesses) occurrence in the evolution of human civilization.

The book Kafka wrote, "The Trial", is the dark story of an ordinary, nondescript man, inexplicably arrested by the police one day, who is then tried, convicted, and sentenced without ever finding out what crime he is charged with, what he is supposed to have done, or hearing the evidence that supposedly damns him.

One day he is just snatched up by nameless policemen, to disappear into the labyrinthine bowels of a faceless, omnipotent, judicial bureaucracy, never to return.

"The Trial" is perhaps the first literary appearance of the concept of a fictional, usually futuristic "dystopian" society (from the Greek, meaning "bad place" – the 1982 science fiction, film noir movie, "Blade Runner" is a more recent interpretation of what a dystopian future would look like) – and the struggles of an individual to survive, or even understand, what is a foreign, hostile, alienating, and indifferent society that he has somehow attracted the attention of, and which now holds complete control over his fate.

A world where he, more than ever, is just a pawn, and where forces are aligned and united against him, and over which he has absolutely no useful information, understanding, or influence.

And though "Josef K." is named so as to represent no one, anyone, or everyone, it is interesting to note that Franz Kafka's middle name was Josef.

* * * *

"In...life one should comfort the afflicted, but verily, also one should afflict the comfortable, and especially when they are comfortably, contentedly, even happily wrong."

-- John Kenneth Galbraith (1989)
interview ("The Guardian", U.K.)

And so, on 22 August 2007, a merely bad day being experienced by Yogi Shan suddenly became as if I had metamorphosed into Josef K.

But I didn't know it quite yet, because the "fun" was just beginning. And at the age of 48, basically not really doing much different than living my usual quietly chaotic personal life, never in my wildest dreams did I expect to be told twice in one day -- in two different Canadian provinces, and for utterly bogus reasons, "You're under arrest."

It was lost on me that **finally**, I would be able to use the term "existential crisis" with authority. (Or even have the "free" time – to use the term loosely -- to contemplate what "existential" actually meant...)

"Whad-da-fuck?" -- I mean, I was **shocked**.

What? How can I be under arrest? I stopped caring about anything **years** ago. Even love has no meaning anymore. Don' cha un'erstan' ?

To my amazement and confusion, they did **not** understand. On the other hand, if I didn't already know that municipal police have no sense of humor, I would have responded with a joke:

"What took you so long?", "Will this affect my career?", or "Can you put me in Lindsay Lohan's or Mischa Barton's – or even Paris Hilton's – cell?"

But there's TV and there's actual reality.

After hearing Wednesday afternoon that the police wanted to talk to me, and calling them up -- having been given the runaround with calls and a visit to the Ottawa Police – I was finally told to go to the Gatineau Police Station, a constable of which said over the phone that I was going to be arrested.

This was to my surprise and distress. I had suspected that something was up, but hoped it would blow over, and had been on the move, and laying low in the meantime. The fact that I was to be arrested, however, verified that the mother of my daughter had indeed made a criminal complaint against me of simple assault and making threats. But the police hadn't even talked to me before deciding to arrest me.

There were no injuries, witnesses or evidence against me, other than her statement, made suspect by her relatively recent diagnosis of mental illness – Borderline personality disorder. The charges were a complete fabrication, a delusion of her mental illness, and otherwise unsupported nonsense.

I surrendered to the Gatineau Police, was detained for a couple of hours, treated respectfully -- other than the general indignity of the situation – given a sandwich to eat, followed by an "extra" sandwich – a nice touch, I must say – and then signed a "promise to appear" document, thus allowing my immediate release from custody.

Since I had fourteen assorted rifles and handguns registered to me – revealed by any standard CPIC (Canadian Police Information Computer) police database check -- one of the standard bail conditions was that I had to surrender these (and any ammunition) to the police until the outstanding criminal charges were resolved.

And so I was accompanied by two Gatineau Police officers – one of whose lame attempts at conversation quickly framed him in my mind as a bozo, and another one who I assumed was a bozo – and we proceeded in a Gatineau Police van to my house on Woodroffe Avenue in Ottawa's west end for the handover of my legal collection of firearms and ammunition.

However, things started to go bad when we arrived at my suburban four bedroom bungalow, in which I lived alone. On arrival, the pigs entered my house, walking past me as I stood at the front screen door, standing on the threshold holding it open.

While I was talking with the talkative cop in my living room, suddenly the other cop started wandering around my place on his own. I loudly called after him, and informed him to get out my dining room and return to my view and supervision.

I was then informed by the other Gatineau pig that they could do what they wanted, and what they wanted to do was search my house. This was *not* part of the bail conditions that I had signed. Without a search warrant, which they did not have, this was completely illegal, and I protested vehemently this gross violation of my rights and abuse of their authority. I knew it, and they should have known it. But being bozos, legally incompetent, as well as armed, further discussion seemed pointless.

I stood down, to wait and see how things played out, and whether I should gun them down like dogs at the first opportunity. They were committing a grossly illegal act, and I knew they could not get away with it in the long run. But I couldn't stop them at the moment.

"I know my rights!" is not just a cliché. The problem, as usual, was that the Pigs didn't -- or more importantly – didn't care. Under color of state law, events would soon result in armed men assaulting me, publicly humiliating me, forcibly detaining me, and proceeding to smash down doors, and ransack my house, without any legal authority or **even** valid suspicion.

They had no informant, tip, inside information, or probable cause -- besides not having the requisite search warrant. They were just grossly ignorant, ill-trained, uneducated, and dull-witted pigs, doing what they felt like, and having the power to do so. And they weren't even in their own jurisdiction.

It was a gross breach of the Charter of Rights and Freedoms, the supreme law of the land, which includes -- under Section 8 – the right of a citizen to be free from warrant-less search. And it was a disgrace and an outrage that these cops didn't know better.

But I was outnumbered, outgunned, facing legally constituted authority with no avenue of timely appeal. I was helpless. Freedom was just within reach, it was a hot evening, and it had been a long, tiring day.

And even though I was well-positioned for a shoot-out with these idiots, I wasn't in the mood for this kind of stuff, and hadn't been for quite a while.

> [A Supreme Court decision makes it perfectly legal to use "reasonable" violence to resist an illegal arrest, search, or other serious police misconduct. If I had pulled a gun and quickly loaded it, and they went for their side-arms, that would have been my legal justification, and it would have been the end of the road for these two cops. ...And a closed coffin funeral, he says slyly smiling.]

These twits were **wasting** my time. And that's what **girls** are for -- and frankly – they're a hell of a lot more fun. If these pigs carried out their threatened room-by-room search, it would be a text-book example of gross police misconduct, so I knew I wasn't in any trouble that would stick. Nothing illegal or alleged to be illegal would hold up in Court in these circumstances.

But, on the other hand, I failed to see the big picture -- that events were snow-balling. First a dubious arrest on false allegations, and now an illegal warrant-less search for absolutely no reason.

I bided my time as I began handing over cases and cases of ammunition, and my rifle and handgun collection. These they were entitled to take, so we got on with it, as I waited to see how the comment about "searching the house room-by-room" played out.

The thing was that -- stashed away in the house -- were two little problems. Unloaded, trigger-locked, and locked in cases -- as required by law -- and stored in locked closets separate from my legal guns and ammo, were hidden two **unregistered** rifles. I had bought them legally – as always – years ago. But when the government demanded that everyone register all the guns **individually** in their possession in 2002 or so, I dutifully complied, **except** for these two rifles, out of a collection of sixteen.

Before registration, the government checked you out, and licensed you, and knew that you probably had firearms, but they didn't know exactly what kind, or how many, in general. And **that** was reasonable. But after years of increasingly restrictive laws, rules, regulations, and harassment, registration was enacted by Attorney-General Allan Rock, of the Liberal Party government then in power.

But enough was enough. So I – like possibly hundred of thousands, and probably millions – an estimated 50% -- of the seven million Canadian gun owners – for right or wrong, committed our little acts of civil disobedience, and broke the law by neglecting to tell the government about some or all of the legal firearms we were licensed to possess.

I considered it "insurance" against a future government deciding to completely ban citizens from possessing firearms, and then confiscating them all. This is what gun registration would make easy, and why it was so feared and vehemently opposed by gun owners. Of course, this would be a gross betrayal of trust, amongst other sins, and I just didn't trust the government, period.

And it seemed that my distrust was valid, as upon me had descended out of the blue a _force majeure_.

Armed agents of the State didn't just break the law, but violated, without the slightest hesitation, the Constitutional protections afforded by the Charter of Rights and Freedoms – the bedrock foundation of all Canadian laws.

The Charter requires that fundamental human rights demand that the police need sound, documented, and detailed justification, **and** a piece of paper signed by a judge – called a search warrant – **before** they can commit the grave and intrusive act of searching someone's house against his will.

A rather minor – though certainly criminal – law says that I should have two pieces of paper certifying that the government is aware and authorizes my possession of these two rifles. I was certainly licensed to own the two rifles, but the government didn't know I had them. Violation of the registration law typically results in a stern talking to by the judge about the importance of obeying even laws you disagree with, and a no-jail, probation or suspended sentence, or similar.

So, ultimately you could say that the whole brouhaha of August 22, 2007 boiled down to the fact that both the police and I had a problem with absent paperwork!

However, **their** criminal misconduct far exceeded mine. But only I went to jail that night, and many nights more, waiting patiently for every single criminal charge to be thrown out -- as finally did happen. Nothing happened to the Gatineau and Ottawa policemen, for their serious and reckless breach of the Charter, along with a host of Criminal Code violations: home invasion, break and enter, destruction of personal property – I later found out they had smashed open multiple locked doors – and the crime of "theft over $2,000".

This police malfeasance established a complete lack of accountability over flagrant and long-standing police incompetence, lawlessness, corruption, and maladministration that I was soon to find had spread and corrupted the entire judicial process and system.

Something was rotten in the State of Denmark. And I was the fall guy selected to take the full guided tour...

> "But how can I explain, how can I explain to you?
> You will understand less after I have explained it.
> All that I can hope to make you understand
> Is only events: not what has happened.
> And people to whom nothing has ever happened
> Cannot understand the unimportance of events.
>
> -- "The Family Reunion" (1939)
> T.S. Eliot

Prison Diary 2: Round Up the Unusual Suspects
==========================

> "Alexander wept when he heard from Anaxarchus that there were an infinite number of worlds. His friends, hearing the commotion, rushed in, asking Alex what was wrong.
>
> Alexander [the Great] responded, ' Do you not think it a matter worthy of tears, that when there is such a vast multitude of worlds, we have not yet conquered one?' "
>
> -- "Miscellanies and Essays" (ca. 100 B.C.)
> Plutarch

> [As Detective Sonny Crockett starts coming towards Evan menacingly, suddenly Evan pulls a handgun on Crockett, and Crockett freezes.]
>
> "Come on Crockett, let's have it out! You and me. Twenty paces. Back to back. What do you say? [Okay, okay.] I'm cool...I'm cool... [Evan lowers his handgun for a moment, then suddenly raises it again.]
>
> **Huh**? One for you, one for me? Let's go together! Hmm? Huh? Huh? Huh?
>
> **Oh –** you don't think I'd do it, do you?
>
> [Pulling out a bullet and holding it out in his hand for Crockett to see] Oh, Crockett -- **here's** one with **your** name on it. Ever think that, Sonny? That there's a bullet somewhere with **your** name on it?"
>
> -- Evan Fried
> "Miami Vice": "Evan" (1984)

There's a Canadian NFB (National Film Board) film that I saw once, on Friday late-night TV. It was entitled, "Tu as crié, 'Let me go!' " (1999) [" 'Let me go! ' you had shouted"] , and I found it mesmerizing.

It was written, directed, and starred Anne Claire Poirier, and it was the story of her search for information and answers about society's hard drug problem after her 26-year-old daughter -- a Montreal heroin addict and prostitute -- was murdered by a person or persons unknown.

The mother talks to – subtly "interviews" – female users, female ex-users, health workers, and others.

One scene I found particularly poignant, was an "interview"/conversation Poirier had with a middle-aged female ex-user, talking about her days as a junkie, and the funny criminal escapades she took part in, to make money for her heroin habit. Like robbing a bank with a plastic gun that fell apart in the middle of the robbery. The ex-user is laughing and all-smiles. And the mother/interviewer lets her talk, and then finally asks her what started her on hard drugs.

And the ex-user suddenly turns serious, briefly pauses, and then responds, "People were always such a disappointment. Ya know?"

* * * *

> "52 Division, handcuffed to a chair.
> I'm joining the lineup to fall down the stairs.
> I tell you, ' I'm innocent,' I try to explain.
> 'We're just making sure you don't do it again.'
> Do **what** again?"
>
> -- "Cherry Beach Express" (1984)
> The Pukka Orchestra / G. Williamson

The Ottawa Police Constable who had placed me under arrest, 'cuffed me – arms in front, rather than behind my back, at least -- then stuffed me in the back seat of the cop car. After leaving me there for about half an hour, he entered the front of the car and without a word, started the car, and began driving.

It was long after dark by now, and I had been trying to chill in the back seat, trying to take a nap, though still handcuffed. When we started driving, I sat up to take stock of the situation, and try to figure out my next destination, which I assumed was the Elgin Street Ottawa Police Headquarters lock-up.

Constable Laplante wasn't the talkative type, and I didn't have anything to say to the arrogant jerk anyway. He had deliberately and with malice aforethought handcuffed me **hard** to cause severe pain to my wrists, when he snapped them on. I knew it was deliberate, because of the hard look on his face, and his complete lack of response or even emotion to my cry of pain – he pressed the handcuffs closed **that** hard. And he completely ignored me when I asked him politely "The 'cuffs are really tight -- could he loosen them a tiny bit, please?"

Our little drive ended up in the bowels of the Elgin Street Police HQ -- the basement parking lot. I was about to enter the Ottawa Police "Temporary Custody Facility", commonly known as a "bucket" by those intimate with the system. There's at least one bucket in every Ontario urban concentration. It's used as a drunk tank, too.

I was briefly registered as an incoming "guest" by the processing unit for new detainees. I wasn't to be photographed or fingerprinted until I was formally charged, but my belt and shoe laces were taken from me, I was questioned as to whether I was suicidal or not, and I was given a list of lawyers to try and contact.

I was not allowed to call my family or any one else. If they were wondering why I had suddenly disappeared, they would continue to wonder and worry for a couple of days. A system where people just disappear doesn't qualify as "a free democracy", and I was rather upset – though not visibly – and I didn't think to ask the lawyer I contacted to call my family. And he didn't think or bother to offer.

After several lawyers I had selected from the provided list didn't answer, I finally made contact with top Ottawa lawyer, Michael Edelson, and we had a brief conversation. My processing was then finished by the police, and I was led to my new accommodations: a single prisoner, cold, barred cell in the basement, with no outside windows -- basically a dungeon.

There was a stainless steel toilet, with no seat, a roll of toilet paper, and a bed whose surface was a flat painted steel plate. There was no pillow, sheets, or blanket. You were supposed to sleep on this torture device – a gray-painted hard steel plate.

How medieval... I couldn't believe that this was how they treated people who had not been convicted of anything. Or even if they **had** been convicted, for that matter. I didn't bother wasting my time complaining, but a prisoner who did got a flip answer from a guard, blowing him off -- "This isn't a hotel."

Apparently "Innocent until proven guilty" isn't a guest at this non-"hotel" either...

Continuing their practice of disgracing Canada and its justice system, the next morning's breakfast consisted of a Styrofoam cup of Tang -- colored sugar water masquerading as fake orange juice -- and a slice of bread spread with margarine. That was **it**. A new standard in "worst breakfast in the world". Not that I was particularly interested in breakfast -- I gave mine away to the guy in the cell next door -- but this was pretty sad.

Other inmates who complained to the guards over various aspects of the conditions of confinement were again given the curt response, "It's not a hotel."

If the tables were turned, I'd be deflecting inquiries about **no** breakfast from arrested guards with, "It's not important, the firing squad is almost ready." Followed by an evil laugh. And I wouldn't be joking, either...

* * * *

"I was staring straight into the shining sun.
Lost in thought and lost in time,
While the seeds of life and the seeds of change were planted."

-- "Coming Back to Life (1994)
Pink Floyd

The end of the Late Republic in Rome was in 44 B.C., the end of five centuries of an early form of democracy, followed by the establishment of Emperor, and Empire.

The decay and then fall of the Roman Empire (the sack of Rome in 410 A.D., and the fall of Constantinople – capital of the Eastern Roman Empire -- in 1453 A.D.).

Sunsets.

With the end of WW1, monarchist empires were out, or on the decline. Hereditary power had lost credibility as its promised stability was undermined by rapid technological advancement and the urbanization that followed.

The new political orders of Fascism and Communism were the new names of absolute power. And after the usually violent upheaval to topple the _ancien regime_ – the old Order – came the new Order's next order of business, the vitally important "cleansing" and re-ordering process known euphemistically as the "consolidation of power".

Ideally it is silent, rapid, thorough, systematic, and step-by-step. It is a brutally efficient, and inexorable infiltration of the new Order into every aspect of the Old Order's hierarchy down to its lowest levels. The public elimination of blatant threats loyal to the Old Order, while systematically identifying all possible enemies for eventual quiet isolation and neutralization, followed by the seamless establishment of parallel or intertwined organs of power of unquestioned loyalty.

Direct comparisons are always problematic -- the Russia that was overthrown was a cruel feudal regime, while the Germany that was toppled was an advanced society, but also a weak and unstable democratic republic. But compared to the crude methods -- order by executive fiat – of brutality and violence that the Russian communists wielded to satisfy their goals – in eliminating all vestiges of the old feudal order -- in a timely, efficient fashion, while under internal and external attack, there was a certain finesse, completeness, and genius in the consolidation of power by Hitler's Third Reich.

And when the Nazis were done, Germany was again – after the humiliating defeat of WW 1 (World War One) – a world power, leader in technology, and a lockstep, unified war machine with a decent chance at world conquest.

How a liberal and democratic nation, the world leader in the arts, literature, culture, medicine, chemistry, engineering, metallurgy – particularly of steel -- mechanical devices, manufacturing, and technology could so quickly and smoothly be forcibly evolved in six years – 1933 to 1939 – into a police state dictatorship is truly amazing. A republic became a totalitarian government that ruled by terror, backed by an iron police fist to ensure total compliance by the German people.

It was so bad that the state made lack of enthusiasm essentially a serious, and sometimes capital crime. And this transformation was effected by a group that essentially began as a bunch of street thugs with vague but extreme political views.

Trouble, there was a-brewing... And what Napoleon referred to as the "butcher's bill" of 15 million dead for WW1 was not the last the butcher would be sending. The next war would be only twenty years later, would involve attempted genocide, and would leave 50 million dead.

But what failed Germany at world conquest in WW2 – the military technology and techniques of ruthlessly efficient internal political control strategies – were certainly not forgotten, but studied and adapted for U.S. -- the new world leader -- Executive branch use in later times.

Good military technology and evil methods of government live on, long past its inventor, adopters, and users. They are just spit-polished and perfected a little more.

* * * *

> "There's more than one way for a generation to be slaughtered. They can be slaughtered in the streets with tanks, with guns. They can be thrown in jail. They can be, as our generation in so many ways has -- and [as] previous generations have been -- demoralized, and emasculated, and beaten down."
>
> -- James Baldwin (1968)
> Chicago TV interview show

At some point Thursday, the day following my arrest, I was taken from my cell and put in a small interview room with a miserable-looking, fat, balding, ugly, middle-aged, Italian son-of-a-bitch pig – I exaggerate not in the slightest – who identified himself as Sgt. Costantini, of the Ottawa Police "Guns and Gangs Unit".

The expression and demeanor of this clown was a sight to see. He looked more miserable than me, and I was the one in big trouble. He looked like he was married to the fattest Italian woman in the world, who beat him with a rolling pin when he got home every night, and who openly cheated on him every day, just out of spite.

The good Sergeant wanted to ask me a few questions, assuring me that the conversation wasn't being recorded, there were no cameras in the room – as his straightened arm swept the room for emphasis that no lenses were visible on the interview room walls. I gathered that the implication he was trying to convey, was that I could answer his incriminating questions without worry.

This didn't fool me a bit, and was a trick to try and get me to say something they could use against me, when I had previously invoked my right to remain silent, and informed the police that further questions be addressed to my lawyer.

This is a police practice of lying to suspects and arrestees on legal issues was ruled totally legal by a bunch of idiot judges on the Supreme Court of Canada who maintain that tricking -- to their serious detriment -- uneducated simpleton criminal suspects into giving up their right to remain silent -- **even** if previously invoked – is a sound and perfectly reasonable pig tactic.

The police can thus lie to you if it suits their purpose, according to the ruling, thus extending routine police lying from their perjury during court testimony, to all areas of public interaction.

(I have four questions about this: Is there an open bar all day at the Supreme Court? Are janitors and McDonald's fast food staff appointed as Supreme Court justices? Can I sign the death warrants? And is that firing squad ready yet?)

I politely listened to all the good Sergeant's questions, because I was bored, and curious about how their investigation was going. I responded occasionally to his questions with answers that were true and which I believed were helpful to me, and would not in any way be to my detriment.

"Why did you have in your possession a 50 lb. bag of ammonium nitrate and a 50 lb. bag of potassium nitrate?" the detective asked.

"Irrelevant," I responded to his question. Because my possession of these two items was completely legal, and was completely unrelated to any criminal charges I was expecting to face, I didn't feel the need to answer questions or justify perfectly legal actions of mine.

"Why did you have 30,000 rounds of ammunition?"

"Irrelevant," I repeated, once again for the same reason. Since I was appropriately licensed, it was perfectly legal for me to possess 30,000 – or 30 million rounds of ammunition, and that was that.

No explanation was required or called for. Basically there was nothing untoward, or unusual to stockpile this kind of quantity of ammo, anyway, besides the fact that since it was legal, it was none of their damn business.

"The deactivated machine guns are legal, we have determined. Where did you get them?"

"Well, well, well," I thought to myself. This policeman had just confirmed to me that Ottawa Police Constable Robert Laplante had falsely arrested me. (He had stated that "You are under arrest for possession of a prohibited weapon." And then refused to respond when my startled question "**What** prohibited weapon?" was posited to him at least two or three times – a violation of proper police procedures and court decisions based on the Charter, that require him to state the **exact** nature of the alleged offense on arresting someone.)

But though, once again, Costantini's question was irrelevant. After confirming that they were legal, what concern of his, was where I had got them? I answered his question with what I considered an appropriately indignant response.

"I would like to inform you that I will be suing **every single** public official who had anything to do with my illegal arrest and detention. Since you have determined that my deactivated automatic weapons are legal to posses, I repeat my statement that it is an irrelevant question, and that I do not feel the need to explain or justify perfectly legal behavior to you or anyone else.

"But I'll answer your last question anyway. I bought them over a decade ago from Peter Kearns of the Edmonton company, "Kearns and McMurchy". Peter Kearns blew his head off for reasons unknown to me, so good luck talking to him. He was an old guy and he had serious heart problems, which I speculate was the reason for his rather disturbing exit."

I declined to answer the rest of the questions, and the interview soon ended. I was returned to my cell. It was late and my buddy Anthony Costantini probably wanted to go home and watch his wife's fat cells divide.

Finally, around 11:00 pm Thursday night, I was again escorted from my cell and **finally** charged, fingerprinted, and photographed. The false arrest had delayed things, and they had to figure out another way to detain me. They had finally got their shit together, I guess.

I was charged with four criminal counts -- two counts of possession of an unregistered rifle (a semi-automatic AK-47 and Colt AR-15), and two other minor weapons counts that were totally false fabrications, that I don't know why they even bothered concocting them.

It was after midnight by the time I got back to my cell.

The following morning, Friday, I was chained up with another detainee, and a bunch of us chained prisoners were loaded into a "paddy wagon" for a brief drive to the Court House on Elgin Street at Laurier. We were unchained and unhand-cuffed on arrival, and locked in a giant holding cell, one of several. It was Friday morning at about 10 am, when I was cuffed and ankle-chained, and led down the hall and up some stairs. My handcuffs were taken off, but not my ankle chains, then I was ushered into what was known as the "prisoner's box" at the side of a small room -- a courtroom. It was my initial -- and a brief -- court appearance, a disgracefully long 36 hours after I had been arrested.

I looked at the Justice of the Peace at the front of the room, the lawyers at tables in the middle of the room, and the many people in the public gallery at the back of the small room. The public gallery was full of people for some reason, and they all looked back at me. I recognized none of them. I was confused at their number and presence.

"What is your name?" the judge soon asked me. I thought it was a strange question at the time. I felt like rattling the leg irons chaining my ankles together in symbolic protest, but restrained myself. I was annoyed that they were not removed for the courtroom appearance, like they had done for my handcuffs.

"Yogi Shanmugadhasan," I responded to the Justice of the Peace in a perfectly calm, clear, audible voice. It was my legal name, though I had used "Yogi Shan" to simplify things for everyone, whenever I could, for the last 30 years.

After a few hushed words to me from the duty (temporary) counsel representing me, the judge "remanded" me (put me back) into police custody until a bail hearing could be prepared for, and scheduled by a legal aid or hired lawyer.

I had now been officially rubber-stamped by a justice of the peace as officially a "guest of the guv'mint". Or, as they put it euphemistically in the U.K., I had been "invited to assist the police with their inquiries."

And so, finally I was ushered out of the courtroom, handcuffed again, and shuffled back, leg iron chains a-clinking, back to the large holding pen to rejoin the other inmates. I found out later that the front page newspaper story described me as having a "shy, but friendly smile" during my brief courtroom appearance. It was an amazingly accurate description, and it kept me laughing, with an amused smile on my face **all** that weekend, somewhat easing the horror of my arrest being plastered all over the front pages of the "Ottawa Sun".

The **front** page of the Sun! **That's** why the public gallery was full. But my more important concern at the time was that every person who knew me or who had known me and remembered me, every person that I had ever worked with or gone to school with, now – no doubt – believed I was some sort of dangerous terrorist, a deranged outlaw kook, a major arms trafficker/gunrunner, or just a generally scandalous and disreputable sort of fellow. And worst of all, it wouldn't surprise them a bit.

After a short time in the holding cell, I was soon chained up again with other detainees who were finished with their court appearances for a paddy wagon trip to my next home.

The rolling snowball was getting still bigger.

> "And in between the moon and you,
> The angels get a better view,
> Of the crumbling difference,
> Between wrong and right."
>
> -- "Round Here (1993)
> Counting Crows

Prison Diary 3:   A Cheap – I mean – Free Press
   =========================

> "The increasingly expert destruction of man's spirit
> by the power of the police -- more wicked...more
> awful than the ravage's of nature's own hand -- is
> another such power, good only if never to be used."
>
> -- B.B.C. Reith Lecture (1953)
> J. Robert Oppenheimer

> "I hear a good deal said about trampling this law under foot.
> Why, one need not go out of his way to do that. This law
> lies not to the level of the head or the reason; its natural
> habitat is in the dirt. ... [H]e who walks with freedom...will
> inevitably tread on it..."
>
> -- "Slavery in Massachusetts" (1854)
> Henry David Thoreau

The Constitution Act, Part 1, proclaimed in 1982, is named the "Canadian Charter of Rights and Freedom". It is the absolute law of the land. In Section 2, under "Fundamental Freedoms" are listed the first of these freedoms, including "freedom of the press". It was over two centuries after that right was guaranteed by the U.S. Bill of Rights, of the American Constitution.

Well, better late than never...

And so, it seems the Ottawa Police Services "Guns and Gangs Unit" had called a press conference on Friday, August 24, 2007, while I had been sitting in jail for a couple of days. My bail hearing was coming up shortly, the police knew, and they had no search warrant for the two unregistered firearms they had found after a false arrest, and a resulting illegal search of my home.

This meant they had no case **and** I was **also** innocent by previous Court decisions -- precedents -- under Section 8, "unreasonable search", and Section 7, "right to...security of the person" of the Charter.

And so, for the press conference, the police trundled out, and happily displayed about 16 legally owned, officially deactivated -- even the police admitted it to the press that they were deactivated and legal -- light machine guns and sub-machine guns (SMG) from WW 2 onwards, the BAR -- the U.S. Browning Automatic Rifle (seen in the movie, "Saving Private Ryan"), a bipod-equipped, box magazine-fed light machine gun -- was the earliest (it was first designed in 1918), and the 1950 U.K. Sterling and an Israeli Army Uzi SMGs were the most recent.

There were British (Lanchester and STEN), American (M3 .45), Israeli (IMI Uzi), Swedish (K SMG), and Soviet (Ppsh-41) gun designs – a complete set of SMGs. There was no duplication -- which should have been a clue to anyone of competence that it was a collection -- actually a small museum of old full auto guns. A museum of one (the SMG's) and two man (for the light machine guns) fully automatic firearms.

I couldn't afford -- well I could, but I didn't want to -- buy the deactivated heavy machine guns -- like the Maxim or Vickers -- or WW2, Nazi Germany weapons, which were much more expensive. Though I drooled over the thought of owning and touching them, cycling their actions, and thinking of their histories and beautiful, but evil designs.

Each of the -- mind you, even deactivated -- cost an average of $500 or $600 dollars or so. The heavy MG's and Nazi guns were over a thousand each, and I would have dearly loved to have some -- especially the Nazi MG-42, a roller block design that revolutionized medium and light MG's and SMG's, and is still in use today in many countries, though renamed to wash away the Nazi taint. But us full auto _aficionados_ know better...

The police also had on display other item. There was a legal starter pistol. There was a legal bayonet. There was a 39 mm flare gun -- perfectly legal, and for which I had no ammo for anyway -- made to look like a highly illegal 40 mm U.S. M203 M16 grenade launcher at first glance, or to a non-expert.

And there were two legally owned, but illegally unregistered rifles, that was the basis of the criminal charges.

But the big story was the big pile of deactivated and legal non-weaponry. And the big question I had -- and still have -- was what right do the police have to display with great fanfare -- or even at all -- my legally owned and illegally seized property to influence the public and inevitably the court process? It was an absolute disgrace, once again.

Both of the two local papers, the "Ottawa Sun", and then the more staid, and detailed, and respected "Ottawa Citizen" had long stories about my case, both complete with big pictures of the guns conveniently displayed just for the press photographers. A picture is worth a thousand words after all. And what a picture a large table full of an assortment of military weapons makes! Deactivated or not.

The police admitted that the guns were deactivated and perfectly legal to own by anyone. Yet what gave them the right to display my legal property in a systematically organized and prepared news conference to the national media, other than as a sensationalized attempt to make a distorted and mischaracterized idea that something bad was going on?

It was the very definition of a smear job by officials of the government, with no one in the media questioning it, its propriety, or their role in taking part in it.

* * * *

"A point of view can be a dangerous luxury
when substituted of insight and understanding."

-- "The Gutenberg Galaxy" (1962)
Marshall McLuhan

I was 48 years old, a Canadian citizen born and bred in Ottawa, and a longtime resident of Ottawa. I graduated high school (Lisgar Collegiate Institute) with a Silver Medal (greater than 90% average), one of two in a school of 1,300 top Ottawa students. I had an Honors Bachelor of Science Degree from the University of Toronto majoring in pharmacology, with a minor in physiology. During third year (a "junior") I had done animal surgery on rats in physiology class, and pharmacology eye research as a university summer job during my Bachelor of Science days.

And I had a Computer Engineering Degree (B.A.Sc.Comp.Eng.) from the University of Ottawa, the hardest thing in my life I ever accomplished. I was an electrical engineer, and a computer engineer, and proud of it. I had worked for the last 13 years as a Member of Scientific Staff at Nortel Networks Corporation, a multi-billion dollar corporate giant with over 100,000 highly paid staff.

I was born and grew up in Ottawa till I was sixteen. I skipped grades twice. I lived for four years in Toronto earning my B.Sc., Honors. I lived for three months in Oakland, California, near Berkeley on the East Bay of the San Francisco Bay Area. I then lived four years in Los Angeles. California after getting married. I also lived for 14 months in Paris, France doing an ex-pat gig for Nortel where I was the liaison in charge of transferring information and answering questions between the French and Ottawa departments.

I spent several times in Spain, near Valencia, on the Mediterranean Coast. I lived in a **real** medieval castle for two weeks in the Departement (a province) of Lot in central France. I traveled all over France, from Arles, Nice, Nimes, and Cannes in Provence, to Normandy to Lille.

I served in the Canadian Forces Reserve, as a private, when I was 16, and was honorably discharged.

None of **that** was published, or brought up in Court, as it should have been, by my idiot lawyers, who didn't even ask.

The "Ottawa Sun" piece -- as characterized the general tone of the rag -- was a sensationalized front page story. A more reasonable follow-up story was on page three of the next day's paper. It talked about how many people in acts of civil disobedience -- just like me -- had refused to register all or part of their guns that they were licensed to own.

It is estimated by Conservative MP (Member of Parliament) Garry Breitkreuz -- and grandfather of MP's on the gun registration issue -- that an estimated **half** of the seven million gun owners in Canada (of population thirty million) have done so.

In interviewing my ex-wife -- who gave a favorable report about me to "Sun" reporters -- they admitted with unconcealed contempt, and interesting candor, that the story and news conference were just in time for the "Guns & Gangs Unit" budgetary approval process.

* * * *

"You have the power to act only. You do not have
the power to influence the result. Therefore you
must act without anticipation of the result, without
succumbing to inaction."

-- "Bhagavad Gita" (ca. 3000 B.C.)

At the end of the Citizen piece was an interesting aside. "The Citizen has learned..." What they had learned but did not disclose was that the police had leaked to them information found during their illegal search of my house of a "youthful indiscretion" of mine that occurred almost thirty years before in the U.S. And they got it wrong, anyway.

The Citizen was tolerating criminal misbehavior and unethical -- disgraceful -- conduct by the Guns & Gangs Unit of the Ottawa Police in leaking confidential information to the reporter. He was more than happy to comply.

It was another part of the smear campaign.

The protection of sources is a hard-fought public policy issue of reporters. Yet it is not meant to act as a conduit for illegal leaks by the police and government sources to smear individuals for the own purposes. It is meant to allow reporters to receive and publish information of public importance -- government corruption and cover-up -- as given by whistle-blowers, not criminal leaks by officials trying to manipulate popular opinion and the courts through a willing and unquestioning press.

When I contacted the reporter who wrote the Citizen article, and quietly and politely explained the background, and privacy, and public policy issues, he blew me off. The free press in action. Tool of the government and pigs if it led to a "good" sensationalist story, even for a staid and supposedly responsible newspaper like the "Ottawa Citizen".

* * * *

"[Lenin] argued that the abolition of the [Czar's] censorship
of the press so much praised by the liberals was not in
fact 'freedom of the press, but freedom for the rich, for the
bourgeoisie to deceive the oppressed and exploited
mass of the people.' "

-- "Lenin" (1971)
Michael Morgan

The Ottawa -- indeed all the big city police departments of Canada -- Police are supposed to be servants of the Court. Not servants of the Crown prosecutors, not public relations machines, not distorted and manipulative press conference holders. And not independent voices on public policy issues, and lobbyists of the government as an independent group, on an equal footing.

They needed to be reigned in, brought to heal, purged of reactionary and resisting elements, and beaten into submission if they don't take the hint of the winds of change.

It was ill-gotten and unallotted police power. They had absolutely no right to do what they were used to doing, and it is both criminal and civil violations of the people's rights. It is an independent, self-instituted usurping -- stealing -- of power from the Constitutional allocation and divisions of such power into legally constituted form.

It was, and is, systematic corruption of the very basis of our Government.

"[Andre] Malraux was the first to understand that the
lie no longer exists in the twentieth century, any
more than truth. He was the first to sacrifice idea
and reality to **image**. All serious power rests on
the imaginary..."

						-- Regis Debray (1976)
						   in "Le Monde"

Prison Diary 4:		And Behind Door Number Two
				========================

> "I'll learn to work the saxophone,
> I'll play just what I feel.
> Drink Scotch whiskey all night long,
> And die behind the wheel.
> They got a name for the winners in the world,
> I want a name when I lose.
> They call Alabama the 'Crimson Tide',
> Call me Deacon Blues."
>
> 				-- "Deacon Blues" (1977)
> 				 Steely Dan

> "We'll be fighting in the street,
> With our children at our feet,
> And the morals that we worship will be gone.
> And the men who spurred us on,
> Sit in judgment of all wrong,
> They decide, and the shotgun sings the song."
>
> 				-- "Won't Get Fooled Again" (1971)
> 				 The Who

Detention, imprisonment, incarceration. Prison, the penitentiary, the pen, a correctional institute, a work camp. Jail, the clink, the can, the slammer, the big house, the hoosegow, the gulag.

With the rest of the unfortunates of the day, I was chained up, loaded into the "paddy wagon" again, and driven to the Ottawa-Carleton Detention Center -- OCDC -- on Innes Road. Intake processing consisted of a strip search, shower, change of clothing into a bright orange jump-suit, one frontal mug shot, classification, and TB test and medical questions by a nurse. It took several hours to process the group.

After processing was finished, I arrived by late afternoon at my new home, "Pod" B, a 32 man range -- one of five -- that made up the maximum security wing. I had been classified as a "maximum security" prisoner automatically since I was charged with a "weapons offense". I was terrified by the possible implications, but a nice, middle-aged French-Canadian guard noticed my emotional unrest, and indicated he would help, if he was around. I was grateful for his interest and compassion.

A heavy steel door was buzzed open at the front of the Pod, and I walked alone into my new "home". I didn't really have a clue what to expect, or what I was supposed to do. There were over two dozen guys, all dressed the same as me in orange jump-suits, on the "range", engaged in various activities, some seated, some walking around, and four on the phone.

The front of the range for a few feet was an empty "no-man's land". Further down, occupying the left side of the range were eight sheet steel tables, on which many of my fellow inmates were seated, four to a table on heavy sheet steel stools bolted to the concrete floor, like the tables.

On the other side of the range, to my front right, was an empty space of concrete floor, and to the extreme front right was a bank of four phones affixed to the concrete wall, above a narrow seating ledge that ran the length of the wall from the range front, to almost the back of the range where two tiers of eight two man cells each were located. Each cell had a heavy, solid steel door enclosing it, with a small window at eye-level, and a lower closed and covered slot for insertion of food trays.

I tried to scan the room, but was overwhelmed by fear of the unknown, and the excessive amount of new visual data. I walked slowly and tentatively forward. Most of the other 30-odd inhabitants were generally seated at the tables, the rest on the phone, or walking the range -- up and down the length of the range, parallel to the phone bank.

An inmate walked nonchalantly up to me, and I froze, unsure of his intentions. He stopped in front of me, and with a friendly smile casually inquired, "You must be the 'gun guy'. The guy in the paper with all those machine guns?"

I'm sure my eyes must have widened visibly, as a shocked and confused look sprang forth from my face. "Yes," I responded with a blatantly surprised and curious tone, "How'd you know?"

I had never seen this guy before in my life. He had never seen me before, I was sure. He didn't know my name, nor I his. I was just an anonymous new prisoner, assigned to an empty cell in a maximum security Pod of OCDC. I was in the bowels of a jail inhabited by hundreds of inmates, people coming and going, usually in chains, all the time: admission, transfers to another jail or the penitentiary, returns from transfers, court (bail or trial), or release.

"We saw the news reports. We knew the name was East Indian, and that you'd be showing up soon. Welcome to Pod B. My name is Mike," he added by way of explanation.

My reality show for real was just beginning.

* * * *

"Bad lands, you gotta live it everyday,
Let the broken hearts stand,
As the price you've gotta pay.
We'll keep pushing' till it's understood,
And these badlands start treating us good."

-- "Badlands" (1978)
Bruce Springsteen

I mean, I had heard of the famous prisons. Kingston Pen in Canada. There's New York City: Rikker's Island. There are the New York, California, and other State prisons: Attica [renamed Ossining, after an infamous riot in the early 1970's], San Quentin, Folsom, Joliet [Ohio]. There's the notorious T.D.C. -- the Texas Department of Corrections. The Mississippi chain-gangs of old. And finally the more famous U.S. Federal Penitentiaries: Leavenworth, Lompoc, and Lewisburg.

I had seen the classic prison movies "Cool Hand Luke" (1967), "Stalag 17" (1953), "Papillion" (1973), and Oliver Stone's "Midnight Express" (1978), but now I was kicking myself for not having been suitably prepared by having watched and picked up any "hostile environment" jail survival lessons to prepare myself from recent TV shows like "Oz", "Prison Break", or even the first "reality show" ever, "Survivor".

"Survivor: Alcatraz" – wasn't it – the last season?

When Mike later passed me the Saturday "Ottawa Sun" piece, it started with a front page full-color close-up photo of a wheeled cart piled full of light machine guns and submachine guns -- all fully deactivated and perfectly legal for me – or anyone – to possess, sell, walk down the street with – legally they were non-guns, just hunks of steel – I was mortified, and astonished all at once.

But I was at least thankful that they had no picture of me. My mother denied **that** request from the "Sun" reporter, naturally. And everyone interviewed for the story -- my mother, son, and a neighbor – said nice things about me. Except the police, naturally, who smeared and slandered me as best they could -- all groundless rubbish, by the way, pigs being pigs. What can you do?

But the best part of the whole media circus was the "Sun's" detailed physical description of me at my brief Friday morning courtroom appearance for arraignment. And I found it absolutely hilarious, in spite of my discomfort, embarrassment, and general unhappy predicament.

Naturally I was having a bad hair day -- after 36 hours without a shower or change of clothes, I was, shall we say, somewhat disheveled-looking -- when I shuffled with my ankle chains warily into the wooden railing-surrounded "prisoner's box" in the small courtroom. The public gallery seated a rather numerous crowd of people, to my surprise. I didn't know any of them -- my family wasn't there, and probably didn't even know about my court appearance. I didn't know at that time that I was all over the news -- TV, radio, and soon newspapers.

And the newspaper did everything *but* say "bad hair day" in describing in detail the appearance of my hair.

But as I mentioned before, their observation of my "shy but friendly smile" had me laughing all weekend long -- one of the few bright points of an otherwise ignominious few days of my unexpected, and unplanned incarceration.

And unintentionally, the big, cheerful smile captured on my OCDC admission-processing mug shot seemed to have saved me from getting my picture in the paper. The smiling terrorist? That wouldn't quite fit the stereotype of the scowling evil-doer, now would it?

And apparently that's what the media circus was about. Somehow the pigs had decided to imply or infer some sort of über-nefarious activities by me, even though what they displayed was perfectly legal, except for two rifles, which they only found by a grossly illegal search of my house without a warrant. They lied to the press about having a warrant. One of many lies to come.

They sometimes like to tell you when unhappy circumstances unexpectedly befall you that, "Life's not fair". Which is not exactly useful, and too cliché to be even minimally helpful, comforting, or even uplifting.

Next time someone tells me that "Life's not fair," I'm going to kick him in the shins for no reason, and quip, "You're right."

But more to the point, I think that life's quite fine. It's **people** that suck. "Life" just got the bum rap...

I had always wanted to be famous, not infamous. Noted, not notorious. Spectacular, rather than just a spectacle. Exalted, rather than exiled. I wanted accolades, not opprobrium, kudos, rather than kicks.

I **did** want to be cremated, but I was kind of expecting to kick the bucket **first**, ya know.

And, God, was I ever shocked when I found out what I thought was a minor criminal weapons offense bust -- a false arrest and an illegal search, let's make that clear once again -- had put me on the front page of the newspaper!

But then again, so has every puppy in Ottawa...

"The only thing worse than being talked about, is not being talked about," said Oscar Wilde famously. I'm going to have to reconsider this. It appears that I'm now an official one-man **scandal**, my reputation in tatters. And it wasn't really much fun at all.

I used to joke that I'd always wanted to be driven into exile, laughing at the irony of fate as I kept one step ahead of a howling mob of social inferiors. Something to tell your grand-kids about and chuckle.

Something that puts you in the company of such historical figures as Voltaire, Oscar Wilde, Pablo Neruda, and Leon Trotsky. Vilified at some point, but eventually resurrected as men of superior caliber and substance, pillars of history, the inhabitants of the pantheon of humanity.

Really, only the unfortunate victims of humanity's forgotten fools or the hopelessly corrupt in power at the time, unable to appreciate greatness – or competition – in human form.

When I was younger, I figured the "best" way to go was to be shot to death in someone else's bed, lying next to his woman. (My last words being, "Relax, I was just borrowing her..." BANG!) But this was more like being tarred and feathered, then pilloried -- unceremoniously clapped into the stocks in the middle of the village square for the entertainment, amusement, and satisfaction of the general population.

I didn't quite realize it yet, but I was the latest 9/11 "scape-goat of the month".

* * * *

"If there were dreams to sell,
What would you buy?
Some cost a passing bell;
Some a light sigh,
That shakes from Life's fresh crown
Only a rose-leaf down.

If there were dreams to sell,
Merry and sad to tell,
And the crier rang the bell,
What would you buy?"

-- "If There Were Dreams to Sell" (1835)
Thomas Lovell Beddoes

The late Dr. Kubler-Ross studied the emotional readjustment to the specter of impending doom in her terminally ill patients. Her ground-breaking research revealed that the ultimate slow-motion emotional earthquake of having to face death could be divided into five discrete, sequential stages.

The five stages are, in order: denial, anger, bargaining, depression, and finally acceptance.

Kubler-Ross is actually mentioned by name, and the five stages immortalized and satirized in an amusing comedy stand-up routine in "All that Jazz" (1979), an entertaining and engaging movie directed by Bob Fosse.

The movie stars Roy Scheider, as a brilliant but morally flawed director/dance choreographer who has a heart attack in the middle of a million dollar Broadway musical/dance production. Confined to the hospital, he reviews his amphetamine tablet and art-driven – but frivolous, ego-centric, and philandering life – playing out, one-by-one, Kubler-Ross' five stages of dying.

But the five stages can actually be applied to any emotionally traumatic event. Being fired, divorced, or, say, being falsely arrested for a serious criminal offense, and tossed into the slammer.

Denial:	You've got to be kidding?
Anger:	Fascist Pigs!
Bargaining:	I'm a mild-mannered, suburban, upper middle-class **engineer**. You **can't** arrest me! My taxes **pay** your salary!
	OK, **fine**! You want to play rough -- my lawyer **kills** a wolverine with his bare hand and eats it for breakfast. When he gets through with **you**, you'll be washing meter maid uniforms for a living.
	Ouch! Stop hitting me!
Depression:	Screwed again. "Oh well, whatever, never mind..."

And, finally -- a drum roll, please:

Acceptance:	I'm fucked...

And now I can add to my resume "black market arms trafficker", or simply "gun-runner." Actually I'd think when armed to the teeth, with modern small arms firepower, you needn't hurry at all. You could walk...or even **saunter** quite casually...

And as Jerry Seinfeld explained memorably in a 1996 (Season 8) episode of his well-loved 1990's TV sitcom, "Seinfeld":

> "Guys with guns don't 'understand'. That's why
> they **get** guns -- too many misunderstandings!"

Thirty years ago I gave up trying to co-exist within an intolerable world seemingly full of assholes, and submerged myself deep into a calming sea of booze. In early 2006, I quit drinking, and as my head cleared, looked around, and to my utter surprise found that the world was **still** full of assholes.

What **was** I thinking?

And some of them had just thrown me in jail without reasonable excuse. Apparently assholes have gotten meaner and nastier in my "absence".

For my high school (Lisgar Collegiate Institute in Ottawa) yearbook graduation picture, over three decades ago, the caption below my picture consisted only of a quote from famed liberal lawyer, Clarence Darrow:

> "I have suffered from being misunderstood. But I would have suffered a hell of a lot more had I been understood."

I guess, I could say that they finally had understood me.

And last, but not least, that old joke about not dropping the soap in the shower, will never quite be the same for me ever again...

> "Why do you have to go and make things so complicated?
> I see the way you're
> Acting like you're somebody else, gets me frustrated.
> Life's like this: you fall, and you crawl, and you break, and
> you take what you get, and you turn it into honesty,
> Promise me, I'm never gonna find you fake it. No, no, no."
>
> -- "Complicated" (2002)
> Avril Lavigne,
> Lauren Christy, Scott Spock, and
> Graham Edwards

> "There is no crueler tyranny than that which
> is perpetrated under the shield of the law
> and in the name of justice."
>
> -- "The Spirit of the Laws" (1742)
> Baron de Montesquieu

Prison Diary 5: A Bullet With My Name On It
=======================

> "Traumatic events call into question basic human relationships.
>
> They breach the attachments of family, friendship, love, and community. They shatter the construction of the self that is formed and sustained in relation to others. They undermine the belief systems that give meaning to human experience.
>
> They violate the victim's faith in natural or divine order and cast the victim into a state of existential crisis."
>
> -- "Trauma and Recovery" (1992)
> Judith Lewis Herman

> "If we inquire historically into the causes likely to transform _engagés_ into _enragés_, it is not injustice that ranks first, but hypocrisy...
>
> [The need t]o tear the mask of hypocrisy from the face of the enemy, to unmask him and the devious machinations that permit him to rule without using violent means...so that the truth may come out..."
>
> -- "On Violence" (1970)
> Hannah Arendt

"Fa, fa, fa, fa, fa. Fa, fa, fa, fa far better. Run, run, run, run, run, run, run away."

"Snake eyes." I was bound to roll snake eyes eventually, after leading -- in a certain sense -- a mostly charmed existence. But this was snake eyes with Cobra fangs and fury beyond reason. I had been dragged into a dark pit, perhaps never to be seen again.

It's hard to express the feelings of disillusionment and betrayal one feels in a situation like this. In my case, it was just the final cynical nails in the proverbial coffin of "I told you so," "I knew it all along," and "Fucked, again!" in a country where the stupid vie for power with the corrupt -- the jackals ruling the jackasses. And where the race to intellectual or moral bankruptcy is the side-show.

What was all this nonsense? Front page news, TV news stories, morning radio reports, murmurings of "terrorism", incarceration in a maximum security hell-hole. What the hell was going on?

Alas, poor Yogi, I knew him well...

> The clusterfuck is coming to town,
> The clusterfuck is here,
> The clusterfuck is up your butt,
> I **told** you it was near!

* * * *

> "[T]he nation's political institutions too often serve up essentially symbolic solutions that fail to resolve deep-seated problems that have over time become worse, and a correspondingly disillusioned and disempowered public is drawn into a culture of consumption and entertainment that provides them with a compensatory but ultimately erosive sense of empowerment."
>
> -- "What Really Happened to the 1960's" (2010)
> Edward P. Morgan

I was born in Ottawa, grew up, and spent most of my life there, except for four years in Toronto, four years in Los Angeles, and fourteen months in Paris, France. I was 48, with **no** criminal record (in Canada or Europe...).

I was a suburbanite, a car owner, a home-owner, a tax-payer. I was a computer engineer, and also had another degree – an Honors Bachelor of Science degree. I was an employee with Nortel Networks for the last thirteen years as a "Member of Scientific Staff" -- a software designer of telecommunications software. I had two children by two ex's, and a solid personal life of decades of troubled, but controlled chaos.

So what exactly the hell was going on? Did they not have enough **real** criminals to prosecute, so now they were turning on innocent civilians? Or more likely, was it too hard to catch the real ones?

"_Sauve qui peut!" ["Save yourself, if you can!"], I cried out to sound the alarm. But it was to no avail, as I was the only one in trouble.

And I had been taken completely by surprise, at that mature stage of life perilously close to the inevitable death spiral of decay. I was way too old for this kind of crap. Forty-eight is **not** the time to deal with the Hurley-burly nightmare world of incarceration, with its violent and dehumanizing environment. Prison is a young man's harsh game. I was way out of my league and I knew it. I was unprepared, outgunned on all fronts, and my feelings of vulnerability were palpable.

Boredom and stress I could handle. (Employment had trained me well...) But an environment of unpredictable and senseless violence, continual institutional abuse and degradation, and open-ended uncertainty, combined into intolerable suffering, I was having to learn to survive.

* * * *

> "And I have never felt, quite this close to Hell,
> All this rock and roll baby, only time will tell
> ...
> Smoke baby, smoke baby, more alcohol, baby.
> Cocaine in Montreal, and back out on the plane, baby."
>
> -- "Smoke Baby" (2006)
> Hawksley Workman

I had been invited by the mother of my daughter to come and stay at their house in Gatineau away from the temptations and personal problems I was having living alone in Ottawa.

I had accepted the offer, after ensuring that we would have sex together, while I stayed there. I was to sleep in a small guest room with a single bed in the basement, though. When I got there, I surveyed my room. It was okay, even cozy, but it was a demotion from the second floor master bedroom where we both used to sleep.

Was I now the troll kept hidden away in the basement? So, to understand and characterize the exact nature of the new paradigm, after looking over my new room, I looked at her, and said, "Let's fuck."

After a brief argument -- or at least an intense discussion -- she agreed, and we stripped, climbed onto the bed, and fucked missionary style on the single bed in my little room in the basement.

But after a week of peaceful co-existence -- including daily sex -- she suddenly went crazy on me, giving me an unprovoked hard slap on the side of the head while I was lying on my side, naked in her bed, totally relaxed, cuddling her clothed body, having invited her to take a work break -- she worked at home on a computer -- by joining me in bed just to hold her.

But then suddenly she was up and standing, yelling at me to get out, and yelling to my daughter in the next room to call "911" for no reason that I could fathom. I had said something, I think, but it was an innocuous joke, or something inconsequential. I don't even remember -- and she went ballistic on me, and -- I was half-asleep -- struck me violently.

I had no idea what was going on, but soon threw on my clothes, grabbed my things, and fled in disarray, totally confused as to what was going on, but clear on the concept that I did *not* wish to deal with the police. I was aware my ex- had been diagnosed as having something called "borderline personality disorder", and was under treatment by a psychologist, but this behavior was way over the top. I've dealt with a lot of crazy people in my time, but this pattern of insanity was beyond my comprehension or experience.

And the principle of the thing. Fabricating an assault complaint against your mate and the father of your daughter? That's crazy! About someone you cared about, and valued. It was absolute treachery. A senseless and unexpected stab in the back. Just because you're angry with a partner, you **never** tap the State's considerable power and manipulate it as a weapon to settle a personal dispute. Talk about an act of dishonor. She knew that was how I felt.

To summarize: "blood -- including your woman -- is thicker than water" and "the state is not your friend".

And when she was believed, with no policemen ever talking to me, no witnesses, no marks or bruises on her, or any other evidence, or other corroboration, and I was called in by the Gatineau Police about a week later, and then arrested and charged, I was aghast.

When I was released after a couple of hours or so, after signing some standard bail conditions, things started to look up. The only catch was that I was going to have to surrender my fourteen registered firearms -- a standard bail condition -- until the criminal complaint was resolved.

They knew about my gun collection, because if you ran my ID or name on the police computer, CPIC, the Canadian Police Information Computer, the printer choked with a list of all my guns. They were registered after all, a development introduced in 2002 or so by the Liberal Party of Canada, amid a lot of controversy.

And when I was driven (without handcuffs or restraints) to my house in Ottawa to effect the "surrender" of my collection of rifles (an M1 Garand, a Simonov SKS semi-automatic rifle, a couple of bolt action rifles, and a stainless steel .22 semi-automatic rifle) and two 12 gauge shotguns, some restricted weapons (handguns: a Colt .45 M1911A1, Beretta Model 92FS 9 mm, and a 5-shot .38 ACP Colt revolver, with a 2" barrel), and prohibited weapons (most military semi-automatic rifles, like an AK-47, AR-15/M16, FN-FAL, MP-5), I suspected nothing untoward.

Also to be surrendered was any ammunition, of which I had dozens of wooden cases amounting, they tell me, to 30,000 rounds of .22 rim fire, and center-fire handgun and rifle ammo. Later, the police and new stories made much of this. It was apparently a cause for much surprise and raised eye-brows, and snide remarks. But the bottom line is that it was perfectly legal for me – or anyone with a gun license – to own 30,000 – or 3 million – rounds of ammo, if they could afford it.

I have never felt the need to justify to the authorities conduct which is legal. This is what is known as innuendo and mischaracterization.

And since it was legal, naturally I faced no charges regarding the ammo. But since it looked "suspicious" to non-gun persons, it was ideal for whipping up public fear and attention to cover up a false arrest and contribute to denying me bail under false pretenses.

And just to quickly explain, since the 14 guns were of 11 different calibers, and thus required 11 different types of ammo, I had something like 2,700 rounds per gun. You can shoot a couple of hundred rounds in one afternoon at the range, it's cheaper to buy in bulk, the price was right, etc., etc.

As Mark Twain said, "Figures lie, and liars figure". Or is it, "There are lies, damned lies, and statistics."?

There are other reasons and justifications. But when public officials feels perfectly comfortable engaging in gross distortion and manipulation -- otherwise known as a smear job -- something is wrong with the system. It was power abused and misused by intrinsically corrupt, overpaid white male _apparatchiks_.

"From the moment of their abduction, the victims
lost all rights. Deprived of all communication with
the outside world, held in unknown places, subjected
to barbaric tortures, [they were] kept ignorant of their
immediate or ultimate fate. ...

They were not mere objects, however, and still
possessed all the human attributes: they could feel
pain, could remember a mother, child or spouse,
could feel infinite shame at being raped in public..."

-- Argentinian "Never Again" report (1984)
Ernesto Sabato

Prison Diary 6:   The Things I Left Behind
        ====================

> "The underlying problem is not men as a
> sex, but a social system in which the power
> of the blade is idealized -- in which both men
> and women are taught to equate masculinity
> with violence and dominance..."
>
> -- "The Chalice and the Blade" (1988)
> Riane Eisler

> "Suddenly there fell upon me without any warning, just as if it
> came out of the darkness, a horrible fear of my own existence.
>
> ... Something hitherto solid within my breast gave way entirely,
> and I became a mass of quivering fear. After this the universe
> was changed for me altogether."
>
> -- "The Varieties of Religious Experience" (1929)
> William James

Ever wonder what jail is like? The view from down here?

I never did. It never concerned, or worried me in the least. In the past, I would just shudder every time I passed the OCDC jail -- that awful, forbidding-looking place of tall chain link fences, topped with coils of razor wire, in the East End of the city, way down on Innes Road -- 2244 Innes to be exact, just shy of suburban Blackburn Hamlet -- and was happy knowing that I'd never end up in that horrid place.

But today, my reality had changed utterly. And here I was, alone, confused, disoriented, and unsure. Only one thing was clear: "I wasn't in Kansas, anymore".

So I did the only thing I could – except panic. I tried to take stock.

And taking stock involved sorting through my emotions, allowing logic and the instinct for survival to take control. I cataloged my resources, attempted to begin understanding my immediate environment, find options, anticipate dangers, and establish a game plan. Taking stock involved establishing that hope was one of the few emotions that was allowed, patience the order of the day, and logic the commander-in-chief.

And it was cold unemotional logic that enabled me to suppress the emotions that normally might have torn me apart. And years of practice suppressing them... Combat mode, once again. I had almost forgotten he was there. My only friend still alive from my "private wars" gone by.

My "private little wars"...

Taking stock was mostly about the price I was already paying. And it was surely a high one. I was ground zero as a metaphorical nuclear bomb detonated above my entire life. My entire belief system had just experienced a nuke attack from my own government, my own country of birth. For no reason that I could fathom, completely unexpectedly and out of the blue.
I had been unceremoniously and publicly stripped of my reputation. Now the process of jailing me was to try and strip me of my dignity.

Suddenly, it was Phoenix Falling.

But would there be a Phoenix Rising? A rebirth or resurrection of any kind? I had certainly experienced major crises before. My personal life was one crisis after another. And I had an air and affect of unflappability, as a result.

But this time was different. Something was way different, and I would need to get my bearings. Fast, and yet very carefully. I did not have any sense of hopelessness at all -- I knew that my arrest had been illegal, and that I would be completely exonerated at the proverbial "end of the day" -- but I was still confused at why things had happened the way they had.

It was "the end of the world as I knew it." And I didn't feel so fine. But uncertainly was a bit exciting, I had to admit. Uncomfortable, highly unpleasant, degrading, humiliating, and tawdry, but exciting. Fascinating to be precise. Informative -- new information, new data. A new world. But, at the same time, I mean, I knew it was the proverbial "all a big mistake" situation. I knew that sooner or later, some official would come to their senses and realize that they had **nothing** on me.

Phoenix Rising.

* * * *

"When sorrows come, they come not in single spies
But in battalions."

-- "Hamlet" (1602)
Act IV, Scene V

Phoenix Falling.

Taking stock. There was a need to catalog what I had lost. The things I left behind. The life I left behind.

Regrets, I've had a few. ...But they were big ones, alas.

I remember when there was a time I could say that I had never hurt anyone. Then one day, I realized that I had to modify my statement. "That I had never hurt anyone that didn't have it coming in spades." Which doesn't sound quite as good **at all**.

The things I left behind. I'm reminded of the conclusion of Justine – the bitter, neurotic, pill-junkie cop's wife – and her "I'm leaving" monologue to her mate, actor Al Pacino, a dedicated, workaholic, hard-driven detective in Michael Mann's hound-dog cop versus super-criminal movie, "Heat" (1995):

"The rest is the mess you leave behind."

In one fell swoop, I had lost my freedom. I lost my right to freedom of movement – freedom to come and freedom to go, or simply to stay wherever I was welcome to stay or live.

I lost my freedom of decision, freedom of association, or non-association. I lost my political rights – disenfranchised, in other words – including freedom of speech, expression, and assembly.

I especially lost my most valued privacy, dignity, self-respect, sense of safety. I lost my sense of comfort. I lost my identity, and became just a number – an entity to be processed, and disposed of at the pleasure of the State, and in the manner of their choice. They had taken away my sense of certainty and serenity. They controlled my here-and-now, and they controlled my future, with the stated intention of destroying it, as their sole choice.

And they took away -- savaged – my precious, most valuable, most treasured right simply to be left alone. And with it, the right to my innermost, most intimate, personal, and private sense of purpose: my own pursuit of happiness and contentment, as I perceived and desired it.

They suspended and took away whatever happiness and contentment I had at present, and cast a shadow over my happy, pleasant, joyful, or humorous memories of times past. And they took away my goals, strategies, plans, or intentions of making for a happy -- or at least happier – future.

In other words, they had snatched away my dreams, and replaced them with uncertainty, fears, dread, and foreboding of dead-ends, and downward spirals. Or just a dead fall onto a sword in the gut that the world would watch, and applaud as a just and well-deserved ending of a bad man.

He got what he deserved, was branded a criminal, stamped a state-certified low-life, and lived unhappily ever after.

It would be the last chapter of my life – a bad ending to a frequently troubled life, already deeply marred and scarred with pain and suffering. My final and "fatal" mistake, exposed to the entire world and branded an incorrigible, irredeemable, deeply flawed man. Twisted and unsympathetic, and best written off and forgotten by all.

* * * *

"Flashback -- warm nights,
Almost left behind,
Suitcase of memories,
Time after --

If you're lost...
Time after time, Time after time,
Time after time, Time after time."

-- "Time After Time" (1983)
Cyndi Lauper, Rob Hyman

I had involuntarily traded the pursuit of happiness with the pursuit of survival when I was unceremoniously tossed into the violent, barren and desolate, claustrophobic concrete and steel human-constructed hell, known as the OCDC, Province of Ontario jail. They shoved me into a hell-hole populated by strangers -- including career criminals, the violent, the predatory, the psychotic, the sleaze-balls, and the immoral. Or some combination of these fine traits.

I lost access to my aging (80 and 88) parents – especially missing my mother – and imagining how awful a feeling it would be – a real horror – if she died with me in jail.

I missed my women "friends", the feelings of pride and joy of my two wonderful children, my friends, the safety and serenity of my house, my daily routine, the comfort of my own bed and clothes, access to **any** reading material.

I even lost such minor daily choices as meals, and the ability to exercise, sleep in, or watch TV when I wanted, to be alone when I wanted, or to be with the people of my choice at my digression (er, I mean "discretion"...).

A legal gun owner for the last twenty-five years or so, I was charged -- after an illegal warrant-less search and then a false arrest for a machine gun that was legally, and obviously, a deactivated hunk of metal -- with possessing two unregistered rifles.

Without a criminal record, and charged only with a paperwork offense – criminal, for sure, but still minor -- I was classified as a "violent" offender – since it was **weapons** paperwork – and placed in the violent and punitive maximum security cell-block: 21 hours a day in a two man cell, with three hours "out" on the range where I could associate with Ottawa's murderers, major drug traffickers, bank robbers, girlfriend beaters, and other assorted violent and career criminals, and other upstanding citizens of the community.

And with the mattress in your cell only two inch thick, and six feet long, on a sheet steel bunk bed, it was always uncomfortable. And God help you if you were very tall, as the mattress was only six feet long.

As a government detainee, convicted or pretrial, treatment was identical. It was my first exposure to Canada's inability to understand and respect the fundamental legal concept of "innocent until proven guilty", that they claimed to uphold. It was actually a total joke -- you had better conditions if you were convicted, or had plead guilty to the crime charged!

But worst of all, the most devastating, debilitating, scary result of all, was not just that I had involuntarily been yanked from my free-wheeling routine, and arbitrarily denied all the options of freedom, and that society's guardians had replaced them for me with a terrible feeling of uncertainty, disenfranchised me without due process, pilloried me and condemned me to disgrace, and being branded a pariah. It was as if I had been sentenced not by trial, but by officialdom's whim, to social and civil death.

But more – much more – than that, was the worst fate of all:

"Naked to mine enemies", was I.

"There are many shades in the danger of adventures and gales, and it is only now and then that there appears on the face of facts a sinister violence of intention --

that indefinable something which forces it upon the mind and heart of a man,
that this complication of accidents or these elemental furies are coming at
him with a purpose of malice, with a strength beyond control, with an unbridled
cruelty that means to tear out of him his hope and fear, the pain of his fatigue and
the longing for rest:

which means to smash, to destroy, to annihilate all he has seen, known, loved,
enjoyed, or hated; all that is priceless and necessary -- the sunshine, the
memories, the future; which means to sweep the whole precious world away
from his sight..."

-- "Lord Jim" (1899)
Joseph Conrad

Prison Diary 7:   A Prisoner of Politics
                                   =================

> "To these I turn, in these I trust
> Brother Lead and Sister Steel.
> To his blind power I make appeal,
> I guard her beauty clean from rust."
>
> -- "The Kiss" (1916)
> Siegfried Sassoon

> "Then the brutal minions of the law fell upon the hapless
> Toad; loaded him with chains, and dragged him from
> the Court House, shrieking, praying, protesting;
>
> across the market-place, where the playful populace,
> always as severe upon detected crime as they are
> sympathetic and helpful when one is merely 'wanted',
> assailed him with jeers, carrots, and popular catch-words;
>
> past hooting school children, their innocent faces lit up
> with the pleasure they ever derive from the sight of a
> gentleman in difficulties..."
>
> -- "The Wind in the Willows" (1908)
> Kenneth Grahame

In one of the few scenes offering social commentary, and not littered with the verbal use of the punctuation mark "motherfucker" (in the 1983 Brian de Palma film -- which became a _fin de siècle_ cultural icon -- "Scarface"), cocaine kingpin Tony Montana -- played by Al Pacino -- finishes his drunken lecture about their hypocrisy to the gawking patrons of a high-class Miami restaurant, following a very loud and public spat between Tony and his wife, the latter whom then contemptuously stalked out.

And finally, after Tony's drunken tirade, he soon lurches towards the restaurant exit, announcing facetiously:

"Make way for the Bad Guy!"

Now, switch gears to one of those 2-minute Canadian CBC – government-funded – public service historical, educational TV ads -- the CRB Foundation's "Heritage Minutes".

The year is 1837. A young French-Canadian _patriote_ has been arrested and is being dragged past the bars of the prison cell of a certain Etienne Parent, who is seated, and quietly writing.

In the background, you hear the new prisoner, angry, and defiant -- those damn "politicals" can be a real nuisance, as well as being so noisy -- denouncing the shamelessly rigged system that feeds simmering political resistance against English oppression of the French-Canadian minority. "This **corrupt** regime!" you can make out the incorrigible activist shouting.

The more reserved Monsieur Etienne has been jailed (briefly) for his writings -- termed by the authorities "seditious schemings" -- expressing the opinion in his newspaper, "Le Canadien", the idea that there was no reason the dominant English and the systematically discriminated against French minority could not both thrive by living together in harmony for mutual benefit.

Now, switch gears again. I'm entertaining a charming but "street-wise" young "lady" of new acquaintance in my house, and I guess she'd managed to formulate an assessment based on various pieces of information that I showed her about me. She looks over at me across the bed, and asks, "Are you an assassin?"

"No, no," I answered firmly, "I'm a 'political'." -- continuing my practice of clarifying nothing with my responses.

I should have added that some people have claimed that I assassinate ideas on a regular basis.

> "The only significance of life consists in helping to
> establish the Kingdom of God. And this can be done
> only be means of the acknowledgment and profession
> of the truth by each one of us."
>
> -- "The Kingdom of God is Within Us" (1894)
> Leo Tolstoy

* * * *

Busted on a Wednesday night, I was only arraigned -- formerly charged -- before a Justice of the Peace on Friday, mid-morning. I suspected the delay was because the Ottawa police realized that I had been falsely arrested for possession of a prohibited weapon. This was presumably my collection of deactivated, and thus perfectly legal, sub-machine guns and light machine guns.

I say "presumably" because the idiot arresting cop failed to respond to my repeated question "What prohibited weapon?", compounding the illegality and impropriety of the arrest, which requires detailed and specific answers to an arrestee's inquiries.

Since even a cursory visual examination would reveal that they were obviously deactivated, non-functional, destroyed firearms it was amazing that this Ottawa pig was called to come and arrest me.

One of the Gatineau pigs had seen the large number of machine guns lying around my gun storage room, and appeared startled. I quickly picked up the nearest sub-machine gun and explained that the full auto guns were all deactivated "non-guns". I showed him the SMG I had picked up, the parts of the gun destroyed by grinding and the extensive welding -- all quite visible -- that rendered each of them non-functional, unusable, and perfectly legal. They were not considered firearms under the law, and required no paperwork -- permits, registration, or licenses -- to own, sell, transport, or anything else. He "seemed" satisfied at my explanation, and didn't ask any questions. I put the SMG down on the floor, and we continued collecting up my **real** working and functional -- and licensed – firearms, all carefully stashed away in locked cases.

The "official" recommended procedure to deactivate a firearm was detailed, extensive, and precise. The breech end of the barrel was plugged, the chamber welded shut. Then the barrel was welded to the receiver -- to prevent barrel replacement. The full-auto triggering mechanism

– the auto-sear – was destroyed by grinding or welding, the bolt face was ground down, and the magazine was welded inoperative.

They were legally just hunks of steel -- non-firearms -- with maybe some plastic or wood attached.

Nonetheless, like all the other errors the police had made at every step of the nightmare, and would continue to make, they happily continued doing whatever they felt like, and always to my detriment, regardless of common sense, the law, or competent exercise of their sworn duty as law enforcement officers.

So now they had to find some charges that would stand up -- to avoid embarrassment, charges of blatant incompetence, and a winnable law suit for false arrest, and other assorted indignities I was put through. Conveniently, two unregistered -- and thus illegal -- rifles had been finally found. But they were found only after a brand new Ottawa police-conducted illegal warrant-less search following my false arrest, which was the result of an illegal warrant-less search by Gatineau police of an Ottawa home. It just seemed to go on and on, as the snowball continued to grow.

The Gatineau officers apparently didn't know what a search warrant was, or looked like, so they assumed that any piece of paper they had was a search warrant. And then Ottawa Police joined the "party". The Gatineau pigs had left with my registered guns and ammo, but now the Ottawa Police started their own search of my home after I had been arrested, publicly humiliated in front of my neighbors, assaulted, stuffed into a police cruiser, and eventually taken away.

The Ottawa Police had no search warrant either. How they justified **their** search, God only knows. They took the word of the Gatineau Police? Who didn't have a copy of the search warrant as required, on them? What B.S.! It was long past going from ridiculous to outrageous.

The Gatineau Police had left with my legal guns and ammo, as that was the only action authorized by the piece of paper they **did** have (a standard no jail "promise to appear" release form). The form had a standard list of release conditions, including that I had to "surrender" my registered firearms and ammo to the police, and which mentioned nothing about authorizing or consenting to a search.

After a stay at Elgin Street jail, and then my Court arraignment, it was off to "OCDC", the [County of] Ottawa-Carleton Detention Center -- the local provincial jail for me. After several hours of waiting, processing, and waiting, at about Friday around 4 pm, I arrived in what the government decided was to be my new home for now: "OCDC general inmate population -- maximum security".

And after a quite extensive period of time in jail, as I got used to things, and took inventory of just how fucked I was, I decided that I needed to get my "head space" into the appropriate place to survive.

And so, I quickly and reasonably decided that I was perfectly justified in thinking of myself as a "political prisoner". And indeed, this belief never left me throughout my incarceration, and provided me with strength and peace throughout my ordeal.

I had been subjected to illegal acts under color of state law at every step of my present troubles, and would continue to be until I was released.

I had not registered two of my 14 firearms as an act of resistance and civil disobedience. I was hurting no one, I was licensed to possess the two unregistered rifles (a beautiful, semi-automatic, Hungarian-manufactured AK-47 and a Colt Sporter H-BAR (heavy barrel) -- the semi-automatic version of the U.S. M-16 -- the two most popular rifles in the entire world).

Yeah, OK, I admit it they were heavy-duty military assault weapons. Well known. Pretty much the best military rifles available to a guy with a "prohibited weapons" designation like I had. Big deal. So what if your average citizen didn't have an AK, and was horrified by them? It was legal at the time I bought them, and still legal for me to own them.

I had bought the unregistered rifles legally, and did not feel it was my job to inform the government about anything further. I was charged with a criminal paperwork violation, and I would normally not serve jail time for such an offense. The maximum they **could** jail me for, was two years.

But I didn't register the two rifles because after much thought, literally months of thought, I had made the "Command Decision" to register all my guns but two. That I could keep unknown to the Government if they instituted a complete gun ban and mass gun confiscation. Now a Command Decision is a formal process within my mind over an issue of extreme importance, where the consequences are bad either way, or bad if the wrong choice is made. A Command Decision is not taken lightly. It is for the hard decisions; where both choices have a cost, and weighing which is the better choice; where you have to take the responsibility for your judgment, and maybe take the consequences.

But knowing that you did your best in the decision-making process.

The two unregistered rifles were my "insurance", because I didn't trust the government. And apparently I was right. They couldn't even obey the basic law of the land, the Charter of Rights and Freedoms, and they expected me to obey a stupid paperwork registration law? They had committed criminal acts, which if I had committed would have gotten me possibly years in prison, such as Criminal Code s.348 "Break and Enter".

The point of course, is, should the state commit far greater crimes to find a little crime, without even an excuse to do what they did? No informant, no information of illegal conduct, no evidence, nothing.

The Canadian Charter of Rights and Freedom says, "Of course not."

* * * *

> "Got a letter from a messenger, I read it when it came,
> It said that you were wounded, You were bound and you were chained.
>
> You had loved and you were handled, You were poisoned, you were pained,
> Oh no. Oh, no. You were naked, you were shamed.
>   [...]
> For a minute I let my guard down, Not afraid to be found out,
> Completely forgotten, What our fears were all about."
>
> -- "The Messenger" (1993)
> Daniel Lanois

I called up my mother from jail -- ever considerate of punctuality and the burden of other minor civilities that differentiate the couth from the unwashed -- to tell her to alert my doctor that I would miss an upcoming appointment. He was the head of my psychiatric team (actually, it's more like a "gallery" than a team...).

I told my mother to give the good doctor a message. "Tell him that I can't make my appointment, because the Government threw me in jail, and that I'm a political prisoner," I told her at least somewhat facetiously.

It was hilarious -- I found out later that she had actually left a message on his office answering machine:

> "Yogi says to tell you that he's a political prisoner of the Canadian Government and will therefore miss his upcoming appointment."

Well, I guess it was about time. I suppose I'd always wanted to take a stand for my political beliefs. Notice that I didn't mention having to suffer, or be martyred as a result...

Controversy or even public opprobrium was as sufficient a storm to weather for expressing an unpopular opinion that needed to be said by **someone** was certainly something I could do and not regret doing, no matter the consequences, or where the chips fell.

And I'm immediately suspicious of beliefs whose proponents expect to be greeted without even a single questioning statement of doubt or criticism. It's not that I like to "ruffle feathers" gratuitously. But sometimes, rather, that chicken needs to be **plucked**!

And if these people think they can slap me down with "outraged" reactions, or venomous attacks in response to my expressing of a reasonable opinion not conforming to their need for support or silence from their audience, they find out they are now wrong a *third* time.

And then they get a lesson I call "Free Speech Can Be Fun(ny)!", as I calmly combine my rapier wit to poke fun and make them appear ridiculous to the audience, as I weave in my counter-arguments and counter-opinion to their original political statement.

When one individual told me one time, to "Get off my high horse," when I criticized municipal police by referring to them as "pigs", I took mild offense. His statement implied that my argument was:

1) utterly devoid of merit and,

2) born of some sort of unreasonable assumption of general police impropriety, and

3) it was okay anyway to ignore police impropriety as acceptable public policy,

that justified his dismissive comment.

Anyway, I responded, "Would I be wrong in suggesting that he was so narrow-minded that I could blind you in both eyes with a pin?" The battle was on.

When he feigned outrage at **my** "personal attack", I spanked him into silence with the remark, "People who tell me to 'get off my high horse' should not stand under its tail and look up."

The audience roared with laughter, as I steeled myself for the next challenger to step up to the plate, and replace his downed team-mate.

"The tongue is the only tool that gets sharper with use," someone once said. The "blind with a pin" comment I borrowed, too.

* * * *

" 'If you had known what "I want mercy, not sacrifices"
means, you would not have condemned innocent people.' "

-- Matthew 12:7

There is a widely held belief that all you have to do when confronted by moral fraudulence is get comfortable with it.

It's no longer "The Truth shall set you free." It's that **you** have to set free the Truth. But don't expect any thanks for doing this.

Just ask Socrates, Gaius and Tiberius Gracchus, John the Baptist, Jesus Christ, Joan of Arc, Voltaire, John Brown, Henry David Thoreau, Gandhi, Bertrand Russell, Ho Chi Minh, Muhammad Ali, Malcolm X, Martin Luther King, Timothy Leary, Nelson Mandela, and Canadian, Dr. Henry Morgentaler.

Persecution and martyrdom as a retirement plan? How's that for irony? To suffer for new ideas, insight, and being ahead of your time. Or just pointing out the obvious... (It was disappointing enough when I found out that my scintillating conversation was utterly useless for picking up chicks. Thankfully, I had a more earthy "Plan B"...)

And when I found out that suffering was going to be my "reward" for seeing the Truth, in my more immoderate days I planned the make other side suffer first. Being ever practical, it would have been fine, I decided, to go down fighting -- as part of a sufficiently large group of like-minded individuals -- as long as I could personally take down at least forty or fifty of the "enemy" along with me.

Exile -- now that was oh-so much more civilized punishment for ideological offenses, which I guess is why it's out of fashion. Exile, preferably to a villa in Switzerland, or even better, Nice, France on the Mediterranean coast. You just can't get enough of Provence, though it's a pity I don't drink anymore, because good red wine is cheap in France.

But I imagine, instead I can do things like ponder the bougainvillea as the sun sets, between half-hearted acts of decadence, as some sort of existential, post-modern paradigm for a happy ending.

And last, but not least, in the meantime, I can add one more interesting thing to my rather unique life history "resumé":

2007 – 2008: political prisoner (Canada) (Charges dropped).

It'll be right below the previous entry, "Intellectual in hiding", and above the entry "Went insane."

Oh, the infamy of it all...

"Where liberty is, **there** is **my** country."
"Where liberty is **not**, there is **my** country."

Prison Diary 8:

-- Tom Paine's retort to
Ben Franklin (ca. 1802)
Confessions of a Fatalist
====================

"If a stone should fall on a jug, it's too bad for the jug.
If a jug should fall on a stone, it's too bad for the jug.
It's always too bad for the jug."

-- The Talmud (408 A.D.)

"I look inside myself, and see my heart is black.
I see my red door, and it has been painted black.
Maybe then I'll fade away, and not have to face the facts,
It's not easy facing up, when your whole world is black."

-- "Paint It, Black" (1966)
Mick Jagger/Keith Richards

Well, here I am in jail, facing somewhat serious weapons offenses. Actually they're not that serious -- criminal certainly, but basically only paperwork violations.

But I'm being held in the maximum security section of the jail -- to my surprise, fear, discomfort, and physical danger -- because all the classification officer sees on my paper-work is "weapons offense", not "criminalized paperwork charge, regarding weapons". So little ol' harmless me -- non-aggressive, serene and sedate, middle-aged, upper middle-class, suburban, easy-going, mellow, and laid-back -- gets classified as "dangerous" and goes to "max" -- which **is** dangerous.

Kafka again?

I don't think it's deliberate torment -- it's just stupidity in rigidity, that's all -- but maybe they want me to confess? So, here goes: I confess. First of all, I confess that I'm a fatalist.

Well, where do I begin?

Fatalist Rule #1: "Friends come and go, but enemies accumulate."

Fatalist Rule #2: "Don't talk about your troubles. Because 80% of people don't care, and the other 20% are glad you are having them." [said by Baseball manager Tommy Lasorda]

Fatalist Rule #3: "Fool me once, shame on you. Fool me twice, shame on me." [said by Scotty on "Star Trek", the original TV series (1967)]

Fatalist Rule #4: "You play the hand you're dealt." No use fretting, whining, or complaining. Just deal with it (no pun intended), and get used to it.

Fatalist Rule #5: "Garbage In, Garbage Out" is in reality, usually the input and output of your mind's information processing capability, unfortunately. Beware. And Be Aware.

Fatalist Rule #6: "You pay your money, and you take your chances."

Fatalist Rule #7: "If it sounds too good to be true, it is."

"Assume everyone is lying, when it's in their interest to do so" is not just a rule, it's a good start to determining the Truth.

Running every human assertion by the "laugh test", is another rule that ferrets out the bogus. Like if someone says, "It's a slam dunk, sure fire win, if you invest in this."

* * * *

>"I do not hope to know again
> The infirm glory of the positive hour
> Because I know that time is always time
> And place is always and only place
> And what is actual is actual only for one time
> And only for one place."
>
> -- "Ash Wednesday" (1930)
> T.S. Eliot

The "marker" for an out-of-the-closet fatalist is his black sense of humor. Morbid, macabre, smuggler's moon dark, pitch black.

A metaphorical black hole from which no light escapes, with an event horizon the exact radius of reality, from which there is apparently no escape, other than in illusions, fantasy, dreams, self-medication, perma-numb emotional catatonia, insanity, or suicide.

But there is an acceptable escape. It is known as "fatalism".

But let's make it clear: fatalism is **not** hopelessness. Fatalism is simply an acceptance of the nightmare as not only true, but likely permanent, and is a way of masquerading and eventually convincing yourself into a mindset that makes this awareness tolerable.

After all, breaking up the word "fatalism", you get "fatal". The fate, the destiny, of everyone is to die one day. It is an absolute certainty, frequently feared, and thought of as unpleasant. But it is a fact of life.

Deal with it, says the fatalist, like the rest of life's down-turns. And so much for Christians, who are supposed to believe in a glorious after-life! Or Hindus and Buddhists who believe in re-incarnation.

* * * *

>"One evening I sat Beauty on my knees.
> And I found her bitter. And I reviled her."
>
> -- "A Season in Hell" (1873)
> Arthur Rimbaud

Again, I bring up Dr. Kubler-Ross' Five sequential Stages of how humans deal emotionally with finding out they have a terminal illness, and are soon going to die (it also applies to how people, in general, react to dealing with going through a foreseeable, life-changing traumatic experience):

Denial, Anger, Bargaining, Depression, and finally, Acceptance.

A confirmed fatalist goes at supersonic speed through the first four stages – sonic boom included for free – and crash-lands in a heap at "Acceptance".

In other words: "You're fucked whatever ya do. Deal with it as best you can." (Or the minimalist version: "Fucked, again...")

The standard evolution of the individual to fatalism, is from innocence to guarded optimism, to universal skepticism and the slow and painful death of hope, to a quick pirouette of _Weltschmerz_, to pessimism.

A down-turn to despair rises up to sardonic humor, then irony-spouting, evolving to bitter cynicism, and finally a downward spiral to biting sarcasm. Then it's on to depressing negativity, to alienation and estrangement, and then on to emotionally numb.

And **finally** to fatalism. The end of the line. Yet a stable equilibrium state, that's not so bad in the final analysis.

In my case, I also took a "couple" of detours -- self-righteousness/moral indignation, dominant-race hate, misogyny, misanthropy, utter moral pollution, and nihilism -- but the evolution was otherwise the same. Ummm. Right...Sure Yogi, whatever you say...

Cues to the evolution of the fatalist are often in his linguistic style: whining and complaining, snide, snarky, indifferent, cynical, sarcastic – and the "3 B's" -- biting, bitter, and blunt.

If optimism is the ideology of the dreamer, then pessimism is the on-going, flat-line of the fatalist. Negativity is the air that he exhales, and a sense of impending, ultimate doom that flows through his veins.

Actually it's ice-water in **my** veins. I paid for a transfusion.

Take the instance I became friends with a fellow inmate at Lindsay CECC (Central East Correctional Center). We began having friendly, thoughtful conversations together -- he had crippled a passenger-friend in a car accident while driving drunk. He was serving an 18 month sentence, and talking himself through his remorse and guilt -- including several intimate talks with me on alcohol and alcoholism. I was two years sober at the time, and a sympathetic ear.

Early one evening, we were on the range just before final lock-up, talking in private conversation. Suddenly, he said earnestly to me in a quietly enthusiastic, cliché revelation, "Think of how many lives alcohol has destroyed?"

"Think of how many lives it's made tolerable..." I retorted instantly, and instinctively, no harm or humor intended.

He didn't talk to me much about alcohol, after that, and was soon after transferred to the sentenced inmate pod...

You see, again, fatalism is not hopelessness. Not even close, though it's a common misinterpretation/misperception. Fatalism is not just a philosophy, not just a splinter faction of cynicism, not just a post-modern cross between depression and a lifestyle, or even just a depressing way of life.

After all, it comes with a free sense of [black] humor all its own!

Some people like to wallow in self-pity. The fatalist likes to wallow in pessimism. (Actually, I've never understood why people dismiss so instinctively "wallowing in self-pity". Would they rather I go on a killing spree? Sheeesh! In other words, it's my party, and I'll whine if I want to!)

Fatalism is a belief system that its proponents, such as myself, claim is a distortion-free view of reality. This claim is violently rejected by non-fatalists, who fatalists view as the deliberately blind.

(There are none so blind, as those I poke in both eyes...)

* * * *

"You come out at night,
That's when the energy comes.
And the dark sides light,
And the vampires roam.

You strut your Rasta wear,
And your suicide poem.
And a cross from a faith
That died before Jesus came."

-- "Building a Mystery" (1997)
Sarah McLachlan, Pierre Marchand

Now let's shift gears. One of my favorite rock bands is, not surprisingly, "Pearl Jam", the heart and soul of which is lead singer and song writer, Eddie Vedder. But with hits like "Wish List", "Daughter", "Jeremy", "Last Kiss", and "Better Man" -- righteous, on-target songs, yet at the same time, songs with depressing lyrics of abuse, pain, and the cycle of mistakes and bad choices repeated.

Songs of loss, unresolved sorrow, and emotional wounds unhealed. Scars old, yet still fresh. The ugly side of reality and the human condition -- unnecessary suffering -- turned into beautiful songs with hard-hitting lyrics that redeem themselves by ringing with the beauty of truth.

And every year, one of my New Year's Resolutions, I like to quip, is to say a prayer that Eddie Vedder feels better this year, from the pain he obviously feels. I sincerely do.

Eddie bears witness.

And I wish I could make the Truth sound as beautiful as he does...

* * * *

"The very tones in which we spake,
Had something strange, I could but mark;
The leaves of memory seemed to make,
A mournful rustling in the dark."

-- "The Fire of Driftwood" (1850)

My favorite Shakespearean play is "Macbeth". "There's daggers in men's smiles," is a typical line in this fascinating tale of power and ambition set in early medieval Scotland. Power has always had a magnetic attraction to evil men. But this play shows the raw evil that ambition for power inspires in the merely morally weak.

And finally, "Macbeth" is an insightful exploration of the bloody, tragic, and terrible cost exacted as Macbeth's regal ambition plays out with human life being reduced to the value of a roll of loaded dice. But Macbeth has been seduced by the three witches' riddles, not realizing that the final roll of the dice is his.

The fatalist would have anticipated the "snake eyes" roll coming up...

> "And all our yesterdays have lighted fools
> The way to dusty death. Out, out, brief candle,
>
> Life's but a walking shadow, a poor player
> That struts and frets his hour upon the stage
> And then is heard no more.
>
> It is a tale, Told by an idiot,
> full of sound and fury,
> Signifying nothing.
>
> -- "Macbeth" (1606)
> Act V, Scene V

* * * *

The fatalist is known to use the comment, "So it goes..." Its origin is Kurt Vonnegut's seminal novel (1969) and movie of the same name, "Slaughterhouse-5" (1972), and the author's emotionally numb comment "So it goes...", every time there is a recounting of a deliberate, human-induced horror or atrocity of some sort.

Vonnegut also wrote "Cat's Cradle" (1963), with everything that touched the synthetic "ice-9", turning into ice. It ends with the whole world dead and frozen, and the protagonist – the very last death – frozen with his erect middle finger giving the finger to God in the sky.

Charles Reich in "The Greening of America" (1971) "argued that the novels of Raymond Chandler, such as his best works, "The Big Sleep" (1939, movie in 1946), or "Farewell My Lovely" (1940), owed their appeal to their portrayal of a "world in which no one could be trusted, where one could only survive by living on one's wits."

Another reviewer (Pico Iyer) said about Chandler, that he was the "classic lonely romantic outsider for our time."

A Mother's milk description for the fatalist...

* * * *

> "What does the radical want? He wants a world in which
> the worth of the individual is recognized. He wants the
> creation of a kind of society where all of man's potential

> could be realized; a world where man could live in dignity, security, happiness, and peace -- a world based on a morality of mankind."
>
> -- "Reveille for Radicals" (1946)
> Saul D. Alinsky

This fatalist's favorite movies include Director, William Friedkin's, "To Live and Die in L.A." (1984). It is a film that superficially classifies as of the criminal mastermind genre. But it's a film that is assembled with a fit that shows a successful -- even brilliant -- achievement for perfection in movie-making, with an outstanding beauty, and complexity rarely seen.

It introduces itself with a desolate original sound track (by the rock group, Wang Chung) and foreboding, red-infused cinematography unveiling a twilight setting of corrosion and decay: the burned out and polluted urban wasteland of the industrial factories of South Los Angeles.

The movie unveils a plot thick with jungle vines curling and twisting to choke each other of light, oxygen, and invariably, life.

A master counterfeiter, it turns out, is guided by what seems to be a twisted and perverse perfectionism. It is his amulet in his search for the beauty in his sophisticated criminal enterprise -- and his true calling -- supreme puppet-master. For it is only in ruthless perfection does he think he will find truth, safety, and completeness of soul.

In his game are a collection of federal agents -- a cowboy cop whose former senior partner the puppet-master has had killed, his new novice partner, and the merely incompetent. The cowboy cop and puppet-master face off against each other, with a cast running interference, of a dangerously beautiful and sexual, avant-garde dancer girlfriend, icy and double-dealing corrupt lawyers, hardened street criminals, shady gang-leaders, and a seemingly innocent female informant working all sides and angles.

And the master criminal manipulates them all in a masterfully choreographed "ballet" dance by letting them think they are actually sleazing their way to success, but instead are merely seduced by the lure of easy money, dog-eat-dog street smarts, and a dedication to success or survival at any cost, while dispensing the least amount of truthfulness. Their dance all guided by their individual, amoral agendas feeding the needs of their corrupt, narcissistic souls.

In the end, it is his perfectionism that is the downfall of the creepy anti-hero counterfeiter, while the rest of the players are all revealed as bottom-feeders screwing everyone around them, until ultimately a literal and figurative – a Hollywood classic -- car crash of a crescendo leaves few left alive.

And the remaining two survivors, the cowboy cop's novice partner and the female informant – who were initially the most guilt-less and innocent -- are now doomed to lose the uncontaminated remnants of their souls, as the cycle of damnation and death starts anew.

Wow! Just incredible.

* * * *

> "He drank himself to death. Which is only another way of living, of handling the pain and foolishness that it's all a dream, a great baffling silly emptiness, after all."

-- Allen Ginsberg (1988)
Speech

"Live and Die" is the finest example of the film genre that arose at the beginning of the 1970's -- starting with "Five Easy Pieces" (1970), and "Clockwork Orange", and was dubbed "nihilism" by film critic Pauline Kael, who gave the 1976 masterpiece of corruption in 1950's Los Angeles, "Chinatown", starring Jack Nicholson, as an example.

Nihilism is a more general form of the dystopian film, but stripped of the latter's cushion for the viewer of a futuristic setting, which allowed for the viewer the comfort of interpreting the dystopia as merely a possible **future** prediction.

This, however, is denied to the nihilist movie viewer, who must confront the in-your-face moral cynicism, and of present-day systemic corruption and decay and desolation, and thus the hopelessness portrayed – or perhaps better described as exposed.

The timing of the entrance of the nihilist movie genre coincided with the end of the Sixties, but more specifically the end of the paradigm shift that it represented: exposition of mass peaceful cultural change, and the explosion of ideas, political thinking, art, and expectations. The uniquely violent revolution/culture war that was mercilessly crushed in the years of 1968 to 1971.

And nihilism in the art of movie-making was the muted cultural scream that the leftist Revolution had been smashed, and the hopelessly corrupt, and decaying reality reinstated.

Every single figure that represented progressive and peaceful thinking and was able to carve a niche in the public eye was imprisoned on politically motivated charges, a fugitive, an exile, or hounded by the authorities. Or they were smeared or radicalized and then marginalized, like Jerry Rubin and Abbie Hoffman and sometimes even killed – the Black Panthers – under color of state law.

The established political order felt threatened by every aspect of the political change and ideas expressed during the Sixties, and used all its resources to crush the revolution, even during its peaceful phase. Though the judicial branch of the government reacted with a resounding rejection of the government's abuses and constitutional outrages, and the legislative branch's persistence eventually disgraced and cashiered President Nixon -- Constitutional rapist supreme -- it came too late.

Nixon drove the secret effort to smash the nascent 60's revolution by an organized, efficient, and systematic campaign of abuse of office and power by commandeering the fascist political machine of a multitude of executive branch and state and local agencies to crush its opponents and their quest for the reformation of American society.

Nihilism (derived from the Latin _nihil_, meaning nothing) is defined as a philosophy, as well as an obscure extremist Russian revolutionary ideology that arose during the late 1800's and damned the social order as so evil that its violent overthrow and complete destruction was justified, even absent any concrete plans for what was to replace it.

The philosophy definition is equally depressing, and denies the existence of all truth, especially moral truth. Nihilism: the philosophy whose doctrine of negativity comprises the absolute rejection of all religious and moral values and principles. It often involves a general sense of intellectual despair, underpinned by a belief that life is devoid of meaning.

It is an extreme form of skepticism maintaining that nothing in the present world has real existence or meaning, and that systematically rejects all values, denies the existence of moral truth, denies belief in existence, and the possibility of valid or useful communication.

The philosophy was nascent and first noted as early as 1793.

It was the doctrine of a Russian extreme-revolutionary group in the late 19th and early 20th C. denying completely, and finding nothing to approve of in, the established order of the Russian Tsar and his feudal empire, its authority and power, morality, or social and political institutions.

It was the doctrine of any of several revolutionary organizations arising in Tsarist Russia that promulgated and upheld the practice of regicide. It is a revolutionary doctrine of destruction for its own sake. It is the abandonment of ideology, focusing only on the complete destruction of the existing government, its institutions, and the social order, with no concrete plans for rebuilding a new social order afterward.

* * * *

"[M]an is equally incapable of seeing the
nothingness from which he emerges and
the infinity in which he is engulfed."

-- "_Pensees_" ["Thoughts"] (1670)
Blaise Pascal

Then there's the magnificent "Seven" (1995), described by one U.K. reviewer as perhaps the "gloomiest movie ever made".

Two homicide detectives are impotent to stop what appears to be a serial killer, played brilliantly by Kevin Spacey, who has decided that the only way he can teach society a moral lesson that will be remembered, is by committing a spectacular series of seven creative and highly newsworthy, meticulously choreographed murders based on the medieval Seven Deadly Sins (the number is bigger now...)., and then surrendering and confessing.

Seven sins, seven killings, in seven days.

The victims that our puppet-master has selected, however, are not simply the superlatively morally flawed, to be made examples of, but people he skillfully manipulates or coerces the death of, by exploiting the very moral failings he's pissed off about. And the mostly gruesomely violent deaths are never really shown -- horrifying and sometimes deeply disturbing only as brilliantly painted in the viewer's mind by director, David Fincher.

But by the end of the movie, as he explains his motivations to the detectives he's surrendered to – because they failed to catch him! -- you're probably convinced that the seriously twisted killer is the sanest person around, for want of anyone better, as the nightmarish moral bankruptcy of the society around him is exposed.

And finally, the credits roll to the inspired selection of the pounding beat of David Bowie's "The Hearts Filthy Lesson", which -- just so you get the point – ends with the lyric, "If there was only some kind of future".

* * * *

"Who hears the fishes when they cry?"

-- "A Week on the Concord
and Merrimack Rivers" (1849)
Henry David Thoreau

It's interesting to note that when the "Unabomber", the highly intelligent Ph.D., Ted Kaczynski was arrested in 1995, ending his eighteen year campaign of mail bombings that resulted in three deaths, the police found in his primitive backwoods cabin a **22,000** page notebook/diary written by Kaczynski. It mirrored the hundreds of notebooks of the thoughts of Jonathan Doe's found in his apartment by the police, in a scene in the movie "Seven", in which Morgan Freeman's detective character reads out-loud a segment, which, though hilarious, shows Doe's complete alienation from society.

Doe realizes that he has puked on a mundane and trite stranger, who starts a conversation out-of-the-blue with Doe, while they are both waiting on a subway platform with Doe nearby to the stranger.

"He was not amused," ends Doe's anecdote.

* * * *

"I go to die, and you to live; God only
knows which is the better journey."

-- "Apology" (399 B.C.)
Socrates

And then, of course, there's Terry Gilliam's dark satire, "Brazil" (1984), starring Robert de Niro as an outlaw heating engineer providing comic relief, and Monty Python's Jonathan Pryce, as a weak-willed dreamer trying to remain anonymous in a dystopian society of the near future, that is decaying around him. He works as a low-level bureaucrat in a monolithic Orwellian police state that tortures its suspected enemies -- and then bills them for the electricity used in their interrogation. Just priceless... (no pun intended).

* * * *

"[B]ut I'll tell you this, outside this campus -- and even inside it --
idealism is under siege, beset by materialism, narcissism, and
all the other 'isms' of indifference. Baggism, Shaggism. Raggism.
Notism, graduationism, chismism, I don't know. Where's John
Lennon when you need him?"

-- Bono, lead singer U2 (2004)
commencement address
University of Pennsylvania

The lyrics to what arguably could be called the fatalist's anthem, is the title cut of Prince's "Sign O' The Times" (1987) hit album (by the way, "horse" is old slang for the illegal narcotic, heroin):

"Sister killed her baby, 'cuz she couldn't afford to feed it,
And they're putting people on the Moon.
Last September my cousin tried reefer for the very first time,
Now he's doing horse.
It's June."

And finally, there's the 1968 "Star Trek" TV series episode title, and also the last words of an alien character in the episode. Before being murdered by a brain implant device – he's killed not for discovering the Truth that their world is the insides of a giant, hollowed-out asteroid, but for committing the mortal sin of trying to spread it – even though he masks it as a double-meaning riddle:

"For the world is hollow, and I have touched the sky."

And it's quite possibly an astounding, triple-meaning paean to the 1960s, and Jimi Hendix's rock classic, "Purple Haze"... (" 'Scuse me, while I touch the sky" goes the LSD-inspired lyric.)

* * * *

"In the 1930s and 1940s, many members of the Nazi Party in
Germany were extremely well educated -- but their knowledge
of literature, music, and philosophy simply empowered them to
be more effective Nazis. ...

[T]hey were still trapped in a web of totalitarian propaganda that
mobilized them for evil purposes."

-- "The Assault on Reason" (2007)
Al Gore

Piles of light machine guns and sub-machine guns stacked in my gun room along with stamped sheet steel-linked belts of blank machine gun ammo. Add three local cops. Well of course, Yogi's GOING to JAIL.

That iconic revolutionary, Che Guevara poster in the gun-room looked mighty suspicious, too. Well, of course, Yogi's GOING TO JAIL.

The "Kill them all, Let God sort them out" shoulder patch tacked to the kitchen memo board – well, of course, Yogi's GOING TO JAIL.

I just sighed, after an unusually deep breath. Here we go... Yogi, you're about to go for the ride of your life.

When they threw the cuffs on and with minimal protocol stuffed me in the back of a police cruiser parked on my driveway for all the neighbors to gawk at, I felt like telling the arresting officer, "It's always something!" or, "Will this affect my career?" Or, even, "Can you call my woman and tell her I'll be late for dinner? If she asks how long, tell her, 'About 5 to 10 years'."

* * * *

"You woke up screaming aloud, A prayer from your secret God,
You feed off our fears, And hold back your tears.
Oh, you give us a tantrum, And a know-it-all grin,
Just when we need one, When the evening's thin.

Oh, you're beautiful, A beautiful fucked up man.
You're setting up your razor wire shrine,
'Cuz you're working, building a mystery.
 Holding on and holding it in."

-- "Building a Mystery" (1997)
Sarah McLachlan, Pierre Marchand

Fatalist that I am, however, I just soldier on. Because the trick is to come to terms with it, accept it, and realize that life is not just about damage avoidance, it's more about "damage control" -- more about the expectation of damage.

And looking for, and grabbing on oh-so-tightly to happiness when it should happen to come your way. Appreciating it, and valuing it, but not expecting it, or expecting it to be perfect, or expecting it to last.

Or expecting happiness to be a right. Hah!

It is impermanent, frequently brief, ephemeral, and transitory. As it's emotions that defeat us, and expectations held too fast to the breast, can be setting oneself up for a terrible fall.

Fatalism, and its expectation of no expectations, initially helped keep me going in jail, until better things like acceptance, rekindled hope, religious enlightenment, keeping busy, psychological resistance, political resolve, steeliness, resistance, and certitude of belief, and genuine serenity, could take hold.

It was sort of like the way the bullet-riddled, but unaffected Terminator cyborg (in "Terminator 2: Judgment Day" (1991)) responds matter-a-factly about his recent "bullet shield" act to protect his human charge, the teenage John Connor. Connor is astonished at the cyborg's apparent indestructibility, and asks him, if it hurts, at least, when he gets shot.

The cyborg, programmed with an affect devoid of emotion, explains simply, rationally, directly, and clearly:

"I sense injuries. The data could be called 'pain'."

"The time for talking's over now,
I guess it's time to let you go,
But I don't mind, No, I don't mind at all.

It's getting so you never know,
When things are better left alone.
But I don't, No, I don't mind at all."

-- "I Don't Mind at All" (1987)

Prison Diary 9:
"Bourgeois Tagg"
Workman/Bourgeois
Why Are These Men Smiling?
======================

"Three hundred thousand Yankees
Lie stiff in Southern dust,
We got three hundred thousand,
Before they conquered us.

They died of Southern fever,
And Southern steel and shot.
I wish they was three million,
Instead of what we got."

-- "I'm a Good Ole Rebel" (1865)
Innes Randolph
Southern drinking song

"Being able to blow somebody's head off at five hundred
meters with an M-16 – that's an interesting skill to have."

-- Dan Farmer (1994)
U.S. Marine, then conscientious objector,
then software designer/Internet hacktivist

Have you heard a gun joke, lately? No? Sure you have! Did you even **realize** that gun jokes -- much less funny ones -- existed?

Remember "Ferris Bueller's Day Off" (1986), the hilarious John Hughes comedy? Young master Bueller -- rebel with a cause -- is an incipient scofflaw who makes a lifestyle of tweaking authority, which he choreographs with a Nobel prize-winning brilliance, and makes being the worst nightmare of high school administrators a brilliant art form of superb mischievousness.

He doesn't just have a pirate heart -- he's the admiral of the fleet!

There's a scene where the truant Ferris, who -- with tongue firmly in cheek -- accuses a high school crony of being so uptight he must have been "toilet-trained at gunpoint".

Or the hilarious 1992 Sam Rami send-up of horror flicks, "Army of Darkness", starring Bruce Campbell, as "Ash", the "S-Mart" department store housewares clerk who falls into some sort of evil, time-and-dimension shifting black hole, and ends up dropping from the sky into a medieval fantasy world of demons, the evil undead, chain mail-clad knights, and peasants.

A surprised knight asks the bizarre new visitor if he's allied with the forces of good or if he's a member of the forces of evil that they're fighting. Ash, who's fortuitously armed with a lever action rifle from the S-Mart sporting goods section, responds cockily -- tongue firmly in cheek:

"Good? Bad? **I'm** the guy with the gun!"

Saving up his _riposte_ for an appropriate moment later on in the movie, the knight comments contemptuously: "Are **all** men from the future loud-mouthed braggarts?"

Then, there's the scene in "The Terminator" (1984), where Arnold Schwarzenegger, as the title robot cyborg enters the slyly named, "Alamo Guns", to load up for killing Sarah Connor. With the social commentary on American gun culture thrown in for free, he buys an Uzi 9mm semi-auto, Colt .45 1911M1 semi-automatic Long slide (a good choice for a .45 Colt!), with Laser sight, and an Italian Franchi SPAS 12 gauge dual semi-automatic/pump action shotgun. The salesman/owner is impressed at Arnold's gun knowledge, and says with unintended humor, "Hey, you sure know your guns, mister. Any one of these is ideal for home defense."

Then Arnold asks for a, "Phased plasma rifle in the 40 watt range."

"Hey -- just what you see, pal." responds the clerk.

Then there's D-Fens, the laid off Los Angeles missile designer. An upper middle class, white collar, anti-hero in the 1993 movie "Falling Down", starring Michael Douglas. After he foils some incompetent Hispanic hoods trying to mug him, they try to shoot him down with a broad daylight drive-by shooting in the middle of a crowded central Los Angeles street. After missing their target, they then proceed to crash their car.

The unperturbed D-Fens calmly walks up to the car wreck, picks up the open-bolt, full auto 9mm Uzi the injured would-be shooter has dropped and shoots the prone, but struggling hood in the leg. As he walks away with the Uzi and a bag full of guns he also picks up, he tells the gang-banger derisively:

"Take some shooting lessons, ass-hole!"

And last, but not least, there's the line delivered by the Australian actor and protagonist of the comedy flick of the lead role's name, "Crocodile Dundee". The Ozzie out-backer is in New York, and one of a pair of low-lifes pulls a knife and tries to rob him.

"That's not a knife," Dundee comments affably, "This, is a knife!" pulling out a rather large hunting knife with a blade almost as long as a machete, stashed behind his back. (Okay, okay, it's **not** a gun joke, it's a knife joke -- but it's still funny)

* * * *

"[W]hat vileness would you not
commit to exterminate violence...?"

-- "The Measures Taken" (1931)
Bertold Brecht

But these joke are just gun humor from the world of fiction. Then there is the **real** world. In the late 1800's, there was a saying. "God made man. Colonel Colt made them equal." Or, "My best friends are Mr. Smith and Mr. Wesson."

Or in 1864, there are the last words of Union General John Sedgewick, at the U.S. Civil War's Battle of Spotsylvania, "Why boys, they couldn't hit an elephant at this distance..." Bang!

Or take the time in April 1866, when a young radical student tried to shoot Tsar Alexander II. But even though shooting from close range, the impetuous would-be assassin missed. A soldier on sentry duty at the scene cried out in surprise, "That is not the way to shoot!"

The failed assassin was executed, the loose-tongued guard got twenty years hard labor.

Or then there's the famous quote from the 1920's Chicago Prohibition-era mobster Al Capone, "You can get much further with a kind word and a gun, than you can with a kind word alone."

Or take the time that Louisiana Senator Huey "Kingfish" Long, controversial populist and demagogue, was fatally gunned down in 1934 in Louisiana by an irate doctor. Nearby was Louisiana Governor O.K. Allen, who grabbed a pistol and cried out,

"If there's shooting, I want to be in on it!"

and then joined the chase after the fleeing assassin -- a surgeon. Talk about medical malpractice -- I think that about takes the cake. And the surgeon? He ended up riddled with 60 bullets -- a physician unable to heal himself.

* * * *

"Powerful lords have their pleasures,
but the people have fun."

-- "Pensees Diverses" (1734)
Baron de Montesquieu

One gun guy -- U.S., of course – that I read on the Internet newsgroup rec.guns -- said that democratic freedom is upheld by four boxes: soap, ballot, jury, and cartridge.

And then there's the acquired taste amongst gun owners of "black humor" -- sometimes pitch black.

For me, it wasn't really an acquired taste -- I just didn't watch myself, and drifted unawares into what physicists' call the "event horizon" of a black hole named "macabre humor". There are worse fates, but some people don't find any "redeeming social value" in this kind of joke at all.

But like the T-shirt says, "Fuck 'em if they can't take a joke."

For instance, there was a fellow named Arthur Bremer. In 1972, he tried to assassinate George Wallace, the racist Governor of Alabama, and candidate for U.S. President. Wallace was only crippled, despite being hit by four .38 caliber revolver bullets fired by Bremer in the impressively short time of 3.7 seconds. Bremer later testified that he had intended to ask Wallace after shooting him, "A penny for your thoughts?"

Apparently, even assassins have a sense of humor. A dry one, at least. Personally, I deplore violence -- unless I'm winning.

Other serious firearm _aficionados_ , I've noticed, are also partial to black humor, and refer to shooting someone as "lead poisoning". This phraseology is meant to be a joke, **not** a euphemism. And getting knifed, or getting your throat cut, I heard one guy refer to as "steel poisoning".

Anyone remember the January 1973 "National Lampoon" teenage adolescent humor magazine front cover -- the most famous cover in the magazine's history? [A side issue: and each magazine had one soft-core porn shot of a woman's breasts! Just one, to thrill the younger than eighteen market segment, but still pass the censor...]

It showed the top of a seated black and white, medium-sized dog. A rather cute mutt. A human arm of an unseen suited man is visible, pointing a shiny, chrome-plated .38 revolver with a 3" barrel, almost point blank to the side of the dog's head. The dog has this hilarious -- sheepish, I'd have to say -- expression on his face in response to his predicament. The caption at the bottom of the magazine cover states:

"Buy this magazine, or we'll shoot this dog."

Then there was the story of Indira Gandhi, who after being elected, seized power after a corruption conviction, converting the world's largest democracy into a _de facto_ dictatorship. After a series of blunders and outrages – like forcibly sterilizing men en mass –, in 1984 she additionally made the mistake of ordering the Indian Army to take by force the holiest site of the Sikh religion, the Golden Temple in Amritsar, in the Sikh homeland of the Punjab -- an unprecedented act of desecration and sacrilege. There were over a thousand casualties, mostly innocent pilgrims held hostage.

The Golden Temple -- equivalent to the Sistine Chapel of the Vatican -- had been occupied by a radical Sikh faction -- a religion known and respected for centuries of resistance against the oppression wrought by foreign invaders. The Sikh people's reputation of ferocious self-defense and militancy against unrighteous behavior directed against their people is also known equally known. The Sikh symbol has two swords in it, okay? And they always carry a (sometimes) symbolic dagger as a religious requirement.

Indira was apparently impatient. However, two of her squad of body guards were Sikhs. At the first convenient opportunity, they avenged the desecration of the bullet pock-marked Golden Temple by emptying their sub-machine guns into her, riddling her body with dozens of bullets.

Indira was apparently now dead.

Well, it may not be your average sort of joke, but *I* thought it was funny and I had even predicted her demise.

"Oh, man! That bitch is dead. What the hell was she thinking?" I exclaimed when I read in TIME magazine that Amritsar had been attacked by heavily armed Indian Army units on Indira's direct orders. The incident was complete with shoot-out, which the also heavily armed, but outgunned Sikh militants lost.

More humorous, was the end of Elena Ceausescu, wife of communist dictator Nicholae. Both were hated by their Romanian countrymen. And when they were overthrown by a popular uprising that established democracy in 1989, in the chaos of the power transition, the couple's well-deserved demise was video-taped for posterity.

When Elena noticed the firing squad getting ready, her last words were the insightful, but rather superfluous and undignified:

"Look! They're going to shoot us like dogs!"

Apparently she didn't avail herself of the more dignified ritual of asking for a last cigarette...

* * * *

"This one, Sister, is a nine millimeter sub-machine gun, used for close-quarter combat fighting -- a very effective weapon. Would you like to see it?"

-- display booth national security
policeman addressing a nun
touring the facility HQ
"Brazil" (1985)

Even **I've** made gun jokes. There was a company computer network chat group discussion where the idea was presented about instituting a "negative" vote -- a democratic election voter being able to cast a ballot that would **subtract** a vote from the ballot count of a candidate the voter despised. (This was in the mid-1990s).

I posted a public response that we already **did** have a "negative" vote: "It's called a Winchester Model 70 bolt-action rifle, with Leupold optics, and firing a 7.62 mm NATO, 150 grain hollow-point Silvertip round."

Some boob posted a follow-up message, informing me haughtily, that he "didn't think my joke was very funny."

I responded, dryly, that, "If he didn't think my joke was funny, he didn't have to laugh..."

Some people have no sense of humor.

Me? If I didn't laugh, I'd have to cry. And that's no lie...

"We had a very nice house and my dad had a whole collection of Nazi war relics: daggers, machineguns uniforms, banners. Growing up and seeing those items, I actually thought the Nazis won the war, and I thought **we** were the Nazis."

-- Thor Sadler (1989)
son of Green Beret singer

S/Sgt Barry Sadler

Prison Diary 10:		Small Arms and the Man
			====================

> "Any form of violence...invokes the bigger energy, the elemental power circuit of the Universe...If you refuse the energy, you are living a kind of death. If you accept the energy, it destroys you. What is the alternative? To accept the energy, and find methods of turning it to good..."
>
> -- "London Magazine" (1971)
> Ted Hughes

> "There are a lot of bleeding hearts around who just don't like to see people with helmets and guns. All I can say is, 'Go on and bleed.' ..."
>
> -- Pierre Elliott Trudeau (1970)
> Canadian Prime Minister
> impromptu TV interview

This chapter is the story of a mild-mannered, licensed Canadian gun collector. Me.

Most people remember the quip that the only difference between a boy and a man is the price of their toys. And mine were not only very expensive -- I had paid $2,000 for a semi-automatic HK MP-5, 20 years ago, for instance -- but the price I was to pay for them was beyond money.

Yet, if I were to be able to do it all over again -- on this score -- I would do exactly the same thing.

Guns had been an interest and hobby of mine since childhood. But, of course, there was more to it than that. There always is, with me. Like most boys growing up, I was fascinated by firearms, and yearned to grow up so that I could be old enough to get a license to buy one, and have the money to afford one.

As a child, of course, with uncooperative, "pacifist" – meaning apolitical -- urban upper middle-class, non-outdoorsy parents, it was a completely theoretical hobby. But so it goes. No pellet gun or BB gun as a pre-teen. No .22 caliber rifle for my 16th birthday. I would have to be more patient. And I was a very patient boy.

But the concept -- what boy couldn't get into the concept of aiming a rifle barrel, pulling the trigger, and "Bang!", you hit something some distance away that you were aiming at. I mean, how cool was that? But not having physical access to firearms, I made due by dreaming of the day. And in the meantime, locating and devouring every book I could find on firearms in the public library, which kept me busy, partially sated my gun obsession, and taught me a host of small arms minutiae.

When I joined the Canadian Forces Reserves for the summer, at the age of 17, it was solely to learn how to handle and shoot military weapons. The guerrilla/counter-guerrilla training was just a bonus, and the night fighting experience invaluable. When I got to shoot a Light Machine Gun, an FN-C2 (Fabrique Nationale -- the Belgium Arsenal that manufactured it) at the firing range, it was a heaven-on-earth thrill for me. Better than sex. But only 'cuz I hadn't had any yet...

It was a magazine fed (as opposed to a continuous belt-fed) light machine gun, with a bi-pod stand near the muzzle, since it was most accurately fired laid on the ground, with the soldier flat out on the ground too.

And I was laid out flat on the ground, aiming it and exercising the "controlled fire" of its full auto capability. The experience was sure as hell fun, but I was about 100 pounds, and firing a 7.62mm light machine gun was like hanging onto the tail of a bucking bronco -- and I'm not exaggerating.

I couldn't hit **shit** -- I just "mowed the grass" in front of the target -- but it was the most fun I'd ever had in my young, virginal life. Firing a *real* machine gun! Wow!

An essential fact of full auto fire in sub-machine guns and light machine guns: the weapon has an almost uncontrollable kick to it. And a soldier requires strength, discipline, experience, and a lot of training to become proficient at using a weapon accurately and efficiently in full auto.

Even firing a "light" machine gun in its most stable position, with the LMG lying on the ground, muzzle supported by a bi-pod, and me -- the shooter -- lying prone, straight and flat on the ground, the recoil was enormous, and the firearm uncontrollable.

Firing three round "bursts", you can learn to control your aim reasonably -- and not use up all your ammo in 2 seconds. It is while reloading, that you are easily killed by the enemy.

And then there was the Sterling. When I cradled a 9 mm Sterling sub-machine gun in my arms and marveled at its beauty, it was love at first sight. It was black, with a metal folding stock, a long, curved magazine sticking out its side, and a really cool looking barrel jacket with peppered with circular holes for air cooling the hot barrel, and preventing you from burning yourself on it

AND for all this training and shooting, I even got **paid** for it. Not much, but at least $25 a week was **something**.

It was a historical and engineering interest, a fascination with their functionality, and intricacy, a love of their paradoxical beauty. Many reasons, many threads of personal connections, woven into a unique fabric of intellectual and emotional solidity. I was beaten up regularly with my brother with instruments by my loving father... For getting 89%, not a minimum of 90% in a test, or coming home five minutes late – I'm not kidding – after a set time playing with the neighborhood kids.

My interest in guns started out with a childhood fascination with these strange mechanical objects known as guns, that seemed to possess so much "power" and influence over human affairs. I had no money and was too young to possess them myself, so my fascination and interest was satisfied by fulfilling an insatiable desire to learn all about them by reading any library book I could get my hands on about firearms.

I fired my first gun in 1977 by joining the Reserves. I fired my first handguns in 1980 by being an out-of-state "contractor" for an outlaw biker group in the Mid-West. I bought my first gun in 1981, from a sporting goods store in California – no license required – a .22 semiautomatic tube-fed rifle with a telescopic sight. I was eventually able to experience as much of the full spectrum of the firearms experience as I could as a hobby.

But I did not want to go to war and shoot people, or become an armorer or gun shop worker – too boring – or other firearms professional. Finally with age and maturity, I ended up where I had started, interested in firearms from a conceptual standpoint, defined by an extremely refined intellectual attraction to knowledge. This was evidenced by the fact that the last time I shot a gun was -- I don't even remember -- about ten years ago, with my teenage son -- who was still in the perfectly valid "bang bang is fun" phase of firearms.

It's a guy thing...

I had evolved into a "pure" non-shooting gun collector. Several of my most prized firearms, I had never fired even once. At least one firearm had **never** been fired by anyone. A virgin, so to speak.

Firearms -- "small arms", to those in the know -- did not define me. They did not obsess me. They were not some sort of fetish, though I suppose I did like to "fondle" some of my firearms every now and then. Guns were a small part of an individual with complex tangents to his simple personality. An individual known for his laid-back attitude, usual unflappability, quiet, polite, and logical affect, and gentle, empathic, compassionate, and kind heart, as much as for my flaws.

But as I grew up, the times changed, and Canada changed. And as I changed and improved, the wind changed direction and never stopped changing direction. And it was an ill wind that blew no good for people who like hunting, target-shooting, gun collection, or even just firearms and the concept of individual liberty they represent in a visibly tangible form.

A harmless computer engineer, with what some people perceive as an unusual hobby, but one that is legal and highly regulated by the Federal Government. Nonetheless, I was apparently swept up by a mindlessly implemented witch-hunt for the "enemy within".

And thus the mild-mannered registered gun collector morphed into the proverbial scape-goat. But that's the end of "my side" of the story, the end of my little whisper of a rant.

* * * *

"The militancy of men, through all the centuries, has drenched the
world with blood, and for these deeds of horror and destruction men
have been rewarded with monuments, with great song, and epics."

-- "My Own Story" (1921)
Emmeline Pankhurst

Ever the consummate collector, there sits on a bookshelf in my place a hefty, roughly rectangular hunk of blued steel. I pride myself on the unique and exotic nature of my souvenirs. It the butte-end of the upper receiver of a Thompson M1A1 submachine gun -- the military version of the 1920's gangster fame Tommy gun, with the circular drum magazine, hanging from the front of the rifle body. The military version used the more reliable stick magazine, rather than the drum, and it's an M1A1 that Tom Hanks shoots in "Saving Private Ryan" (1998), if you'll recall.

My souvenir piece has been what's called "demiled" -- demilitarized -- to U.S. BATF (Bureau of Alcohol, Tobacco, and Firearms) specifications, by having the main gun part -- the "upper receiver" (for the Thompson) cut into three with a oxy-acetylene torch, and discarding the center piece.

The demil specs were modified (the middle piece used to be allowed for sale) because some jokers used to hacksaw, then re-weld the three pieces back together. The BATF -- who enforce U.S. Federal gun laws -- did not find this particularly amusing, even though it was a rather precise, meticulous, exacting, and thus impressive technical and philosophical (Yes! Philosophical!) feat.

It is also a defining example of the lengths people will go to arm themselves, with, or without government approval. A well-established principle that is utterly beyond the grasp of people and policy-makers who don't approve of and seek to limit, the private ownership of small arms.

I used to keep my demiled M1A1 hunk of steel as a nice little souvenir -- unique and exotic. But now, alas, it's pretty much all that's left of what used to be a decent weapons collection. Military, or military-type weapons, exclusively. A .22 semi-automatic "survival" rifle that could be disassembled into smaller pieces, another .22 semi-auto rifle with a scope, 2 bolt-action antique military rifles, six semi-automatic military rifles: two AK's, an AR-15, a U.S. a 1944 M1 Garand, an Indian ("Ishapore") FN, and a Chinese SKS. I also had an HK MP-5 sub-machine gun (the civilian semi-automatic version), a Ruger mini-14 semi-automatic rifle, two 12 gauge shotguns (pump riot gun, and double-barreled), and 3 handguns (a Colt .45 semi-auto, a Beretta M1, and a .38 police "throw-away" or ankle, snub-barreled revolver).

There was an array of specific military functionalities represented: from state-of-the-art offensive/defensive, to concealability, to high ammo capacity, to close-quarter or house-to-house combat, to defensive breakout, to robust battlefield/jungle/forest firepower that my collection represented.

Oh, and let's not forget a complete array of available caliber ammunition, in case of "shortages".

I did not collect "sniper" type military weapons -- .30 or higher caliber, long range, accurate bolt-action rifles. (For example, the Winchester Model 70.) They were expensive, and it really wasn't my thing. My thing was raking and clearing at short or ultra-short ("belly gun" -- just kidding) range.

Every kind of problem requires a specific tool, such as target practice requirement, concealable, reliability/robustness, rate of fire, target number and range, etc., but I have no more to say on the matter. Trade secrets, ya know.

Of course some weapons were just so cheap, that I couldn't resist buying them. Who could resist a brand new semi-automatic Chinese SKS for $130 plus tax (at the time), that is a superior, yet cost-effective, defensive weapon, with an interesting history, to boot. An eclectic, but decent collection. Long guns (rifles and shotguns), restricted firearms (generally handguns), and prohibited firearms, *all* legally owned.

An impressive collection, I guess. Impressive enough that in spite of a stack of permits, it apparently spooked the nitwit pigs, and got me thrown in jail for no reason. Book collector, gun collector, weapons reports, and data collector. I have other collections and hobbies, too. I take my hobbies with an obsessive -- even fanatical -- seriousness.

Science, Religion and the Search for the Truth. Religion comes off the rails pretty early in the game. That leaves us with Science and the Search for the Truth. We can always come back to salvage Religion.

Out of chaos, order. Out of a black abyss: greyness, shades, and nuances. Out of infinity, only manageable and understandable complexity. It is truly staring into the abyss. The darkness of the human soul, collectively, historically, and in the present.

But once again, reality cannot and should not be denied. The result of denying reality is the folly of deliberate ignorance. Reality is, what it is. If you disagree with reality, rise up and steel yourself to begin to attempt to change it, if it is human-caused. But reality cannot be denied, and the messenger of reality is not the evil that reality may appear to be.

Oh, and by the way, current young Canadian criminal slang for a handgun is "a burner". As in, "Did you have time to hide the burner when the police raided your house?"

The things you pick up, in jail, eh? I guess it's time to retire the slang words, "gat", "roscoe", and "piece". And my favorite -- "hardware".

"Say 'hello' to my little friend!"

-- Tony Montana,
 "Scarface" (1983)
  getting ready to fire the integral
  M203 40mm grenade launcher
  on his M16.

"Marley was a Rasta, Moses was a Jew,
Jesus was an outlaw, Just like me an you,

Let love be our religion, Until this life is through,
The next best thing to Heaven, Is a sinful life with you."

-- "A Sinful Life with You" (1994)
Timbuk Three

Prison Diary 11:   I Love the Smell of Cordite in the Morning
====================================

> " 'Fire!' cried the Sergeant-Major,
> The muzzles flamed, as he spoke,
> And the shameless soul of a nameless man,
> Went up in the cordite smoke."
>
> -- "The Deserter" (1984)
> The Pukka Orchestra
> Duggan-Smith

> "[T]he simplest surreal act consists of dashing down into the street, pistol in hand, and firing blindly, as fast as you can pull the trigger into the crowd.
>
> Anyone who, at least once in his life, has not dreamed of thus putting an end to our petty system of debasement and cretinization, in effect has a well-defined place in that crowd, with his belly at barrel level."
>
> -- "Second Manifesto of Surrealism" (1929)
> Andre Breton

...It smells like....Freedom.

I'm not kidding. I **do** like the smell of gun smoke -- burnt double-base smokeless powder -- when you fire a rifle outside, in the country, in the cool autumn air.

When I used to go target shooting -- the informal target shooting known as "plinking" -- in a gravel pit deep in the forest or woods, away from houses or people, that whiff of cordite smoke emanating from the muzzle of your rifle, just after you've fired a round, combined with the crisp, cool and nippy, clean fall air combines to form an absolutely unique and wonderful smell.

In the smoke of burnt powder, I assume, is a hint of unburnt nitroglycerine vapor, and NG is a heart stimulant of sorts used medically for angina. The double-base powder -- cordite was an old trade name -- the propellant of a modern center-fire rifle cartridge consists of two ingredients, cellulose trinitrate, and nitroglycerin, mixed and extruded through a small hole into a thread-like strand or "cord", which was then chopped up into small lengths, before loading into the brass casing. From cord was derived the name "cordite".

But I digress.

So, every once in a while, typically on a lazy Sunday summer afternoon, I used to take my son, and head out in the wilderness to shoot at paper targets, clay targets (for the shotguns) or anything, like a tree trunk. Plinking is just an excuse to haul out the hardware, head out into the countryside and do some shooting, occasionally at an actual paper target.

Plinking. To go out just to "bust caps", using the cool ghetto vernacular.

I'd take a rifle or two and some 20 round boxes of military surplus ammo -- it's cheaper -- to feed the rifles. Ammo is *expensive*. We're talking a dollar a round for commercial center-fire rifle ammo, and this was fifteen years ago.

And so a valid and historical expression of masculinity, and father-son bonding was established and enjoyed as a controlled, yet relaxing excuse for an outing and adventure from city to the great outdoors.

Tom Diaz, in his pro-gun control book "Making a Killing", wrote that guns represented "a tangible symbol of such fundamental American values as independence, self-reliance, and freedom from governmental interference." But it's more than that. Firearms are a symbol of freedom, self-defense, and manhood as embodied in the concept itself, and the ability to protect the safety of one's family.

Oh, come one! You've got to admit that it's a unique pass-time, a rather exotic hobby? Would you rather collect stamps or take an AK-47 to the range and blast some rounds? Empty its thirty round magazine as fast as you can. That kind of thing.

I have never considered myself a regular target shooter, and never considered myself that great a shot. All I've ever wanted to do is hit what I'm shooting at, at a hundred yards or so. If I can get a good score or even just hit the paper target that the bull's-eye is printed on -- well that's gravy -- a bonus.

But I do remember something that makes a good story. The first time I fired a semi-automatic AK-47, it was the late 1980's, in a gravel-pit out in the country, in the eastern, maritime, Canadian Province of New Brunswick, far off in the woods, I hit a bull's-eye on the paper target at 80 meters. Which was pretty cool for an untested rifle, an unfamiliar weapon and round, a heavy recoil thirty caliber round, and my mediocre marksmanship, in general.

But I guess it was luck, because my subsequent shots weren't so accurate, and my second shot was, alas, not a bull's-eye.

It's like an adult version of kids playing "cowboys and Indians" or "war", except with real hardware, a real "bang!", instead of a make-believe "bang-bang". It's more realistic, but still nobody gets hurt.

And it's a well-known father-son bonding ritual. The father teaching gun safety. The son graduating from sling-shot to BB-gun, to his first .22 rifle, and then finally to a center-fire rifle. An inter-generational male act -- a changing of the guard. A handing down of the mantle of masculinity. Before sex takes the mantle. And drugs...

Drugs were illegal for a century, and guns are under increasing and continually restrictive attack. Will they try and ban sex next?

I remember one time my young son informed me enthusiastically that I was considered the "coolest dad in Grade 6" by his [male] classmate associates. I was surprised, my curiosity piqued, and I asked him on what basis I should be accorded such an important honor.

It seems that it was that one time, when the thought occurred to me to include my son in the adult fun, and at the same time teach him a little lesson. I wanted to let him fire an AK-47, my favorite weapon for many reasons, and the most common firearm in the world, with hundreds of millions in existence the world over. But it fires a .30 caliber bullet (like the .3030, .308, and 30-06), and the recoil is significant -- it has a tremendous kick when fired. Even with a muscular shoulder such as I had, as I fired with the rifle stock against my shoulder, the horrendous jolt of firing it would leave a number of broken blood capillaries visible through the skin of the front of my shoulder from the recoil shock, by the end of the afternoon.

As well, there was no way my son was big enough to even **hold** the AK rifle: it was too big for him, and too heavy, even if the recoil wouldn't knock him on his ass. So, I got on my knees, held it for him, and instructed him to stand on the other side of the rifle, and carefully pull the trigger backwards.

"Bang!" The bullet went down range, there was smoke and flame out of the muzzle, the empty brass casing ejected on the other side of the rifle as the AK action cycled and reloaded another round into the chamber.

I clicked on the safety, and lowered the rifle. I turned away from Oz, removed the AK's banana clip, clicked off the safety, and jacked the new round out of the chamber, just for extra safety. I figured a lesson was now due, after the fun was over for my kid.

"Remember that sound, son. That's the sound of Freedom," I intoned gravely. "And you've just fired your first AK-47, the **best** and most popular rifle in the world! There are something like 70 million AK-47's in the world. I'll bet you're the only kid in Grade 6 who's **ever** fired an AK-47!"

And then I filed it away into the back of my memory, and promptly forgot about the whole incident.

But **he** was certainly impressed that he had got to fire a real adult rifle. So when – naturally – he told all his friends at school, indeed, I was apparently acclaimed the "coolest dad of all the Grade 6 kids."

I wasn't the greatest of fathers, I'll never win a Nobel Prize, and Claudia Schiffer will never drop by one night and force herself on me. But "coolest dad of the Grade 6 boys"? Well, what can I say?

"High five!"

"And what's wrong with being an
angry prophet denouncing the
hypocrisies of our time?"

-- Howard Beal
"Network" (1976)
Paddy Chayefsky

Prison Diary 12:   "The Prophet Armed"
================

> "Interesting. Your Earth People glorified organized violence for forty centuries, **yet** you imprison those who employ it privately."
>
> -- Mr. Spock
> "Star Trek" (1966):
> "Dagger of the Mind"
> Shimon Wincelberg

> "Pass me a bottle, Mr. Jones.
> Believe in me,
> Help me believe in anything.
> I want to be someone who believes."
>
> -- "Mr. Jones" (1993)
> Counting Crows

" 'Charlie' don't surf!" shouts Colonel Kilgore of the Air Cavalry, in the middle of a battle, summarizing neatly the cultural difference between Americans and the North Vietnamese.

It's since the name of a "Clash" song.

And the lead singer of the rock band "Marillion" wears Col. Kurtz's camouflage makeup.

"Apocalypse Now" (1979), directed by Francis Ford Coppola, is my hands-down favorite movie of all time. A spectacular movie-making extravaganza. A brilliant cinematographic achievement. An all-encompassing tour-de-force of the American experience – and loss – in the Vietnam War, 1962 - 1973, hung on a scaffolding as a modern re-write of Joseph Conrad's _fin-de-siècle_ classic, "Heart of Darkness".

A Captain Willard, an alcoholic Special Forces assassin is tasked, much to his surprise, to assassinate an out of control, former star high-ranking U.S. **officer** that the CIA and Army high command wish to secretly and quietly – partially out of embarrassment – put out of commission. Like at the real War's end, American forces turned on themselves and were frequently in open mutiny against their officers and each other, instead of fighting the enemy.

"Apocalypse Now" was only the second movie made on the War (the first was "The Deer Hunter"), and both years after the War's end, so devastating was the loss on the American psyche. "Apocalypse Now" is brilliant for many reasons, but one of note is that it covers every single cliché I can think of, of the Vietnam War.

> [For example, in Sept. 1969 a Special Forces Colonel was charged with the first-degree murder of a South Vietnamese double agent. The charges were dropped

by the intervention of the Secretary of the Army.

In "Apocalypse Now", Special Forces Colonel Kurtz derides the fact that the Army wants to charge him with murder of several South Vietnamese double agents that he uncovered. Capt. Willard is equally astounded, stating that Kurtz's predicament with the Army was like someone "handing out speeding tickets at the Indy 500."]

And cliché, or not, it handles them perfectly.

It is interesting to note that the title phrase "Apocalypse Now" is never said in the movie. It appears ever so briefly as Capt. Willard, the Green Beret hit man passes it, spray-painted on a stone wall deep in the Cambodian jungle where the army of Hmong native tribesmen led by the renegade American Colonel are ensconced.

The lone graffito says merely: "Our motto: Apocalypse Now".

> "In a war, there are many moments for compassion and tender action. There are many moments for ruthless action.
>
> What is often called ruthless, but may in many circumstances be only clarity. Seeing clearly what there is to be done, and doing it.
>
> Directly. Quickly. Awake.
>
> **Looking** at it.
>
> -- Colonel Kurtz
> "Apocalypse Now" (1979)
> John Milius, Francis Ford Coppola

* * * *

Not with a whimper, but a "Bang".

It was one of those magic moments. It was a 5-10 second video segment -- what journalists call a "sound-bite" -- a short piece of video they can use in their two minute TV news piece that is interesting, preferably poignant, and captures the sense of a certain side's argument and position, but without complex, logical, or fact-encumbered arguments.

But this "sound bite" was the shot heard 'round the world, as TV network news broadcasts picked it up and showed it all over the U.S. that evening.

It was former actor, Charleton Heston -- "Ben-Hur", "The Ten Commandments", "Planet of the Apes" -- holding with one hand high over his head, an antique flintlock rifle, as he thundered to his rapt audience the words:

"From my **cold, dead** hands!"

He was standing behind the podium, as the then NRA President, as he dramatically ended his speech for the 2000 National Rifle Association annual convention, held that year in Charlotte, North Carolina.

The flintlock was the firearm used by the citizen-soldiers in the American Revolution. 1776 and all that.

The NRA conventioneer audience loved it -- and he brought the house down. It was an inspiring, straight-from-the-heart, electrifying, crystalline pure message -- harmless, yet paradoxical. It was honest, back-to-the-roots simplicity, taken from a modern but dated, crude anti-gun control statement of militancy, and recycled into devastating minimalism.

And it was only Heston, who could have pulled it off so brilliantly, taking the last phrase from one of the oldest -- and crudest and extremist -- anti-gun control slogans ("You can take my guns away, when you pry them..." -- you now know the rest), and reinventing it as potent symbolism.

It just warms the cockles of your heart, doesn't it?

Sorry, it's a guy thing.

* * * *

> "The central conservative truth is that it is culture, not politics, that determines the success of a society.
>
> The central liberal truth is that politics can change a society and save it from itself."
>
> -- Godkin Lecture (1996)
> Harvard University
> Daniel Patrick Moynihan

Disarming the respectable, the vetted, the approved, the cleared. But implementing an easy solution against an easy target, agreeable to the government for not completely honest motives, well...

The latter of which is extremely bad public policy, and an affront to the honesty and trust that is essential to the Social Contract -- an idea written in a book of the same name by Rousseau in 1762 -- that is supposed to exist between citizen and government, and effectively balances and reconciles individual liberty with governmental authority, by requiring each to respect the other. Voluntary citizen compliance as the basis for good governance.

The result, a society in harmony, where citizens comply with state authority out of logic and the collective good, and governmental authority is likewise dispensed with reason, consideration, and care. Thus is produced the efficient result of voluntary and willing compliance, as it is viewed not as authority by edict, but change that is sensible to the citizen, promulgated for the collective good, with respect for and absent the abuse of minority concerns.

But instead, we have what we have. Rousseau would roll in his grave (which would surely upset tourists visiting the Pantheon in Paris).

The government itself has no qualms about being armed to the teeth, be it the Armed Forces or the internal security forces (who seemed to be more and more armed, and more and more trigger-happy, as citizens are more and more restricted and disarmed).

A hobby for sure. Unique, fascinating, different, exotic, exciting, fun, masculine identity/rugged individualism-stroking.

The "guy thing". A sense of security. A sense of knowing you're ready for any threat. Self-defense and to defend your home and family -- genetically ingrained as a man's responsibility, right, and duty.

" 'Vengeance,' sayeth the Lord, 'is mine." says the Old Testament. But some take issue with the speed and timeliness of the Big Guy's schedule.

There are many reasons and arguments for private ownership of firearms -- rifles and handguns. And the horrendous U.S. toll of civilian firearms deaths is a price that quite reasonably appalls gun ban proponents, and any reasonable person. But it is collateral damage worth enduring to gun people who form a large, significant, and well-organized lobbying and voting group in the U.S., where privates ownership of firearms seems to be enshrined in the Bill of Rights, as the Second Amendment, right after free speech and other idea-related actions (free press, religion, and peacefully demonstrate).

However, in countries with more equitable wealth re-distribution *and* private ownership of firearms, the same firearms death toll is only a tiny fraction of the U.S. death rate, and the civilian ownership of guns -- usually hunting rifles and shotguns -- in the First World is widespread -- from Canada to Scandinavia to Switzerland to New Zealand.

Gun possession is highly illegal in dictatorships and other authoritarian regimes. Something that should give cause to ponder. The smart money, in good times or bad is that a cheap, semi-automatic, center-fire rifle and some ammo is good insurance given the stability of governments, democracy, and the inconsistent morality of those who govern.

And goddammit, if I have a say in the matter, if such horrors come to pass, it'll be a Bang, not a whimper.

In 1776, and in any of the successful revolutions that overthrew despotic tyranny, and brought liberty and freedom, or other significant improvements to the masses -- American, French, Russian -- it was the spontaneous mass action of the citizenry that achieved that goal through violence and armed attack. And it was the only way that the success of these celebrated revolutions could have been attained.

And an armed populace -- respectable, established citizens of stable background, good character, and sound judgment -- exerts a deterrent effect and is the last ditch insurance policy of a free and democratic society against a government that should it decide it can slip well outside of its proper bounds and no longer respect the will and desires of its true master -- the people -- it cannot, without the possibility of an effective armed backlash if enough people are outraged.

And "enough" is the key word.

But there is another more concrete and less conceptual and speculative reason why the government should stop the folly of gun control and gun bans. The reason shows just how incompetent the government and our leaders are. It's almost infantile in their reasoning process as to expose just how poorly we are served by our government in failing to realize the depth of men's need to arm themselves -- legally or not.

The U.K.'s "Telegraph" newspaper reported in 2008, that in England and Wales -- where handguns are banned, and rifles severely restricted -- there were 400 knife crimes per week. In other words, 20,800 per year, not including Scotland. So much for gun control.

When the Canadian government mandated under threat of criminal sanction, that citizens (something like 7 million Canadians) already approved and licensed to own firearms report exactly how many and what guns they possessed, it is anyone's guess how many didn't.

Well, at least one -- me.

People have a right to effective self-protection, and a right to self-defense, a right to the feeling of security. People have a right to carry on with their life and normal pursuits free from intrusive -- and especially creeping -- government interference.

And with gun control, one has to pause and reflect on what amounts to a concerted attempt by the government to essentially shift the balance of power by leaving only itself and its security forces armed. Heavily armed, actually.

And their primary purpose, if you didn't realize it, is the protection of ruling officials, the preservation of their power, and the social order that established it and -- once these people are ascendant -- corrupted to prevent effective social change that would diminish their power, perks, positions, and economic control.

Mass killings by lone gunmen, deranged, or disgruntled and unhinged, are a recent social problem. However, despite the disturbing and sensational aspects of such incidents, the sudden urge to run amok and engage in mass murder for no apparent reason is not common thinking. [I wish people would stop thinking "TV reality" is actual reality.]

And since there are around two hundred million firearms -- including seventy million handguns -- in civilian hands in the U.S., and numerous problems of social stability -- though tragic -- the number of incidents is amazingly low.

Yet anti-gun advocates paint gun owners as "demented and blood-thirsty psychopaths whose concept of fun is to rain death upon innocent creatures both human and otherwise." [National Institute of Justice (1981)]

And if we used the same logic, well, cars cause many thousands of deaths a year in North America. Planes crash, sharks attack, and lightning bolts fry people. Just as an example, in the U.S. in 2008, almost 15,000 people died from opioid painkiller overdoses.

The fact of whether a death is caused deliberately or accidentally is of concern to the Law and one's emotional response only. The person is still dead.

We live in a world whose nations are armed to the teeth. We live in a world that solves its problems with violence. Its external problems as well as its internal ones.

We live in a world that in the 20th Century killed over 200 million fellow human beings in war and internal conflicts.

It says a lot about us, doesn't it?

* * * *

"Where were you when I was burned and broken?
While the days slipped by from my window, watching,...
While you were hanging yourself on someone else's words,
Dying to believe in what you heard,
I was staring straight into the shining sun."

-- "Coming Back to Life" (1994)

Pink Floyd

Starting in April 1994, between one-half and a million Rwandans of the minority Tutsi tribe were killed by extremist members of the majority Hutu tribe in an 8 month period -- an astoundingly fast attempt at genocide. 38% were killed by machete, 17% by clubs, and 15% by firearms.

The Hutu government had, the year before, imported around one million machetes -- weighing around 500 tons, at a price of less than a dollar a piece. It was around one machete for every three adult males in the country.

It was the first mass killing by machete in world history.

* * * *

"My holy of holies is the human body, health, intelligence, talent, inspiration, love, and the most absolute freedom -- freedom from violence and lying, whatever forms they may take."

-- Anton Chekhov (1917)
letter to a friend

I've legally owned guns for almost twenty-five years, and I've never shot anyone, obviously. I don't actually believe in violence. I've seen enough of it in my lifetime... It disturbs me profoundly, depending on the circumstances.

You see, violence is a choice, I like to say. And I choose *not* to be violent. And to anticipate and avoid finding myself in a situation where I have to even make a choice. But fate is unpredictable, so, in that case, I say "No," to violence, except as a last resort, or when it's unavoidable.

Or the best of two bad choices.

" 'But I could say that the decision was made as soon as I heard my own voice shouting.' "

-- "The Manticore" (1972)
Robertson Davies

Prison Diary 13: "Isn't it Ironic?"
==========

> "I have lived a great deal among grown-ups. I have seen them intimately, close at hand. And that hasn't much improved my opinion of them."
>
> -- "The Little Prince" (1943)
> Antoine de Saint-Exupéry

> "Our apologies, good friends, for the fracture of good order, the burning of paper, instead of children, the angering of the orderlies in the front parlor of the charnel house.
>
> We could not, so help us God, do otherwise. ... The time is past when good men can remain silent, when obedience can segregate men from public risk, when the poor can die without defense."
>
> -- "Meditation" (1968)
> Daniel Berrigan

In modern democracies, the rich are entrenched, as they always have been. Their status and position and power remaining secure by the availability of a convenient, stabilizing buffer: a large, comfortable, and generally satisfied – or at least distracted – middle class.

The poor and sick struggle along on their welfare tokenism. The adults are anesthetized into non-voting apathy, and the children just suffer in the poor-man's public housing ghettos.

A stable societal equilibrium has been established and has made "capitalism is theft" into "climbing the social -- and economic -- ladder" the new paradigm. Thus, has been diverted the struggle for survival in an exploitation-based society, to one of the -- usually -- a fantasy of attaining a better life through effort, talent, and persistence.

Meanwhile, the forces of foolish liberalism, say, "I don't need a gun in a peaceable society such as ours." **So you don't need to own one either.** Who should need to be armed in such a stable and comfortable society as ours?"

In the city, the hunters and target-shooters are marginalized by being out of place in their urbanized environment.

And those in power – with a circle of expensive and armed body-guards, federal agents, municipal police, and loyal army – just smile.

* * * *

> "Being a lover of war and of wisdom, the Goddess
> chose the place that would bear men most closely
> resembling herself, and there made her first settlement."
>
> -- "Timaus" (ca. 360 A.D.)
> Plato

It's what Napoleon called the "Butcher's Bill" -- war's cost in human lives. Some well known examples of white people killing white people in the last two centuries:

U.S. Civil War	(1861 - 1865) -- 620,000 soldiers dead in 5 years (U.S. pop. 20 million)
World War I	(1914 - 1918) -- 15 million dead in 5 years
Spanish Civil War	(1936 - 1939) -- 1 million dead in 4 years
World War II	(1939 - 1945) -- 50 million soldiers and civilians dead in 7 years

And finally:

World War III   (Not yet) -- With an all-out nuclear war, there would have been something 150 million U.S. dead (even in they "won") and 600 million "enemy" civilians dead in 30 minutes, 1 billion dead in less than a day from the resulting fire-storms, 1.25 billion total dead from fall-out in six weeks, 2 billion dead in less than a year from radiation, disease, and starvation, and -- the final act – nuclear winter, and human extinction.

It could be said that it was human war -- World War Two -- that signaled man's beginning conquest over nature, when war deaths began to vie for killing power with the maximum of deaths from disease. The Black Death -- bubonic plague -- of the 1340's killed an estimated 30% of Europe's population – 50 million dead in four years.

*   *   *   *

> "A consensus was forming in the periodic surveys done by American
> historians that [G.W.] Bush was the worst President in American history,
> the patron of unnecessary pre-emptive war, of unilateral abrogation of
> treaties, of ignoring allies, of torture, of imprisonment without charges
> or representation, of illegal surveillance, of unaccountability to Congress,
> of economic mismanagement, of incompetence and cronyism and
> corruption..., and on and on."
>
> -- "Bomb Power" (2010)
> Gary Wills, referencing a 2006 "Rolling Stone"
> piece by Sean Wilentz, "The Worst President
> in History?"

Dedicated fanatics with really cool uniforms. They were the Nazis.

It is hard to believe the United States, with their revolutionary foundations and ideals, carefully planned republic and constitution, have become the **new** Nazis.

Since the world was dominated by the United States, who inherited the role in 1945 with the defeat of Nazi Germany and Japan, and the devastation of Europe, they were to adopt the German invented, and Nazi-perfected guiding policy of _realpolitik_.

This is loosely defined as making all government actions, particularly government foreign policy, solely on a practical basis of self-interest and expediency, ignoring completely ideals and moral or ethical considerations.

Realpolitik is a political strategy that maximizes future success and growth -- but not goodness -- and was actually probably the oldest example of the application of Game Theory, invented by von Neumann -- and disclosed in a book published in 1944 -- while working on the U.S. development of the atomic bomb.

But stalemate and huge U.S. casualties in Korea in the early 1950's, and defeat in Vietnam and Cambodia in 1975. showed there were limits to "containment" on the door-steps of China and the USSR's own spheres of influence.

And the War on Terror, through the lies instigated by President George W. Bush, led to Iraq in 2003, and a quagmire of deaths *greater* than "9/11", the original start of things.

But realpolitik was not the only thing adopted by the U.S. as "last man standing" in the post-World War Two world. It was a whole plethora of Nazi technology and strategy that were adopted. Everything but the Concentration Camps.

But instead of Concentration camps, they were replaced by a huge prison network, the American _Gulag_ for the underclass. The black minority was kept poor and under constant police harassment and criminalized as often as possible. And constant wars kept going that employed the poor and the blacks to fight for the rich and the powerful.

One particularly effective technique of Hitler's, he dubbed "_Nacht und Nebel_" -- "Night and Fog" -- the sudden disappearance from an occupied country, without formal arrest, of an enemy of the state, with no notice, attention, or information. The families never knew what happened to those kidnapped by the Nazi state. One day, the targeted person in a conquered or annexed country just disappeared, never to be seen or heard from again -- having been quietly grabbed, and shipped off covertly to Germany for forced labor.

This technique was taught by the Americans to their client military dictatorships in Latin America, mostly, and made famous by the heart-breaking, public -- but quiet -- protests in Argentina by the "Mothers of the 'Disappeared' " for their sons and daughters "disappeared" by the American-instigated military dictatorship of the mid-1970's.

And then there was the "Night of the Long Knives", when governments were overthrown by military coup d'état -- again most well known in Latin America. Bloodless -- the public could stand no chance against the military plotters. The "Night of the Long Knives" was the name given to the 1934 decapitation of the Nazi S.A. -- _Sturmabteilung_, "Storm Troopers" -- by the professional elite Nazi S.S. -- the Schutzstaffel. The S.A. -- amateurs, petty criminals, grown-up delinquents -- had grown too demanding once Hitler rose to power in 1933.

There was only so much power, and the spoils of victory, to go around. The S.A. -- street thugs -- had outlived their usefulness. So Hitler secretly ordered the quick and secret round-up and liquidation of the S.A. leadership.

Forty to sixty officers in all were executed without trial in a day or two, and the S.A. threat was neutralized. It was an effective means also for ensuring success in a military take-over of a government for a troublesome country America wished to dominate.

There was the concocted _casus belli_ to justify the Vietnam War (the 1964 Gulf of Tonkin "incident"), the 1983 invasion of the island of Grenada (to rescue American medical students), and the Second Iraqi War (Al Qaeda, and non-existent Weapons of Mass Destruction).

There was "the Big Lie", when U.S. State Department head Colin Powell gave his phony presentation on alleged Iraqi WMD to the U.N and the world. And all the experts were silent that it was all B.S. But, being an expert on Chemical Warfare -- I knew.

The Nazis introduced loyalty oaths to professional employment, a practice adopted by the U.S. in 1948, preceding the anti-communist, McCarthy era of the early 1950's.

There was the use of a pantomime of legal methods to give the cover of legality to patently illegal governmental behavior.

There was Ronald Reagan, whose greatest crime in his eight years in office, was the savaging of the federal judiciary with corrupt, biased, and incompetent extreme right wing appointees who began the roll-back of the judicial affirmation of the Bill of Rights that reached its pinnacle in the "1960's Spring".

George W. Bush mimicked Hitler's breaking of treaties with shocking ease -- treating them as mere pieces of paper. The Geneva Conventions, the Hague Convention, the 1984 Convention on Torture and Treatment of Minors. And Canada was knowingly complicit in these war crimes, making us equally guilty. We **are** war criminals, too.

The surviving Nazi leadership was hung after the post-war Nuremberg trials for similar crimes.

I'd like to have a ticket to see (or hear) Canadian Prime Minister Stephen Harper's neck snap.

* * * *

> "Look! Those two specimens are worth millions to the Bio-Weapons Division -- right? Now, if you're **smart**, we can both come out of this heroes, and we will be set up for **life**."
>
> -- Burke
> "Aliens" (1986)
> James Cameron

Then there was Nazi weaponry, tactics, and other advancements snapped up and adopted for American usage:

The first use of paratroopers in combat (Crete).

There was development of a superior (than the Allied one) proximity fuze for anti-aircraft use. Proximity fuzes made AAA ten to twenty times more effective at killing crewmen and downing bombers than regular fuzes.

There was the revolutionary MG-42 medium machine gun, the best machine gun design of all time, its roller-locking firing mechanism pure genius. Cheap and quick to manufacture, it was robust, reliable, and deadly effective, with a high rate of fire, and a quick change barrel. It was made from sheet metal pressings, castings, and plastics, instead of expensive and labor-intensive forgings and machined parts.

A variant was later developed and adopted as the U.S. M-60 machine gun. It functioned poorly.

The design of the first assault rifle, the Stg44 (Sturmgewehr 44), with a 30 round magazine, 16" barrel length, and full auto selective fire, was the predecessor of the U.S. M-1 carbine, and M-16 assault rifles (of which "Sturmgewehr" is a loose translation).

The M-1 cartridge had inadequate lethality and even "stopping power", and the M-16 jammed profusely unless kept scrupulously clean.

The development of the MP-40 ("Schmeiser") submachine gun (preceded by the MP-18 drum-fed WW1 SMG), was the predecessor of the CAR-15 SMG.

The development of the first infrared night vision rifle scope.

The superior armor, gun, and general design of the Panzer and then Tiger tanks.

The legendary and feared Flak 88 anti-tank gun: a tungsten high density, kinetic energy shell, that first used high quality steel to use the squeeze bore principle to attain a high velocity projectile in a gun of not excessive size.

The _Panzer Faust_ disposable anti-tank weapon provided superior armor penetration of 5.5" armored steel, and was much shorter, smaller and cheaper than the American bazooka.

The flame-thrower, first invented and used during WWI.

The first fast jet engine fighter plane: the Messerschmitt Me 262.

The V-1 "buzz bomb" was the first cruise missile.

The V-2 ballistic missile was the predecessor of the modern ICBM. And to guide the new rocket-powered V-2, was the new inertial guidance system, necessary for the ICBM's accuracy. The German's U-boat submarine warfare was the precursor of the modern U.S. ballistic missile-launching and enemy attack submarine naval force.

The steam catapult for launching the V-1 was adapted of usage for Naval aircraft carriers to launch short take-off jet fighters.

The invention of stealth technology -- now used with fighters and heavy bombers -- was another Nazi first, with the installation of anti-radar coated submarine snorkels.

The first development of nerve gas, a revolutionary new, and much more toxic, agent for chemical warfare than previous chemicals.

The first development of a synthetic morphine substitute: methadone.

The development of large scale use of the German-invented Syn-Gas Process to produce a gasoline/diesel fuel substitute from plentiful German coal stocks

* * * *

gnome: "Oxford Dictionary", definition 3: a person with sinister influence

"[General LeMay] impressed upon his forces the need to be vigilant, warning them about the 'gnome' in the Kremlin basement who was waiting for the perfect moment to attack. Most days the gnome would say, 'No, we won't attack today.' But some day, the gnome would announce, 'Yes, the correlation of forces is right today. Let's go.' "

-- "The Wizards of Armageddon" (1983)
Fred Kaplan

The collapse of the Soviet Union, and the end of the Cold war in 1991 **finally** ushered in a new era of arms control -- actual nuclear arms reduction -- treaties. Better late than never, I suppose...

1991   START Treaty   6,000 strategic warheads   1,600 ICBM's/SLBMs/bombers

2002   Moscow Treaty   2,200 strategic warheads   500 ICBMs/14 Trident II subs/115
                                                         B2 and B-52 bombers

2010   New START Treaty   1,550 strategic warheads   800 ICBMs/SLBM's/bombers

As of September 30, 2009, the U.S. stockpile of nuclear weapons was 5,113 warheads -- with about 3,000 in storage -- the nuclear lay-away plan -- the "inactive stockpile", which are not mated to a delivery vehicle -- B-52, or ballistic missile.

The 5,113 warhead number is down from the maximum of U.S. warheads in 1967, totaling *31,255*, and down from the stockpile number of 22,217 warheads held in late 1989, when the Berlin Wall fell, and the Cold War started its end.

There are currently about 3,200 U.S. nuclear warheads in service, about 2,000 of which are strategic warheads -- missile warheads aimed at Russia, either the one thousand 8,000 mile range Minuteman III, or the twelve (with twenty-four missiles each) sub-based Trident II missile with a range of 7,500 miles.

The rest are battlefield "tactical" rockets or artillery shells -- the U.S. has only 180 of these at the ready in Europe. The Russians, by treaty -- the Moscow Treaty of 2002 -- also has about 2,000 strategic warheads aimed at the U.S. or its allies, and 3,000 - 5,000 plus tactical nuclear warheads

* * * *

"Let everyone sweep in front of his own
door and the whole world will be clean."

-- Johann von Goethe (1832)
on his deathbed

In the 20th Century approximately 231 million people died in wars and conflict and, "by human decision" (deliberate famine, etc.).

At the beginning of the 21st Century, approximately 95 wars were being fought, the world over.

The Russian standing army consists of 1,000,000 men, and 20 million in reserves.
The U.S. standing army consists of 1,600,000 soldier, and reserves of 865,000.
China has a standing army of 2,300,000 with reserve forces of 500,000.

All standing armies of the world total 20 million soldiers.

There are about 70,000 soldiers in the Canadian Forces, and 24,000 in reserve units. This is out of a total population of about 30 million. This is proportionally **half** the percentage of active duty soldiers on a per capita basis as the U.S., the world's only Super-Power, and thus the military and political leader of the entire world.

World military expenditures in 2009 are an estimated $1.53 trillion, or approximately $225 for each man, woman, and child in the world. According to one estimate in the "Economist" magazine, 10% of the money involves bribery of government officials.

The U.S. accounts for just under 50% of the total world military budget. The First World, consisting of North America and Western Europe -- the NATO countries -- total almost 70% of world military spending. (Ref: SIPRI, the Stockholm International Peace Research Institute.)

Of 184 member-states of the United Nations, there is a U.S. presence of 141 countries.

In 1984, the University of Oslo, and the Norwegian Academy of Science, determined that since 3600 BC, there had been almost 15,000 wars, resulting in 3.6 billion deaths. There was peace in the world for 292 out of 5600 years studied, probably due to inadequate historical records

At the end of 1989, forty to one hundred, mostly internal wars, were going on in the world, including 700,000 insurgents fighting their own governments.

There are 670 active duty military personnel – but only 85 doctors – per 100,000 people world-wide.

* * * *

> "When the third victim was brought back, dead or unconscious, the elder of the _Guardia Civil_ shrugged his shoulders with a glance in my direction; it was an unconscious gesture of apology.
>
> ... In it was expressed an entire human philosophy of shame, resignation and apathy. 'The world's like that,' he seemed to be saying, 'and neither I nor you will ever change it.'
>
> The shrug...is more vivid in my memory than the screams of the tortured."
>
> -- "Dialogue with Death" (1943)
> Arthur Koestler
> on his capture by the fascists
> during the Spanish Civil War

Then there is assassination as government policy, a contradiction in terms -- an absolute abortion of the principles of a democratic republic.

Assassination, political homicide, a "hit", termination with extreme prejudice ("Tweep", a CIA term dating from the 60's), a "sanction", and "executive action" or "left hand operation" (both expressions from the 1970's). Or simply "wet work".

There was religious assassination. One of the last of these was Joseph Smith, founder of the U.S. religion, Mormonism, who was assassinated in 1844.

(In 1981, there was a brief, unsuccessful revival of religious assassination, when Pope Jean-Paul II was shot four times with a 9 mm Browning Hi-Power semi-automatic pistol by 23-year-old would-be assassin, the fascist "Grey Wolf" organization member, Mehmet Agca. The attempt was a failure, and the Pope recovered completely.)

Then there was revolutionary assassination. _Riscus sardonicus_, the smile of death, caused by muscular contractions after death.

There was a string of assassinations of European royalty and French and Italian political leaders in the period of 1890-1905. Damn nihilists and anarchists!

The assassination of the Arch-Duke Ferdinand in 1914 by a Serbian nationalist started World War One, leading to 15 million deaths of young men, and leading twenty years later to World War Two and another 50 million deaths (with a far greater percentage of civilians killed).

Rafael Trujillo, dictator of the Dominican Republic for thirty years was assassinated by an ambush of his car, and riddled with sub-machine gun fire.

Anastasia Somoza, ex-dictator of Nicaragua was assassinated in 1980 in Paraguay, by bazooka and AK-47 fire, the target of a 7 person team of Sandinista commandos

JFK, who tried to overthrow Castro, and introduced U.S. military "advisers" to Vietnam in 1962, was killed by communist Lee Harvey Oswald in 1963.

There was the AK-47 assassination of Anwar Sadat, Egyptian President, by Islamist radicals in 1981, for signing the 1978 peace treaty with Israel for an outright bribe of billions in dollars in aid from the U.S.

There was the wide smile observed by the compound guards on the face of the suicide driver of the truck bomb in the 1983 Beirut U.S. Army barracks bombing that killed 246 U.S. Marines.

There was the massive bomb that killed Benazir Bhutto in 2007 on a political parade route.

But those were just small potatoes compared to government killings.

* * * *

> "[They] found a kind of serenity in resolute acceptance and engagement with the violence in our nature, deriving from it what I would call the only true solution...the pursuit of virtue, the only proper pursuit of a human being.
>
> Such is an uncomfortable route towards a solution, since it demands individual self-examination and renounces the consoling ideologies of a collective resolution of the human problem."
>
> -- "The Bullet's Song" (2004)
> William Pfaff

There was J. Edgar Hoover, head of the FBI, who focused on the illegal neutralization of the Black Panthers in the COINTELPRO Operation of the 1960s. Through the Chicago Police Department he murdered Fred Hampton and Mark Clark of the Black Panthers, firing something like 200 shots into his apartment and bedroom, as he slept when the concocted police raid began.

Fred Hampton was a rising star and effective organizer and rhetorician of the Black Panthers silenced forever.

The was AIM (American Indian Movement) leader Leonard Pelletier framed in the early '70s for the killing of two FBI agents on the Pine Ridge reserve in South Dakota; extradited by the U.S. from Canada where he had fled, based on a perjured FBI affidavit and Canadian acquiescence.

He remains incarcerated to this day.

There was the 1975 Church Committee Senate Report "Alleged Assassination Plots Involving Foreign Leaders", which examined U.S. government assassinations and plots against Patrice Lumumba in the Congo, Fidel Castro in Cuba, Trujillo in the Dominican Republic, President Ngo Diem in South Vietnam, and General Rene Schneider, a strict democratic constitutionalist, whose obstruction to U.S. interests in Chile resulted in his assassination in 1970.

In the late 1970s to 1980s the white racist South Africa _apartheid_ government ran assassination squads using toxins and investigating biological warfare.

In Chile, after a right-wing coup d'état in 1973, the Pinochet dictatorship established Operation Condor, a campaign of foreign assassinations of both exiled guerrilla leaders and moderate politicians.

In Iran, after Ayatollah Khomeini overthrew the dictatorship of the Shah of Iran and took power in 1979, he died in 1989, but not before compiling a list of 500 "enemies of Islam". By 1992 sixty of them had been killed in foreign countries, a Special Operations committee having been set up after Khomeini's death to carry out the hits. Killings of Kurds in Germany were traced back to Iran, who followed by the rest of the EU recalled their ambassadors and broke relations with Iran, for six months, then reestablished relations. The hits stopped.

In 1993, an armed assault on the Branch Davidian religious compound in Waco, Texas killed four U.S. Federal BATF agents, and six Branch Davidians. There followed a botched assault with a tank that started a fire that killed 70 people, including women and children, and David Koresh, the religious cult's leader, and the target of the BATF.

In Northern Ireland, the English government's SAS commandos began a covert "Dirty War" against the Provisional IRA -- the "Provo's" -- in the 1980s. It was approved all the way to the top of the U.K. Government, by Prime Minister Margaret Thatcher (a right-wing fascist, friend of fellow fascists Ronald Reagan and Pope Jean-Paul II, and the murderous Chilean dictator, Augusto Pinochet).

* * * *

"Our mental life is governed mainly by a cauldron of emotions,
motives, and desires which we are barely conscious of, and
what we call our conscious life is usually an elaborate post
hoc rationalization of things we really do for other reasons."

-- "Illusion of Body Image" (1996)
Dr. Vilayanur S. Ramachandran

And people wonder why * I * want to be armed.

Isn't it ironic?

"Civilization and violence are antithetical concepts."

> -- Nobel Peace Prize
> acceptance speech (1964)
> Martin Luther King

"I've got to stand up for those who won't stand,
and I've got to stand up for those who can't."

> -- Martin Luther King (1967)
> answer to athlete John Carlos,
> following MLK speech in New York

Prison Diary 14:	Is It Possible That **We're** The "Bad Guys"?
================================

> "They despise me as an upstart, I despise their worthlessness. They can taunt me with my social position, I, them with their infamies. My own belief is that men are born equal and alike: nobility is achieved by bravery."
>
> -- Gaius Marius (107 BC)
> Consul, Roman Republic

> "The beasts of modernism have mutated into the beasts of post-modernism -- relativism into nihilism, amorality into immorality, irrationality into insanity, sexual deviancy into polymorphous perversity."
>
> -- "On Looking into the Abyss" (1994)
> Gertrude Himmelfarb

The winners write the history. And the winners, of course, are always the "good guys" who won legitimately. Well, I got news for you, about the sound application of logic to that statement...

"A safe estimate is that several hundred times every day... [CIA] officers engage in highly illegal activities." So wrote a 1996 (Clinton era) report of the unassailably credible House Permanent Select Committee on Intelligence.

So it goes.

*   *   *   *

> "A winner will emerge, but there's often no reason to expect that winner to be truth."
>
> -- "The Moral Animal" (1994)
> Robert Wright

There is routine lying on major issues by the Government. There is the _casus belli_: Cuba, Vietnam, Nicaraguan contras, Grenada, Columbia, 1% of drugs interdicted instead of the phony value of at least 10%, the use of street prices to inflate drug seizure values, rather than the wholesale value of the load intercepted.

It was Republican Party's Ronald Reagan, an actor and fascist who began the practice of systematic judicial stacking of the Federal courts, beginning with the Supreme Court on down.

The classic example was in one Supreme Court case where speeding cars was ruled grounds for stopping and searching them by the Border Police. **Then**, a short time after, the Supreme Court ruled that it was hunky-dory for the Border Police to stop and search cars going the speed limit! The Supreme Court had ruled that an otherwise illegal stop under the Fourth Amendment of the Bill of Rights was legal if a car was within a hundred miles of the Mexican or Canadian border if it was moving at the legal speed limit.

It was an unbelievably outrageous ruling.

Look at what the U.S. has become: the prison business, media conglomerates, the national security state (the alphabet soup of agencies: the CIA, NSA, NSC, NRO, FBI), the military (AF, Navy, Army, Coast Guard, Reserves, National Guard) and its corporate subcontractor support structure, the justice system (FBI, DEA, BATF, Customs, Secret Service, federal, state, municipal, police, prisons, lawyers, corporate lawyers, judges, and prosecutors), bureaucracy; manipulation of the truth with selective leaks, no accountability, and a phenomenal National Debt

And then there is the death of democracy, and the rise of corporocracy: election stealing, a painfully low voter turn-out, black disenfranchisement, lying attack ads that are successful, TV ads only for the super-rich, voter ignorance of issues of importance, responding only to canned ideas: "tough on crime", "starting wars".

* * * *

"Terrorists don't have a chance against the United States.
Terrorists will never destroy America. Greed might."

-- "Let's Blow Up Our Brand" (2002)
Karl Idsvoog in "Into the Buzz-saw"

Then there's "TV reality". Take for instance the super-rich and famed author, Tom Clancy.

With trade deficits, an exploding national debt, financial scandals, recessions, unemployment, the media feed us soft, feel-good stories, and celebrity scandals, but everyone loves a Tom Clancy book's happy ending.

Tom Clancy's jingoistic, right-wing stories are great. The trains run on time, and rah-rah happy endings makes us feel that all is right in the world. His "the U.S. is great" B.S plots, that promotes the military superpower status of the government, and the righteousness of its leaders, the courage of its covert soldiers, while serving up a one-dimensional, blatantly racist portrayal of its enemies.

He enhances, placates, and sedates the upper echelons of society with a feel-good view of world reality, with the U.S. rightfully at the top of the heap. And forget about its tottering.

* * * *

"We don't even have bad ideas."

-- "The End of Science" (1996)
Noam Chomsky

George W. Bush's election stealing when he was first elected in 2000 is a matter of record. His brother, Florida Governor Jeb Bush illegally (and then ignored court orders) ordered the removal from voter roles of felons entitled to vote: 50,000, mostly Democratic voters. Also there were 58,000 removed as felons, most of whom were not felons, and 54% were black, again mostly

Democratic voters (95% at least); with also the tampering of election machines in black districts to "eat" mismarked ballots, instead of spitting them back out, as in white districts.

The U.S. Civil Rights Commission held dramatic and damning hearings on the evidence.

The U.S. has 5% of the world's population, and 25% of the world's prison inmates. It has per capita, five times more inmates than Canada, and seven times more inmates than **all** of Western Europe combined. 90% are imprisoned for non-violent crimes, mostly for drug offenses.

> "Activists fall into three basic categories: radicals, idealists, and realists.
>
> The first step is to isolate and marginalize the radicals. They're the ones who see inherent structural problems that need remedying if indeed a particular change is to occur...
>
> The goal is to sour the idealists on the idea of working with the radicals. Instead get them working with the realists.
>
> Realists are people who want reform but don't really want to upset the status quo."
>
> -- Consultant speaking to corporate executives, quoted in "Organizing for Social Change" (2001)
> Kim Bobo, Steve Max, and Jackie Kendall

Prison Diary 15:   Armed Right vs. Armed Left
=====================

> "[T]errorism comes in two forms, revolutionary and reactionary, and it is engaged in by both non-state and state actors. ... 'War' is a polite term in contrast to 'terrorism'; it refers to legal boundaries.
>
> But when war becomes illegal, it merges into war crimes and terrorism. And terrorism is the more primal form of warfare, the means of social domination to which states as well as non-state actors frequently revert when they drop the mask of legality."
>
> -- "In the Name of God and Country" (2010)
> Michael Fellman

> " 'Tsk, tsk,' I said, not moving at all. 'Such a lot of guns around town and so few brains.
>
> You're the second guy I've met within hours who seems to think a gat in the hand means a world by the tail.' "
>
> -- Phillip Marlowe
> "The Big Sleep" (1939)
> Raymond Chandler

Anti-gun ideologues often express pride in admitting they know nothing about guns. It is part of the polarization of the extreme. Polarization intrinsic to the mind, an unchangeable dogma, unresponsive to rational argument.

It is a polarization seen in many political positions that divides the proponents into two diametrically opposing, irreconcilable groups, with the resulting loss of any possibility of the sensible compromises that are necessary to solve complex societal problems.

Similar "hot button" issues adopted by politicians, and popular with the electorate, like gun control, are going to war, defense spending issues; passing illicit drug laws with harsher penalties, and hard on crime issues, such as "three-strikes" laws .

Liberals and moderates are held in contempt or even despised by the revolutionary left as gradualists and believing in the false hope of reform under a regime beyond hope. Not surprisingly, the liberals and moderates (what I call the "soft left"), are the chief supporters of gun banning, masquerading as gun control, because they reject the idea of any possibility of violence.

Which is unfortunately a hope, rather than a reality, as the right wing is heavily armed and will remain so; it is them that are the stalwarts what is called "reactionary". The citizen "soldiers" prepared to fight on the streets to maintain the unjust social order, in addition to the regular security forces.

The start of any revolution begins with what I call, "clearing the deck." The wiping out of the moderates.

The trouble with moderates is that when the shit hits the fans, their usefulness is at an end and they're just a troublesome nuisance, eminently expendable, their moral righteousness no match for a mass grave and the final blessing, "Good riddance."

* * * *

"[T]errorism [is] a phenomenon of the nineteenth-century breakdown, and an evidence of political despair... associated with retrograde nationalism and utopian anarchism."

-- "The Bullet's Song" (2004)
William Pfaff

It is the demonization of each other of two groups. The destruction of rational argument and conversion into bigotry, and then dogma. It is the destruction of "the trust that is essential of compromise and accommodation".

It is a one way street from regulation, stricter and stricter, more and more enforcement mechanisms, and leading inevitably to absolute prohibition. Could it be that stricter and stricter laws on an issue that do not solve the perceived problem are simply irrelevant? Wrong? If an act does not have any effect on a problem, then why would stricter acts?

It is a civilized version of "war" – a political, non-violent, civil war. A take-no prisoners "**war** of different ideologies"

Israel, New Zealand, and Switzerland are countries where a large percentage of the population is armed, two with machine guns. Yet they have no problem with guns. Then there are the democratic countries where gun ownership is severely restricted: Britain and Canada (between 1976 and 2011). (I say democratic, because civilian gun ownership is non-existent in dictatorships and authoritarian regimes. A matter that should give one pause.)

And then there are countries where gun ownership is both severely restricted, and unlawful ownership severely punished, sometimes including the death penalty (Taiwan, South Africa, Jamaica), which have far higher apolitical (criminal) murder rates than the U.S.

Sri Lanka – interestingly, since I am half Sri Lankan – has the highest non-political murder rate in the world, but severely restricts gun ownership. 30 million people on an island smaller in area than New York State, is the problem.

And on the political front, a bloody 25-year internecine civil war broke out after total systemic minority discrimination aligned itself with government policy: hundreds of thousands died, mostly minority Tamils.

And what was covered up was that the final government push that crushed the Tamil Revolt came with funds stolen from Western-donated tsunami relief funds and diverted to artillery and military rifle ammunition.

The 8" artillery shell is a magnificently engineered and perfected piece of technology. And it annihilated the Tamil separatist struggle, and overthrew real democracy and installed an extremist, right-wing, "death squad" government.

The 15% Tamils were Hindu, the majority Sinhalese were Buddhists. So much for peaceful religions, again. Another human tragedy unfolded...

And, by the way, New York City has banned handgun ownership for nearly a century. NY City police estimate that the number of handguns in NYC is 2-3 million... [Kates, Don B. "Gun Control: A Realistic Assessment". San Francisco: Pacific Research Foundation (1990).]

* * * *

"[T]he principle that the end justifies the means -- all means without exception... Whoever did not understand this necessity had to be destroyed.

Whole sets of our best functionaries in Europe had to be physically liquidated. We did not recoil from crushing our own organizations abroad when the interests of the Bastion required it...

We did not recoil from betraying our friends and compromising with our enemies, in order to preserve the Bastion. That was the task which history had given us, the representatives of the first victorious revolution. The shortsighted, the aesthetes, the moralists did not understand.

But the leaders of the Revolution understood that all depended on one thing..."

-- Rubashov
"Darkness at Noon" (1940)
Arthur Koestler

What the pro-gun control liberals don't seem to realize is that the right is more effective than left, when it comes to violence. And the right will never give up their guns.

1650: By this date, 90% of indigenous peoples (20 million) slaughtered in Latin America by the Spanish _conquistadors_.

1871: Paris Commune crushed and 25,000 Communards (men, women, and children) killed by French troops. The war was with Germany, who defeated France. Figure that one out.

1890: 12-15 million Native American Indians slaughtered in the U.S., leaving 200,000 forced onto reservations – shitty land that nobody wanted.

1919: Leftist Spartacist Uprising in Berlin, Germany. Lasted ten days: a general strike, barricades in the streets, and armed battles. It was brutally crushed in ten days by the German Freikorps, disaffected, rightist, and heavily armed WW1 veterans of the defeated German Army, who had reconstituted as violence-seeking, roaming private militias. Casualties were 17 Freikorps members and 100 civilians and leftists killed.

1936-1939: Spanish Civil War. 150,000 executions by Franco's forces during war, 50,000 after the war. Untargeted mass political killings.

1939-1945: Nazis (50 million dead; including an estimated 10 million Jews, Slavs, gypsies, communists internally, as well as in conquered territory). The Soviets lost 20 million men. It was Stalin's efforts that defeated Hitler, and then had to deal with the American-instigated Cold War. Hitler wasn't crazy, he was a genius, though a little egomaniacal, so he didn't take good advice when it was proffered.

The Nazis, I describe them, were "dedicated fanatics, with really cool uniforms".

1957-1961: In 1957, Bernard Fall, combing through Vietnamese police records, discovered that in ever- increasing numbers the Viet Minh guerrillas assassinated 7,000 village officials in nighttime operations, controlling and confining police to their stations and barracks.

1968: 30,000 CIA-trained South Vietnamese cadres of the Operation Phoenix counter-terror program assassinated more than twenty thousand suspected Viet Cong members and supporters.

1976: Argentine dictatorship was the most bloody of all Latin American dictatorships, excluding , of course, civil war situations. They "disappeared" 30,000 civilians.

1973-1989: Chile under Pinochet and his secret police organization DINA "equivalent to the Gestapo" "and the KGB" as described by U.S diplomats, set up Operation Condor with three other Latin American dictatorships, for the purpose of exterminating the "leftist threat" out of Latin America. 13,000 people killed, or tortured.

1976: Orlando Bosch and Luis Posada Carriles, anti-Castro exiles living in the U.S., bombed out of the air an Air Cubana flight from Caracas to Havana, killing all 73 passengers and crew. The first terrorist bombing of an airliner. Random killings.

1960-1989: South Africa. The white apartheid Afrikaner's string of atrocities is long. Along with assassinations, and fighting the armed wing of the ANC, they engaged in bio-war, submarine-delivered commando raids, the Sharpeville massacre, the killing of Steven Biko (made into a movie), and the jailing of Nelson Mandela for decades. They annexed Namibia, and engaged in war with Angola (500,000 dead), and Mozambique (900,000 killed). Racist, imperialist war.

1965-1973: U.S. Black Panthers vs. FBI, Fred Hampton and Mark Clark shot dead, Timothy Leary, Ken Kesey, AIM (American Indian Movement) leaders jailed. Hampton was a targeted assassination.

1980: Right-wing extremists: Bologna, Italy bombing of central train station; 85 people killed, >200 wounded. Random killings.

1984: The "Stalker Affair" confirms the existence of Northern Ireland police "death squads" in use against unarmed IRA suspects, since at least 1982. Targeted killings.

1994: Rwandan genocide of between one half and one million unarmed minority Tutsi s, along with moderate Hutus by extremist Hutus. Genocide.

1995: Right-wing extremist Timothy McVeigh bombs U.S. Federal Building in Oklahoma, killing 168 people. Targeted killings.

2011: Right-wing Norwegian extremist, Anders Behring Breivik with a bomb and shooting attack kills 77 Norwegian civilians, mostly teenagers. Random killings.

* * * *

"[C]overt, as well as overt, political repression
reached massive levels during the Vietnam war era."

-- "Political Repression in Modern America:
From 1870 to the Present" (1978)
Robert Goldstein

Compared to the left and other groups:

1950: Stalinist Russia; deliberate mass starvations, Stalin era foreign assassinations; Trotsky as the most famous; cyanide spray; ricin in umbrella platinum-iridium pellet; Polonium-242, TCDD of Ukrainian political leader; chloracne.

1968: Two thousand government official, administrators, bureaucrats, soldiers, and other government collaborators killed and buried in shallow graves at Hue, taken temporarily by Viet Cong forces, during the "Tet Offensive".

1970's: The infamous Carlos "the Jackal" – the Venezuelan, Illich Ramirez Sanchez – the legendary free-lance international terrorist is the only "rival" in killing effectiveness ever.

Killer of two French intelligence agents, the bloody 1975 OPEC Hostage-taking, the Entebbe Air France high-jacking, associated with the PFLP (the Popular Front for the Liberation of Palestine) he remained free for twenty years. He once bragged that he was responsible for the deaths of eighty-three people.

He was finally caught in Sudan in 1994, and remains imprisoned in Paris. He was accused (but denied) four French bombings that killed eleven people and wounded nearly 200.

1975: Cambodian communist dictator Pol Pot killed mostly by starving and working to death 700,000 "collaborators" of the old U.S. backed dictatorship of Lon Nol.

1983: Beirut U.S. barracks suicide truck bombings: 241 Marines and over 50 French soldiers killed by Hezbollah (Party of God), prompting withdrawal of U.S. forces from Lebanon by Reagan; U.S. retaliated by random, terroristic 16" Naval gun shelling by U.S.S. New Jersey off Lebanese coast of Shi'ite mountain villages.

Anybody notice that when it comes to killing, the right-wing extremists are much more effective, as well as being random, rather than engaging in targeted "liquidations" (as the Stalinist killings were referred to)?

"You wonder what I am doing? Well, so do I, in truth. Days seem to dawn, suns to shine, evenings to follow, and then I sleep. What I have done, what I am doing, what I am going to do, puzzle me and bewilder me.

Have you ever been a leaf and fallen from your tree in autumn and been really puzzled about it? That's the feeling."

Prison Diary 16:   "Pig!"

-- T.E. Lawrence (1922)
to his friend Eric Kennington

===

> "While evil may manifest itself obviously,... it rarely does so. More commonly by far, its manifestations are seemingly ordinary, superficially normal, and even apparently rational. ...
>
> [T]hose who are evil are masters of disguise...The evil always hide their motives with lies. ... They [are] the people of the lie."
>
> -- "People of the Lie" (1983)
> Dr. Scott Peck

> "[T]his social gelatin composed of the mass of all those who just want to live their little private lives at a distance from history and its tumults.
>
> This swamp is predisposed to be the champion of false consciousness, half-asleep and always ready to close its eyes on the war that rages all around it."
>
> -- "The Coming Insurrection" (2007)
> The Invisible Committee

In 1989, the rap group NWA (Niggaz With Attitude) had broken through with their debut album, "Straight Outta Compton" [an L.A. ghetto]. They also broke through other barriers, creating a minor flurry/ controversy with their cut, "Fuck the Police". The album went triple platinum (3 million copies sold).

Rap musician, Ice-T, who did the Billboard Top 100 hit, and title track "Colors" for the 1988 Dennis Hopper movie, starring Sean Penn, first achieved widespread fame (or at least notoriety) over his 1992 track "Cop Killer" which was nationally pilloried as justifying and glorifying the killing of municipal pigs by blacks, adopting the mantle of the militant Black Panthers of the 1960s.

The first of its kind, back-lash campaign against Ice-T's "Cop Killer" title and lyrics was formed and led by Charleton Heston, with Al Gore's wife, "Tipper" Gore, joining in.

So much for the First Amendment, huh?

The rapper community fought back, with one rapper writing a song fantasizing about, and extolling the rape sodomizing of Tipper's two teenage nieces.

Oy vey! And people complain about **my** sense of humor!

But in a final irony, in 2000 Ice-T reappeared on the public scene, when he became a regular cast member, as of all things, NYPD Detective Finn Tutuola, in the hit Dick Wolf NBC show, "Law & Order: SVU" (Special Victims Unit).

* * * *

> "To no longer wait is...to enter into the logic of insurrection. ... Because...every act of government is nothing but a way of not losing control of the population."
>
> -- "The Coming Insurrection" (2007)
> The Invisible Committee

How many people know that Bob Dylan's 1965 song, "It's All Right, Ma (I'm only bleeding)" is about police brutality? Or that the Rolling Stones hit "You Can't Always Get What You Want" includes a lyric couplet about police brutality?

Once established and consolidated, the first and only goal of tyranny is to defend its power from being eroded. For external threats, there's the military. For internal threats, up step the municipal, urban police forces as the hammer of state repression and brutality against its citizens.

Even before the intelligence agencies, federal police, national security agents, and secret police.

The primary means to defend a monopoly of power is the most simplest. It is to extinguish hope of change in any form. Reform, improvement, change. There are the security forces for the maintenance of internal security, and the detection and quashing of internecine conflict, rebellion, overthrow, revolution, insurrection.

Extinguishing hope; to accomplish it, to act for it to occur, to organize against it, to say it, to believe it possible, to even hope for it.

This is why you can't say "pig" to or about municipal policemen without eliciting a violent emotional and verbal counter-reaction from older white -- generally racist -- white men and their ilk. Because to say it is to allow discussion, leading to acceptance of the possibility that something is wrong. And the acceptance that something is wrong is the only possibility of beginning the long, slow change in reining in a corrupt, violent, illegal, and ultimately totally evil organization.

And by definition any organization that resists any reasonable and effective process for change, audit, oversight, and reform **has** to be evil.

That they are violent is beyond dispute. That they commit illegal acts routinely is beyond dispute. That they are out of control is manifestly without question.

* * * *

> "[T]he justification for this behavior...lies in the words 'saving American lives'. Any action can be condoned, any excess tolerated, any injustice justified, if it can be made to fit this formula. The excess valuation on American life, over any other life, accounts for the weapons and tactics we feel entitled to use..."
>
> -- "New York Review of Books" (1968)
> Robert Crichton

Fred Hampton, Chicago Police. A genuine political homicide. Asleep. 200 shots.

Mark Essex, 1973, riddled his already long dead body with a total of 200 bullets, after his anti-police shooting spree.

And they were both black.

* * * *

"My wife, children, and family were beside themselves with worry and fear. Their respect for the police was rapidly evaporating, and although they love me as a husband and father, their pride in me as a policeman was dying."

-- "The Stalker Affair" (1988)
Former Deputy Chief
Constable John Stalker

Pigs have become society's bureaucratic department of harassment, fear, and violence directed at the discards of society. The legal system is a bureaucracy of pantomime justice, nominally transparent, but so slow, which most people can't follow it, or are put to sleep by.

Pigs are the eminently visible interface between the acceptable middle class and upward, and society's rejects, the working class and below. It is the fighting edge of a racist, corrupt system, and its unending war against society's weak, powerless, defenseless, and disenfranchised; the politically unrecognized, the criminal class, the outlaws and those on the fringe, socially and economically marginalized -- society's rejects.

Pigs are the essential core of fascism, and corrupt political power. Violent, criminal, evil, this panoply of horror is known, and accepted by the leaders, and upper and middle classes, since they are not exposed to its depredations. Or even aware of it, frequently.

Police are effectively a state-sanctioned criminal gang fighting all other smaller, non-state sanctioned criminal gangs. A devil's bargain for sure, but one acceptable to supposedly free and peaceable democratic states and the comfortable components of their citizenry.

It is a spectacular, blatant, and obviously unacceptable outrage that belies the true, fundamentally corrupt nature underlying such "free" democratic states.

It is merely the most obvious example of the class war of the wealthy upper classes versus the lower classes formalized, and invisibly codified, and mutated into an orderly, systematic form.

* * * *

"Whenever we encounter such a purely evil Outside, we should gather the courage to endorse the Hegelian lesson: in this pure Outside, we should recognize the distilled version of our own essence."

-- "Welcome to the Desert of the Real" (2002)
Slajov Zizek

U.S. and Canadian municipal police – the security forces – are over-armed. They are armed to the teeth, aggressive and belligerent, travel in packs, and are set to hair-trigger brutality and lethal aggression. In other words, they are brutally violent, out of control, and butchers.

Its perfectly clear and well-documented, and easy for everyone but the blind to see: the police have throughout all history been the enforcers of government oppression and repression against the middle and low classes for the aristocracy, oligarchy, plutocracy and ruling classes, and upper classes and wealthy.

The police of modern times have been intimately involved in all overthrows of democracy, as well as the willing and enthusiastic enforcers of the continuance of brutal dictatorship, and the instigators and practice of mass torture of dissenters and ordinary civilians.

The police of today – and since the 1960s -- are the hammer of society's deliberate and sub-conscious war against the poor and the different, the lower and working classes, immigrants, ethnic minorities, especially the non-white, the unemployed, the mentally ill, the homeless, street alcoholics, suffering drug users and petty criminals, the adolescent young, students, the gay and sexually different, political radicals and dissenters, and union and political protesters exercising or not exercising their right of peaceable protest, and the fringes and drop-outs of society.

They target the marginalized and disenfranchised of the human menagerie. Society's rejects, discards and outcasts; the obsolete, the cashiered, the internal economic exiles, consigned to the back-alleys, gutters, outskirts, and fringes of society.

Pigs are evil shit, pure and simple. The pigs are corrupt. The pigs are fascist. Society's faction of modern day fascists that society tolerates as a society-approved, self-contained criminal organization of white men, as long as the middle and upper classes are free from interference.

And the few pigs that aren't violent and criminal and corrupt, cover-up for, and tolerate those that are.

The word "pig" originated in the 1960's, yet a half century later, it is the worst thing you can call a policeman, who knows it very well still, and from which it provokes a violent criminal reaction – either a false arrest, or a violent physical assault, or both.

But even just criticizing police behavior, or talking bad about cops must be suppressed – immediately and violently. For talk is the first step in reform and accountability, and change. Or exposure of their systemic misconduct to others.

* * * *

"My life is violent, But violence is life.
Peace is a dream, reality's a knife."

-- "Colors" (1987)
Ice-T and Afrika Islam

And their "make work" programs are laughable, useless and unending: the unending and intractable social problems intrinsic to modern society: "drunk" driving, false domestic abuse arrests, computer child porn, vice squad activities such as prostitution. Make work for police and the social engineering industry, instead of attacking the real or root problems.

Drunk driving reflects the social problem of widespread alcoholism. Domestic abuse is typically either vengeance by the female partner, or the dysfunctional dynamics of a troubled couple forced to live together out of poverty. Prostitution is a reflection of the hard drug problems endemic to modern society. None of these severe, as well as exaggerated issues are made better by the hammer blow of the criminal justice system, which does such a bad job in the social worker function.

* * * *

> "A young man of lower-middle-class origins who
> is both aggressive and conventional, for instance,
> would be quite likely to seek a position on the
> [police] force."
>
> -- "People of the Lie" (1983)
> Dr. Scott Peck

The fact that pigs are rarely ever charged, prosecuted, and convicted for brutality except when caught on video or camera is _de facto_ evidence that there's systemic pig brutality and a corrupt process in place.

It's incontrovertible based on the simple law of probability.

* * * *

> "[A] day will soon dawn when our sleepy and lazy society will wake up and be
> ashamed that it allowed itself to be humiliated for so long; that it has not
> resisted when its brothers, sisters, and daughters were taken away and
> destroyed for no greater crime than being true to their convictions. And
> when this day comes, society will avenge us.
>
> "Go on persecuting us. Physical force is still on your side; but moral force
> and the force of historical progress is on ours. Ours is the power of ideas.
> And ideas...cannot be impaled on the points of bayonets."
>
> -- address to the Court (1877)
> Sofya Bardina

Regarding the homicide of cops, the deadliest decade for pigs was the 1970's, when over 200 pigs were killed annually. High five!

In 1973, 268 pigs were killed, an all-time high. Right on! The crushing of the peaceable 1960's Revolution got smart, and took up the gun.

Since the 1970's, there's been a steady decline. According to the DoJ, in 2004 fifty-seven were killed that year, almost all by guns.

* * * *

> "Many of these men have made a mockery of the soldier's oath of
> obedience to military orders. When it suits their defenses they say
> they had to obey; when confronted with Hitler's brutal crimes...they
> say they disobeyed.
>
> The truth is that they actively participated in all these crimes, or sat
> silent or acquiescent, witnessing the commission of crimes on a
> scale larger and more shocking than the world has ever had the
> misfortune to know. This must be said."

-- "Judgment of the International Military Tribunal" (1946)
[Nuremberg Trial]

If evil were obvious, it could and would be confronted immediately and head-on. Evil requires stealth, manipulation, and subtlety to thrive and prosper unchallenged. Its weakness, however, it that it is lazy. This isn't surprising.

Hard work is the mark of honest, not malignant, effort. Truth requires insight and the effort to gain it. It requires honesty, and self-criticism. It's as simple as that.

"A predominant characteristic...of the behavior of those I call evil is scapegoating. Because in their hearts they consider themselves above reproach, they must lash out at anyone who does reproach them.

They sacrifice others to preserve their self-image of perfection."

-- "People of the Lie" (1983)
Dr. Scott Peck

Prison Diary 17:		Witness
		======

> "[I]t was the first experience of freedom that I had. I was free for an hour and a half because during that time the repressive forces couldn't put their hands on me."
>
> -- Eldridge Cleaver (1968)
> commenting on his shootout
> with Oakland, California police

> "For [Reinhold] Niebuhr, Americans could never overcome the love of possessions because sin is ineradicable: the flesh and spirit are in a permanent state of war.
>
> The theologian argued that the dualisms of religion -- good and evil, spirit and matter, freedom and fate -- cannot be reconciled because each is a part of our humanity and each requires the other to convey the meaning of Christianity."
>
> -- "Why Niebuhr Now?" (2011)
> John Patrick Duggins

Bearing witness.

Bearing witness as an act of revolutionary truth. A polite, impartial, unarmed, neutral observer his only function. A private, citizen "truth reporter". Of non-violent, revolutionary change. The mass demonstration. The right to freedom of assembly, the modern mass demonstration expressing displeasure with the government, its performance, plans for the future, or untenable recent policy, act, or law passed or just introduced

> "But...resistance, while sometimes effective in raising the costs of State violence, is of limited efficacy as long as it is not based on [an understanding] of the forces at work and the reasons for their systematic behavior..."
>
> -- "The Manufacture of Consent" (2002)
> Noam Chomsky

*   *   *   *

To replace with a non-violent, independent monitoring of police malfeasance, and the dissuasion, by their mere presence of such malfeasance, and their challenging, or recording of it:

1) the abandonment of responsibility of reporters to the principle of independent, impartial and accurate reporting,

2) the incomplete coverage of the entire action, due to the limited numbers of reporters,

3) the reporter's general focus on demonstrator misconduct, rather than police malfeasance, playing to the TV viewer's innate expectations or biases,

4) the government's general strategic and tactical plan for the overall police action,

5) to replace the government's lack of oversight to their original plan of police actions,

6) and indeed the government's accountability to change in the face of overwhelming abandonment of Truth, replacing it with bias, manipulation, distortion, and outright lies, or propaganda.

* * * *

> "And the perverted fear of violence
> Chokes the smile on every face.
> And common sense is ringing out the bell.
> This ain't no technological breakdown,
> Oh no, this is the road to hell."
>
> -- "The Road to Hell (Part 2)" (1989)
> Chris Rea

Neatly, and maybe well-dressed, or maybe, in a T-shirt labeled front and back, in large letters, "WITNESS".

Preferably another member of the team armed with video cameras hidden, or from a distance. Ideally in groups of three or more witnesses, ideally two of which will approach an incident involving a policeman, identify themselves as "official witnesses", and calmly and rationally question the propriety of any police misconduct, attempting to convince the cop to stop his violence or the unjustified arrest.

> "In his 'Commentaries', [William] Blackstone described how...
> confiscation of property without accusation or trial, although a
> a sign of despotism so extreme as to herald the 'alarm of
> tyranny throughout the whole kingdom,' is not as serious an
> attack on personal liberties as 'confinement of the person, by
> secretly hurrying him to gaol, where his sufferings are unknown
> or forgotten.'
>
> For imprisonment, being 'a less striking' punishment is
> 'therefore a more dangerous engine of arbitrary government.' "

-- "Commentaries" 1:132 (1765)
William Blackstone

Prison Diary 18: "The Investigation of a Citizen Above Suspicion"
=====================================

"I declare my independence from the fool and from the knave,
I declare my independence from the coward and the slave,
I declare that I will fight for right, and fear no jail, nor grave,
That's why we keep marching on."

-- "Move on Over" (1965)
Len Chandler

"The fact is that what politicians and newspapermen and police consider as great problems, don't bother a single citizen in the Province.

There was unanimous condemnation in Quebec of the [James Cross and Pierre Laporte 1970 political/"terrorist" hostage-taking] episodes, but all of the Quebec juries refused to become hysterical about it -- refused to be pushed around by the Crown [prosecutor], or by the extreme and dramatic measures that were taken against the FLQ [French-Canadian leftist urban guerrillas]."

-- Claude-Armand Sheppard (1973)
Dr. Henry Morgentaler's lawyer

Everyone complains about the government, but nobody does anything about it.

Well, democracy, in its short existence (less than 250 years in its current run) has pretty much established itself to be utterly worthless in our modern technological world. Its fine principles and good intentions have been subverted and corrupted completely, having failed to not only deter the forces of greed and injustice, but even to reflect the reasonable will and desires of the people in general, while protecting the fundamental rights of the minority -- read, rich and powerful -- segments of its populace.

Democracy has no problem catering to its leaders, and the rich, and the unreasonable desires of the people -- and so it seems like it's time for a change.

By any means necessary.

I used to fly quite a lot between Europe and Canada. Destination France. A lot of lay-overs in other European countries when I tried other airlines after Air Canada pissed me off with the racist passport harassment by their staff at CDG [Charles de Gaulle] International Airport in Paris.

I carry EU -- European Union -- and Canadian passports. I was never hassled in the slightest until the World Trade Center towers fell down. Ever since 9/11, I have been stopped and hassled 25% of the time, when crossing national borders. Randomly, unpredictably -- but *never* by French passport control. (I'm a French national through my mother, as well as a Canadian citizen by birth.)

Air Canada security in Paris CDG Airport, the Germans in Munich, the Dutch, and Toronto passport control *all* have done it. The Swiss at Zurich airport *almost* did it, but then the Customs officer suddenly had second thoughts, and asked me how long I'd be in Switzerland. "Zwei Horen" / "Two hours," I responded in German. He then realized I was just waiting for a connecting flight, and waved me through, instead of the BS he was planning for me, until he realized he should ask one more pertinent question.

And, as a European Union member state passport holder this harassment is unacceptable as a matter of law.

It's pure racism, and I'm sick of it. It's the color of my skin, a possible (but ridiculous) resemblance to a "Middle Eastern" person, and a last name that might be mistaken as Arab -- not that it should matter, I add vehemently -- if you're an incompetent border official or uneducated airline security guard or Canadian city pig. I no longer fly Air Canada, by the way. I give the Europeans my money, instead. I'd be damned if I was going to put up with this crap from my *home* country.

*　*　*　*

>"Delusion arises from anger. The mind is bewildered by delusion. Reasoning is destroyed when the mind is bewildered. One falls down when reasoning is destroyed."
>
>-- "Bhagavad Gita" (ca. 3,000 B.C.)

"**No one** expects the Spanish Inquisition" goes the repeated tag line for a notorious and hilarious U.K. Monty Python comedy skit from the 1970s.

On Friday afternoon of August 31, a little more than two weeks after I was jailed, I was unexpectedly summoned from the range by a guard. I assumed it was visit I was expecting from my lawyer, as I was escorted to one of the interview rooms, where two men I had never seen before sat. I was confused, and assumed I was in the wrong room, or something.

But it turned out that the two casually dressed, serious looking gentlemen -- one looked Middle Eastern, the other looked Vietnamese -- were indeed my visitors.

No one expects the RCMP, National Security Unit!

I was impressed, fascinated, excited, and curious -- all at once. The "big boys", I thought to myself -- this should be interesting. I mean how could it *not* be exciting to be able to say that one had actually been "interviewed" by a couple of secret policemen?

Assuming you had done nothing wrong, they sort of believed you, and it was a civilized country, as opposed to a "We have **vays** of making you talk!" Gestapo sort of place.

I settled in mentally for this unexpected and exciting development. (There was **no** possibility of useful or intelligent conversation with a municipal pig -- a dumb white boy, who barely passed high school. Even a passing encounter or interaction of any sort was highly likely to be unpleasant with local cops. But I didn't have such a virulent hatred of the RCMP. They seemed to be quite pleasant and civilized to deal with – and I appreciated the contrast, and the possibility of an actually polite, civilized police force.)

The flip-open leather badge and photo-ID card carriers came out, and I studied them with interest as they lay on the desk for me to peruse, and verify that the two plain clothes "spooks" were, in fact, spooks.

"First off, is there anything you'd like to tell us?" the lead spook opened up the interrogation with.

"Yes. I don't like the government!" I announced.

"Why is that?"

"Because they threw me in jail!" I stated, as if it was obvious, throwing out my hands for effect. It was a joke, but I wasn't smiling.

They asked if I traveled a lot. I said I used to go to France on vacation a lot, years ago, and worked a gig in Paris for Nortel for fourteen months. I mentioned that I visited Spain on vacation several times during the Nortel ex-pat gig to get drunk with English-speaking friends I had met, who lived near Valencia.

They asked about my having books on the IRA (twenty or so, out of a total personal library of almost 2000 books). I said that I bought them to celebrate "Freedom to Read" Week...

The books were widely available, historical accounts and political analyzes, and legal. I explained that I do not, in general, feel it necessary for me to explain legal behavior on my part, but I would anyway.

I commented that if I had a copy of "_Mein Kampf_" ("My Struggle"), written by a certain jailed Austrian extremist -- before he became a rather infamous German dictator -- would that make me a Nazi? No, I didn't think so.

And then there was the picture of James Joyce on my kitchen wall memo board. That was clearly subversive! [Strangely, the two women who bore my two children are the only people who had ever ID'ed the photo as that of author James Joyce -- or even noticed and commented on the black-and-white newspaper picture of Joyce -- a dorky looking pencil neck white guy who's in a suit and fedora -- but still looks dorky.]

I wanted to tell them that I believed it was sometimes justified and necessary to kill people over ideology, but never ever religion, power, money, resources, or conquest. But even though it's the accepted practice of all governments pretty much, somehow I had the feeling that saying so to people who had thrown me in jail for doing nothing would not like to hear this.

They asked about a few non-working and disassembled cell phones that had been noticed in my house. I had just picked them up for free from various people and thought I might screw around, and take them apart or such, being an engineer working on cell site software and all. I didn't mention that I knew that GSM cell phones were convertible to remote bomb triggers, since I didn't know how to do it, they weren't GSM cell phones, and I certainly was not interested in making bomb triggers anyway. I use light-able homemade fuse...

They asked if there was anything they would be interested in on my IBM laptop computer at my house. Since my laptop had the temerity to die a few weeks before my arrest, they couldn't easily search it. It contained my writings and files, I said -- about girls and stuff, I said, a little embarrassed. I hadn't gotten very far on my book on how to overthrow the government. But I didn't mention this. (Actually, I consider writing on such a topic "relaxation" in my twisted universe, where detailed fantasy plans and stratagem is much more pleasant to think about than actual reality.)

I mentioned that as a person that used to be called a "Net personality" for their prolific (1994-

1999) and interesting contributions, there were about 400 "postings", and a few monographs of mine on the Internet available for anyone to read. I told them to do a Google search engine search on the term "Yogi", as well as "sarin" and "BZ" (chemical warfare agents), "clandestine", and "moonshine" for what I regarded as my best, and most detailed technical posts, archived by others.

"Do you have a web site?" one of the agents asked. I answered, "No." I think they asked the exact same question two more times in the ensuing conversation. They were confused because I indicated that my writings were all over the Net, not because I was an ego-maniac, but because people thought they contained useful information, and kept them publicly available on **their** websites.

> [I'm "like a rash all over the computer system", as the "Illinois Nazi" comments about John Belushi's character, in my favorite comedy movie, "The Blues Brothers" (1980), directed by John Landis.
>
> A movie, in which, incidentally, Carrie Fisher periodically pops up to try and shoot John Belushi using an assortment of small arms, bombs, and a bazooka -- for having left her standing at the altar.]

They asked me for my email address. I politely refused to give it to them for privacy and irrelevancy reasons. "Sorry, but I assume you can find it, anyway. But it's private -- personal emails to girls and...stuff [I blushed at the "girls" comment; I'm sure they noticed my embarrassed facial look, and body language] -- so I'm not going to make it any easier for you." I was polite, calm, and laid-back in demeanor. I shrugged my shoulders to sort of say, I was helpless, because I couldn't cross a certain line. I was being cooperative, but I wasn't no push-over.

Finally, they threw me a curve ball. "If you could change the government of Canada, what changes would you make?"

I sat their stunned. I couldn't believe it. I was speechless -- stumped. They had perplexed me with the question. I also had some sort of vague, disembodied, and rather disquieting feeling of -- well -- embarrassment.

"I don't know..." I stuttered. "No one has ever **asked** me before..." I admitted finally, a touch embarrassed. No one cared, I didn't want to add.

"Well, **we're** asking," he reiterated, with enthusiasm. As if they were genuinely curious.

"Well, I'll have to think about it. I've never really thought about it, since no one has ever been interested in my views on the subject," I confessed. "Let me think for a minute..."

After considering the matter, I cobbled together some loose thoughts and launched into a lecture of how the basic U.S. republic, with its "Declaration of Independence", and Constitution, including the "Bill of Rights" and "checks and balance" system was the form of government I approved of, since the latter helped prevent executive branch abuses by either legislative branch or judicial branch veto (though the fascist pig Ronald Reagan had begun the traitorous sabotage of the system, which was later continued to completely destroy the intent of the Founding Fathers. But that's a whole new book...)

The Canadian parliamentary system should be scraped and replaced by this, I said. Since it was worthless, I didn't add... It was derived from the English system and laws, and like most ideas and people from the U.K. were historically garbage.

You've got to wonder about this country, where the leaders are so stupid they have to import their dumb ideas...

"Have you ever made a pipe bomb?" he asked.

"Well, I know how to make one, if that's what you mean," I answered, trying not to appear evasive.

"But have you ever **made** one?" he persisted, zeroing in on my evasiveness...

"Well, I was on CBC [TV] news show in 1994 as an explosives 'expert' of sorts." I continued, "I **did** build and detonate a *concussion* device for the show. They had wanted me to build a pipe bomb, but I told them it was too dangerous, so I would only agree to make them some sort of improvised "Flash/bang" type device, after they agreed to pay for its construction."

"Actually, I believe my use of the term 'IED' on the show was its first public use in the media" ("IED" -- Improvised Explosive Device -- was a term later made famous by the 2003 Iraq insurgency, following George W. Bush's disastrous initiation of the Second Gulf War.)

"Really?" he responded, sounding, or trying to sound, impressed and intrigued. I wasn't sure, but I think he was sincere.

"Why did you have 30,000 rounds of ammunition?" he inquired nonchalantly.

"Several reasons. I had sixteen different firearms with different calibers. That amounts to something like 2,000 rounds per gun. That's about four afternoons of plinking in the country. Plus they were military surplus ammo, and cheap, and I knew the price might go up a lot, or the availability...**And** it's perfectly legal, and I don't feel I should have to explain legal behavior to anyone."

"Why do you collect firearms?" he asked.

"Because it's legal, and I generally don't feel it necessary to explain my legal behavior. However I will expand on 'why' anyway, just for your edification. Lots of reasons. Because it's a hobby and interest that I began almost 25 years ago, it's legal, I can afford them, and they're fun to shoot. Because I like them, and like to study them and 'play' with them.

And because they're beautiful." I must have involuntarily smiled a big smile. "It's hard to explain, but as an engineer, you become so immersed in (usually a boring) technology, that even on my **own** time I can only relax and unwind with something technological. I find studying a technology of **my** choice that I find fascinating and enjoyable, entertainment -- what people I guess would call a 'hobby'.

And after years of reading and studying about firearms and other weapons, I eventually realized that there's a certain 'beauty' to guns. It's hard to explain, but they **are** beautiful. Not how they are used for evil, but technologically – their functioning, for instance."

But we humans seem to have a problem with beauty and ugliness. "Only **bad** witches are ugly," Glynda, the Good Witch, informs Dorothy, in the movie, "The Wizard of Oz."

Mr. Spock's comment is: "The idea promulgated by your Ancient Greeks, that what is beautiful **must** be good, and vice-verse." It's in the original "Star Trek" TV series, in the 1968 episode entitled, "Is There in Truth No Beauty?"

I continued, "A well-designed firearms is a work of art, and a thus a thing of beauty. Oh! -- I know. There's a song that comes to mind..."

And then, after a brief pause to collect myself, I began reciting in cadence (without singing) a slightly modified version of part of the lyrics of what I consider Leonard Cohen's best song, ("First We Take Manhattan" (1988)) which I thought explained things quite well:

> "I'm guided by a signal in the heavens,
> I'm guided by the color of my skin,
> I'm guided by the beauty of our weapons,
> First we take Manhattan, then we take Berlin."

I pointed at the sky for the "heavens" line, and I laid two fingers lightly on the skin of my forearm for the "color of my skin" line, and smiled.

They didn't clap at the end of my performance, but they didn't interrupt or laugh, either. And the interview soon ended. I guess they had had enough. I imagine they thought I was a rather unusual fellow, but **not** an "Islamist terrorist" -- their main concern -- not a "terrorist", and not a "threat". I was thus of no interest, and no concern to them. Their time had been wasted by stupid Ottawa cops, who had requested a security investigation of me to try and bolster their smear job.

The lead agent commented that it was clear that I liked women. Apparently I mentioned the issue several times. (Well, after two weeks in jail, I was really missing them...even though – fortunately -- I had gotten laid the evening before I was arrested. Thanks, Lorna!)

[With regards to sexual needs, a lot of women seem to be like camels and water. Men are a **little** more impatient than that, as far as their sexual "time-table" goes...]

And, as should be noted, they not **once** asked if I was a Muslim. (And **I** didn't use the cop-out in saying "I'm not a Muslim", of which I'm quite proud.) In fact, not once did any police or investigatory official ask. Interesting, hmmm? The question was blatantly obvious in its absence.

They could commit all manner or atrocious human rights violations, ruin people's lives, be accessories to torture and war crimes, but could leave no evidence that they had been "religiously profiling" or otherwise "politically incorrect" by asking a directly relevant question, that was obviously of interest to them.

I was hoping they'd asked me, "Who I took my orders from?" To which I'd answer, "H.L. Mencken" or "my mother", depending on my sense of levity at the time.

As they prepared to wind things up, I complained about the lack of RCMP accountability -- nobody being fired over the Maher Arar Syrian torture scandal. Which should have been a new scandal all its own. No punishment, no change. So it goes...

The next guy to suffer was a Polish immigrant, arriving in Canada for the first time to start a new life, and being Tasered to death by the RCMP at Vancouver International Airport for his "Welcome to Canada" _fete_.

As they were leaving, I asked if I could use them as a reference -- which for the first time **did** elicit a laugh from the two agents, as they continued leaving. Civil servants are just such a pain to deal with...

* * * *

"Vain the ambition of kings

> Who seek by trophies and dead things
> To leave a living name behind,
> And weave but nets to catch the wind."

-- "All the Flowers of the Spring" (ca. 1623)
John Webster

There's a bumper-sticker popular amongst libertarian types in the U.S. It says simply: "Question Authority".

But as one wag commented: "Question authority, and the authorities will question you."

"When Socrates and his two great disciples [Plato and Aristotle] composed a system of rational ethics they were hardly proposing practical legislation for mankind.

One by his irony, another by his frank idealism, and the third by his preponderating interest in history and analysis, showed clearly enough how little they dared to hope.

They were merely writing an eloquent epitaph on their country.

They were publishing the principles of what had been its life, gathering piously its broken ideals, and interpreting its momentary achievement."

-- "The Life of Reason (1954)
George Santayana

Prison Diary 19: "Bad Moon Rising"
                  ==============

"Mother of mercy! Is this the end of Rico?"

-- Rico
"Little Caesar" (1930)

"As Gregor Samsa awoke in his bed from uneasy dreams one morning, he realized that he had been transformed into an enormous bug."

-- "Metamorphosis" (1915)
Franz Kafka

September, bloody, September.

As every revolutionary knows,
Christmas comes early,
Or death, the hard blow,
The winds of change,
Or change too slow.

It was to be an easy bail hearing. Nothing the prosecutor said should have any effect on a formal, honorable legal judicial ruling, I knew. I was sure of it. Pretty sure of it. I was hoping for it...

To keep my spirits up. To keep from crashing into a deep depression...

The future you have no control of is not worth worrying about. It is pointless.

But, dreams of cynicism and horror, nightmares and disturbing dreams of dystopia and misery, lamb's blood hurriedly smeared on thresholds, daydreams of violence and casual unpleasantness. Mental amulets of protection.

But it was not to be. Holy Mother of God, the Long Knives are upon me!

They call it "Screw" -- the So-Called Real World. And I've lived in a carefully constructed fantasy dream world most of my life to avoid it. Successfully for the most part. Until one day, it appears, I was swept unawares into a paradigm shift -- or something like that.

By the end of the day of my bail hearing, I was so depressed, even my dog was on anti-depressants. But after the vet examined him, he said my dog wasn't depressed – he just didn't like me.

Life was always an adventure, for me. And the entire world was my stage -- my playground, actually. "May you live in interesting times", and I certainly did. But, boy oh, boy, I certainly wasn't ready for the roller-coaster ride of my life that was getting worse and worse with each day, or so it seemed. It wasn't the sudden onset, it wasn't the severity or horror, or anything even close -- because, really, it wasn't physically a severe threat, or the end of my freedom, or anything like that.

It was just so unexpected. It was just seemingly without justification, so illogical. The Canadian government put people in positions of power, backed them with armed force, and paid them well, assembling them into a well-ordered and organized system.

Yet it seemed that they were all united in concert against my interests, for no particular reason, and contrary to the basic legal rules and rights they were supposed to uphold.

It was just that one day, at the age of 48, after two university degrees, and twenty years after abandoning a flagrant and illegal lifestyle, and having settled into a calm, suburban, upper middle-class life of a sedate, but opinionated computer engineer, I suddenly went from a position of status, value, and respectability, in accordance with my modest professional status to a downgraded classification of criminal dirt, *and* threat to the state. I went from stability and predictability to the dark and unknown in a matter of hours. And it just kept getting worse.

> "Then Jesus said to the chief priests, temple guards, and leaders who had come for him, ' Have you come out with swords and clubs as if I were a criminal?
>
> 'I was with you in the temple courtyard and you didn't try and arrest me.' "
>
> -- Luke 22:52-53

* * * *

No matter what, however, throughout, I never lost faith in the fact that if I held firm, eventually I would be absolved completely. The fact remained, I was guilty of nothing, because my home had been searched without being under suspicion of any criminality, and without a search warrant. And I faced two actual, but minor charges, as a result, which would be tossed because of the lack of a warrant. And if they were not, I would get a suspended, or probationary sentence at worst.

Yet there's faith, and there's "here and now", and I just couldn't shake the overwhelming feeling of utter confusion. Was I being "humbled" for some reason of supreme cosmic balance? Or just a "victim of random chance"?

My first "real" court appearance -- where a decision would actually be made -- was a bail hearing to decide whether I could be released from jail. I didn't know what to expect, but assumed I would get out -- a 48 year old, upper middle class, Canadian born citizen, an employed computer engineer with a lengthy work record, and no criminal record, facing minor criminal charges.

It was Friday September 7, 2007 -- over two weeks after my arrest -- by the time I got my shot at release from jail. But what I got instead was a monument to judicial cowardice, cowardly playing to the crowd, judicial idiocy and gross incompetence, casual stupidity, and rubber-stamp indifference, and acquiescence to shrill and abusive, and quasi-criminal prosecutorial excess.

It was played out by a casual playing up to sensationalized police-instigated news head-lines that resulted in a full public gallery, the trotting out of every terrorist stereotype and every possible malevolent speculation, in a fairness-discarding atmosphere of witch-hunt politics, with no actual supporting evidence whatsoever, completely beyond the actual charges I was facing.

It was as if the bail hearing for someone charged with simple assault had been charged with a bloody knife 2nd degree murder of a nun.

The police calling a press conference, and hauling out a trolley-cart full of malevolent looking machine guns and sub-machine guns -- all legally and obviously deactivated -- had done its job.

What exactly then was the point of calling a press conference and displaying legal property of mine erroneously seized by the police other than as a subterfuge and smear campaign against me? The answer was simple: to have me appear dangerous – and get a front page story with a full page picture of legal but evil-looking, blatantly criminal-looking guns -- and thus cover-up their illegal search of my home and false arrest of me.

It was an outrage, and evidence of cover-up by smear, and raised the policy issue of government manipulating -- essentially using -- the press for its own hidden agenda.

A life in pursuit of the exotic, the arcane, the obscure, the bizarre, and the horrific -- as a tourist. A hobby infiltrator, as it were, a fugitive from the endless deadly drip of the boredom of conventionality and the conformity it demands.

Yeah, yeah -- I know. Fools rush in, where angels fear to tread. Curiosity killed the Yogi. The candle flame was just too irresistible. The Siren Song of observed and experienced Reality just too magnetic.

And I realized one day that though I seemed to stumble into hot water with amazing regularity -- and usually could have avoided the crisis in the first place -- my talent was in almost always being able to extract myself from the trouble I'd walked into -- though usually with a mustering of all my resources and abilities and energy. Not very sensible, but it made for an interesting life, for sure.

> "I had a dream that my house was on fire.
> People laughed while it burned.
> I tried to run but my legs were numb.
> I had to wait till the feeling returned.
>
> I don't need a doctor to figure it out,
> I know what's passing me by.
> When I look in the mirror,
> Sometimes I see traces of some other guy."
>
> -- "Till I Am Myself Again" (1990)
> Blue Rodeo

* * * *

The foundation of a system of justice in a modern free, democratic state consists of the fundamental concepts of the "Rule of Law", "presumption of innocence", and the general idea of a process guided by "fundamental justice". These simple, yet noble, concepts are the pillars of a system meant to ensure fairness, prevent injustice, and enshrine society's commitment to the search for truth as the underlying goal. They are the Rule of Law, the guiding principle of justice in a democratic society with an independent judiciary.

> "Let us inspect the dream that has brought to the
> climactic years of the twentieth century the
> assurances and rewards of the madhouse."
>
> -- "The Social Contract" (1970)
> Robert Audrey

I had two witnesses to testify on my behalf, that I should be released on bail. My father, 88, who received a Ph.D. from Cambridge University in England in the mid-1940's, and who had worked since then at the National Research Council on Sussex Dr. in Ottawa as a researcher in theoretical physics -- a physicist -- an impeccable reputation if there ever was one.

My second witness was a co-worker at Nortel Networks, who had known me for about fifteen years, worked beside me, was a personal friend, and just so happened to be a member of Canada's National Shooting Team. Talk about the *perfect* defense witness! He had shot competitively for Canada around the world, including in the Pan-Am Games, the Commonwealth Games, and other matches. He was as ideal a witness as I could ask for, and he had nothing but good to say about me, my competence as an engineer, my historical knowledge of firearms, and my character.

But I should have known better. My tip that none of that mattered in the face of a Crown attorney (prosecutor) determined to keep me in jail was my brief conversation with my first lawyer. She informed me that due to the "political climate" that bail would be problematic.

"Political climate"? I responded in wonderment. "What are they going to do? Ship me to Syria?" I asked rhetorically, referring to the place where a totally innocent Syrian-Canadian electrical engineer. Mahar Arar was sent and tortured, as a direct result of Canadian official incompetence, and American evil.

To my utter amazement, she responded in a serious tone, "They might..."

I did not like that answer one bit. I was a Canadian-born citizen. They were only able to ship Arar and others to Syria because he was a foreign-born Syrian citizen.

And she hadn't even heard my side of the story. She had only talked to the Crown Prosecutor, I guess (a prejudicial state of affairs that is routine, and which I have issues with). The conversation went down-hill from there. Within a few quick questions, I zeroed in on one that sealed her fate. " *Exactly* how many bail hearings have you handled?" I asked simply.

"A few..." she answered pathetically, and with embarrassment. I had cornered her, with no wiggle room at all. I ended the conversation. She was *gone*. I was firing my first lawyer. And I didn't even bother to inform her personally, to symbolize my utter contempt for her, her statements, and her attitude.

But indeed, it was a prelude of things to come...

* * * *

> "Why has government been instituted at all? Because
> the passions of men will not conform to the dictates
> of reason and justice without constraint."
>
> -- "The Federalist Papers" (1787)
> Alexander Hamilton

The dog-and-pony show -- the press conference -- that the Ottawa Police had called had done its dirty work. They had trundled out a cart full of perfectly legal deactivated guns for the press photographers, and laid everything out on a big table for more pictures. Oh, there *were* two unregistered rifles, but everything else was perfectly legal.

I have yet to receive a reason as to why the Pigs are allowed to display for the press legal, non-weapons for the purpose of smearing the reputation of a citizen. Items seized without benefit of a search warrant, no less. And the citizen is a registered gun collector for decades with no criminal record, obviously.

But there was method to the badness. The press conference ensured wide publicity, destroyed my reputation, and guaranteed there would be a full public gallery and reporters and press coverage at the bail hearing. And publicity counts against you *big* time, when you have a petty official -- called a Justice of the Peace -- who is a patronage-appointee that is usually not a lawyer. And on the spot with a full gallery watching -- he starts to sweat, and makes sure he can't be exposed as an incompetent by appearing "soft".

The press conference and cart full of legal guns had done its dirty work. The "New Rules" of 9/11 and the Islamic terrorist witch-hunt would do the rest.

If they don't catch you red-handed, they catch you engaged in "terrorist training". If they can't make that stick, you're associating with "known" terrorists. If they have no evidence at all, you're a "sleeper agent". And if that won't fly, you're a "potential threat".

A tan skin, and non-white vaguely Middle Eastern look, and it's a slam dunk.

> "[John the Baptist] also spoke out against [King] Herod for
> all the evil things he had done. So Herod added one more
> evil to all the others; he locked John in prison."
>
> -- Luke 3:19-20

* * * *

> "You can't expect a chap with such
> a name as Dick Dead-eye to be a
> popular character -- now can you?"
>
> -- "H.M.S. Pinafore" (1878)
> Gilbert & Sullivan
> William Gilbert

But it was all for naught.

An ordinary compound (long) bow, became a nefarious "cross-bow" – remember that this is a police officer testifying and being grossly wrong.

A very small collection of knives, including a very expensive, collector's item, triple-ply steel Japanese tanto, and a Marine Corps combat knife, became "killing" knives, a name, I have never heard before. I had two **combat** knives, and that was that.

Combat fatigues become suspicious and malevolent evidence against me, as the pig kept a straight face. His pants were themselves camo pants, making his testimony a joke...

Popular books on the Irish War of Independence, 1919 - 1921, became terrorist training manuals (Leading to my comment, that apparently the Government of Canada was planning on breaking relations with the Republic of Ireland, and the U.K., which signed the treaty with the I.R.A. recognizing the independence of the 26 southern counties of Eire.) in 1923.

A couple of unused paint ball guns became "terrorist training gear" (even though I had an AK-47, and an MP-5 -- wouldn't you train with real guns if you had them?).

A perfectly legal, but obsolete, Kevlar bulletproof vest became the ominous-sounding "body armor". I'm a collector, OK? I collect things.

A perfectly legal, rather eclectic collection of military and police rifles and handguns -- fourteen in all, and each with the appropriate up-to-date permits, became an "unusual" and therefore suspicious collection, according to the sworn testimony of the Constable -- the testimony that resulted in my bail being denied.

It was even brought out that I had a lot of books on nuclear weapons, as if this was somehow suspicious. Yes, indeed, Yogi the latest nuclear power!

And even the fact that I had a Bachelor of Science degree, made me a threat to public safety. That didn't even pass the "laugh test" for my lawyer, who openly ridiculed that one.

My legal collection of deactivated sub-machine guns was repeatedly trotted out for no discernible reason, even though they acknowledged that they were legal.

My 30,000 rounds of ammunition was brought up, even though this is not illegal (or even unusual). And my legal, registered collection of fourteen military and police rifles and handguns was described as "suspicious" and "unusual", which was interesting, since the federal and provincial government firearms authorities knew all about my collection, and didn't mind.

> " 'John [the Baptist] came neither eating nor drinking,
> and people say, "There's a demon in him!"
>
> The Son of Man came eating and drinking, and
> people say, "Look at him! He's a glutton and a drunk,
> a friend of tax collectors and sinners!" ' "
>
> -- Matthew 11:18-19

* * * *

A recognized part of the legal system, is the "adversarial process". Basically the Crown prosecutor is the accused's enemy, and tries to convict him by interpreting evidence in an unfavorable way, and never giving the accused the benefit of the doubt. The defense attorney is in the accused's corner, and interprets the evidence in a favorable light, giving the accused the benefit of the doubt.

However the adversarial process is subject to such reasonable limitations as not tolerating the introduction of known perjury, and the requirement for the crown to disclose all its evidence to the defense, especially favorable evidence. There's also the matter of "fundamental justice" which doesn't seem to fit distorting evidence (and especially mere suspicion) to the point that it misleads the Court.

Apparently they canceled the class in Law School on the Adversarial Process, and substituted an extra class on "Billable Hours".

What really got my goat was:

> 1) the bringing up of things that were legal for me to do. I don't believe I have to explain or justify conduct which is perfectly legal.
>
> 2) the fact that almost everything they brought up had nothing to do with the charges I faced. It was a bail hearing over a couple of unregistered guns, not an excuse to hold me while the authorities conducted a fishing expedition for their claimed terrorist suspicions. The charges over the two unregistered gun – found in my house after a false arrest, and an illegal warrant-less search -- were relatively minor criminal charges since I was licensed to possess them if registered.

* * * *

> "The bitter frosts, with the sleet and rain,
> Destroyed hath the green in every field..."
>
> -- "The Canterbury Tales" (1387)
> Geoffrey Chaucer

And so a petty official with the title of "Justice of the Peace", in the guise of a small man named Herb Krelling, denied me bail. He relied on a steady stream of gross mischaracterizations and distortions, innuendo, inflammatory statements, and prejudicial errors of fact. He accepted damning speculation lacking any evidence that would justify painting the extreme picture of criminal terroristic malevolence that it did.

He discounted the calm, solid testimony of two pillars of the community that I was just a normal, intelligent, reliable citizen of long-standing reputation who had collected firearms legally for years, and was very knowledgeable on the subject. Who had always been stable, and of good character and judgment in the past, and otherwise suitable for release on bail.

I was denied bail as a "threat to public safety". I was so shocked, I didn't even think to be disgusted. But I tried to hide my emotions. I had a pleasant smile throughout the hearing. "What fools these mortals be," I might have thought, but didn't.

The rolling snowball was now humungous in size.

> " ' Stop judging by outward appearance!
> Instead, judge correctly.' "
>
> -- John 7:24

And to add insult to injury -- Krelling looked like such a pompous and self-satisfied man, the way he strutted out of the courtroom -- he declared that I also could not be granted bail because releasing me would likely "bring the administration of justice into disrepute."

"Too late" was my response to that.

It's a reason that the Supreme Court has ruled is to be rarely and specially used for serious cases. Not two gun registration charges. So Krelling was an ignorant fool, breaking the law without consequence.

The Canadian police and judicial system had slipped into an astounding pattern of what amounted to police state practices and tactics administered incompetently to no useful effect, senseless damage, and great mischief. And the icing on the cake, at huge taxpayer expense. The municipal police had proved my long-held belief that they were _de facto_ a state-sanctioned criminal organization.

They were not the independent and impartial enforcement arm of the Courts that is their only authorization and purpose. Instead, they were corrupt tools of the Prosecution that I was soon to find are comfortable in engaging in routine criminal misconduct, perjury, and fraudulent manipulations of the Court process. They abused their office by covering up malfeasance as much as possible, resisting and sabotaging effective oversight, control, and accountability, and even engaging in manipulating the media to further, justify, and protect their corrupt aims.

A majority of the police were a clear disgrace to their office, their uniforms, and themselves. To them obeying the Criminal Code was optional, and the human rights guaranteed by the Charter were just suggestions and, in general, just a nuisance to be forgotten or ignored.

And no one had noticed or realized -- including me -- that this had become firmly established and accepted within the government. No one seemed to realize the grave threat that the police organization's criminal and ethical misconduct posed to the Rule of Law and our society itself. It was open sabotage of the search for Truth and Justice by the group on the front lines of enforcing the law, and sworn to uphold it.

Who the hell was in charge of this ridiculous situation? It certainly didn't happen overnight.

"My later experience with police methods and the use of lies and insinuations confirmed my suspicion that it was a blatant lie, or a grossly exaggerated story."

-- Dr. Henry Morgentaler (1973)
Canadian legal abortion pioneer

Prison Diary 20:   "One Day at a Time"
================

> " 'Father said a man is the sum of his
> misfortunes. One day you'd think
> misfortune would get tired, but then
> time is your misfortune, Father said.' "
>
> -- "The Sound and the Fury" (1928)
> William Faulkner

> "I have wasted time, and now doth time waste me."
>
> -- "Richard II" (1597)
> Act 5, Scene 5

Tell my friends: "Alas! The House of Shan has fallen." But after a brief pause, add: "His flair for the dramatic, however, still stands."

The slings and arrows of outrageous fortune. And a sea of troubles, as well. A fall from grace, inner turmoil, and **especially** real, pervasive, unsettling tension and fear, as I found myself alone in the valley of the shadow of death. And I was afraid.

Ruin had been visited upon me, and all it had taken was a phone call from a mentally disturbed woman. "God, what had I done to deserve this?" I wondered again and again.

"What had I ever done to deserve **this**...?"

> "The Pharisees asked Jesus when the Kingdom of
> God would come. He answered them, 'People can't
> observe the coming of the Kingdom of God. They
> can't say, 'Here it is!' or 'There it is!'
>
> You see, the Kingdom of God is within you."
>
> -- Luke 17:20-21

* * * *

So let me try and explain what a typical day is like in the maximum wing of OCDC Ontario Provincial Jail. In a word, "medieval". In a sentence, something you can only deal with, "One day at a time." (one of the helpful slogans that came out of A.A. -- Alcoholics Anonymous -- the self-help, stop drinking group).

If you thought -- like I did -- that we lived in a liberal, progressive, well-meaning, and morally superior society with a civilized rehabilitative criminal justice and prison system, then you -- like I was -- have fallen for the "snow job" propaganda that the government has fooled us with.

They have certainly made excellent improvements. They keep more people -- those who they think deserve it -- out of prison. They have reduced sentences to make the punishment fit the crime, and *not* turn the person into an angry, vicious, unsalvageable animal upon release. And they had added helpful inmate programs to the Federal institution network (where convicts with sentences of two years or more are sent).

But these were obvious improvements that were simple to implement, long overdue, and mostly saved enormous amounts of money. And they were obvious, because they were to the benefit of both society *and* the convict beneficiary.

But the core problem is unresolved: provincial jails like OCDC are an abomination to "Canadian values" -- in fact, to civilized values, period. And since the government should know, and doesn't care to have resolved this, I can find no satisfactory explanation. Damn them to Hell!

* * * *

> "How horrible it will be for you scribes and Pharisees! You
> hypocrites! You give God one-tenth of your mint, dill, and
> cumin. But you have neglected justice, mercy, and faithfulness.
>
> These are the most important things in Moses' Teachings.
> You should have done these things without neglecting the others.
>
> You blind guides! You strain gnats out of your wine, but you
> swallow camels."
>
> -- Matthew 23:23-24

A dreary, Spartan environment, privation, boredom, long-term uncertainty, regular mistreatment, abuse, and degradation -- all these I could handle, even though they are an outrage against the unconvicted. It was the environment of random inmate violence from which no one could escape that I found intolerable. Our enemy in not each other, I repeated endlessly, our enemy is out there, I said pointing to the outside of the range. Our enemy was the guards, the prison system, the justice system, etc.

Beating up each other was pointless, and flawed in theory and practice, beyond redemption.

It was Friedrich Nietzsche who said that what doesn't kill you makes you stronger. Yet, I'm not sure what to make of that statement, since the 19th Century German philosopher and author -- including "Beyond Good and Evil, "Human, All Too Human", and "Thus Spake Zarathustra" -- went completely insane in 1889. Either way, jail does not make you stronger, and that is the fact whether you are a criminal, corrupt and got caught, broke the law, or someone like me – innocent, unfairly detained, and unconvicted.

In "max" the inmates -- mostly pretrial detainees recently arrested, or denied bail as people who have missed previous court dates, violated previous bail conditions routinely, have extensive criminal records, or are considered a danger to society -- are kept locked in their two-man cells for 21 hours a day, with only three hours a day outside the cell onto "the range" where they can mingle with the 30 other inmates of the "range" for conversation, chess games, playing cards, regular bloody fist fights and serious physical assaults, and theoretical access to a pay phone (collect calls only, of 20 minute duration maximum).

You are allowed two twenty minute visits a week, with the visitor separated from you by thick glass, talking to the visitor by **recorded** telephone.

I was incarcerated with a great many so-called "career criminals", usually with drug problems, making for a rather unsuccessful career. As it was max, I got to mix with people such as alleged first degree murderers, bank robbers, armed robbers, medium to large-scale drug dealers, home invaders, street gang members, and other violent, or assault-prone hoodlums.

But the treatment of them, most with lengthy criminal records, and easily provoked to violence was interesting. First of all, most of them were here because of bad behavior. I did not think that by subjecting them to mistreatment for many months was going to get them to behave any better. A regime of daily mistreatment, degradation and emotional abuse by fellow inmates and prison staff in a predatory environment of constantly impending violence and collective punishment -- of people who had no privileges worth mentioning -- was certainly *not* going to positively influence these already fucked up people in *any* way, except to reinforce or increase their maladaptive personality traits.

You mistreat badly behaving people, and all that happens is that they get meaner, more resentful, and overall, worse.

And they become more certain that their charming, but useless idea, that having *no* moral code is better than having society's corrupt and hypocritical moral code is valid.

Thus the circle of stupid government policy is made complete by an ever-increasing population of worse and worse criminals.

This is obvious and it is well-established. A sad reflection on those who administer this system, as well as the government officials who could reform it, since it is outside public view, and thus outside effective oversight and accountability. No one really knows what goes on in the prison system except the inmates -- who have no credibility for obvious reasons -- and those petty officials and guards who are part of the prison system.

So, let me get this straight. Strategy No.1. I reiterate: mistreatment of people is well-established as a failure. Yet state-sanctioned abuse is visited on jailed people who are mostly educational failures, social deviants whose moral code is totally warped -- typically by some sort of childhood abuse, ill-treatment, or neglect -- and whose dysfunction has manifested itself in becoming mentally impaired with drug or alcohol abuse, and engaging in criminal behavior, which apparently they're a failure at as well.

And abusing these people is supposed to make them shape up and respect the law on release?

So finally they tried a more liberal approach -- Strategy No. 2 -- but this did not produce significant *identifiable* improvement in levels of criminal behavior, though it did probably show why the welfare rolls, suicide rates, and number of untreated mentally ill increased over time (with Strategy No. 1) wasting enormous amounts of money. *This* proved that longer sentence do not do anything but waste millions of dollars, since imprisoning one person, requires an enormous amount of correctional, transportation, administrative, parole and court staff -- all well paid.

So, enter Strategy No. 3, the solution of which was to return to the same modern update of the medieval-type punitive system (Strategy No. 1) that is known to not work, and wastes enormous amounts of tax dollars.

The only comment one can make is: "How stupid can you get?"

Obviously the problems existed with previous governments, who snow-jobbed the public with propaganda about an entirely progressive type of prison system. But Prime Minister Stephen Harper is a leader proud of his intellectually retrograde ideology, that's perfectly clear.

If a clown with his intellectual capacity and striking pattern of total dishonesty, vindictive, narrow-minded moral fraudulence, and intellectual bankruptcy somehow came to power in many countries of the world, you'd hope he'd be either assassinated, deposed, or "shot while escaping lawful detention".

And then his family would be "strongly" advised to avoid public exposure or statements, quietly move to a rural town, change their name, and destroy or withdraw all known genealogical records and family biographies.

* * * *

" 'I was in prison and you visited me.' "

-- Matthew 25:36

During the day-time "lock-down" period, most of the detainees just sleep in their cells to pass the time, stay stoned, play cards, or talk to their cell-mate about their lengthy and totally bogus history as master criminals.

When outside their cells, inmate fights and assaults would break out unpredictably, but regularly. Over drugs, debt non-payment (usually related to drugs or gambling), stealing, phone access, extortion of weaker inmates, being gay, being found to be an informant or alleged sex criminal, any personal conflict, or a host of unwritten rules that were completely without merit, and just an excuse to pick a fight with a weaker person.

There was almost no reading material -- especially books -- available, which I couldn't believe was the situation. So apparently the penal strategy included boring to death poor and uneducated welfare recipients -- mostly drug addicts or non-white immigrants, who believe that committing crimes is their only available strategy for economic betterment and social advancement in the dreary and barren world of urban poverty.

And I thought it was an absolute disgrace. You'd think that they'd realize that it was a good idea to provide books and reading material to inmates, since some of them may actually want to better themselves intellectually, so that they could improve their logical skills, moral skills, and understanding of why their previous conduct was a failed strategy, and so that they could escape the dead end cycle of crime, poverty, and welfare. Instead they largely sleep a lot, get stoned a lot, and are bored the rest of the time.

The one bright spot -- and a big relief -- was that homosexual rape is not an issue **at all**. Apparently it's only an issue in U.S. prisons. Thank God for big mercies.

"One doesn't arrest Voltaire."

-- Charles de Gaulle (1968)

"Society secretly wants crime, needs crime, and gains definite satisfactions from the present mishandling of it! The crime and punishment ritual is a part of our lives.

We need crimes to wonder at, to enjoy vicariously, to discuss and speculate about, and to publicly deplore.

We need criminals to identify ourselves with, to envy secretly, and to punish stoutly."

Prison Diary 21:   "In the Belly of the Beast"
-- "The Crime of Punishment" (1968)
Karl Menninger

==================

> "I despise **you**. I despise your social order, your laws, your authority propped up by force. **Hang** me for it!"
>
> -- Louis Lingg (1887)
> anarchist, to his trial judge
> Chicago Haymarket 8 Trial

> "Society takes upon itself the right to inflict appalling punishments on the individual, but it also has the supreme vice of shallowness, and fails to realize what it has done."
>
> -- "De Profundis" (1905)
> Oscar Wilde

"Rage, rage, against the dying of the light," said Dylan Thomas. But rage is pretty worthless against concrete and steel. You don't need to be an engineer to realize that.

In here, a place of crowded unhappiness, loneliness, sorrow, suffering, misdirected anger, and the resulting unpredictable and senseless violence, it would be easy to wither into a broken, embittered, hollow shell of a man.

The air was stuffy with official indifference, the day punctuated by routine degradation, abuse, and humiliation, and the occasional beating and bloody fisticuffs. And boredom filled all the gaps and holes. I had been buried alive. I foresaw my own "death" in captivity, and saw the seeds of hopelessness begin to germinate in my soul.

First you have hope. Hope of release, hope of bail, hope of getting out soon. Hope that this horror that has been thrust upon you will soon be over. Hope of mistakes corrected. Then hope withers, fades, and appears to die... Kubler-Ross, again.

> "Lost in a Roman wilderness of pain,
> And all the children are insane."
>
> -- "The End" (1967)
> Jim Morrison

I was left with two choices, I realized. I could allow myself -- and in this predatory environment, becoming an easy target as a result -- to follow my emotions and be dragged down into the pit of deep depression, *or* I could rally all my energies into somehow keeping it together emotionally. Keeping my spirits up through force of will, distraction, diversion. Trying to stay upbeat.

Body, mind, soul. In this barren environment stripped of almost everything, muster your internal resources and creativity to substitute whatever you can, if nothing is available that you would normally use. No exercise equipment -- then walk, use push-ups, and calisthenics. Beat them at their own game of deprivation, and keep hope burning, and to keep your soul alive and beating.

Everything starts with the body. Exercise to get fit, and lose the excess weight I brought in with me. Examine and modify my diet to help with weight loss and health. The food was high in fat (gravy), and sugar (dessert), and carbohydrates (bread), to make up the calories cheaply. Trade or give away as much of these fattening empty calories as possible.

I would force myself to shave to keep my spirits up by looking my best -- totally clean shaven. I would keep body and mind busy and active, and seek to use or improvise with whatever resources were available.

I began by keeping busy by establishing a schedule and routine, the discipline and self-regimentation helping maintain emotional stability, filling up the time anyway I could, chewing it up anyway I could to keep boredom and depression at bay.

I would work to always keep everything organized and filed, or stored neatly. Daily I would clean at least part of my cell to take control and command of my personal space -- my personal environment. Laboriously cleaning my cell -- the floors. walls, and bed of hair, dust, and dirt. And doing it by laboriously and slowly hand-cleaning the cell with toilet paper. Because it took longer I adopted using this deliberately slow and inefficient method to nibble away at my surplus of time, and as a change of pace from reading or writing. And there was a certain sense of accomplishment and achievement, because the hand-cleaning was certainly thorough and effective.

Then there was the basis of maintaining my psychological well-being. My three mental pillars:

1) the Serenity Prayer ("God grant me the serenity to accept the things I cannot change, the courage to change the things I can, and the wisdom to know the difference."),

2) the concept of "work your own program" (focus on yourself, and do not let externalities or the problems or situations of others distract you from that self-focus that is the only productive use of your attention), and finally,

3) taking reality "one day at a time", an invaluable survival strategy to avoid being overwhelmed by the magnitude of your problems, and which makes you realize that you cab *only* hope to survive "a sea of troubles" by accepting them a manageable nibble at a time.

As far as I was concerned, these were the three pillars of recovery and emotional survival in times of crisis, I had realized from my past experiences. These three concepts calm the rage, allow you to disengage -- or focus as need be -- and accept and deal with your reality. The reality being that survival is only possible by simply accepting and working as best you can with the "hand that you've been dealt", whatever that may have been.

Survey the situation and environment with a view to identifying and marshaling available resources and collecting the tools available to stay alive and thrive as best as I could. Look at the bright side, to keep depression at bay, and focus on any progress and gains and improvements inside and with the legal process.

Looking at the bright side. It's hard, and sometimes in the day, super-hard, and energy-consuming. But you have to do it, or just maintain an emotional "holding-pattern", until your spirits can lift up again to minimal positivity, and thus functionality. And as a long-time fatalist, I've had enough practice, surely. Readying me for this experience? Who knows?

I've always found people to be more or less, such a disappointment. But you need to look on the bright side. It's just that you have to be patient waiting for the good ones to appear and enter into your life.

I gobbled up whatever sparse reading that I could ferret out. Whatever was available: borrowing the daily newspaper some inmates subscribed to, magazines, and the occasional paperback book, since there was no longer any library, I was told.

Back to basics and time nibblers: hand copying things of interest, hand calculations to chew up time; weapons calculations done by hand, that I could think of that might be interesting, legal strategies and analyzes for my case.

My interpretation of Zen is fundamentally that it is the practice of the realizing and defeating the natural human reflex expressed by the quote, "It's emotion that defeats us."

The Serenity Prayer, I mentioned -- "The things we cannot change..." It's logic in action, but it's also the philosophy of Zen Buddhism in practice.

The goal of Zen is to disengage -- to detach from the unmet needs, pain, discomfort and suffering we allow our emotions to subject ourselves to. The practice of Zen teaches us to separate our mental processes and processing from the emotions that insert themselves in between reality and our mind to consciously and subconsciously interfere and distort our feelings and observations. This interferes with our actions, reactions, and responses to our distorted view of reality, and causes us the fail in our goals as a result.

We end up by wasting energy, or using it self-destructively, taking mistaken paths or choices, and we take ourselves away from serenity and constructive actions and thinking, thus generating more destructive emotions such as unhappiness, resentfulness, and hopelessness. The fears that defeat us -- the insecurities, the over-confidence, the low self-esteem, the self-sabotaging pessimistic thoughts and behaviors.

The stripping away of emotions to leave only reality observed free of distortion, with emotion-less logic to then guide us rationally through life, situations, and problem for optimum results with what we have. The serenity and sense of control and stability of logic and the feeling of liberation as you free yourself from the wasteful and unnecessary pain of destructive emotions.

* * * *

> "Unless we...recognize that television, in the main, is being used to distract, delude, amuse, and insulate us, then television, and those who look at it...may see a totally different picture too late."
>
> -- Edward R. Murrow (1958)
> speech

The soul. As foolish as it may sound, adopting and believing that I was a "political prisoner" of the Canadian Government was a cloud of dignity and serenity on which I could rest. I was certainly *not* a criminal. My crime if anything, was an act of civil disobedience, as I told my latest -- third -- set of lawyers. And even that was no crime, since that State had no right to search my house, and had committed the greater crime of warrant-less armed home invasion with absolutely no probable cause or concern when they found the two unregistered rifles.

So I was a civil resister. A political prisoner, awaiting a judge to see that I should be freed without further ado. Patience, with a smile.

After much thought, and observation, I adopted and practiced what I called my philosophy of "maintaining an air of quiet dignity". No matter what indignity or abuse or provocations were heaped upon me by some of my jailers periodically, I would take it with neither anger, or getting upset. No reaction at all to any provocation. Quiet dignity. And it worked for me! I soon found there was serenity and dignity in not reacting to insults from guards. Soon they just rolled off me like water off a duck's back.

My up-close interactions were with the guards -- the correctional officers -- as well as the paddy wagon transportation officers, and the Court-house special constables. And I was always quiet, relaxed, civil, well-mannered, and polite. When the guard unlocked and pulled open the heavy steel door to return me back into my pod, I said, "Thank you."

I never reacted or showed that they had got to me, if they were abusive. And I *always* made a display of gratitude for the occasional display of kindness or flexibility that a guard was decent enough to bestow upon or allow me, without having to.

I would immediately take notice of the kindness, and say to the guard: "Blessed are the merciful. For they shall be shown mercy." And I would say it with complete sincerity, after thanking the officer while looking him right in the eye to show that I had observed and appreciated his gratuitous decency. Then I would flash a brief smile, and then it would be back to the regular programming...

And finally there was a single act of resistance I realized I should take. As a "political", it was my duty to live out my beliefs. Prisoner. Solidarity with the poor. Denying myself the one luxury the jail system provided: "canteen". But I realized that there were inmates who had no funds in their account for whatever reason. And it was yet another thing noted with disdain by other inmates -- poor, no family that cared, no girl, etc.

So out of silent solidarity with the destitute inmates who had no money to buy canteen items, I vowed to never buy anything, and do without such extras.

There's a sense of victory in the control you exert successfully on yourself by the self-imposed denial of the things you don't need -- the luxuries, the extras, the perks, the unnecessary. Lived frugally. Adapted to worse-tasting skim milk, walked or biked rather than took the car, whenever I could. Washed my hands in cold water. When asked at someone's house if I wanted something to drink, I'd opt for cold water. "Ice water, if you have it. That's all." I was adaptable to privation, and just the basics. Even the uncomfortable, I tried to put up with. I appreciated luxury, but I didn't need it or crave it. The liberation of asceticism.

That bastard Crown Attorney Phillips would think I was used to my nice upper-middle class life of creature comforts, and would be dying to get out, horrified and miserable at the conditions of regular boring jail-house life deprived of everything.

He was dead wrong. All I needed was something -- anything -- to read. And if that wasn't available, just a pencil and paper. And *that* was available, if I dashed quickly enough when the head guard dropped a pile off on a table once a day, and there was usually a mad scramble.

So, I could proudly say, head held high: the fire still burns.

The Long War was to begin anew. They had fired the first shot. They had drawn first blood. And all I required was steely faith and my usual adaptability to the new and less than ideal.

And patience.

Patience, bloody patience. The Long War. And I was here without having fired a single shot.

> "The real roots of crime are associated with a constellation of suffering so hideous that, as a society, we cannot dare to look at it in the face."
>
> -- Speech by David L. Bazelon (1981),
> Senior Circuit Judge, U.S. Court
> of Appeals for the D.C. Circuit

Prison Diary 22:	Torture of an Innocent by His Own Government
=======================================

> "There's an absolute ban on torture for a very good reason.
>
> Torture taps into the deepest recesses -- unexplored recesses -- of human consciousness, where creation and destruction coexist -- where the infinite human capacity for kindness and infinite human capacity for cruelty coexist..."
>
> -- Prof. Alfred W. McCoy (2006)
> interview

> "There are no magic answers...to overcome the problems we face, just the familiar ones: ... action that raises the cost of state violence for its perpetrators, or that lays the basis for institutional change..."
>
> -- "Turning the Tide" (1985)
> Noam Chomsky

I understand dogs are re-thinking whether they wish to be referred to any longer as, "Man's best friend".

There is apparently an uproar within the canine community over, what's been termed by spokes-dog, Fido, as "a raising of dog standards". Scattered outbreaks of barking and snarling have been reported near fire-hydrants (and ex-U.S. President, and fugitive war criminal, George W. Bush's leg) across North America.

When reached for comment, Spot, Dog Federation President, merely responded, "Arf, arf..."

* * * *

> "I heard the people who live on the ceiling,
> Scream and fight most scarily.
> Hearing that noise was my first ever feeling,
> That's how it's been all around me."
>
> -- "Lost in the Supermarket" (1979)
> The Clash

"Memories! You're talking about memories!" exclaims Rick Deckard in "Blade Runner", in the theme that runs throughout the movie – the role of the memory as integral to a man's personality. To his identity.

An almost perfect memory, in my case was a gift, as well as a curse.

I always considered myself a soldier – a hardened veteran of the five pillars of Samson's Temple. *All* of them. Within my family, as a child, from bullies and other children at school, by a stranger in the dark, predatory girlfriends, etc., etc.

Physical abuse, verbal abuse, emotional abuse, and finally the abuse that has no name.

The fourth one, I don't like to talk about. One of the reasons I don't really like men, and prefer socializing with women. 'Nuf said

And last, but not least, the fifth pillar of Samson's Temple, the overt and covert attacks of devastating and unpredictable systemic and systematic racial discrimination.

Starting as a young child, and continuing, on and on, as I grew into adolescence, and then adulthood. And I never talked about it. What was the point, when there was no one who would listen, or care, or just dismiss it out of hand?

This was just the final blow to a marred and bruised struggle of finding a life, and an identity, while fighting against society and its hypocrisy and injustice. And failing.

But like I've said, the way I finally dealt with it, was to realize that you just play the hand you're dealt, as best you can. And with that realization, came an incredible inner strength, to join the wall I had erected around myself, without knowing.

But I also realized the fault was with my tormentors, and not me. The evil was all theirs. So why should I let it continue to bother or affect me? It was done, and it was over, after 25 years, and my final integration into society's upper middle class, and an engineering education and then career.

> "The grew up in a family where they watched their
> father beat the mother regularly. ...
>
> They grew up in a family where they were slapped
> and pummeled and belittled for paltry affronts."
>
> -- "Shot in the Heart" (1995)
> Mikal Gilmore, brother of murderer,
> Gary Gilmore, executed by firing
> squad in 1978.

Until the final blow, as the hands of whom I least expected it from – the State – and for no reason. Unconscionably.

* * * *

> "I have sailed over a somewhat stormy sea for nearly half a
> century, & have experienced enough to teach me thoroughly
> [to] buckle up & be prepared for the tempest. ...
>
> [L]et our motto still be Action, Action; as we have but one life to live."
>
> -- letter to his family (1846)

John Brown

Torture is wrong. It is *simply* morally wrong. An absolute abomination, and a sin against God and man.

Torture is a violation of the Charter of Rights and Freedoms. Torture is a violation of the Fifth Amendment of the U.S. Bill of Rights. Torture is a violation of international treaties – the Geneva Conventions – that Canada promoted and signed of all things -- and thus a violation of International Law.

Torture is a war crime if committed during war, and certainly in times of peace a very serious matter, by inference. And most important of all -- from a legal perspective -- torture is an abrogation of Canada's legally *binding* treaty obligations forbidding the use of torture.

And if you put morals aside -- God forbid -- it's *just* common sense. You don't fight barbarism with barbarism.

And when you see the next Canadian soldier -- or abducted diplomat -- come home in **three** body bags, perhaps you will begin to understand *why* torture is not something we as a nation wish to be engaged in.

"Oh, the humanity!"

I got news for the Canadian Government -- a dung heap is *not* the moral high-ground.

* * * *

> "I still believe that one day mankind will bow down before the altars of God and be crowned triumphant over war and bloodshed, and nonviolent redemptive goodwill will proclaim the rule of the land.
>
> I still believe that 'we shall overcome'."
>
> -- Martin Luther King, Jr. (1964)
> Nobel Peace Prize acceptance speech

I was tortured, pure and simple. It is calmly, and without exaggeration, that I make such a charge, against my own country.

I was a forty-eight year old -- on the cusp of old age – an upper middle class engineer, a suburbanite, with a suburbanite's conservative car and a home, both fully paid off. I was without a criminal record, polite and non-aggressive in my daily dealings, and moral and ethical in my personal dealings to a fault.

I lived a quiet life, read a lot, had a coterie of just a few close and dearly loved friends, and was a long-term mid-level employee in a respected and highly paid position of a major multinational corporation, with nothing but good employment reviews – though I hated my job, and the corporate management. Nothing unusual there...

And I was guilty of nothing but being the victim of an illegal warrant-less search by Gatineau and Ottawa Police officers, using the excuse of a bogus arrest on minor domestic incident charges.

Yet I was thrust without cause into a barbarous, predatory, and uncivilized jail environment of unceasing violence by inmate-upon-inmate, being full of hair-trigger brutes. I was denied basic human needs and essential human rights, even though I was convicted of nothing yet. I would be imprisoned for seven months for a criminal offense -- two unregistered guns -- that would merit on conviction of a non-jail sentence of probation, or a suspended sentence at most.

My imprisonment before trial was based on a false arrest, followed by the perjured bail testimony of one police constable who wasn't even there for the search of my home, an evidence review revealed.

My bail was denied by a worm of a Justice of the Peace -- a non-lawyer, and political appointee -- awed and cowed by a full public gallery, and public attention focused by newspaper articles that were a smear campaign orchestrated by the police. The insinuations of terrorism played right into their hands in their cover-up of misconduct.

I was imprisoned for seven months to cover-up police misconduct and prosecutorial malfeasance to coerce a guilty plea to cover it all up, for an offense that they knew would be thrown out immediately as soon as it got to trial.

And if they got their way and squeezed a guilty plea out of me, the cover-up would be solid. They would say, "Oh, he was convicted." "Oh, he has a criminal record." "Oh, he admitted guilt for his crimes." Case closed.

I was tortured deliberately for this purpose, at the deliberate and knowing behest of Crown attorneys based on lies to the Court by them, and the police.

It was a disgrace and an abuse of process that was apparently so routine behavior by Crowns that they didn't think anything of it in the slightest. It was the gradual erosion of absolute morality that ends with the absence of moral standards, or the establishment of a moral "standard" flexible to any personal whim, immediate need, or prosecutorial long-term goal.

It is the destruction of any moral standard and its replacement with a sense moral propriety known as hypocrisy. A state under which any manner of mischief is possible, justified, and rationalized, utterly free from restraint, limitation, and – most importantly – guilt or troubled conscience.

* * * *

"My aim then was, to whip the rebels, to humble
their pride, to follow them to their innermost
recesses, and make them fear and dread us.

'Fear of the Lord is the beginning of wisdom.' "

-- "Memoirs" (1886)
William Tecumseh Sherman

It's a scandal. A national scandal.

But a scandal requires moral outrage, righteous indignation. "Have you no shame?" It requires a state of awareness and concern.

It requires either a shrill vocalization of men of reputation, or the focused attention of the news media sensing the smell of blood, or the upset of their veneer of impartiality by blatant moral indecency. It requires a public galvanized by collective disapproval, blossoming into anger.

It requires a metaphorical lynch mob enraged in their living room arm-chairs, or surrounding their workplace water-cooler, the entertainment of gossip pushed aside by the tongue-wagging indignation of the cubicle zombies aroused from their laptop graveyards.

Aroused into rage, and from rage to action, no matter how small. Doing something.

> "Any Christian who blindly accepts the opinions of the majority, and in fear and timidity follows a path of expediency and social approval is a mental slave."
>
> -- "Strength to Love" (1963)
> Martin Luther King, Jr.

Prison Diary 23:   A Friend in High Places
                               ===================

> "A robin redbreast in a cage
> Puts all Heaven in a rage.
> A dove-house fill'd with doves and pigeons
> Shudders Hell thro' all its regions.
> A dog starv'd at his master's gate
> Predicts the ruin of the State."
>
> -- "Auguries of Innocence" (1803)
> William Blake

> "Oh Egypt, Egypt!
>
> Your religion will be no more than a fable which not even your own children will think true! Nothing but words will remain, chiseled in stone, to tell of your acts of piety. [...]
>
> Gods and men will be at odds -- diminishing both, with the wicked angels victorious.
>
> A time of cynicism will come, where it will appear the epitome of folly to imagine that Egypt ever thought that reverence for the divine was supreme over all."
>
> -- Hermes Trismegistus (ca. 350 A.D.)
> Egyptian seer

One of an engineer's favorite quips in response to a technical question from a fellow engineer who's just a new, young pup engineer, or just being lazy because the answer is easily found, is simply the comment "RTFM", which stands for "Read the Fucking Manual".

So I took my own advice to educate myself on my religion -- Christianity -- and got a plain, readable modern English bible -- free, from a really nice girl from the Salvation Army -- and starting from the beginning of Matthew, read the four accounts -- the Gospels -- of the story of Jesus Christ. The Gospels -- the "Good News" -- according to four out of thirteen of Jesus' apostles: Matthew, Mark, Luke, and John.

It was the first time I had read them, the first time I had read them complete from beginning to end. It's never too late to change, goes the religious cliché. And it pisses me off no end, to have to say that I never thought I'd have to agree. But I do...

After reading the New Testament in OCDC jail, I became what they call a born-again Christian. A heavily loaded term, but one that technically applies to my change.

Once again, it is a cliché -- real, nonetheless -- of becoming a Christian convert in jail. But this I deny strenuously. I am *not* a bible-thumper -- I just *got* the message, complete and unadulterated, of Jesus Christ. And I liked it, learned it, was amazed at what I had never heard before, and it changed -- Boom! -- the way I thought and behaved, and gave me the full story of Jesus' teachings and message.

I've always subscribed to Christ's general message that I had picked up, and considered myself a Christian -- though there was a rather deep chasm dividing my belief and my practice -- not to mention my simultaneous beliefs of political violence, killing evil people, and righteous bloody vengeance.

Jail just gave me the time, and inclination to look for answers. And it was a well thought out acceptance of His impeccable logic that brought me to His acceptance. His words ring true and valid after *two millennia*. I simply could not reject them, having read all four Gospels completely, and in order. And it only took the short time of two or three weeks.

Each Gospel is a bite-sized chunk of *only* 20 or 30-odd pages. *Just* 20 or 30 pages, Goddammit! Enough for anyone to snap up, in next to no time. And I wonder how many Christians have actually read it, short as the Gospels are? Now, that's incredible, because I think only a small percentage have. Or at least have read it and understood its simple story and message. Four versions of the same story by four different apostles -- members of Christ's self-chosen "team" of close followers.

> "I do not believe in the creed professed by the Jewish church, by the Roman church, by the Greek church, by the Turkish church, by the Protestant church, or by any church that I know of.
>
> My own mind is my own church."
>
> -- "The Age Of Reason" (1794)
> Thomas Paine

* * * *

> "...I continued a discussion with my lawyer [in the 1950's] about the Broadway Theatre, which I said was corrupt, the art of theatre had been totally displaced by the bottom line, all that mattered any more. Looking up at me, the notarizing lawyer said, 'That's a communist position, you know.' I started to laugh until I saw the constraint in my lawyer's face, and I quickly sobered up."
>
> -- "Are You Now or Were You Ever?" (2000)
> Arthur Miller
> "The Guardian"

There once was a time that I would say, "I've never hurt anyone." And then one day, I realized a change was necessary. "I've never hurt anyone who didn't have it coming in spades." And then, one day later, I decided, I'd be better off dropping the line completely, and not saying anything...

The road to hell, indeed is paved with good intentions. And my road to hell was a three decade old committed dedication to "righteous" vengeance against anyone who crossed me in a way that required or begged for retaliation, added insult to injury, or otherwise required a crushing response, just on general principles.

And I dispensed my own brand of ruthless justice that I considered an art to behold, an eloquence of revenge that was in a sense beautiful in its planning, choreography, execution, and sense of equitable, retaliatory justice.

Thirty years of studying, improving, practicing, experimenting, and honing to ultimate perfection the science of bloody, ruthless, righteous vengeance, to those who made the mistake of thinking that they could cross me, get away with it, and then amble close enough for the claws of pay-back to rake their souls to shreds, as I smiled with an intense look on my face that must have been a sight to behold -- or chill to the bone.

* * * *

"[T]o observe guerrilla warfare is to find yourself everywhere
and no where, witness to a particularly fragmented reality."

-- "Report from Afghanistan" (1981)
Gerard Chaliand

The meaning of the word "exegesis" is an explanation or interpretation of Scripture. But the words of Jesus Christ in the Christian New Testament are so crystal clear, so plain, and so obvious, that they need no analysis. One wonders why there is even a word, "exegesis".

" 'Absolutely pointless!" says the Preacher.
Absolutely pointless! Everything is pointless!' "

-- Ecclesiastes 1:2

Or re-interpreted to the vernacular:

"Bullshit!" says the Preacher. "Bullshit! It's all bullshit!"

And people think the Bible is boring!

In the Old Testament -- the stories of Abraham, Moses, and the prophets before the birth of Jesus -- the chapter "Ecclesiastes" is quite short, and quoted all the time -- though you may not realize it.

And it's rather interesting, down-to-earth message is that what people think is meaningful, heroic, or successful is just a waste of time.

Hard work, and productive effort -- it's ultimately meaningless and forgotten when you die. Then it talks about having fun, getting drunk, and enjoying life's simple pleasures, and *their* lack of meaning.

And so he concludes when everything you can possibly think of, from enjoyment to hard work, the sensible to the lack-luster, is just meaningless bullshit, he concludes, when the world is just a bunch of empty endeavors whatever you do, or try, irrespective of values, there's is one thing of eternal value.

Only the practice of the word of God, and his worship is meaningful, the unnamed Preacher concludes, simply, because **that** leads to eternal life. The worldly is of no value, in other words. Only the practice of Good has any real meaning.

* * * *

"It's a lonely ol' night,
I don't know,
I'm just so scared and lonely all at the same time.
No, no, no, no, no.
...
Can I put my arm around you?
It's a lonely ol' night –
Custom made for two lonely people like me and you.
Radio playing softly some singer's sad, sad song.
His singing about standing in the shadows of love.
....
And it's a sad, sad, sad, sad feeling,
When you're living on those in-betweens,
But it's okay."

-- "Lonely Ol' Night (1985)
John Mellencamp

Quotations of hope and logic by Jesus Christ from the New Testament. And like J.C.'s entire message, as relevant today as it was 2,000 years ago in the Ancient Roman Empire:

" '[D]on't ever worry about tomorrow. After all, tomorrow will worry about itself. Each day has enough trouble of its own.' "

-- Matthew 6:34

" 'Can any of you add an hour to your life by worrying? If you can't do a small thing like that, why worry about other things?' "

-- Luke 12:25-26

" 'Come to me, all who are tired from carrying heavy loads, and I will give you rest. Place my yoke over your shoulders, and learn from me, because I am gentle and humble.

Then you will find rest for yourselves because my yoke is easy and my burden is light.'

-- Matthew 11:28-30

* * * *

I was rather disappointed and disillusioned *way* too young about organized religion. Because when I grew up there was no warning taught -- or it otherwise made clear -- that most Canadians and Americans, especially, are useful to help you learn about Christianity by showing an example of what **bad** Christian practice looks like.

And how it's part of human nature that must be understood and addressed, of how people can go so far off the rails in their behavior and belief system, and not realize it, doing consistently the opposite of Christ's teachings. It is so bizarre that it defies logical explanation to the point of incomprehensibility.

* * * *

Priest 3: "No riots, no armies, no fighting, no slogans."

Caiaphas: "One thing I'll say for him: Jesus is cool."

-- "This Jesus Must Die" in
"Jesus Christ Superstar" (1970)
Tim Rice

My first specific and effective tutelage and interest in the message of Jesus Christ, was sparked by "Jesus Christ Superstar", lyrics by Tim Rice and music by Andrew Lloyd Weber. I was eleven years old.

A couple of decades passed uneventfully from a religious perspective. I was angry, cynical, bitter, vengeful, and hateful. I was lost, and I didn't even realize it.

Then came a bolt from the blue -- a small bolt, actually more of a religious milestone. A growth in my understanding of Christianity. I happened to notice an mid-1990's Nortel Networks-internal Internet posting from a Nortel "bible-thumper" of sorts, that was different that the usual BS American Nortel bible-thumper. It was a message posting from an educated RTP guy -- an engineer -- who made the interesting comment -- the first time I'd ever heard anything as intriguing as this -- about the fact that he liked and was amazed by the Bible, because he read it, and found that there was a passage, or story covering every possible aspect or situation in life.

Now *that* was a pretty heavy take on the Bible, and my interest was piqued. *Every* aspect of life! I "filed" it away mentally, as a new, surprising, and rather interesting comment. It was worth checking out sometime, with no specific date on "sometime".

Every possible situation in life. It didn't sound like something a guy would make up, and I was intrigued for sure. A seed had been planted. I don't remember the guy's name, or details, but I thank whoever he was.

Another few years passed, and I started being visited by Jehovah's Witnesses on Saturdays. I had nothing better to do Saturday morning, so I would let Wally in, and listen and talk to him. I didn't have a Bible and eventually told him so, because I was interested in seeing the "Manual".

A fellow Jehovah's Witness showed up soon after, another Saturday and I was

a) given a Bible for free,

b) most importantly, a Bible translated into everyday, understandable English, and

c) given the important advice that it was OK -- it was *not* desecrating it -- to underline, or otherwise annotate any part I found interesting. I began occasionally reading parts here and there. And underlining short parts I found interesting and meaningful. More importantly, the stage was now set.

I occasionally read bits and pieces, but no coherent picture developed. And usually I was drunk, it was something like 6:00 in the morning, I was alone, and I was bored.

Years later, again -- the year 2007 to be exact -- I found myself with some free time. In jail, to be precise...

And so, I did what I had always wanted to do, and amazingly did it: I started reading in order the four Gospels – the four accounts of the life and teachings of Jesus Christ by four of his Apostles -- of the New Testament (Matthew, Mark, Luke, and John) one after another, and then continued on, until I had read the entire New Testament, except for the last book, Revelation. (I wanted to keep the ending a surprise.)

All jokes aside, I was amazed at what I read. I was amazed at what I had *never* heard before. His bitter railing against those in power. And hypocrisy. And his real and total message.

> "They were more concerned about what people thought
> of them than about what God thought of them."
>
> -- John 12:43

But the hardest of all of Jesus Christ's teachings to practice -- or even consider practicing -- is, "Love thy enemies." I mean, can I *at least* talk bad about them? Or revel in their fall or misfortune? No. Sheesh. Talk about a tough nut of a teaching to follow!

And, last, but not least, the message got through, including: that when you fight evil with violence, even if you win, you inevitably become the very thing -- evil -- you fought to end. And so the cycle of evil and violence continues.

The struggle of Good against Evil is the struggle of light against darkness. And you can only end darkness with light, not with more darkness. It was really quite simple.

Bingo!

That I was it, when I became convinced and accepting of the full message of non-violence of Jesus.

* * * *

> "We either live together as brothers, or
> we are going to perish together as fools."
>
> -- Martin Luther King

But there was still one stickler of a point that had always barred me from full acceptance of Jesus. I agreed with Jesus' message of peace, forgiveness, and love – but the Son of God, no way. But one of several stories convinced me that Jesus was, indeed, the Son of God – such as John 7:32-46.

Two temple guards are sent by the chief Jewish priests to arrest Jesus. They find him preaching to a crowd in another temple, and they begin listening to his preaching as they wend their way through the large crowd towards him to execute the arrest. But then the two temple guards turned around before arresting Jesus and returned to the chief priests empty-handed, telling them, "No human has ever spoken like this man."

And indeed, after reading all four Gospels, I decided the guards were right -- Jesus was the Messiah. Jesus was the son of God. It was the resolution of the hardest part I had difficulty accepting all these decades. (Not that that was as important as his message.)

* * * *

> "Happy days are here again,
> The skies above are clear again,
> Let us sing a song of cheer again,
> Happy days are here again."
>
> -- "Happy Days are Here Again" (1929)
> Jack Yellen

Sometimes I wonder about this Christ fellow, though. His words are immortal, priceless, perfection itself. But I wonder about his employment practices. Take his apostles. There's Judas -- a thief and traitor. There's Peter, who betrays His memory with three denials. There's "Doubting" Thomas, and Simon the Zealot (fanatic). Jesus was frustrated because his Apostles continually could not understand his teachings. And they all ran away when Jesus was arrested.

Messiah, he was, but Human Resources Director -- he would never have lasted.

And he's constantly criticizing his Apostles. Quite bitterly and often -- right up until just before his arrest in the Garden of Gethsemane.

* * * *

> "Virtually all the members of the Convention were followers of
> natural religion; they believed in a Supreme Being conveniently
> distant, dwelt on man's abundant capacity for natural virtue,
> and regarded priests as charlatans and revealed mysteries as
> a delusion.
>
> [Organized] religion, they would say, was in its death throes,
> soon to expire before the searching eye of reason."
>
> -- "Twelve Who Ruled" (1941)
> R.R. Palmer

And last, but not least, I look to the words of Jesus Christ, Prince of Peace, for guidance regarding private ownership of weapons.

> " 'The person who doesn't have a sword
> should sell his coat and buy one.' "
>
> -- Luke 22:36

> "The disciples said, ' Lord, look! Here are two swords!'
> Then Jesus said to them, 'That's enough!' "
>
> -- Luke 22:38

So there you have it -- from the "Big Guy" himself -- love they neighbor and love thy enemy, certainly, but *be* careful. "Turning the other cheek", loving your enemies, accepting Christ's message of peace, love, and non-violence, does *not* mean being defenseless

It doesn't mean one needs to become an easy target. Or, in the vernacular, "He who fights and runs away, lives to fight another day."

Given Christ's views on peace, love, and turning the other cheek, does not mean not trying to avoid being hit on one cheek if possible, being an easy target for the predatory, or letting people stab you again, if they stab you once.

So. I wonder? If firearms, rather than swords were around in Christ's time:

> What would Jesus pack?

"History is long; even the most terrible political events are passing ones and eventually self-rectifying. It is art and intellectual work that ultimately count, and morality and virtue are the measures of human worth and accomplishment."

-- "The Bullet's Song" (2004)
William Pfaff

"Beware the beast Man, for he is the Devil's pawn.
Alone among God's primates, he kills for sport, or lust, or greed. Yea, he will murder his brother to possess his brother's land.

Let him not breed in great numbers, for he will make a desert of his home -- and yours.

Shun him -- drive him back into his jungle lair -- for He is the harbinger of Death."

-- 29th Scroll, 6th Verse
"Planet of the Apes" (1968)

Prison Diary 24:   Writer's Retreat
                                    ================

> "I wish to be read for Art's sake, not for my notoriety."
>
> -- Oscar Wilde (1897)

> "Within a foreseeable future, men will cast off their useless chains. They will realize their full individual potential, according to the necessary order of spontaneity -- in splendid anarchy."
>
> -- "Le Refus Global" (1948)
> Paul-Emile Borduas

Sic transit gloria homo. So passes the glory of man.

It was the fourth week, when I stabilized enough emotionally from the shock of my totally unexpected incarceration -- initial and continuing -- and settled down enough to realize what I had to do. I needed to seize an obvious opportunity, and possibly a lifeline with maybe my only chance of maintaining emotional stability, and thus sanity.

A way to keep at bay terminal depression, the suffering of mind-numbing endless boredom, and the surrender to hopelessness, the harbinger of madness. A way to make something permanent and useful out of what would otherwise be a waste of my mind and my talents (like employment).

Keeping busy and keeping my spirits up, were the obvious things to do in such a circumstance. Keeping the *mind* busy, specifically. In common parlance, I needed to devise a good "time-waster". But perish the thought, I find it difficult to waste time. But my previous bosses filled the void...

It's not commonplace knowledge that the brain can be considered to be like a muscle. One is born within a certain range of mental ability, and with specific aptitudes, strengths, and weaknesses. One is born with a certain learning curve, curiosity, focus and persistence, and precision in memory and logical information processing. And awareness of limitations. Lastly, the size your memory, and the ability to memorize is a function of short-term memory, and forgetting a function of long term memory.

With proper use and exercise, the mind strengthens, grows stronger, and achieves stamina and robustness. With special training it can be polished to exact precision, trained to perfection in a specialized skill, or molded into an analytical engine that is a towering tribute to logic, inference, extrapolation, modeling, and experimental design. Add a memory trained to cross-reference the information stored within and the mind rounds out to as close as one can get to perfection of intellectual functioning.

It goes to seed if wasted or unused. If it reaches its full potential, it is man's most useful -- or dangerous -- "muscle".

From Paris Diary to Prison Diary. It seems just a short while ago that I was writing the "Paris Diary" for the amusement, via e-mail, of my colleagues back home as I did a 14 month ex-pat gig for Nortel in Paris, France in 2001-2002. And years earlier, as a respected "Net personality" from 1994 - 1999 or so, as I wrote extensively for public consumption and the amusement of myself and others. Besides, writing is the only way I can get my thinking out without being interrupted.

It seems in a sense, almost a seamless transition, to now be writing my "Prison Diary". Though the expense account seems to be a little late. Yes, talk about a reversal of fortune... So, let's consider the "prison diary" genre. It is well established that some great writings have come out of prisoners, or ex-prisoners, criminal or political.

Samuel Pepys was the first -- and most famous -- diarist of note, chronicling his observations of mid-1600's England, including the Black Plague, and the Great Fire of London. Though not a prisoner, he put the diary/personal journal on the literary map.

The autobiographical prison experience books, "Papillion" and "Midnight Express", were both made into movies. Oliver Stone's "Midnight Express" (1978) captures the hopelessness of lengthy imprisonment under harsh conditions disturbingly well.

As a writer in jail I'm merely following in the footsteps of inmate writers Jack Abbott, celebrated Canadian convict-writer Roger Caron, and 1950's California death row inmate Caryl Chessman.

Political prisoner writers include St. Paul, John Donne, the Marquis de Sade, Oscar Wilde, Henry David Thoreau, Gandhi, Ho Chi Minh, Bertrand Russell, 1960's radicals Eldridge Cleaver, Malcolm X, Abbie Hoffman, Timothy Leary, and Jewish-Argentinian newspaper editor, Jacobo Timerman (falsely imprisoned by the Argentinian fascist junta in the mid-1970's).

Now, I have a question. Everyone who has ever challenged the social order in any way, violently or peaceably has been thrown into jail, if not murdered, by state authorities. Why don't they teach *that* in school, I ask? But that would just either depress young people, or encourage them to vow to change society. And the social order is perfect.

So perfect that there is no efficient, reliable, timely, and effective method for peaceable change. So perfect that there is no efficient, effective auditing of those in power to deter and detect corruption -- malfeasance, misfeasance, and nonfeasance the triumvirate of power abused, from soft to hard corruption.

And since they've already thrown me into jail, what else can they do should they dispute my firebrand writings? So, no rotting in jail for this lad! Lemons and lemonade, and all that. Sanity demands it. Development or decay, the only choices. Courage under fire. Accept and wait out the darkness before the dawn.

And forget the plaintive whimper: "I want my mommy!"

It's just a new adventure, a new opportunity to be embraced, utilized, and taken advantage of.

It's not jail, I have to convince myself, but a vacation, a retreat, from the world. Time for my head to clear, unencumbered by the demands, wants, needs, desires -- certainly creature comforts, basic comforts, and luxuries -- and all the vicissitudes of modern life -- in the sunset days of yet another tottering _ancien regime_.

But there was more. I realized that the sensory deprivation of the stark, sterile, gray concrete cell was the ideal germinating ground for thoughts and creativity. It's time for a voyage personal discovery, or re-discovery. A time for reflection and revision, and reconsideration. Time to recharge. Insights to be gained or reinforced. Solitude and quiet to be taken advantage of, to provide a fertile environment for calm introspection and serene self-analysis. Between the violence, abuse, and degradation outside the confines of my cell...

And with 21 hours a day locked in my cell, my cellmate generally sleeping, writing is the perfect way to pass the time. And with meals delivered three times a day to your cell, you can say that the room service is great!

It's meant to trigger boredom, depression, misery, and despair, leading to submission or behavioral changes to avoid jail. It is to avoid these unpleasant emotions that keeping busy is one counter-measure.

But when I picked up that pencil and began writing down everything I was thinking about, that I thought worth preserving, that I noticed an amazing thing.

In fact, I can say from experience that this sort of "solitary confinement" can be an environment that the writer finds to be not only comfortable, but that is phenomenally productive for serious writing. The sensory deprivation of the barren isolated environment, and freedom from distraction is there for sure, but it's more than that. The complete cut-off of any sensory or mental input of any significance puts the active mind in an environment that is static, and plain, devoid of sensory input of any sort.

Almost a complete sensory vacuum. With almost a complete cessation of external inputs to be picked up and processed. Complete cessation of data flow inwards. And in reaction, my brain turned its focus entirely inward to my own thoughts and ideas, as if keeping itself busy. And then, the "tide" reversed, and ideas and information, and data began flowing *outwards* from my brain -- output taking over the empty "channel", that was normally dominated by input data coming in.

My brain -- turned inwards on itself -- was not just keeping busy, but sort of keeping itself entertained, as it explored my memory recalling interesting caches of data -- ideas, concepts, events, experiences, readings I have remembered -- looking for "interesting" data to re-run or data that had been filed away into memory that likely could be re-processed to reveal further information or insights.

Potentially interesting data that was "re-run" brought a memory into my conscious thoughts. It would play like a tape, as I was amused or looked for explanations or insights that would provide closure or a lesson to learn for the future.

Incidents that bubbled up from memory as something that might prove "re-processable" were usually memories I had stopped thinking about for some reason and filed away in memory with a "note". The note said, "Processing suspended before completion. Future data collected may allow this file to be re-processed to reveal answers or conclusions thought to be hidden in file, but not uncovered by last attempt at file analysis".

A river started flowing full of ideas. Soon it was a torrent flowing out of my mind, and through my pen and onto the paper. It was amazing! In less than seven months, I wrote continually: a day-to-day diary, a personal journal of ideas, jokes, observations, and philosophies. Much of this "Prison Diary" -- my experiences, reactions, thoughts, emotions, and everything else I felt. And lastly I buried myself deeply within my historical escape from the tedious annoyance of thinking about humanity and its folly (well, not exactly). I also disappeared in a project related to the writing of a book on a certain highly technical subject which had been my hobby on the outside. The exact subject of which will have to remain a surprise for the moment.

But all in all, during seven months in detention, with only a pencil and paper, and a sheet steel dinner table in my two-man cell to write on – and my beloved coffee -- I wrote over 850 pages, single-spaced! Even I was surprised at my productivity.

And I believe it was probably only possible because of the large amount of information stored in my brain, and my extensive writing experience -- writing skill -- to provide a suitable outlet for the torrent, allowing data and ideas to flow out in a random yet coherent and articulate form suitable for committing to paper.

I have to admit it, the barren cubicle they keep you in when you work for the government or big business is a good idea to keep empty. No radio and empty, or at least uncluttered desk would possibly reproduce the same barren environment.

So what should I call my first book? I know: "A Man of Conviction". No -- too prosaic – and joke to obvious.

What about "_Urbi et Orbi_" [To the City, and the World"]?   Nah -- too pretentious.

Or -- OK, I give up -- "Will Write for Attention".

"There is a side...that develops better and stronger by [writing] than contact, especially with some people who can get their thoughts clearer when they see them written.

Another thing -- that beastliness, self-consciousness, is left out -- shyness, shamefulness in exposing one's inner self there, face to face before another, getting rattled and mislaying words.

The absence of the flesh in writing, perhaps brings souls nearer."

-- "Hundreds and Thousands" (1935)
Emily Carr

Prison Diary 25:   Prisoner of the Heart
                        =================

> "P.C. [Police Constable] 31 said, ' We caught a dirty one',
> Maxwell stands alone,
> Painting testimonial pictures, ohh, oh, oh, oh.
>
> Rose and Valerie, Screaming from the gallery:
> ' Say he must go free! Maxwell must go free!' "
>
> -- "Maxwell's Silver Hammer" (1969)
> John Lennon, Paul McCartney

> " ' We knew that: "Hey, I could be dead tomorrow. So
> why not live every moment?" The insanity, lies, cruelty
> all around us, all of the time, made us want to run to
> the soft gentle flesh of a woman accepting, without
> any hint of danger...to make life intimate and worth
> living again...' "
>
> -- "Odysseus in America" (2002)
> Dr. Jonathan Shay

When all is said and done, as The Cult sang: "She sells sanctuary".

It's prison time, when a young man's thoughts turn to love... And the only "spring" time will be if my lawyer springs me, by getting me bail.

Well, I'm not so young anymore. And -- at my age -- I've become a little jaded and cynical about love. Face it, I'm a **cured** Romantic. If you told me that love was a crap shoot, I'd respond that you were half right...

The evidence: my current favorite song (it changes every few years) is "The Sweetest Thing" (1998), a "love" song by "U2". The beautiful melody is accompanied by lyrics suggesting a young man in a difficult relationship, struggling to remain untroubled as his girlfriend seems to be rather careless of his emotions, indifferent about his genuine and undemanding love. The opening lines:

> "My love, she treats me like a rubber ball.
> Oh, oh, oh, the sweetest thing.
> She won't catch me, or break my fall."

*   *   *   *

Bumped into an ex-girlfriend on the street recently. Some low-key chit-chat led to a no-pressure, but low-key suggestion by me. Imbued with an interesting tone and way -- direct, to the point, blatant. "Take it or leave it", but not dismissive or devaluing. Firm and deadly serious. And all these emotions, simultaneously, and unmistakable to any woman who's been intimate with you for any length of the past.

"Ex-sex", I call it. It's convenient and the fact that there are no surprises makes it predictable, with no possible misunderstandings, disappointments, or expectations that cannot be fulfilled.

"The devil you know...", and all that. Actually, it's more like an angel for an evening, or a date with a script including a happy ending, everything mutually agreed upon in advance, if only by virtue of knowing each other that well. Isn't that beautiful, in a limited sort of way?

It's a known level of satisfaction on both sides, both of us relaxed for a low pressure, no risk few hours or evening of nonsense-free physical enjoyment. The key concepts are low-key, "static"-free, low maintenance, realistic, and up-front expectations, if any.

All in all, a no hassle, enjoyable get-together, because everyone knows the score, and truly, no strings are attached -- no expectations to complicate or disappoint, or ignite any feelings of betrayal, or being used.

And so, for an hour or so of glorious and exhilarating physical intimacy, my surrender obvious, my needs, motives and vulnerability unconcealed, my pain palpable, and my motives bereft of ulterior motives or slippery skullduggery, I could lay down beside another soft human being, and forget about my pain and troubles just for a little while.

And then we'd lie next to each other in bed, skin touching skin, until we both drifted off to restful and untroubled sleep, next to each other and in the same bed.

I treasure the feeling of intimacy when I wake up and realize that a woman has accepted her safety and given me the honor and comfort and unconcerned trust of the simple act of spending the night -- sharing my bed until we wake up to each others smiles the next morning. To me, that waking up is one of life's joys and most beautiful, purest pleasures.

But what I particularly treasured was as I lay there in post-coital bliss the night before, as I waited to drift off, I could pretend for a time that I was lying next to someone who actually might care about the warm body lying in bed next to them.

And feel almost happy.

* * * *

> "When I was one-and-twenty
> I heard a wise man say,
> 'Give crowns and pounds and guineas
> But not your heart away;
> Give pearls away and rubies
> But keep your fancy free.'
> But I was one-and-twenty,
> No use to talk to me.
>
> -- "When I was One-and-Twenty" (1896)
> A.E. Housman

I think I first began to suspect that male-female dynamics were not so straightforward was when I realized that there were lots of little blue cartoon smurfs, but only **one** smurfette. A suspiciously bouncy young thing. I no longer have a crush on the smurfette -- that little blue cartoon slut.

But nonetheless, women are one thing I miss in here, including one, my beautiful 15-year-old daughter, "A.C." Like Louis Armstrong used to sing, "Thank God for little girls".

And Hallelujah for the bigger ones, too!

For the cult of heaven and hell, aka, the "cult of earthly female pleasure" is one that no man can resist, forget, or ever leave, except by death.

Like when by chance, I happened to meet some sweet young thing for the first time -- admittedly way too young for me -- but she's friendly, gorgeous, has a body to **die** for, maybe big brown eyes and blonde hair -- or a thick mane of black hair -- capped off by a wonderful comely smile -- directed straight at me. Reaching down to my very soul.

Of course, I'm captivated. And should she happen to ask me how old I am, I'd respond with a mischievous smile, "Old enough to know better..."

My problem is that when I was a kid, I broke all my toys. It seemed a prophetic pattern and apparently not much has changed. I don't understand it. I don't know why. But my relationships tend to be rather stormy, to say the least. And it's the one problem, the one regret that I've always felt that has always demoralized me to the bone.

As I like to joke, the only problem with chasing women, is that occasionally you catch one.

My favorite woman's hair color, you ask? Blonde, brunette, and redhead.

How do you "bag" any woman you want? Find out what she **really**, **really** wants. And act like you're going to give it to her...

My favorite type of women? A woman who says, "Yes."

Well, enough of that...

I don't like to admit it, but occasionally I **do** find enjoyment in the chase, when I can classify it as a "glorious failure". Like the time I was waiting in line at the cash of a convenience store in the Agora at the University of Ottawa. The clerk at the cash was a university student – a really hot chick about twenty: cute, stylishly dressed, very attractive, and **stacked**.

When I advanced in the line, finally to present her with my purchase, she looked at me and seemed to recognize or remember me from somewhere. "Do I **know** you?" she asked, furrowing her eyebrows.

Without missing a beat, I responded, "No...do you **want** to?"

Apparently not, I immediately found out, when she turned away without a word. Some women have **no** sense of humor...

You see, I'm basically a "free agent" when it comes to love (the first half of our life is ruined by our parents, the second half by our ex-wives), and as far as I'm concerned **all** women are my friends.

They just don't necessarily realize it or accept it quite yet. But I'm working on it. One recruit at a time. And, as I like to say, it's an all-volunteer army. After all, "there ain't no future dating magazine girls."

And I definitely consider myself a mature professional. In other words, I'm not looking for one-night stands, I'm looking for "repeat business".

But let's make it clear: I'm no Don Juan, Lothario, or Casanova. Though I do like to stretch serial monogamy to the limit. I just don't hang around when it starts to become obvious the future is starting to look down-hill. Sure, let's stick around and torture each other for a while... I learned a **long** time ago that it's better to be unhappy alone than unhappy with someone else.

Some women I've been intimate with -- hard souls themselves -- have commented about **me** having a wall as my "armor".

One time we were in bed together and I responded, "Yeah. I have a wall 1000 meters high, a hundred meters thick, and do you want to know what's behind it? An ocean of tears." And then I choked up.

God, I hate doing that...

But there's no point in recriminations, hard feelings of any kind, or wasting each others time any further. It's really nobody's fault, and people should just realize that and move on with a minimum of fuss and a maximum of civility. It is really the only rational efficient thing to do.

People are just so different. Requirements, attitudes, physical needs, compatibility, and many other traits need to match, and you have to figure that out real fast. And like they say, many are called, but few are chosen...

* * * *

"Jump up, bubble up, what's in store,
Love is the drug and I need to score.
Showing out, showing out, hit and run,
Boy meets girl, as the beat goes on.

Stitched up tight, can't break free,
Love is the drug, got a hook on me."

-- "Love is the Drug" (1975)
Roxy Music
Bryan Ferry

I'm not a shameless womanizer, philanderer, or cad. Though I encourage any sort of negative reputation in this area. The curiosity factor brings in a lot of traffic. I **am** somewhat of a rake. Raffish, decadent, a dissolute sybarite, a libertine in some people's eyes.

It's just that you don't really get to know the woman behind the mask, until you've jumped that low wall, that hoop of **physical** intimacy. After all, she doesn't need to hoard it. It's **not** a weapon or a tool. And after the sex is over and done with, I have found that **every** single time, you see the real woman, and one quite different than with her mask **on**.

Or at least, you get closer to seeing the real woman. Sharing each other's body in glorious nakedness, all the sticky and uncomfortable formality melts away, and you can really see and get to know the woman behind the mask, the woman behind the wall.

The camouflage and smokescreen dissipate, and you finally should be seeing the bullshit-free individual with all the possibilities, the inner beauty to discover, or the inner flaws or ugliness that we can either deal with, negotiate, improve, accept or just need to know to make an informed decision about the possibility of building a more permanent bond of intimacy, or accept that it'll never work, and it's no one's fault.

I see nothing wrong with this, and everything right with it.

And it's just that I love being around women. They're just so **soft**, and sweet, and oh-so-cuddly.

And they can be so much fun, if you can "connect" mentally and emotionally somehow. If you can -- and first you have to **try** – great. And if you can't, you simply move on.

NEXT!!

* * * *

> " ' Never shall a young man,
> Thrown into despair
> By those great honey-colored
> Ramparts at your ear,
> Love you for yourself alone
> And not your yellow hair.' "
>
> -- "For Anne Gregory" (1930)
> W.B. Yeats

A lot of women can't stand to be alone. Get to know me, and **I** can cure you of that, I like to joke...

And why do so many women seem to think that the world revolves around them? Even 400 years ago, Galileo established that **that** was bullshit.

And do you know Yogi's definition of "safe sex"? Don't tell her your real name.

And what's the definition of "promiscuous". I don't know, but I only hear the word used by people who don't seem to get a lot...

It's an interesting dichotomy, this label of "promiscuous". It's the goal of every man, and the goal of every woman to keep secret...

I was being examined by a resident [trainee doctor], and she asked me, "How many sex partners had I'd had?"

I answered, "You mean, like, last week, or my **entire** life?"

Apparently she took that as a flip response, and copped an attitude. **Not** a good idea to treat me, or to get me to cooperate with seriousness. I felt like saying, "Every girl but **you**!", or "Do I count the women if the sex was lousy?"

* * * *

> "One's own free, unfettered choice, one's own caprice, however wild
> it may be, one's own fancy worked up at times to frenzy... What man
> man wants is simply **independent** choice, whatever that
> independence may cost and wherever it may lead."
>
> -- "Notes from Underground" (1864)
> Fyodor Dostoevsky

Do you know **my** definition of "foreplay"? "Hi, my name is Yogi."

A lover told me one time, that she just wanted to be "friends" from now on. So I said, "Sure." So, since we're just friends now, would she mind introducing me to that really hot friend of hers? And the answer was a curt, "No!"

Some friend, huh?

Q: How do you make a woman come?
A: Women can come??

Q: How do you make a woman come?
A: Who cares!

Now what kind of a moron – Yogi waves his hand from the back bench – could think they could **not** be torn apart by women when they find out he thinks he can define "sexual immorality" as he sees fit – on-the-fly, as it seems?

Well, um, you see.... "Not exactly" is my response to that charge.

You see, I treasure the tryst, adore the assignation, and love the liaison. The surprise, the spontaneity, the pleasant unpredictability, the rawness, the innocent impulsiveness, the validation of life, and joy. A prelude to nothing, but perhaps a fond memory, or an entree to a more long-term garden of earthly delights.

Dating is for young guys. I'm a mature adult now. I don't date anymore -- I just go straight to fucking.

And I don't apologize. I'm old enough to know what I want, and a reasonable game plan on how to get it. But that kind of lifestyle has had its costs and disadvantages and dangers, unfortunately. And I finally had to raise what I considered my reasonable "standards" as far as women go. The standard used to be: they had to say "yes".

I remember one chick I linked up with one afternoon, out of the blue, After hanging together every day for less than a week, I found out she had a vicious, aggressive, and threatening temper that spooked the hell out of me. It was the first time a woman had ever actually made me physically afraid -- from a voice mail message she left me, of all things.

I called her up immediately, and directly and diplomatically told her that it was over, and why. Honestly, and straight up, with no candy-coating, or blame.   Sincere discussions are hard to fake, making their truth and believability obvious. There is absolutely nothing to be gained from any standard other than the full, clear, honest, and open truth, delivered with sensitivity, and understanding, and the distortion of self-interest either pointed out honestly, or filtered out, or cleansed.   There is no substitute for honest, effective communications, no games, no bullshit, and the valuing and respect for other people's feelings, no matter what the problem.

All problems can be negotiated to common ground and worked out, or identified clearly as insoluble or insurmountable differences that are not deliberate, or malicious, but must be identified clearly, talked out, and either addressed or accepted as a ticket out, better uncovered now that ignored, allowed to grow and fester, or denied to no one's ultimate benefit. There's still the possibility of fun, though, and no honest connection between two people is ever a waste of time, no matter how doomed its future may be.

I later joked to friends that at four days long, it was either the shortest relationship I've ever had, or the longest one-night stand.

You have to accept things. You grab anything positive you can. You value what is valuable, and you accept that you did your honest best, until calling it quits was obvious to both of you as the next step. Nothing is permanent, anyway. It's all temporary. And you're gonna wait a long time and waste it as well, if you think perfection is the only sensible choice.

That woman was just a private in the Army of the Walking Wounded. And I taught her to salute a superior officer. So I spanked her ass, and sent her on her merry way. I wish you all the best, but frankly, good luck with your strategy...

It's not disloyalty or betrayal or selfishness to learn the art of knowing when it's appropriate to cut your losses...and **run**.

* * * *

"To love for money all the world is prone:
Some love themselves, and live all lonely:
Give me the love that loves for love alone --
I love that love -- I love it only!"

-- "The Sorcerer" (1877)
Gilbert & Sullivan
W.S. Gilbert

It's not "love" out there, anymore. It's war. And I seem to have found myself a prisoner of the heart in a war that takes no prisoners, and seems to have few winners, in a war with no useful purpose or direction. Replacing loneliness with self-interest, shameless manipulation, predatory behavior, and an astounding ability to switch from love and caring, to pointless and damaging attacks on the undeserving.

You'd think at my age, with all my experience, maturity, acquired mostly by "learning the hard way", that love and relationships and all that would be a piece of cake. But sometimes I think they should replace the word "love" with "war". But there's no surrender for me in this war. I just find it hard to believe that me – charming, laid-back, and accomplished fellow that I am – in my entire life I can't seem to have been able to convince a **single** female of my enduring value, and worth. Which seems a little pathetic to me. And brings tears to my eyes, if I dwell on it too long in my consciousness.

The mother of my daughter, when we were together, constantly and persistently gave me static about every aspect of my character and personality. I wondered that if that's what she thought of me, why she stayed. And I told her that constantly, out of sheer frustration, if nothing else: "If I'm so bad, go find someone else more to your liking."

Then one day she tells me, "I love you, Yogi".

I immediately cried out: "**Love** me? You don't love me. You don't even **like** me!"

"What's enough love?"

"Always a little more than anyone ever gets."

-- Alice Kinnian answers Charly
"Charly" (1968)

* * * *

But all in all, it was Proverbs 4:23 that said it first, well over two millennia ago.

"Guard your heart carefully,
because it is the source of all
that is good in life."

Or, as "Cheap Trick" sang in 1978, in the classic "Surrender" song:

"Surrender, surrender -- but don't give yourself away."

Or, more simply:

Love, but love carefully.

"If I had a souvenir,
Just to prove the world was here,
And, here is a red balloon,
I think of you, and let it go."

-- "99 Red Balloons" (1984)
Sung by Nena
Lyrics by Carlo Karges

Prison Diary 26: "The Hearts Filthy Lesson"
====================

> "Give me an underground laboratory, half a dozen
> atom-smashers, and a beautiful girl in a diaphanous
> veil, waiting to be turned into a chimpanzee, and I
> care not who writes the Nation's laws."
>
> -- S.J. Perelman (1950)
> "New Yorker" article

> "My deathless love will die an early death
> If, scorning me, you stay a faithful wife;
> Why should your husband hoard your loving charms,
> While I get none, grow gray, and spoil my life?"
>
> -- Squire at Arms
> "The Festivals of Bacchus" (ca. 1650)

When ever I see a woman I think I'd like to get to know, and I have the time, and the self-confidence, I approach her and try to start a conversation. Typically, I'll introduce myself, and ask her what her name is.

But it's really strange... Did you know how many chicks in this town are named "Fuck Off"?

* * * *

> "She sees the world through rose-colored glasses,
> Painted skies and graceful romances.
> I see a world that's tired and scared,
> Of living on the edge too long.
>
> Where does she get off telling me
> That love could save us all? Save us all."
>
> -- "Rose Colored Glasses" (1987)
> Blue Rodeo

As well as -- laughably -- thinking the world revolves around them, many women are uncommunicative with their thoughts and wants, and the need to resolve disputes through negotiation. Their mouth seems to be only a "complaint orifice". I inform such pathetic individuals that I can do many things, but reading their minds is not one of them, alas.

I finish off my brief discourse with the words, "I used to be able to read people's minds, but I gave it up. It was too boring..."

* * * *

> "He who does not bear within him
> the seeds of the demonic will never
> give birth to a new world."
>
> -- "Magie: Geschichte, Theorie, Praxis" (1923)
> ["Magic: History, Theory, Practice"]
> Ernst Schertel

I used to have a "dream catcher", that hanging Native Indian decoration of woven string. But eventually I just threw it away.

It wasn't working...

I hear women intimates, friends, and acquaintances periodically complain about being "hit on" or propositioned by strangers, co-workers, and other men. My answer to that is, "The only thing worse than being sexually harassed by men, is **not** being harassed by men..."

My second comment is -- which is meant in all good intentions -- "I thought that women learned early in life how to fend off unwanted approaches."

My third comment is that I thought the woman were used to unwanted approaches by men after a certain point in their life. That they just accepted them as unavoidable, and a minor nuisance. Nothing to get upset about, in general.

Repeated harassment, or crudity, or insults, being an entirely different matter, **of course**.

My fourth comment is, "Let me introduce you to the English word, the minimalist, 'No.'. Then there's the direct, 'Sorry, bud -- not interested...' "

The statement, "Sorry, I have a boyfriend, and I love him." (frequently untrue, or a pathetic exaggeration, I might add, cynically), or the crude, but effective, "Fuck off." delivered with a firm, voice of normal volume, and dismissive, bordering on contemptuous tone.

Of course telling the unwanted guy, "Drop your underwear right here and now, so we can all have a good laugh." **is** close to the ultimate insult. But that's a wild card that could lead to further unpleasantness, or unpredictable results.

And the idea of the first three of my suggestions is to cut the conversation short, the unwanted interaction brief, and the return to personal serenity, and peace of mind as **quickly** as possible.

Your call, baby.

> "Kylie wont kiss my friend Cassandra,
> Jessica won't play ball,
> Mandy won't share her friend Miranda.
> Doesn't anybody live at all?"

Prison Diary 27:
-- "Porn Star Dancing" (2010)
My Darkest Days
No Laughing Matter
================

> "I want you to put up with a little foolishness
> from me. I'm sure that you will."
>
> -- 2 Corinthians 11:1

Wife: "Have you ever **killed** anyone?"

Husband: "Yes...But they were all bad."

> -- "True Lies" (1994)
> Wife on finding out her
> husband is a secret agent.

Alas, poor Yogi. I knew him well. A man of infinite jest...

In this chapter, I hope I don't annoy too many people. But if I do, remember, I'm just returning the favor... In fact, I keep threatening to change my name to "Pearls T. Swine"!

Oscar Wilde once said that the worse thing than being talked about, is **not** being talked about.

Unless it's **me** that's talking about you...

In this chapter I make fun of fools, bosses, shrinks, white alpha-males, and women. Yet of all five categories, I'll only just be accused by **some** women of being anti-woman. See what I mean about women being self-centered...?

> "The girl is gone, and now I'm forced to see,
> I think I'm on my way, Oh, it hurts to live today.
> And she said, 'Don't you wish you were dead like me?'
>
> And I remember the day you left for Santa Monica,
> You left me to remain with all your excuses for everything."
>
> -- "Santa Monica" (2005)
> Theory of a Dead Man

* * * *

The genre of the satire has been around since Ancient Rome, at least. Jonathan Swift, who in 1726 penned the biting -- though now dated -- political fable, "Gulliver's Travels", used the medium of literary political satire. It is still a famed work – though mostly as a children's cartoon utterly bereft of its political pointedness.

Satire – social and political commentary wrapped up as amusing humorous writings – is an interesting and entertaining literary art-form. It is akin to the verbal form of humor known as "wit", pioneered by Oscar Wilde, in the late 1800's.

But in the early years, the use of indirect, implied, and sometimes not-so-subtle biting political commentary or criticism was mandatory for political writings, because it avoided the imprisonment, exile, or other such unpleasantries visited upon an author who was suicidal enough to attempt a direct attack on the ruling elite, in a time where execution required only a parchment, a signature, and a gallows.

A direct attack would be deemed to have dishonored, insulted, or otherwise been criminally rude in criticizing in a society run by rulers with absolute power, or the cherished social orders that kept them in luxurious idle uselessness, and thus would not be tolerated. And so would be justified severe, and immediate retaliation. Thus, the era of self-imposed exile came into existence, amongst those outraged authors who could not restrain their vocal and pointed criticism of outright infamy.

This threat was ever-present in absolute monarchies and other societies unsympathetic to freedom of speech, or liberal political ideas in general. And political writings could only be of a fawning, obsequious sort of nature.

How dreary...

Surprisingly, it was even terminally stupid – included on the negative side of the balance sheet of political offenses – to pen polite, soft-peddled, and reasonable, non-violent political statements **critical** of the government administration -- especially the king and the monarchy system, as well as the social inequities and suffering that resulted.

Most people don't realize that "Gulliver's Travels" was political satire. Apparently satire – which includes many poems we call "nursery rhymes" today – doesn't always age well. It does however seem to recycle into children's books and movie cartoons, and nursery rhymes, centuries later quite well.

Lewis Carroll's "Alice in Wonderland" and "Through the Looking Glass" are interesting in their perspective. Carroll was a mathematician and logician, and he wrote a realistic portrayal of a child's (and a girl, at that) innocent, and uncontaminated perspective – the simple logic of a child – to satirize and skewer the adult world with its bogus and contradictory rules, conventions, and social rituals. And in particular, holding up to ridicule the legal system and its judges for their unfair, arbitrary, capricious, and ultimately meaningless exercise of authority.

But with playing cards as adult human stand-ins – and being a children's book – no one is ever actually harmed, as the Queen willy-nilly orders the execution of anyone inconveniencing or annoying her in any way.

Rabelaisian satire is where instead of poking fun at the target, the reader is the one being made a fool of. Like the Canadian legal system, with the promise of justice, but the joke is on the Canadian public, and the accused, while the Judge sleeps, the Crown and the defense lawyer trade trivialities and misrepresentations, and all three laugh all the way to the bank.

> [And as a footnote, science fiction, particularly apocalyptic and dystopian SF – John Brunner's "The Sheep Look Up" (1972), and the movies "Blade Runner" (1982), and "Brazil" (1985) being prime examples – perform the same modern-day function as satire performed in yesteryear, warning us of psychologically unpalatable horrors presented in the safety and reassuring point-of-perspective

of a warning, a distant future unknowable and changeable. And as
to be expected, a warning unheeded.]

With the introduction of freedom of speech, freedom of expression, and freedom of the press (if you owned one), satire's original purpose evolved into simply a genre of literature that poked fun in an original way at the rich and powerful, the system, or the social mores of the day.

Or if you were a Charles Dickens novel, just serve to bore students, and cause the high school drop-out rate to skyrocket.

Paradox that I am, I like to mix humor and blunt criticism at the same time. So let me start by being blunt, before I let loose with the humor and satire.

In Canada, the maladministration and systemic moral and intellectual bankruptcy of our leadership and institutions is fairly easy to uncover. But I realized that this pathetic situation was not just _de rigeur_, but an entrenched and dangerous threat, more concerned with defending its exposure as an illusion – a mirage – by abusing the powerful existing laws to protect its reputation, rather than carrying out the less difficult demands of its fundamental responsibilities – slowly improving the quality of Canadian society.

Beware the networks of a society's institutions and their claim of perfect harmony, yet who are unwilling to cede the ability to unleash extreme, oppressive, and repressive legal powers against the People, whom they are sworn to bow down before.

And at the same time they dare to begin disarming the people, without disarming their own security forces to the minimum required. An act which threatens the very democracy they are supposed to defend. And they cannot in any way fool many us to believe otherwise.

* * * *

"Idiots (n): a member of a large and powerful tribe,
whose influence in human affairs has
always been dominant and controlling."

-- "The Devil's Dictionary" (1906)
Ambrose Bierce

Amen, Ambrose, my bro', wherever you are. _Via con Dios_ [Bierce disappeared in Mexico around 1906.]

I admit, therefore, that it is Me versus humanity. One issue is that I bore easily. And people don't help. The sleep of reason begets morons...

They say that "insanity" is "doing the same thing and expecting different results". That's **not** insanity, that's stupidity.

One time I went to a psychiatrist for a session. We talked a bit. After I took the knife from his throat, he straightened his tie and announced, "Mr. Shan. It is my professional opinion that you are completely insane."

I responded indignantly, while I started cleaning the gun I had had in my shoulder holster, "Listen doc. I have a hot car, I own a big house, I have a fat bank account. I have two university degrees and a good job."

He looked at me and said, "I said you were nuts, **not** stupid."

I went to another shrink. He was German and his name was Dr. Hans Inyourpocket. I entered his office, and sat down. I looked at him seriously, and said, "Doctor, I'm crazy."

He tells me, "You're not crazy. Every crazy person I've seen usually says something like, 'I'm not crazy! You're the one who's crazy!' "

So I said, "Okay. Good. So I don't have to pay you, because you didn't have to treat me."

* * * *

"A great deal of confusion arises when people decline
to classify themselves as we have classified them."

-- "Public Opinion" (1922)
Walter Lippmann

Ever notice that when you tell a fool he's "painting himself into a corner", all he does is paint faster?

But I swear, I call myself a "moron magnet", because they just seem attracted to me like a magnet, putting themselves in my presence for no reason that I can figure out. They zoom in out of nowhere to darken my world with their idiotic spoutings, like some sort of swarm of mosquitoes spawned in the fetid swamp of intellectual vacancy.

As best I can define it, a moron is like some sort of mobile "black hole" in which all knowledge, wisdom, insight, and learning is snatched a hold of within its swirling vortex, and disappears, never to be seen again.

It's not that they're incompetent. It's just that they're manifestly unsuitable to do anything useful.

After years of annoyance, irritation, frustration, and hand-wringing, I finally thought I'd found the philosophy to help me cope and learn toleration and help restore my serenity. "Everyone's good for something, if only as a bad example," helped me learn tolerance for a while. But they were just too many, and I abandoned this rationalization as impractical in practice.

Then I came up with another guiding principle to achieve serenity in the face of the inexorable march of human stupidity.

"It'd be a lot harder to make a decent buck in this town if most people weren't so stupid".

It was simple and logical appeal to my own sense of avarice and self-interest, once again in the interests of serenity and sanity. I justified any possible charges of "greed", with a curt dismissal of such an idea: I'm sympathetic to the poor, but I just don't want to be one of them.

This worked quite a bit better, until it wore thin too, after some personal misfortunes and down-turns, including this one.

One of the big issues, is that unfortunately there's a time limit to how long I can tolerate being in the **presence** of abject, vocal, egregious stupidity. Stupidity that has absolutely **no** redeeming social value.

I received one commentary from some dud, "Get a life!"

So I punched him in the nose, and said, "Get a face!"

Male stupidity, alpha-male stupidity, management stupidity, female stupidity, gender non-specific common stupidity. I could go on. There was minor amusement in classifying and briefly examining the sub-categories of it, but that was as short-lived as the existence of a mayfly.

And the solution to the problem always seemed to wind up with satisfying fantasy daydreams involving machine guns or other extreme and impractical _quod erat facienda_. Though, it did result in horrifying and uneasy speculation as to the positive aspects and prevalence of war in modern history. Like it being a human version of Darwin's "Survival of the Fittest". Survival of the bulletproof? Survival of those who can duck faster?

Except if this is evolution, it seems to have resulted only in helping perfect war-mongering militaristic societies. Clearly the birthrate of idiots must be enormous, or there's lots of money in the arms trade -- or both.

I mean I have nothing against stupid people in small manageable numbers: village idiots, _fou de roi_ court jesters, clowns -- funny or just amusingly foolish.

But what bothers me about idiots, is that they seem to require an audience.

> [Some even have the utter temerity to be both stupid **and** ambitious. I have been unable to explain this from a logical or scientific point of view.]

And I get really annoyed when I get chosen for the "honor" of their attention. So like a loud guest who's not only an uninvited party-crashing stranger, but rude to boot, I try to have some fun by converting the moron's soliloquy into an uncelebrity roast in the imbecile's honor. And thus, I am a living example of the quote, "The tongue is the only tool that gets sharper with use."

With an imaginary drum roll, the curtain slowly rises, the audience hushes. The play bill today trumpets opening night for – wait for it -- "the slings and arrows of outrageous Yogi." A play in one scene. Followed by the target moron scampering away.

People like that are just so incorrigible, sometimes I can't resist the urge to make a target of them – the butte of my jokes. I mean, it's a public service to checkmate nitwits and send the honorary _piñata_ slinking away from the scene in confusion and disarray after a non-stop onslaught of clever jabs.

A good verbal spanking. Maybe then they'll learn to just stay silent in the shadows, or simply show respect, instead of being a loud nuisance and constant irritation.

As well, the humiliation of an obnoxious nitwit must be done in a public way for maximum effectiveness, alas. I mean otherwise they don't take your criticism seriously, and it just goes on and on...

The trouble is that they just love to talk: not just drivel and inane comments, but stupid suggestions they expect you to immediately acknowledge as brilliant, or implement forthwith as obviously meritorious improvements. Ill-mannered by their brashness, and incorrect, and patently idiotic in their suggestions and assertions, they expect you to laud them with verbal accolades and, more to the point, express agreement with them.

The human brain weighs only three pounds and has between fifty and one hundred **billion** neurons – the gray matter of the brain. The number of inter-connections – the internal white matter of the brain is the axon "wiring" – between these nerve cells consists of trillions of synapses, each firing up to hundreds of times per second to a neighboring neuron. Each neuron is connected to about ten thousand neurons through the axon's synapses.

Out of the interactions between multiple neurons stimulating neighboring nerve cells through an extended and staggeringly complex sequence, produces what we call consciousness.

Or stupidity, as the case may be...

> "The danger is, you'll become something like a moron. You'll be incapable of learning from most people in the world, no matter how much experience they may have in their particular areas that may be much greater than yours."
>
> -- "Secrets" (2002)
> Daniel Ellsberg

* * * *

I **love** women. I enjoy their company. And they're just so soft and cuddly, as I've said. A friendly, innocent, sweet female just makes me melt. "Sugar and spice," and all that. I love being around women, in general. And I don't just mean in a sexual way -- okay?

Women are perfect -- well, except that they nag and needle, are self-centered (which makes them thoughtless, insensitive, inconsiderate, and sometimes impolite), communication-deficient (in other words, they expect you to read their minds), bitchy, because that's their baseline psychological state, or bitchy because another man has mistreated them (**not** me, get it?).

Other than that, they're just **perfect**...

And I admit it. I like cheap sex. As opposed to marriage, which I would call "expensive" sex... When people ask me if I'm married, I respond, "No. I tried it once – I didn't like it." There's a reason they called the marital state "wed**lock**"! Locked in and locked out.

My marriage was so bad, I wouldn't even call it a marriage. It was more like an occasional one-night stand...

When a friend was getting married, I sent the groom-to-be a sympathy card. That's when I found out that brides have no sense of humor.

"You hate women!" goes one unimaginative, slanderous retort. "No, I **love** women. I hate **bitches**," is my dismissive retort. And I have no time for women who have just made a clear statement indicating that they're not interested in getting down with me...

"You think women are stupid!" goes another _non sequitor_.

"No, I think **you** are stupid. Your gender has **nothing** to do with it."

"Now why don't you go somewhere and find a man you can cook something for?" I add contemptuously.

"Men are just interested in **one** thing." goes another common dead-ender of a line.

To which I retort, "No doubt to get away from **you** as quickly as possible..." With the follow-up crack, "By the way, have you ever had a **second** date?"

I'm older, mature and experienced, know what I want, and what I'm looking for, and sort of expect the same in a woman of the same age group that I'm attracted to, as I take the risk of rejection that hitting on them entails.

Rules of the Game, bro'. Alas...

So I follow up with a joking remark -- which I know is subject to serious misinterpretation and don't really care at this point if it is.

"I'm too old for tiresome games, like 'playing hard to get'. Listen honey, we're both **way** past our 'Best Before' dates, so let's get real. I know I'm not perfect, but I have certain qualities and abilities. But I know what to get you for Christmas – a mirror!"

Followed by the _coup de grace_: "Is it something about **me**, or have you just decided you're OK with the fact that you're going to die alone?"

Sigh...

And if they still haven't stalked off in disgust, then it's obviously time for the atomic bomb:

"Ya know, with a pick-up like you, I used to need you to wear a paper bag during sex. But then I realized that 'doggie style' is simpler and much more fun! But seriously, I am attracted to you, and sex is not a tool or weapon – which is **so** adolescent. So how about we cut the crap, and get down?"

And finally, crudely put, but well intended in meaning:

"Fuck or fuck off, princess."

* * * *

> "If men are to be precluded from offering their sentiments on a matter which may involve the most serious and alarming consequences that can invite the consideration of mankind, reason is of no use to us...and dumb and silent we may be led, like sheep to the slaughter."
>
> -- General George Washington (1783)
> speech to officers

In a slow, romantic cadence, to the tune of "Are You Lonely Tonight?":

> "Are you lonesome tonight?
> Are you lonely tonight?
> Are you easy, will you please me, toniiiiight!"

Have I ever cheated on a girlfriend? It doesn't count if the sex was bad, right?

I've always just wanted the love of a good woman. Or ten bad ones.

And then they're the women that keep apologizing that they're "so" fat, when you're fucking her,

her legs spread wide. Listen, don't keep reminding me, it's gross enough. Let me try not to think about it and try to develop erotic feelings about sex with you, and then figure out an escape route after the sex is over."

One girl, she says, "I'm sorry I'm so fat. It must be like fucking the Titanic..."

"Nah, thousands of people wanted to get inside the Titanic. I was thinking more along the line of the size of the iceberg that sank the Titanic. And don't talk with your mouth full, OK?"

* * * *

"It just shows that there are poets
everywhere and practical jokers, too."

-- Phyllis
"Monkey Planet" (1963)
Pierre Boulle

Then there's my list of pickup lines.

"Hello. Having a good day? It would be better with **me** in it. [Pause] How about thinking about running with me if you're single? Or want to be?"

or,

"Hi, there, sweetie! I'll bet your father told you **never** to talk to strangers. But, I'm not a stranger – I'm just **strange**..." [Smile, chuckle, and twist your face to give a funny smile at this point.]

Gauging from the reaction, and looking for the laugh or -- even better -- a smile, you _ad lib_ it from this point. "So, if you think I'm cute, and that I might have a chance, let me take a few seconds and introduce myself. How about we go get a cup of coffee on me, no strings attached, and no expectations except politeness, and friendly talk? So how about it, sweetie?"

I don't hustle women under 25. No, guys, not **age** -- I.Q.

Actually, it **is** a joke, in that I don't reject "simple" girls, or seek out "intelligent" woman. I find "simple" girls delightful in their innocence – joyful and happy. There's a minimalist beauty in a personality that is uncomplicated, straightforward, and "real".

I don't mind if a woman finds **me** fascinatingly intelligent, though...

* * * *

"All the vampires walkin' through the [San Fernando] Valley,
Move west down Ventura Blvd.
And all the bad boys are standing in the shadows,
All the good girls are home with broken hearts.

And I'm free! I'm free fallin'. Yeah, I'm free. Free fallin' "

-- "Free Fallin' " (1989)
Tom Petty

"I would recognize the top of your head **anywhere**!" I joked to one lady-friend. She took it well, as the joke it was intended to be. She also knew me well.

"You're beautiful when you're stupid," was a line I sand-bagged one girl with, who **started** the insult-fest first, and thus took my counter-attack well.

Or you ask a woman if she uses make-up. When she says, "Yes," you respond: "When does it start working?"

But women who deride men as, "Only thinking with their little head," not only are carping an ancient, tired line that men don't disagree with, and is nothing novel or new. The ludicrous part is that they are basically expressing resentment that the little head ignores **them**.

Who would complain about being head-turning attractive to men to the point that men are helpless to resist? But all humor aside, physical attraction based on looks is how men first perceive a woman. ("The eyes have it," he quips.) Sure, it's superficial, but our reaction to the visual image is initially all we have to go on. So what?

When I first met a new girlfriend of my son's, I took one look at her – she was absolutely gorgeous – and soon asked her nonchalantly, "How many car accidents have you caused?"

Without missing a beat, she answered, "Eleven."

She **knew** exactly what I was talking about, and didn't even react as if it was a strange question. A model, she looked hot, and I didn't particularly think she was some sort of stats freak...

Which reminds me: there are some women who confuse "discrimination on the basis of sex", with "discrimination on the basis of ugly."

Do you remember when the Canadian government sponsored an anti-AIDS ad, with the slogan for women to use: "No glove, no love."

I never got to hear in person that rather rude, insensitive remark from a prospective bed-mate, but if I had, I would have retorted, "No lay, no stay"!

I swear, it just goes on and on. You'd think we could just declare a permanent truce in the gender war and get on with something useful and interesting – like fucking more often or better! I'd consider surrender, just for that!

But at my age, I'd rather just curl up with a good book or a bad girl.

* * * *

> "For he can prophesy, With a wink of his eye,
> Peep with security, Into futurity,
> Sum up your history, Clear up a mystery,
> Humor proclivity."
>
> -- from "The Sorcerer" (1877)
> Gilbert & Sullivan
> W.S. Gilbert

And then there are the alpha-male "know-it-alls". Ah yes, a Fool's thoughts -- the briny well that never runs dry. **They** act like rude ass-holes to women, and us normal polite guys have to take the flak.

These alpha-male "know-it-alls" are either wannabee "know-it-alls" who missed the starting gate, or delusional ego-maniacs. And they're naturally verbally loud and aggressive male **white** guys with some sort of ingrained racial or gender-based sense of status entitlement, and an unquenchable need for attention, adulation, and self-promotion, rather than the ingrained sense of meritocracy and earned accomplishment that they **should** have.

And when you contradict them gently, politely, and diplomatically, instead of discussing the issue calmly and reasonably, they get mad, insulting, and loud. Well how about that? Ill-mannered, unreasonable, loud, and obnoxious -- what an **impressive** combination...

Frankly, I consider such Neanderthals as an uncouth splatter of rancid ketchup in the glorious _bouillabaisse tableau_ of post-modern intellectualism.

They're a gratuitous reminder of humanity's limitations, individually, and collectively. God's making fun of us, the sick bastard.

And to add insult to injury, I remind the reader of the morons you **also** have to contend with – and one of the two is going to be your boss.

Yes, it's true, I **do** like a good, well-crafted insult. And I insult friends for the humor value only, obviously. Insulting enemies invites an immediate beating. No future in that...

That's why I needed to write this book – sort of a mass production way to insult the deserving. One-on-one was just not spreading the message fast enough. And renting a stadium was too expensive.

Filling it! Now, **that** would be easy.

So, here I sit. In jail, of all places. At my age, the last place I would have imagined ending up in. Me being such a stand-up guy, and all.

But then again, as Oscar Wilde said, the only thing worse than being talked about, is not being talked about.

I've always modeled myself to be the witty and funny heterosexual version of Oscar Wilde – and apparently the arrest and disgrace Oscar suffered has been included for good measure, alas.

However, even in dark times like these, I feel it's important to maintain a sense of humor – dry, like a good martini. Dry, black, deadpan, even a little macabre. The cynicism of a man who's survived too much.

And yes, it's true: I'm a bite – I mean a bit – of a wit, and have a somewhat sharp tongue. And combined, they make for an interesting combination, and an interesting social and spiritual dynamic.

But the pay needs improvement...

* * * *

"In human beings, attachment is the driving force of behavior...
For most of us it is present throughout our lives... We may also
also attempt to satisfy the lack of the human contact we crave
by various other means, such as addictions, for example. ...

Much of popular culture, from novels to movies to rock or country
music, expresses nothing but the joys or the sorrows flowing from
satisfactions or disappointments in our attachment relationships."

-- "Scattered Minds" (1999)
Dr. Gabor Mate

There are no stupid questions. There are however, stupid questioners...

When someone calls me a "know-it-all". I snort contemptuously, "No, I'm not a know-it-all, I'm just a 'know-more-than-you' ".

"Yogi -- you're arrogant, argumentative, and opinionated," said one fool. "I'm not arrogant," I retorted deadpan.

"You have an answer for everything, don't you?" said one inferior, as a rather pathetic excuse for losing an argument with me by running out of retorts. I answered simply "Yes," with a big smile, after a brief pause, for effect. He just couldn't admit defeat, being of higher rank...

Or in terms more easy for the powerful but incompetent to understand, "I stand on my reputation. You, sir, stand on your feet".

Or: "If nature abhors a vacuum, how come your ears are so far apart?"

I was once in a "Mexican stand-off" with someone over a monetary matter. I made a suggestion to resolve things that would prevent a "fast one" or "sleaze play" on his part and, ensure a reasonable, negotiated, and fair resolution of the problem, rather than continuing the stand-off we were stuck in.

Rejecting my reasonable suggestion out of hand, he said, "Do I look stupid?"

"No," I responded. "You **look** ugly. You **act** stupid."

The sad part is people don't mind being stupid, they just mind when you notice....

And some questions are just guaranteed to get you in trouble, no matter what you say. Like when your woman asks you, "Does this dress make me look fat?"

"No, baby, it's your body that makes you look fat. Don't blame it on the dress."

* * * *

> "In fact, the final episode (a much-underrated encapsulation of the ["Friends"] show's recurrent themes) has an almost Waughian valence in its savage mockery of the loss of self and soul among the cosmopolitan heathen."
>
> -- Michael Hirschorn (1999)
"Slate" on-line piece

Did you hear about the Canadian Forces Intensive Foreign Language Course? It's a day long, and teaches you how to say "We surrender!" in 87 different languages.

Not including French.

Or did you hear about the Ottawa Valley country boy. He lost his virginity to Daisy the Cow. And he thought "Moo! Moo!" from the cow, was "More, more!" Actually, it was cow talk. She was saying, "I've had better."

You know that because of the desolateness of the permanently frozen Arctic, polar bears have a huge territory. Something like the equivalent of a 75 miles square. And a polar bear wanders around his icy, snow-covered territory looking for anything that moves to eat, in a lonely, solitary existence.

Did you know that Canadian Arctic research scientists after years of study of the solitary creature, have found out that there are only two words in the entire polar bear vocabulary: "Eat" and "Fuck"?

So what do you do if you're up North, and you run into a polar bear?

Hope he's hungry!

* * * *

"I am going to put myself to sleep now for a bit longer than usual. Call the time 'Eternity.' "

-- Jerzy Kosinski (1991)
suicide note

You've heard about the legend that there are no snakes in Ireland because St. Patrick chased them all out?

Nah. They were just looking for better company...

I once had a ceramic coffee cup. On the bottom was stamped "Made In Ireland", with "Ireland" spelt wrong...

* * * *

"It's clear that you are Christ's letter, written as a result of our ministry. Your are a letter written not with ink, but with the Spirit of the Living God, a letter written not on tablets of stone, but on a tablet of the human heart."

-- 2 Corinthians 3:3

I looked at the Crown Prosecutor, and said calmly, with icy contempt, "Your remarks are illogical and foolish, and provably untrue. Why don't you shut the fuck up, you unethical, incompetent, corrupt boob."

"Please direct your remarks to me – the judge – only," the Judge quickly interceded.

"Well, I would, Judge, but I'm trying to avoid a contempt of court citation"

I wanted to become a stand-up comic. But everyone keeps telling me to sit down...

* * * *

"Irony is a particularly useful stance in therapy.

> Seeing the world with ironic detachment is similar to what the Buddhists tell us to work toward – a giving up of attachments or rigid beliefs that get in the way of directly experiencing the world."
>
> -- Dr. Liz Margoshes (2003)
> "Village Voice" interview

Being a jail-bird, I have a lot of celebrity company: Socialite Paris Hilton, young actresses Lindsay Lohan and Mischa Barton, rock singer, Amy Winehouse, and public spectacle, Mary Kay Letourneau.

Andy Warhol got it wrong. In the future, everyone will not be **famous** for 15 minutes. They'll be **in jail** for 15 minutes.

Times up! Someone let me out!!!!!

And then there's an intellectual "celebrity", whose name was Dr. Henry Morgentaler, a Canadian, and a hero now, but not when the Quebec Provincial Government, was persecuting and jailing him relentlessly and vindictively, in the early to mid-1970's. It was a shameful and disgraceful chapter in Canadian judicial history documented in detail, and remembered by few.

He was acquitted by three juries on the **same** charges before someone in a position of authority realized Quebec Provincial – biased Roman Catholic -- judges were exposing Canada to disgrace and ridicule. The law was changed, making a jury acquittal the absolute end of legal proceedings.

What was perfectly legal behavior in Canada was outlawed hundreds of years ago in the United States as not "legal process", but a relic of dictatorship, not qualified to exist in a modern democratic republic, or even any government operating under the Rule of Law.

The officials who were doing this kind of thing in the U.S. at the time, had to flee for their lives to Canada, or were summarily executed when the American people realized that a musket ball through the head was the only way to stop this kind of outrage by government officials, in the process of the replacement of the entire Government, criminal code, and system – the American Revolution, celebrated every July 4th.

It soon after became well-established that government corruption and injustice could be discussed for hours using reason and logic without any effect. But a bullet in the brain stopped official misconduct in seconds.

It was a simple matter of the efficient use of time, and the relatively small number of policy-making government officials in a highly centralized absolute monarchy that made cleaning up government misbehavior related simply to how long it took to reload a flintlock rifle a few dozen times, followed by the voluntary mass resignation and flight of the rest of the retrograde government officials.

Most of these former officials, unfortunately were never found again, but were last heard about attempting to leave the U.S. without settling outstanding criminal charges of treason and other capital crimes. The modern prison term, "Dead Man Walking" referring to convicted prison inmates soon to be executed, was apparently a linguistic evolution of the original 1776 term, "Dead Man Running".

* * * *

> "But some, besides allegiance to their original error, possess I know not what fanciful interest in remaining hostile not so much towards the things in question as toward their discoverer. No longer being able to deny them, these men now take refuge in obstinate silence, but being more than ever exasperated by that which has pacified and quieted other men, they divert their thoughts to other fancies."
>
> -- Galileo Galilei (1615)
> letter to Grand Duchess Christina

And then on the other hand is your boss. Your manager. A master of expressing with great flourish the trite or the trivial, masquerading as insight!

When a newly promoted manager took over our department at work, she got her new team together for a meet-and-greet to introduce herself. At some point she mentioned that we should be aware that she had difficulty accepting compliments.

"That's okay," I piped up, "We won't give you any!"

Then, there was this other boss I had at work – a totally unqualified technically obsolete aging white male boob, who had the connections to keep his job. When I was late with a report that he had to pass onwards to **his** boss, he informed me that I was making him look bad to his boss.

I assured him that he didn't need **me** to look bad. That's when I re-confirmed that bosses have no sense of humor.

The second best compliment I've ever received came from a guy with a computer science degree, a fine man, and a former boss with enormous talent himself – he wrote a thousand lines of telecommunications software code in one day!

He told me after a group restaurant outing, privately, one day that, "I was the smartest person he'd ever met."

"Yeah, me too...", I would have jokingly responded, if it wasn't such a gracious compliment. I thanked him profusely. You'd never get a complement like that from a white boss or colleague.

* * * *

> "How in order that we may harvest some fruit from the unexpected marvels that have remained hidden until this age of ours, it may well be if in the future we once again lend ear to those wise philosophers..."
>
> -- Galileo Galilee (1612)
> letter to one of his patrons

I used to (and still do) consider myself an intellectual, ideologue, and polymath. But the audience simply wasn't there, there was not much demand or respect, and the pay was lousy. So I switched to being a wit, or a wag. The pay is still lousy, but the audience is slightly bigger, and the emotional rewards are so much greater.

Then again, maybe you should just call me (shades of Oscar Wilde, again) Sebastien Melmoth...

Then again, literary humor is **so** _passé_. "Machine Gun Shan" has a nice ring to it, and sounds so much cooler. I think I'll stick to that.

I have to make a joke of everything. In my life, it's either laugh or cry...

> "Give them the third-best to go on with. The second best comes too late, and the best never comes."
>
> -- Robert Watson-Watt (ca. 1950)
> Chief Scientist, U.K. Radar Project

Prison Diary 28: Ode to the Sacred Bean
               ====================

> "We are **all** in the gutter, but some
> of us are looking at the stars."
>
> -- "Lady Windermere's Fan" (1892)
> Oscar Wilde

> "A spider dosed with a little alcohol weaves a drunken
> web. If stimulated with caffeine, she will build one
> which is a model of engineering precision. With
> mushroom [hallucinogenic] drugs, she builds one
> circular strand with a couple of spokes, then hangs
> in the center, a spider god in a spider universe."
>
> -- "The Ring of Ritornel" (1968)
> Charles Harness

Since a century or so ago, many blues, jazz, and finally rock musicians believed that smoking marijuana improved their instrument playing skill in live performances. There's a well known link between writers and their art -- Malcolm Lowry, Hemingway, Fitzgerald, Capote, and the Lost Generation -- and excessive, and even abusive, alcohol consumption.

But after a career of three decades of heavy drinking, I finally quit in early 2006 -- and haven't missed it a bit. I was a happy drunk, but I **did** miss out on a lot of reality. But ya can't cry over spilt beer. And more importantly, it's never too late to change.

And boy, it's a pleasant surprise, to find that it wasn't so bad dealing with the ups-and-downs of life, the pains, and disappointments, rather than just sleep-walking through them with the help of booze. And I never really considered boozing as an aid to writing, though I **have** written while drunk -- my well-received "Paris Diary" series comes to mind.

But I've drunk coffee since I was 22, and never found anything particularly special about it, other than it helping me to wake up, and stay awake, for work in the morning.

Phoenix rising. Phoenix ascendant. The unexpected discovery of a simple pleasure in the barren, desolate environment of my provincial jail-cell home. For me, the discovery was my simple morning coffee. My morning home-brew, improvised "espresso", to be exact.

Oh, magic espresso! My espresso magical!

Need becomes habit. Habit becomes reflex. Reflex becomes instinct. Instinct becomes ritual. Combine ritual and instinct, and you have the sacred. And so was born the sacred jailhouse ritual of my wonderful magic espresso, writing creativity aid _extraordinaire_.

To be a balanced "whole", you need to take care of body, mind, and soul. To my ever-lasting astonishment, my morning jail-cell espresso ritual, fed and nourished my writing soul -- the author hidden within me.

Detail and attention to detail. Elaborate, precise, and obsessive. The ritual. With every meal, every jail inmate received a square blue foil-lined packet of about a teaspoon of instant coffee. Hot water was available in the morning, brought around by the inmate server in a pouring jug (in OCDC – the Lindsay jail had a hot water dispenser).

The instant coffee package was rather stingy in coffee powder contents -- perhaps a level teaspoon of coffee powder. So being a dedicated espresso _aficionado_, and inveterate improviser and experimenter, one morning I stumbled upon the magic formula, itself. I had saved five instant coffee packets.

So one morning before eating breakfast, I added them *all* to about a quarter cup of hot water, quickly stirred to mix, and then downed the home-brew espresso substitute on an empty stomach for maximum caffeine absorption effect. Then I waited five minutes before eating breakfast to prevent the food from absorbing the caffeine liquid in my stomach, and allow it be absorbed as quickly as possible, unhindered.

It was a quite tolerable substitute for real espresso. And being in jail, it was the only game in town. But the real surprise discovery was when I sat down at the table in my cell after breakfast and began my normal pass-time of writing. But this time after consuming my first home-brew espresso, I kept writing, and writing, and writing.

Something was different. The ideas flowed from my brain in an unstoppable, continual, torrent. Yet manageable. Not so fast that I'd lose track of them. Out from my brain, through my hand, and onto the paper. On and on, one after another, subject after subject, concept after concept, idea after idea without cease, without end. It was amazing. My caffeine-electrified brain ground out its intellectual product continually for the entire morning without stop, as page after page filled up with my pencil scrawlings without pause.

I was forty-two in Paris, France when I first discovered the delight of espresso, and had loved it ever since as the ultimate form of coffee: very hot, undiluted caffeine-concentrated coffee. I was now forty-eight in a jail-cell in Ottawa when I discovered how solitude, quiet, and a barren environment combined with my improvised, caffeine turbo-charged, super-espresso synergized to turn me into a veritable writing **demon**.

And thus was born my daily espresso ritual, and my transformation into the Coffee King of Cell-block D (or whatever cell-block I happened to be residing in at the time). I would trade with the other inmates practically **anything** for their coffee packets – usually deserts -- and would get two to five coffee packets depending on the desert, and the competition if there were any other coffee lovers on the range.

And soon the sacred coffee ritual was born and perfected. Instant coffee has the lowest caffeine content of all forms of coffee, which is why I needed five packets per espresso.

(Brewed coffee has the highest caffeine content – about double that of instant coffee – other than espresso – which requires a special purpose machine, which is not particularly common in North American households.)

Holding it by the top, I would shake the square packet, then tap it at the bottom end of the foil square, to loosen, then settle the instant coffee powder to the bottom of the packet.

Then I would slowly, and with excessive carefulness, tear a thin strip off the top edge of the coffee pouch, which was a foil-lined paper packet. I would empty the instant coffee powder into my cup of hot water, I would brush the empty packet of every last particle of instant coffee. I would repeat with the next packet – it would go on-and-on, as your typical ritual is want to do.

And it kept me busy, focused, and, most delightful of all – **writing**.

Long live Super-espresso!

* * * *

> "Triviality destroys at once robustness of thought and delicacy of feeling. No enthusiasm can flourish, no generous impulse can survive under its blighting influence."
>
> -- "The Right to Privacy" (1890)
> Samuel Warren and Louis
> Brandeis, "Harvard Law Review"

There is caffeine (1,3,7-trimethylxanthine), theophylline (1,3-dimethylxanthine), and finally theobromine (3,7-dimethylxanthine). All in different combinations and amounts in coffee, tea, and chocolate. And they vary their effect, and synergy by varying combinations.

Caffeine is water-soluble, quickly absorbed, and evenly distributes itself around the entire internal human body, and easily crosses the blood-brain barrier. Maximum concentration in less than an hour. Taking vitamin C (ascorbic acid) immediately before drinking coffee helps absorption. And the amino acid taurine is supposed to potentiate the effect of caffeine. About a gram a dose – it's an ingredient in carbonated energy drinks like "Red Bull".

Theobromine (in cacao powder, cocoa butter, and thus chocolate) inhibits caffeine metabolism, potentiating its action. Theobromine is either an additive or synergistic effect combined with caffeine.

It was fun to experiment with combinations, but I wasn't sure of any conclusive results or definitive observations.

Tea: leaf powder, like in tea bags is twice as strong as full-leaf tea; black tea infused for four minutes has double the caffeine (up to 100mg) than tea infused for three minutes.

Agonist, and antagonist modes of pharmacology. Agonist (like amphetamine), more addictive. Caffeine believed to be an antagonist adenosine re-uptake inhibitor.

> "Sometimes the System goes on the blink
> And the whole thing, it turns out wrong
> You might not make it back and you know,
> That you could be well -- oh -- that strong,
> And I'm not wrong. Aaaah!
>
> ...
>
> 'Cause you had a bad day,
> You're taking one down,
> You sing a sad song, just to turn it around."

-- "Bad Day" (2005)
Daniel Powter

Prison Diary 29: The World According to Yogi
=======================

"['T]is an unweeded garden
That grows to seed; things rank and gross in nature
Possess it merely. That it should come to this!"

-- Hamlet
"Hamlet" (1602)
Act I, Scene II

"If suffer we must, then at least
let's suffer from the heights."

-- "Les Malheureux" (1855)
["The Unhappy"]
Victor Hugo

To attempt to fully, or even partially, understand someone is a pretty tall order at the best of times.

To understand *ourselves* is difficult enough, much less the examination of such an elaborate, intimate and, by definition, infinitely complex task of probing, exploring, examining, dissecting, and analyzing the intimate thoughts, feelings, emotions, mental processes, reactions, and personality of a different human being.

The determination of an individual's way of thinking and reacting to others, the way they respond and interact with other individuals, groups, or their environment, their feelings, overt and covert, hidden, or expressed comprise their behavior, personality and identity.

Emotions are just one – and a particularly complex one, at that – of the human psyche. To break down and characterize an individual, one must examine not just their present state, but their history and evolution over time. The complete story comes from your history and your experiences. To do otherwise, is to read the last page of a book and expect to know the full story. But that's really all most people do in judging someone.

You must look at their actions, reactions, and experiences -- of infinite possible length and complexity. In documenting accurately a person, and exploring and analyzing their personality, your efforts are degraded by the given limitations of time, interest, and importance, availability of data, the flaws and limitations of memory, insight, self-delusion, denial and dead-ends in the attempt to get inside their head, their heart and the personality.

A person is the sum of all their actions, their experiences, their thoughts, emotions, ideas, and knowledge, and their morality, as judged by their speech and actions.

* * * *

"It is no use trying to sum people up. One must follow hints,

> not exactly what is said nor yet entirely what is done."
>
> -- "Jacob's Room" (1922)
> Virginia Woolf

My favorite color is red, with black a close second – it's a political thing. My favorite season is spring -- the end of damn winter.

I hate camping – even the idea of it. Throwing away 60 centuries of civilization for a few days, just so you can wipe your ass with leaves and get covered with bug bites. No thanks.

I'm a city boy, born and bred. And I love it. Thus my original attraction and espousal of urban guerrilla warfare to counter a racist, anti-leftist society in deed if not word, as it emerged from the 1950s, the date of my birth being 1959.

I used to like crunchy peanut butter, but in early 30's I changed for some reason, and now prefer smooth. I like my steaks very rare -- "blue". I love smoked salmon and _baguette_. I love shrimp and lobster. My favorite dish is cubes of beef – eye of round – marinated in olive oil and lemon juice, with onions and tomato chunks, spiced with _herbes de Provence_, salt and pepper, and broiled to perfection. I love curries, especially chicken, with rice.

I like ginger ale soda pop. My favorite ice-cream is chocolate, or rocky road. Or vanilla. Or…hell, I like it **all** in a pinch! I like to read – only non-fiction now. History or technical, are my favorites. I like to bike. I like to lift weights.

I'll always turn my head for a blonde – I don't know why. I may turn away immediately, but I'll always turn my head and give her a scan by reflex – strangely enough.

My fave sexual position is now "missionary" -- so I'm looking right into her eyes throughout, face-to-face, from inches away. My favorite position used to be "doggie-style" when I was younger, and less experienced.

But enough of that.

The following are the important influences that affected and formed my -- mostly -- early personality -- my late childhood, and adolescence, and early adulthood.

* * * *

> "If you reveal your secrets to the wind, you should not
> blame the wind for revealing them to the trees."
>
> -- "The Wanderer" (1932)
> Kahlil Gibran

Influential Books

Adolescent/children's classics and fantasy book series:

"Listen and I'll Tell You" (1962), "The Wind in the Willows" (1908), the 7 "Chronicles of Narnia" series (1950-1956), "The Hobbit" (1937), the "Lord of the Rings" trilogy (1954-1955), "The Black Cauldron" (1965), the "Tintin" series of cartoon story books by Hergé (translated from French), "Bored of the Rings" (1969) (satire), "The Outsiders" (1967), "A Connecticut Yankee in King

Arthur's Court" (Mark Twain, 1889).

Evolution from Children's fantasy, to my teenage Science Fiction phase:

"Dune" (1965), "A Canticle for Leibowitz" (1959) (post-nuclear apocalypse), many of SF great, Robert A. Heinlein's books, including "Farnham's Freehold" (1965) (post-nuclear apocalypse SF), "The Red Planet" (1949), and "The Moon is a Harsh Mistress" (1966) (a lunar colony revolts from Earth's oppression), "Alas, Babylon" (1959) (post-nuclear apocalypse).

On to non-fiction: the education of a revolutionary:

"The Trial of the Germans" [Nuremberg],
"The Fire-Bombing of Dresden",
"The DuPont Blaster's Handbook",
"The Politics of Heroin in Southeast Asia" (1972) (Alfred W. McCoy)
"The Anarchist Cookbook" (in Grade 7 -- 1970 – got it from a friend of Finnish extraction whose father had given him a copy),
"Seize the Time" (Bobby Seale, one of the leaders of the Black Panthers),
"Explosives and Homemade Bombs" (a technical, "serious" book)
"Total Resistance" (1965) (Swiss Army guerrilla warfare "manual"),
"Steal This Book" (1972) (Abbie Hoffman -- counterculture leader),
"Do It!" (1970) (Jerry Rubin, counterculture leader),
"In the Spirit of Crazy Horse", (Peter Mathieson)
"The Night of the Long Knives" (1972), [the Nazi internecine purge of the S.A. by the S.S.]
"Serpico" [Fighting NYC police corruption.].

Grade 12: the first Shakespeare play I understood – thanks to a great teacher's explanations – and consequently loved:

"Macbeth". And then (the second, by far, though) "Julius Caesar".

The beginnings of a weapons expert:

"The Saturday Night Special" (1973), Robert Sherrill
"Improvised Modified Firearms" (1975), J. David Truby and John Minnery
"Improvised Explosives Handbook" (1969), Dept. of the Army
"Small Arms of the World", Edward Ezell
"The Technology of Killing" (1995), Eric Prokosch
"The Secret that Exploded" (1981) (H-Bomb secrets uncovered by Howard Morland)
"Born Secret: The H-Bomb, the Progressive Case, and National Security" (1981) Alex DeVolpi et al.
"Nuclear Weapons Databook" Volume 1 (1984), Ted Cochran et al.
"Weapons of Tomorrow" (1984), Brian Beckett
"The Samson Option" (1991), Seymour Hersh.

The making of an "underground chemist":

"Cannabis Alchemy", by D. Gold
"Snow Blind", by Robert Sabbag
"The Big Brother Game" and "The Street Drug Game", both by Scott French
"Marijuana Chemistry", by Michael Starks

"Psychedelic Chemistry", the classic, written pseudonymously by Michael Starks as "Michael Valentine Smith",
"Phenethylamines I Have Known and Loved", by Alexander Shulgin, Ph.D.
"Tryptamines I Have Known and Loved" (Dr. Shulgin again) (in which my name appears in the Footnotes section).

Adult readings, and the consolidation of my principles:

"The Perfect War: Techno-war in Vietnam" (1986)
"A Bright Shining Lie" (Neil Sheehan) (Vietnam)
"The Provisional IRA" (1989)
"The Dirty War: Covert Strategies and Tactics Used in Political Conflicts" (1999)
"The Peter Principle" (1969), the "way of the moron" in modern life
"The Curve of Binding Energy" (1974), 1950's nuclear weapons designer tells (almost) all.
"The Secret History of Nuclear Weapons" (1988) (by the late Chuck Hansen)
"And the Band Played On" (1987), a fascinating analysis of the emergence of AIDS, made into an excellent movie by HBO in 1993
"The Making of the Atomic Bomb" (1986), Nuclear weapons
"Dark Sun: The Making of the Hydrogen Bomb" (1995), includes detailed Fat Man A-bomb plans".
"The Swords of Armageddon" (1994), nuclear weapons doc; on CD or microfiche by the late Chuck Hansen
"Shoah: The Script" (1985): Script of the documentary film on the Holocaust.
"People of the Lie", brilliant, incisive book by shrink Scott Peck on modern life and lies
"Odysseus in America" (2002), brilliant. By another shrink: Homer's "Odyssey" as an example of Post-Traumatic Stress Disorder.
"Blank Spots on the Map" (2009), the Great Satan: the American Economic Empire triumphant

The four Gospels of "The New Testament" of the Bible. My happy re-birth.
The chapter "Ecclesiastes" in "The Old Testament" of the Bible.

Highly Influential Newspapers and Periodicals

"New York Times" (now seriously in decline and decay)
"Washington Post" (the new leader in newspapers, surpassing the NYT)
"Los Angeles Times"
"The Economist" (on the verge of decline)
"The New Yorker" (for the intellectual in me scratching to get out)
"Esquire"
"New Scientist" (the current view on science; entertaining and presentable to a wide audience)
"Scientific American"

All-Time Favorite Rock Albums

"Jesus Christ Superstar" (1970) (my **first** introduction to rock music at age 11 – and I loved it!)
"Tommy", the Who (the second of four "rock operas" ever)
"Sgt. Pepper's Lonely Hearts Club Band", the Beatles (1967)
"Tubular Bells", Mike Oldfield
"Hot Rocks 1964-1971", the Rolling Stones (1971)
"The Wall", Pink Floyd (the third [of only four that I know of] rock opera)

"Strange Animal", Gowan
"Aja", Steely Dan
"To Live and Die in L.A.", movie sound track, by Wang Chung (1985)
"Wide Awake in America", U2 (1985)

Rock Groups

The Beatles (from 1966 on), "The Rolling Stones" (1964 - 1980), "The Who", "Pink Floyd" (the four greatest rock bands of all time), Bruce Springsteen, "Steely Dan", "Chicago", "The Sex Pistols", Tom Petty and the Heart-Breakers, "U2", "Pearl Jam", "The Tragically Hip", "Blue Rodeo", John Mellancamp, "The Cars", "Lighthouse", Terry Jacks.

   Honorable Mention:

Max Webster, Gowan, Cyndi Lauper, "Chrissie Hynde and the Pretenders", Harry Chapin, "The Pukka Orchestra", "The Clash", "Kiss", Alice Cooper, Billy Joel, Jim Carroll Band, King Crimson, Todd Rundgren, Phil Collins, Daniel Lanois, Alanis Morrisette, Avril Lavigne, Steve Earle, Billy Talent, Eric Clapton ("Layla", Derek and the Dominos, Eric Clapton, and Jim Gordon).

Favorite Music Videos (youtube.com)

U2 – "One", "The Sweetest Thing", "Sunday, Bloody Sunday"
David Bowie, Freddie Mercury – "Under Pressure" (Live)
Sarah McLachlan – "I Will Remember You" and "Building a Mystery"
Alanis Morrissette – "You Oughta Know"
Chrissie Hynde and the Pretenders – "Brass in Pocket"

All-Time Favorite Movies

#1: "Apocalypse Now" (1979);
#2: "Blade Runner" (1982) (original release with Harrison Ford narration);
Tied for #3: "Brazil" (1985) and "Highlander" (1986).

Favorite Movies

[You'll notice that I stopped going to movies in 2003.]

Best of the Best:

"A Beautiful Mind", "Seven", "Heat", "Pulp Fiction", "Terminator 2", "Point Break", "Goodfellas", "Colors", "Predator", "Aliens", "To Live and Die in L.A.", "Gandhi", "Network", "Dr. Zhivago", and "Lawrence of Arabia".

2000's: "A Beautiful Mind" (2001), "Lord of the Rings" trilogy (2001, 2002, 2003).

1990's: "Sunshine" (1999), "American Beauty" (1999), "Tu as crié, 'Let me go!' " (1999), "Hurly-Burly (1999), "Saving Private Ryan" (1998), "Men in Black" (1997), "Trinity and Beyond: The

Atomic Bomb Movie" (1995) "Seven" (1995), "Heat" (1995), "12 Monkeys" (1995), "Pulp Fiction" (1994), "Ace Ventura, Pet Detective" (1994), "True Lies" (1994), "Mr. Jones" (1993), "And the Band Played On" (1993 HBO cable), "Falling Down" (1992), "Army of Darkness" (1992), "The Unforgiven" (1992), "Terminator 2: Judgment Day" (1991), "Point Break" (1991), "Pump Up the Volume (1990), "Europa, Europa" (1990), "Goodfellas" (1990).

1980's: "The Cook, the Thief, His Wife, & Her Lover" (1989), "Talk Radio" (1988) (Dir. by Oliver Stone and starring Eric Bogosian), "Colors" (1988), "Predator" (1987), "The Princess Bride" (1987), "Full Metal Jacket" (1987), "Aliens" (1986), "Ferris Bueller's Day Off" (1986), "To Live and Die in L.A." (1985), "The Terminator" (1984), "The Blues Brothers" (1984), "Starman" (1984) (starring Karen Allen and Jeff Bridges), "The Falcon and the Snowman" (1984), "The Day After" (1983 -- made for TV movie), "Scarface" (1983), "Gandhi" (1982), "Fast Times at Ridgemont High" (1982) "Outland" (1981).

1970's: "Alien" (1979), "All That Jazz" (1979), "Time after Time" (1979), "Midnight Express" (1979) (Screenplay written and Dir. by Oliver Stone), "Wizards" (1977), "Network" (1976), "Three Days of the Condor" (1975), "Roller-ball" (1975), "Shampoo" (1975), "Chinatown" (1974), "Deliverance" (1972), "The French Connection" (1971), "The Omega Man" (1971), "Five Easy Pieces" (1970). "Jesus Christ Superstar"

1960's: "Charly" (1968), "Planet of the Apes" (1968), "Dark of the Sun" (1968), "Cool Hand Luke" (1967), "Alfie" (1966), "Dr. Zhivago" (1965), "Von Ryan's Express (1965), "Dr. Strangelove" (1964), "Laurence of Arabia" (1962), "Spartacus" (1960).

< 1960: Jimmy Stewart's movies, Bob Hope's comedy movies, the "Bedtime for Bonzo" series of movies, "Ben-Hur" (1959), "Cat on a Hot Tin Roof" (1958), "Scrooge" (1951), "_Quo Vadis_" (1951) "Monsieur Verdoux" (1947) (starring Charlie Chaplin; talkie, not a silent movie), "The Lost Weekend" (1945).

Actors:

Sean Penn ("Fast Times at Ridgemont High", "Taps", "Colors", "The Falcon and the Snowman", "Thin Red Line", "Hurley-burly", "The Assassination of Richard Nixon") is god.

Kevin Spacey ("Seven", "American Beauty", "Hurley-burly") is god.

Directors

Oliver Stone ("Talk Radio", "Midnight Express", and the rest forgettable),
Michael Mann ("Miami Vice", the 1980's *TV* series, "Heat"),
John Hughes (too many to list; consistent in his plentiful work of satisfactory, above-average
              quality),
Ridley Scott ("Alien", "Blackhawk Down"),
David Lean ("Dr. Zhivago", "Lawrence of Arabia", and the lesser "Ryan's Daughter").

Movie Endings:

The original "Planet of the Apes" (1968), starring Charleton Heston, has the best movie ending of all time -- with Heston, playing "Taylor", a cynical U.S. astronaut finding an atomic blast-scarred Statue of Liberty half-sunken in the sand, and realizes to his horror and rage that his spaceship has actually returned earth, an earth destroyed by nuclear war, where humans are mute savages and apes rule the world, now at the top of the evolutionary ladder.

Sequels or Remakes of Movies and Songs That Were *Better* Than the Original:

"Aliens" was much better than "Alien", which was just "good",
"Scarface" (1983 version over original 1932 version),
"I Shot the Sheriff" (Bob Marley song, re-recorded by Eric Clapton),
"Blinded by the Light" (Bruce Springsteen song, re-recorded by Manfred Mann),
"My Way" (Frank Sinatra song, "re-recorded"/butchered by the "Sex Pistols": "Tonight I
    killed a cat").

Movie Miscellanea

The Difference Between a "Great" Movie, a "Classic" Movie, and a "Masterpiece".

Great: a great plot, with good acting, **and** one great memorable or insightful line (e.g. "Alien",
    "Army of Darkness", "Talk Radio", and "Mr. Jones")

Classic: memorable line after memorable line (e.g., "Aliens", "The Blues Brothers", "The
    Princess Bride", and "Cat on a Hot Tin Roof")

Masterpiece: superb script, with abundant memorable and quotable lines, a spell-binding
    plot, social commentary or insight, great acting; and if possible, a great soundtrack, with
    cinematography as the icing on the cake (e.g., "Apocalypse Now", "Colors", "Scarface",
    "To Live and Die in L.A.", "Gandhi", "Pulp Fiction, "Seven", and "Planet of the Apes",
    "Network", to name a few).

    You might be aware that except for "Colors", none of these movies have a sex scene in
    them.

Comic Strips

Calvin & Hobbes ("Spaceman Spiff!")
Doonesbury (political satire at its best, and in comic strip form, of all things)
Dilbert (the high-tech corporate world satirized brilliantly)

Pornography

The best porn scene is the pool table scene in the 1970's movie, "Garage Girls".

"Debbie Does Dallas" stars the most beautiful blonde porn star – the "Debbie" -- ever. It is also from the 1970's.

From the mainstream Hollywood world, "Sunshine" (1999) has the best, most erotic, sex scene in a mainstream movie that I've ever seen – in the office sex scene in communist Hungary.

TV shows:

Childhood & Adolescence: "The Beverly Hillbillies", "Get Smart", "Star Trek", "Kung Fu" (starring David Carradine), "The Man from U.N.C.L.E", "The Wonder Years".

Adulthood: "Alf", "The Wonder Years", "Miami Vice", "The Simpsons" (until about Season 10), "Law and Order" (until about the time the actor playing Lenny Briscoe died), "Homicide: Life on the Streets", "Arrested Development" (2003-2006, Fox), "War" (1983) (documentary; 7 part mini-series by Gwynne Dyer), "Seinfeld", "Friends", "Fresh Prince", "Cheers", "The Family Guy".

Funniest "The Simpsons" scenes:
----

1) Prof. Frick rolls his latest invention into a conference room: a sarcasm detector. Obese collector's item store owner comments derisively, "Oh, now that's **really** useful...", at which point the sarcasm detector promptly explodes.

2) In the second most hilarious Simpson's scene, (one of the "Halloween" specials, and a parody of the Charleton Heston SF movie, "The Omega Man") nuclear war has just broken out. As the obese collector's item store owner, is lurching along the side walk stuffing his face, a Russian ICBM equipped with a neutron bomb suddenly appears heading directly along the sidewalk towards him. A second before it is about to hit him, he blurts out "Oh, no! I've wasted my life."

One Shot Wonders (but *still*, I'd call them "glorious meteors"):

Jim Morrison only wrote **one** worthwhile song -- and it's a masterpiece. It was "The End".

John Carroll and his classic, "People Who Died".

Wang Chung and their beauty, "To Lie and Die in L.A.", doing the soundtrack of the movie of the same name.

Alanis Morrisette's classic of bitter love lost, "You Oughta Know."

Billy Talent and his outstanding song of Vancouver heroin addiction, "Fallen Leaves".

The Clash and "Julie's in the Drug Squad" (though they did do other good stuff, this one song is far above the rest).

Director Oliver Stone, of his many, many movies made, only made two good movies -- and they're masterpieces -- "Talk Radio" (1988) and "Midnight Express" (1979)). Every other of his many movies that I've seen either sucks, or is mediocre.

Berg ("Talk Radio") was an actual Jewish talk radio personality who was machine gunned to death by the Order, a right-wing white supremacist urban guerrilla group (in the mid-1980's), using an Ingram MAC-10 machine-pistol (a good choice of firearm). The scene at the end, when Berg in a soliloquy, berates his radio audience, is stellar.

Favorite Title of an Academic Article

"SIGINT Used by Anti-State Forces" a 1983 academic paper on the IRA vs. the British Security Forces and the war/counter-war on IRA radio-controlled bombs.

Music:

Walking paradox that I am, I must admit that I'm a sucker for rock duets. I don't know why, but I just am and I admit it. There's just something about this musical form that I find just **magically** beautiful.

"Under Pressure" (David Bowie/Freddie Mercury; live version), "_Sous Le Vent_" ["In the Wind"](Celine Dion/Garou).

I am also embarrassed to admit that I like **two** Celine Dion songs ("My Heart Will Go On" (1997); "The Titanic" movie theme song) and the French language "Sous Le Vent" ("In the Wind") a duet with another French-Canadian singer, Garou).

Fiction Books I've never read, but decided in late middle age, that I must, by the time I die
----

"Ulysses" by James Joyce
"Gravity's Rainbow" by Thomas Pynchon
"The Catcher in the Rye" by J.D. Salinger
"Zen and the Art of Motorcycle Maintenance" by Robert Pirsig
"Under the Volcano" by Malcolm Lowry
"Atlas Shrugged" by Ayn Rand
"A Confederacy of Dunces" by John K. Toole
"Stand on Zanzibar" and "The Sheep Look Up" by John Brunner
Philip K. Dick's SF short stories
The "Foundation" trilogy (actually 4) by Isaac Asimov
The "Hitchhiker's Guide to the Galaxy" trilogy (actually 5) by Douglas Adams

Religion

Jesus is **cool**.
Zen Buddhism

My Motto

1977-1980:   "We are stardust. We are golden. And we've got to get ourselves back to the Garden." ["Woodstock" (1969), by Joni Mitchell]

1980-1984:   "_Viva La Muerte!_" ("Long live Death!")
             [Slogan of Spanish fascist Falange Party (ca. 1934)]

1985-2006:   "Oh well, whatever, never mind."
             ["Smells Like Teen Spirit" (1991), Nirvana, Kurt Cobain, Krist Novosolic, Dave Grohl]

2007–present: "My motto: Apocalypse Now" ["Apocalypse Now" (1979), John Milius]

Favorite Fish

I don't care what anyone thinks, I **adore** anchovies – those salty little critters. I eat them plain, right out of the can or jar. But they're **so** expensive...

But it's salmon and mayo, for a cheaper – and affordable – sandwich lunch.

Older U.S. Presidents I Consider Particularly Totally Evil

Interestingly enough, most of these assholes (and the next set) are celebrated as great Presidents...

Andrew Jackson: Violator of the constitution, genocide of the Indians

Theodore Roosevelt: Imperialist dog, genocide of the Indians; described by Noam Chomsky as "one of the most extraordinary racist, raving lunatics of contemporary history".

Eisenhower: the Butcher of Iran, the Butcher of Guatemala, began wholesale butchery around the world using the Cold War as an excuse, that made the U.S. the Great Satan.

> However his parting speech about the military-industrial complex being a profound threat to democracy were valid and insightful, but were ignored, and came true with a vengeance.

Recent U.S. Presidents I Consider Particularly Totally Evil

They all waged evil wars and financed huge arms build-ups (only employing white people!) costing hundreds of billions to trillions of dollars while calling the opposition Democratic Party "tax and spend" liberals. All three topped the evil of their predecessors, as hard as that is to imagine.

The end of the Bill of Rights, and the end of the American Republic, and the sacrifice of American Democracy on the altar of fascist capitalism.

Oh, God, no...

Richard M. Nixon: (1968 - 1974)

The butcher of Vietnam, the war criminal of My Lai, the butcher of Argentina, the butcher of Chile; resigned in disgrace with a popularity of 15% (or was it 5%?), to avoid certain impeachment for criminal acts and treason to the Constitution. Most of his Cabinet and top officials were indicted and jailed for politically-motivated criminality.

> "You must remember this: Nixon was evil. If we forget that, his crimes will cease to be crimes; evil will become instead the way of the world."
>
> -- "NYT Book Review" (1996)
> Gary Taylor

Ronald R. Reagan: (1980 - 1988)

Dumb as a beagle; supported Cambodian dictator Pol Pot, armed the evil Iraqi dictator Saddam Hussein and supplied him with a chemical warfare capability. Armed and trained Osama bin Laden and many others who would later form the Al-Qaeda anti-Western guerrilla group.

Reagan was described as "a technically ignorant and incompetent [President]" by Swedish Government adviser Milton Leitenburg (1984), and numerous others.

Reagan was the Butcher of Grenada, the Butcher of El Salvador, the Butcher of Nicaragua, the Butcher of Guatemala, the Butcher of Afghanistan, the Butcher of Libya, the Butcher of Iraq (the Kurds and the Shi'ite Marsh people murdered by Saddam Hussein), and the Butcher of Iran.

He trained Osama Bin Laden and Al-Qaeda's future membership during the 1979-1988 Soviet-Afghan war that cost 1.3 million civilian lives, just to punish the Soviets for Vietnam. He was thus directly responsible for the 2001 "9/11" World Trade Center slaughter, and the even higher death toll of American soldiers in the Second War on Iraq.

Falsely took credit for "defeat of communism" (news, no doubt, to Red China) which was known by the U.S. to be on its way down to economic collapse since the 1970's, and which finally occurred in 1991 **after** Reagan left office in 1988.

Towards the end of the Reagan administration in 1988, the "Washington Post" published a 22 page list of Reagan administration and appointees convicted, under indictment, charged with illegal, unethical acts, or corruption, or resigned as a result. As a reporter in a "Doonesbury" cartoon asked: "Have you no shame?".

It still confounds my sense of logic and reality that out of a population of 250 million Americans armed with two hundred million firearms, only one person tried to murder him out of office, and that guy, Warren Hinckley (with the Jodie Foster hang-up) was just completely nuts -- that is, it was a non-political assassination attempt.. (And to top it off, Jodie Foster turned out to be a lesbian!)

Which doesn't really count from my purist ideological perspective. Though I wouldn't quibble about the result itself, if it was successful, at least...

And on it went: Reagan *tripled* the National Debt with his arms build-up.

He systematically packed the federal judiciary with corrupt, unqualified incompetent (including James M. Ideman of the Central District of California) evil flunkies, and right-wing moral frauds with no respect for the Constitution or Bill of Rights, in attempt to sabotage judicial independence and impartiality, and roll-back the freedoms rightfully gained in the 1960's under the Supreme Court Chief Justice Earl Warren. Packed the Supreme Court with absolute fascist pigs: Warren Burger, William Rehnquist, Antoninus Scalia, Clarence Thomas.

He failed with the Supreme Court nomination of fascist pig Robert Bork, a whore remembered forever as the dirt-bag of Watergate's "Saturday night massacre". Unbelievably, Nixon had ordered Special Prosecutor Archibald Cox – whom he had earlier appointed -- fired for doing his job investigating Nixon too well, rather than doing the "white wash" that was expected of him.

The two highest ranking members of the Justice Department as honorable men, in turn, one after another refused, and resigned for their refusal.

Bork was third in rank, and gladly did Nixon's biding, and fired Cox.

The moral elite of the Democratic Party remembered the Son-of-a-Bitch Bork over a decade later, and handily put the kibosh on Bork's nomination as Supreme Court Justice. The Good Guys won for once, and Bork had to slink away, a humiliated, embittered slug of a man.

George W. Bush: (1996 - 2004)

Dumb as dog shit, a lazy, worthless ex-drunk, he was a draft-dodging coward, a moral and intellectual fraud, liar, and war criminal. Mass murderer of thousands of U.S military personnel. Authorized mass torture. Evil personified in the human shell of an utter moron.

Front man for fellow war criminals and traitors to the Constitution, Vice-President Dick Cheney, and Sec. of Defense, Donald Rumsfeld (both ex-Nixon officials), Condinigger Rice, amongst other cabinet and high official appointees.

Doubled the National Debt. When George W. Bush was inaugurated in 2001, the National Debt stood at $5.7-trillion. He ran up more debt faster than nearly all of his predecessors combined: just under $4.9 trillion in his eight year dictatorship. The National Debt stood at $10.6-trillion on the day Barack Obama took office. [Source: CBS News]

Lost the 2nd Iraqi War (2003-2011), begun on his lies and deceit, and cost 4,500 American lives on a deliberately phony and concocted Al-Qaeda linkage (whose World Trade Center attack only claimed 3,000 lives!) and his phony Weapons of Mass Destruction _Casus Belli_.

But who was dumber or more evil?: him or the people who elected "the worst president in the entire history of the U.S." not once, but twice!

I fuck your mouth, George. I fuck your mouth.

"Civilization secures three chief advantages: greater wealth, greater
safety, and greater variety of experience. Whether, in spite of this,
there is a real -- that is, a moral -- advance is a question impossible
to answer offhand, because wealth, safety, and variety are not absolute
goods, and their value is great or small according to the further values
they may help to secure. [...]

An animal's existence is not improved when made safe by imprisonment
and domestication; it is only degraded and rendered passive and melancholy."

-- "The Life of Reason" (1954)
George Santayana

Prison Diary 30:   The "Commie from Mars" says, "_Via con Dios_"

> "Listen, Jesus, do you care for your race?
> Don't you see we must keep in our place?
> We are occupied --
> Have you forgotten how put **down** we are?
> I am frightened by the crowd,
> For we are getting much too loud,
> And they'll crush us if we go too far."
>
> -- Judas Iscariot
>   "Heaven on Their Minds" (1970)
>   in "Jesus Christ Superstar"
>   Tim Rice

> "The Truth is simple, but the Way
> of Man is hard. [...] Thou art God.
> Know that and the Way is opened."
>
> -- Michael Valentine Smith
>   "Stranger in a Strange Land" (1961)
>   Robert A. Heinlein

Via con Dios.

The Spanish _Adieu_; the Spanish "Farewell". "Go with God", literally. Or "May God be with you [on your trip/journey]."

I think that the Bible is the "coffee-table book" of all coffee-table books. Everyone in your typical Christian-majority country of the Western World has one laying about, or on a shelf in their abode somewhere. But – so it seems – just for looks or appearance's sake, because nobody ever seems to bother reading it – or even *ever* having read it -- pretty much like your average "just for show", display-to-impress-the-company coffee-table tome.

There are 30, 20, 33, and 22 pages in Matthew, Mark, Luke, and John, respectively – the Gospels that give four versions – written by four of his thirteen apostles – of the life of Jesus Christ. Four versions of his life and one version of his message.

The reason I say how long each gospel is, is because it doesn't take that long to read one – or all four, one after another -- and not long at all, if – like the Bible I got in jail from a nice – and lovely – young lady from the Salvation Army – you get a Bible translated into modern, every day English.

Older Bibles, like the King James Version, are totally incomprehensible, unreadable – in archaic English – and thus useless, as far as I'm concerned. Though it used to be even worse, since no commoner could read the Latin bibles that used to be the only available (until the King James Version), and widespread literacy is a relatively recent phenomenon.

And, by the way, reading the cartoon strip or sport sections only doesn't count...

* * * *

> "How much is enough, when your soul is empty?
> How much is enough, in the Land of Plenty?
> When you have all you want, And still feel **nothing** at all,
> How much is enough?   Is enough?"
>
> -- "How Much is Enough?" (1991)
> The Fixx

The Bible used to be hand copied by monks in the early Middle Ages. It was slow and arduous work, and the number produced was small, and of restricted distribution, eliminating access by the common people.

It was also written in Latin -- not to mention that most people in the Middle Ages were illiterate. To remedy the first problem, followers of John Wycliffe began the first English translation of the Bible around 1380, near the end of the medieval period.

Johann Gutenberg, who by chance happened to have studied the metallurgy of soft metals, such as lead, came up with the idea of movable cast lead alloy type printing, which lead him to invent the first printing press in 1455 in Maintz, Germany. The first book he printed was the (now called) Gutenberg Bible. In Latin, unfortunately.

But hand-copying was out, and with Gutenberg's invention of the printing press, widespread availability of knowledge, and translated Bibles were soon in.

Martin Luther had a simpler idea: his "press" was a nail that he used to hammer his "95 Theses" to a German cathedral door in 1517, an act of bravery that he survived only by good luck and the right timing, but which started centuries of bloody religious conflict in Western Europe.

All due to the irredeemably corrupt, rotten-to-the-core Catholic Church. Personally, I've always thought killing for religion was not only ludicrously hypocritical, but atrocious and primitive. Killing for ideology, on the other hand, I used to consider quite admirable, and much more advanced.

As I used to put it, you can argue with someone for hours, and convince them of nothing. Or put a bullet in the back of their head, and not have to. I no longer subscribe to such rather harsh ideas.

In 1521, Martin Luther, in hiding, translated the Bible into German from Latin. The English King James translation, ordered by the English King James I, was completed in 1611, and finally broke the monopoly of Latin bibles.   Several revised versions were produced over the centuries.

Still, no one seemed to pay much attention to Christ's message. War got bloodier and bloodier, and the majority of people eventually lost all faith in organized religion.

In 1961, Robert A. Heinlein, one of the greats of science fiction, wrote the novel, "Stranger in a Strange Land", the title an Old Testament quote, about a Martian who comes to earth spreading strange ideas, but comes to a bloody end, like the Jewish one.

In 1969, The Who wrote the first "rock opera", a smash hit entitled "Tommy". Both "Tommy" and "Stranger" were thinly disguised versions of a messiah and his message of truth, that is ultimately rejected by the people it is meant to save.

And finally in 1970, Tim Rice (lyrics) and Andrew Lloyd Weber (music) wrote the second rock opera, "Jesus Christ Superstar", another smash hit, consisting of a modernized interpretation of the final days of Jesus Christ, with stellar – I refuse to say "heavenly" – lyrics, set out to absolutely lovely soft rock music. (And **NOT** the nauseating soft rock composed since then.)

It spawned a movie, with Uzi submachine gun-carrying Roman soldiers wearing shiny chrome-plated helmets, an Israeli Air Force Star fighter jet fly-over, a complex black Judas, plagued by doubts, a rag-time singing King Herod, and a bare-bones set, all filmed on-location in Israel.

* * * *

I may compare myself to Saul, who went
to seek his father's ass and found a Kingdom."

-- Heinrich Schwabe (1843)

But what most people don't realize is that Jesus was the world's first **communist**.

Don't believe me? Listen to the Man himself:

The Rich

"Jesus said to his disciples, ' I can guarantee this truth: It will be hard for a rich person to enter the kingdom of heaven. I say **again**: it is easier for a camel to go through the eye of a needle than for a rich man to enter the Kingdom of God.' "

-- Matthew 19:23-24

" 'Stop storing up treasures for yourselves on earth... Instead, store up treasures for yourselves in heaven...' "

-- Matthew 6:19

" 'No one can serve two masters. [...] You cannot serve God and wealth.' "

-- Matthew 6:24

" ' Other people are like seeds planted among thorn bushes. They hear the Word, but the worries of life, the deceitful pleasures of riches, and the desires for other things take over. They choke the Words so that it can't produce anything.' "

-- Mark 4:18-19

"Mary said, ' My soul praises the Lord's greatness! ... He sent rich people away with nothing.'

-- Luke 1:46 & 53

" 'One day the beggar died, and the angels carried him to be with Abraham. The rich man also died and was buried. He went to hell, where he was constantly tortured. .... Abraham [said to the pleas for mercy from the rich man] "Remember, my child, that you had a life filled with good times, while [the poor man's] life was filled with misery. Now he has peace here, while you suffer." ' "

-- Luke 16:22-23 & 25

Income Redistribution: Sharing with Others

"Jesus said to him, ' If you want to be perfect, sell what you own. Give the money to the poor and you will have treasure in heaven. Then follow me!' When the young man heard this, he went away sad, because he owned a lot of property.

-- Matthew 19:21-22

" 'Give to everyone who asks you for something.' "

-- Luke 6:30

" 'Sell your material possessions, and give the money to the poor. Make yourself wallets that don't wear out! Make a treasure for yourself in heaven that never loses its value! In heaven thieves and moths can't get close enough to destroy your treasure. Your heart will be where your treasure is.' "

-- Luke 12:33-34

" '[W]hen you give a banquet, invite the poor, the handicapped, the lame, and the blind. Then you will be blessed...' "

-- Luke 14:13-14

Possessions & Materialism

" 'Whoever has two shirts should share with the person who doesn't have any. Whoever has food should share it too.' "

-- Luke 3:11

"He told the people, 'Be careful to guard yourselves from every kind of greed. Life is not about having a lot of material possessions.' "

-- Luke 12:15

Anti-capitalist: "Capitalism is Theft !"...and Capitalism is Lies!

[Not inventors, scientists, artists, music composers and rock musicians, and other deserving folks. Corporate bureaucrats and executives, politicals, rip-off sellers who make more than 1% profit, jack up prices with air conditioning, advertising, excessive features or variations, etc.)

"Jesus went into the temple courtyard and threw out everyone who was buying and selling there. He overturned the moneychangers' tables and the chairs of those who sold pigeons. He told them, 'Scripture says, "My house will be called a house of prayer," but you're turning it into a gathering place for thieves!' "

-- Matthew 21:12-13

" 'The master praised the dishonest manager for being so clever. Worldly people are more clever than spiritually-minded people when it comes to dealing with others.' Jesus continued, ' I'm telling you that although wealth is often used in dishonest ways, you should use it to make friends for yourselves.' "

-- Luke 16:8-9

"Class War", Leading to Equality Throughout Society

"Jesus called the apostles and said, 'You know that the rulers of nations have absolute power over people and their officials have absolutely authority over people. But that's not the way it's going to be amongst you.' "

-- Mathew 20:25-26

"All the tax collectors and sinners came to listen to Jesus. But the Pharisees and the scribes complained, 'This man welcomes sinners and eats with them.' "

-- Luke 15:1-2

"Jesus said to [the Jewish high priests and the leaders of the People] ' I can guarantee this truth: Tax collectors and prostitutes are going into the kingdom of God ahead of you."

-- Matthew 21:31

J.C.s condemnation of rulers, the powerful, the wealthy, government officials and bureaucrats, lawyers, and the "pious" elites – such as the Pharisees – are quoted in other chapters of this book.

" 'Peace on earth!' was said. We sing it,

And pay a million priests to bring it.
After two thousand years of mass
We've got as far as poison gas."

-- "Christmas: 1924" (1924)
Thomas Hardy

Prison Diary 31:   The Lawyer Game
                                                  ===============

> " ' Woe to the Lawyers! They stand in the Doorway
> of Knowledge, but they won't unlock the door,
> and they block anyone else from entering.' "
>
>   -- Luke 11: 52

> "You start a conversation, you can't even finish.
> You talk a lot, but you're not saying anything.
> When I have nothing to say, my lips are sealed.
> Say something once, why say it again?"
>
>   -- "Psycho Killer" (1977)
>   Talking Heads
>   David Byrne, Chris Franz,
>   Tina Weymouth

This missive was hand-written with my own blood as ink, on toilet paper, and smuggled out of the jail's maximum security wing inside my lawyer's ass, then typed onto a TEMPEST-secured laptop and burned onto CD-ROM, which was transferred to an ex-colleague, before being sent out by email to a select list of suspected dupes of my literary powers and ideological magnetism.

OK, OK, I'm exaggerating.

It wasn't blood...

The life and message of Jesus Christ, as covered by the four Gospels (Matthew, Mark, Luke, and John) of the New Testament, discloses that besides preaching love for one's neighbor and fellow man as a way of life -- the *only* way of life -- he also railed and condemned loudly, bitterly, and repeatedly the flagrant hypocrisy of those in power. Those he declared *should* know better, and therefore those from whom more was expected by God.

And what also doesn't seem to be well known, is that he named names: the elite classes of society. Damned by Christ are: the rulers, judges and holders of power, the government administrators and bureaucrats, the religious leaders, the "devout" pious religious community at large, and the rich. And last, but not least, *lawyers*.

And Christ's withering criticism of lawyers is thorough, accurate, and in no uncertain terms. He verbally savages them individually, and as a group, not once, but twice. Only the rich, and those who pursue money are condemned more often than lawyers. Though -- really -- lawyers are just a subset of the latter, aren't they? Thus, they face double damnation.

And, by the way, if crime doesn't pay, how come lawyers are so expensive and many are rich?

Twenty centuries later, how little has changed about the ruling elites and administrators of power, and the pious frauds and their hollow humility and phony morality. And the hypocrisy of lawyers, as they go through the motions -- pun intended -- is manifest, held in contempt by even ordinary non-political citizens.

* * * *

"Jesus said, 'How horrible will it be for you lawyers!
You burden people with loads that are hard to carry.
But you won't lift a finger to carry any of these loads.' "

-- Luke 11:46

You've probably heard this one:

Q: What do you call 10,000 lawyers on the bottom of the sea?
A: A start.

I add a couple more:

Q: How do you tell if a lawyer is lying?
A: His lips are moving.

Q: How do you tell if a lawyer is **thinking** of lying?
A: He's breathing.

* * * *

"All thieves who could my fees afford,
 Relied on my orations,
And many a burglar I've restored,
 To his friends and his relations."

-- "Trial by Jury" (1875)
Gilbert & Sullivan
William Gilbert

It was Clarence Darrow, famed and – exceptionally – a righteous U.S. lawyer of the first half of the 20th Century, who said, "The first half of our lives is ruined by our parents, and the second half by our children."

The second half of **my** life is being ruined by lawyers.

J.C. -- the Big Man, himself -- got it right with his swipe at the legal profession, quoted above. Lawyers are truly the used car salesmen at the car lot of Truth. They always focus on dotting the "i's" and crossing the "t's" of trivial, inconsequential points, while avoiding substantive issues and real work, all the while billing you at $250 - $400 an hour, and trying not to let you see them as one step away from disbarment, should some of their behavior be noticed and exposed.

In fact, the minimum standard for proper lawyering and a faithful practice in service of justice is given by the Prophet Isaiah (1:17), six centuries before Christ:

"Seek justice, stop oppression. Defend
orphans. Plead the case up of widows."

Which part of this quote do lawyers who claim to be Christian or Jews **not** understand? So by claiming to be Christian or Jewish are such lawyers liars, hypocrites, or merely complete moral frauds? Or all three?

Or did they somehow stumble metaphorically into a kind of quicksand that engulfed all their good intentions?

When they can soar with the angels by fighting injustice, most just profit at the pig trough of misery, as the malevolent Power of the State mishandles and abuses those unfortunates caught in its evil clutches.

> "The lawyers, of whose art the basis,
> Was raising feuds and splitting cases.
> Opposing all registers that cheats,
> Might make more work with dipped estates;
> As wer't unlawful that one's own
> Without a law-suit should be known.
> They kept off hearings willfully
> To finger the refreshing fee;
> And to defend a wicked cause
> Examined and surveyed the laws,
> As burglars shops and houses do
> To find out where they'd best break through."
>
> -- "The Grumbling Hive" (1705)
> Bernard Mandeville

* * * *

One pleasant and sunny afternoon, I strode purposely into the lobby of a West End apartment building lobby on my way to drop in on a female friend. On the way to the lobby elevator, I was engaged in conversation by a short blonde with a winsome smile, and a twinkle in her eye. As we chatted amiably, she followed me into the elevator. Just as the elevator approached my floor, she inquired as to whether I wanted to "fool around".

Well, naturally, I did. I mean, such rather generous offers do indeed come in from time-to-time -- though not lately. However, alas, when we arrived at her place, I was dismayed to be informed by the young lady that her entreaty was actually a pitch. And she wanted **money** for her "charms". Well, **that** was not "fooling around", it was fooling **me** around".

Classify that one under Fatalist Rule No. 6: the "Too Good to be True" statute.

But the most I remember about the incident is the come-down from generous offer to sordid commercial proposition. I wasn't just disappointed, I was rather irritated.

It's the kind of feeling you get if you have to hang around lawyers for any length of time... Having a lawyer is like having a whore for a girlfriend. They sort of leave a permanent bad taste in your mouth. Of course, **their** mouth always has a bad taste...

* * * *

> It is Tellurian who is credited with the motto, _Credo,
> qiua absurdum est_: I believe because it is incredible.
> Needless to say, he began life as a lawyer."

-- "Treatise on the Gods" (1930)
H.L. Mencken

But being a man of substance, a man of action, a man of arguably poor judgment and outrageous behavior, a genuine full moon, living-large icon of indiscretion, an ideological and intellectual gypsy, lawyers have been a part of a lot of my life. It's either that, or bad luck...

I used to joke about always having three lawyers on retainer. One to keep me out of jail, one to keep me out of the loony bin, and one to inflict terror on my enemies, pursuing them to the ends of the earth with harassing and bankrupting law suits.

Now in my mind, even a lousy lawyer should be able to get an innocent man off. By definition, a legal system that is fair should acquit an innocent man. Routinely. Preferably speedily. And uninjured.

A *good* lawyer is one who can get guilty guys off. *That* is skillful lawyering, moral questions aside, if there are any. (It is not the purpose of the legal system to convict the guilty. It is to determine whether there is *evidence* proving guilt. Case closed, on the moral question.)

And another issue I've noticed is that they have to "like you" before they'll make any effort to seriously help you; a general problem of Canadian "professionals". It's overdue for Canadian lawyers and doctors and teachers and other such so-called professionals to look it up in the dictionary and start finding out exactly what it means to be called a "professional", and that what they're doing isn't even close to making the grade.

In his most acclaimed song, "I Shot the Sheriff" (1973), by the late Bob Marley, revered reggae singer, the protagonist is caught in a jam by circumstance, and fated to lose in a society that exercises one aspect of its injustice by branding him as undeserving of the benefit of any doubt.

Bob Marley's lyrics explain the outlaw's predicament as a problem many of us eventually find ourselves facing:

> "Reflexes got the better of me
> And what is to be, must be,
> Every day the bucket goes to the well,
> And one day the bottom will drop out,
> Yes, one day the bottom will drop out."

Perhaps a similar idea is expressed in the movie "Falling Down" (1992), starring Michael Douglas, as the anti-hero, an upper middle class, laid off missile engineer, who after a life of conformity, finally decides he's had enough of modern society's bullshit, and decides to live life on his own terms as he takes a long walk across from down-town Los Angeles, westward to Santa Monica, and ultimately to Venice Beach. On his journey, he encounters a string of losers, living in their own meaningless little worlds, seeking to impose on him the very bullshit he has disavowed.

But in disavowing bullshit, he has also unleashed his own uncontrolled anger and resentment, as blithely unaware, he gets into more and more trouble as he walks unknowingly to his doom.

After all is said and done, he is branded an outlaw, but his crime is not malice, or breaking society's rules and mores, but simply the intolerable act of rejecting them completely and substituting his own, flawed judgment for that of society's.

In August of 2007, surely the bottom fell right out of my private little world, and I needed a damn good lawyer – and FAST. After all, the reality was this: front page news, terrorist allegations, piles of automatic weapons trundled out for press conferences, rumors of grenade launchers, heavy machineguns, and 50lb. bags of explosives.

All a complete legal mirage, no question, I knew. But not a mirage I was willing to entrust to a legal aid bottom-feeder or "dump-truck" – as the worst were called by inmates – to dispel. I needed an experienced, aggressive, come-out-fighting attorney. One who was either up-and-coming, or one with a solid reputation and proven track record.

These were dangerous times, and the government was playing hard ball. It was crazy, in my mind. Absolutely ridiculous. But it was suddenly clear to me, I was now a victim swept up in witch-hunt, a confused public uneasy with fear over half-real, half-imagined images of invading terrorist hordes screaming for infidel blood, and their reward of 72 virgins. Shee-et – they ain't gonna find them girls here...

Press sensationalism had combined with a government and police belief that they needed a public display of steely, alert, whip-cracking efficacy on the issue of non-state Islamic political violence to use me as the patsy. I was the **first** non-Muslim caught in their witch-hunt. I was innocent, and they knew it, and now I had to suffer for the cover-up of their mistake. A press conference and a display of a pile of machine guns by the police was the first step of the cover-up. The prosecution took the ball from there. And on it went.

Innocent or not, there was a lot at stake for me, and I needed to get ready for the fight of my life, if that was the game the government was going to play. I needed the Hounds of Hell fighting for me. And I wasn't going to settle for the Beagle from Bavaria.

My second lawyer failed to get me out on bail, so he wanted me to plead guilty to avoid six months in jail wasting away – waiting for my trial – by pleading guilty and getting probation. This was one of the many reasons he was "out", including failing in the outcome of the bail hearing.

I don't plead guilty just to avoid suffering. "Do they still teach 'principles' in Law School? Or did you miss that class?" I asked, a faux-earnest look on my face. He didn't answer, except with a sour look on his mug.

As they say, "Fire 'em, if they can't take a joke", or something. So he was **gone**. And thus my third lawyer was soon hired and on board, and I kept him until the end of the case.

* * * *

> "So I call up my preacher,
> I say, 'Give me strength for Round 5.'
> He said, 'You don't need no strength, you need to grow up, son.'
> I said, 'Growing up leads to growing old, and then to dying.'
> And dying to me don't sound like all that much fun,
> So I said, 'I fight authority, authority always wins.' "
>
> -- "Authority Song" (1985)
> John Mellencamp

"_Fiat justitia ruat caelium_" ["Let justice be done, though the heavens fall."]

I only work when the boss is watching. He turns his head, and I immediately stop what I am doing. It's called it "workplace fucked-over employee catatonia". It's a medical disability. It's just not recognized as one yet, especially when I brag about it to my management.

The reason was that my boss was incomprehensible, inarticulate, useless, and overpaid. I thought he should become a criminal defense lawyer.

I **do** know of at least three classes they have in Law School. "Billable Hours", "Billable Hours", and "Billable Hours".

They **had** a class called "Ethics", but they canceled it because everyone who showed up cheated on the final exam. The others called in sick on exam day, and used a forged doctor's note.

Oh – wait – there's another class I know of, "Eliminating 'Yes' and 'No' from Your Answers".

Ever wonder why you can never get a straight answer to a legal question? I never thought that "fence-sitting" or speaking an obvious "smokescreen" or verbiage was a useful or productive skill. Apparently evading responsibility, by refusing to commit to even a qualified "Yes" or "No" response is considered a worthwhile and acceptable response to a client.

They say someone who defends himself in Court has a fool for a client. But better a fool for a client, than a fool for a lawyer, I like to say.

"It will not...be said, here stood a temple of vast antiquity; here rose a Babel of invisible height; or there a palace of sumptuous extravagance; but here, Ah, painful thought!

The noblest work of human wisdom, the grandest scene of human glory, the fair cause of Freedom rose and fell."

-- Thomas Paine (1796)
open letter to Pres. Washington,
published in a Philadelphia
newspaper

Prison Diary 32:   The Way the Wind Was Blowing
================================

> "The dissenter is every human being at those moments
> in his life when he resigns momentarily from the herd
> and thinks for himself."
>
> -- "In Praise of Dissent" (Dec. 16, 1956)
>   "New York Times Book Review"
>   Archibald MacLeish

> "It is easier to say what loyalty is not that what it is. It is not conformity.
> It is no passive acquiescence to the status quo. ... It is not an ostrich-
> like ignorance of...other countries and other institutions. It is not
> the indulgence in ceremony – a flag salute, an oath of allegiance,
> a fervid verbal declaration. ...
>
> It is a tradition, an ideal and a principle. It is a willingness to
> subordinate every private advantage for the larger collective good. ...
>
> It is allegiance to...the traditions of freedom, equality, democracy,
> tolerance, and the tradition of Higher Law, of experimentation,
> cooperation, and pluralism...born of revolt, [and] flourished on
> dissent..."
>
> -- "Who is Loyal to America" (1947)
>   in Harper's Magazine
>   Henry Steele Commager, Columbia
>   U. historian

"You don't need a weatherman to know which way the wind blows". ["Subterranean Homesick Blues", Bob Dylan (1965)].

The above Bob Dylan line was adopted by the spectacularly unsuccessful young, white, upper middle class, U.S. national urban guerrilla group, who took the name, "The Weathermen" in 1970, as they broke away from the non-violent SDS (Students for a Democratic Society), fed up with the failure of society to change for the better in response to massive, repeated, non-violent Vietnam War protests and action. But the wind had changed direction, and the Weatherman had failed to notice.

And failed to note that it was the wrong direction...

Despite their bombing and arson attacks on government, military, and financial buildings ("Don't Bank on America!" was one slogan as they targeted BoA institutions), but only when they wouldn't injure anyone, their efforts were in vain.

The middle and upper class people turned hard right, Nixon was re-elected and escalated the Vietnam War to unprecedented heights, and they faced a huge military and police monolith that they could not even begin to dent.

* * * *

> "The parting on the left, is now a parting on the right,
> And now the trains no longer run overnight."

sang The Who, in the quintessential and insightful political song, "Won't Get Fooled Again" (1971), with its memorable "screeee-aaam!" of disillusionment, at the end of the song.

And so I faced the changed world of 9/II. Canada had never been attacked by terrorists, but was a staunch ally -- satellite -- of the U.S, who had.

Many Muslims were rounded up over the years in Canada since 9/11, charged, held without bail, and then released completely innocent. One was sent to Syria to be tortured for months, while RCMP personnel released secret documents and phony information to the media, condemning him as a legitimate terror suspect. A campaign that multiplied the sins against Mr. Mahar Arar multi-fold, and were illegal, unethical, and a gross violation of RCMP regulations.

Arar was finally exonerated and compensated with $10 million dollars by the Canadian government. He is still an unhappy man, blighted by his horrendous experience.

And there is still the fact of the gross manipulation of an unwitting -- but compliant -- media, that became an arm of a government campaign of disinformation against a completely innocent Canadian who had done nothing to deserve his torture, much less his total betrayal by the federal police force sworn to uphold the truth and the protection of the innocent.

There have been a couple of *real* home-grown terrorists caught, but they were easily caught, and in the conspiracy stages of their planning. You could say it was unlikely they would have ever amounted to anything devastating, but in their dreams and fantasies. Amateurs.

Though they made for good headlines to keep the terrorism propaganda fires burning. To show that the police were competent and well on top of things. To increase police funding requests, and their high profile.

And to bow down to the American War Machine and Police _apparatchiks_, as part of the -- hint, hint -- successful War on Terrorism. We're doing our job. We're doing our part. We're low-key **and** successful, and keeping things in check.

* * * *

> "It's the terror of knowing what this world is about,
> Watching some good friends screaming, 'Let me out!'
> Pray tomorrow gets me higher, higher, high.
>
> Pressure on people, people on the streets.
> Turned away from it all, like a blind man,
> Sat on a fence, but it don't work,
> Keep coming up with love, but it's so slashed and torn.
> Why, why, whhhhhy?"
>
> -- "Under Pressure" (1983)
> David Bowie, Freddie
> Mercury, and Queen

And then there was me. A name that was Hindu in origin, it could be mistaken for Islamic by ignorant police unfamiliar with an obviously Tamil Sri Lankan name. Tamils comprise one of the largest groups of recent Canadian immigrants. Tamil names are very distinctive and they're very long, *unlike* Arabic names.

Mine I call "fourteen letters" as a joke to whites astounded by its length. But the last letters are "hasan" -- "obviously" Muslim. Really though, the last syllable of my long name is "dhasan" with the "h" silent.

But though Hindu in origin, I'm half-white – from my mother – and **Christian** since my preteen years – by my own choice – impressed by Christ's message of love and peace, though not impressed by the Christians I had to deal with, at **all**.

There was no Koran in the house that they searched -- there *is* a well-thumbed bible, and a small collection of Christian-related books.

There was a poster of Che Guevara on the wall -- leftism is no favorite of Bin Laden -- and many other clues that I was a cultural Canadian of flagrant and obvious leftist political bent. I'm proud of my extensive collection of leftist books on the Sixties Era, its history, personalities, its nature as a cultural phenomenon, and most importantly, its left-wing politics...

There was a bottle of French wine -- not likely in the house of an Islamic extremist.

I could go on, but that's enough to make the point.

* * * *

"Political power, properly defined, is merely the
organized power of one class for oppressing another."

-- "The Communist Manifesto" (1848)
Karl Marx, Friedrich Engels

My skin is light brown -- easily mistaken for Middle Eastern, like any of the thousands of Canadian brownish-hued people. But I've been mistaken for every non-white race under the sun, except Chinese. Black, Mexican, Iranian, Hispanic, you name it.

One thing was definitely true however. I was, in fact, the first non-Muslim swept up in -- a badly run -- witch-hunt for Islamic terrorists. And it was all based on criminal acts through-out the judicial process. A terribly flawed, perversely manipulated legal and judicial institution. The Gatineau Police Force, the Ottawa Constabulary, the Ottawa Police Guns & Gangs Unit, and most particularly, the Crown Attorney's Office, especially two disgraceful, unethical and probably criminal Crown Attorney's.

Along with several unethical criminal lawyers. And a boob of a judge who slept briefly through part of an aborted trial proceeding.

The Ontario Legal System is a laughable sham and a fraud, I say, but the rot is too deep. And no one listens.

"To think of the future and wait, was merely another way of saying one was a coward; any idea of moderation was just another attempt to disguise one's unmanly character; ability to understand the question from all sides meant that one was totally unfit for action; fanatical enthusiasm was the mark of a real man

... anyone who held violent opinions could always be trusted, and anyone who objected to them became a suspect..."

-- "History of the Peloponnesian War" (ca. 410 BC)
Thucydides

Prison Diary 33:   Pantomime Justice
                              ===============

> "Midway upon the journey of my life,
> I found myself within a forest dark.
> For the straightforward pathway had
> been lost."
>
>   -- "The Divine Comedy" (1320)
>   Dante Alighieri

> " 'I will ask my Father, and he will give you another
> helper, who will be with you forever. That helper
> is the Spirit of Truth. The World cannot accept
> him, because it doesn't see or know him. You
> know him, because he lives with you and will be
> in you.' "
>
>   -- John 14:15-17

Errors, flaws, imperfections. They are the essential, inescapable components of human frailty since time immemorial. And so, they are the invariably part of all human endeavor itself -- our ideas and concepts, our concrete plans, our implementations, our constructs -- in other words, all our creations.

The various nuances of ill-thought out design or uncorrected implementations that disturb the serenity of the trained and experienced engineer with their "screams" -- the blatancy of their incongruous corruption of the beauty of perfection -- begging to be corrected, to conform and be a part of the beauty of ultimate, smooth-flowing, sleek, efficient, operational, and functional perfection.

And so it was with the legal system, that like a modern day Jonah I had been swallowed up by a man-made, belching, evil Leviathan.

A legal system that had somehow been permitted to evolve into a mere theater of "pantomime" justice. With the external look, the sound and apparent workings, the mimic and appearance of a sound system of judicial propriety -- that was anything but.

And with the unique perspective of a victim, that was to make its corrupted workings obvious to me, in this unique time, when its inadequacies and weaknesses combined and synergized with the spectacle of invented or exaggerated public crisis and the government's irrational response to it, to produce anything but that which it was supposed to -- justice, through a thorough, thoughtful, and careful process of logical and fair evaluation.

      *   *   *   *

> "Those who, having no passions of their own
> because of no intellect, have spent their lives in
> curbing & governing other people's."

-- "A Vision of the Last Judgment" (1880)
William Blake

The government had decided that it had to be effective and decisive, but the result was that it only had to be seen by the public as such. And so as it manifested itself so pitifully, and in my case, so painfully, in a hollow edifice of a pretend, shadow-boxing, illusion of a justice system, the legal system went through the motions.

The road to hell, they say, is paved with good intentions.

But I submit, that is the **least** expected surfacing one will find today. The road to hell is also paved with bad intentions, surfaced with a culture of lying, varnished with hypocrisy, and the denial that goes hand-in-hand with smug self-satisfaction. As things go to hell, its individual operators' main concern withdraws into self-interest -- financial security, blame avoidance, and cover-up -- and so there develops a culture or lifestyle of contractual avarice, look-the-other-way indifference, and contented, unconcerned feather-nesting official decadence.

The road to hell is paved by forgotten values, empty morals, an abandoned or discarded sense of duty, debased honor, and heroes whose deeds are but dim or forgotten wisps of memory, and half-forgotten dreams.

* * * *

"He who would do good to another must do it in
minute particulars -- 'general good' is the plea of
the scoundrel, hypocrite, & flatterer."

-- "Jerusalem" (ca. 1820)
William Blake

So, as usual I guess, the choice is between sense or nonsense, the sensible or the foolish, the hard road or the easy road, and finally, dedication to the Truth, or addiction to lies and enslavement to hypocrisy.

It was all irrelevant to the charges I faced. And all legal. And all distorted. A mirage. An illusion. And it was the last thing I expected was for the system of justice to be a mirage.

They say, to a hammer, everything looks like a nail.

The justice system is society's hammer, and it becomes readily apparent to the dispassionate observer that the criminally accused person is the nail. But that is the fundamental purpose of having "innocent until proven guilty" as one of the bedrock principles of fundamental justice. And how *that* can be dispensed with is a question of the ages. A question for the _philosophes_, if there are any here.

The hammer of justice is appropriate only upon the formal adjudication of guilt, as the punishment aspect of Justice is something that should strike fear in the wrong-doer, for purposes of deterrence, and compliance by fear, or wariness, if mere morality is insufficient.

As well, Justice should be seen as powerful for the good of heart and the well-behaved, as their insurance and assurance of society's vigilance, and knowledge that there is an ever-present sentinel of an orderly society, whose fundamental values and mores are encouraged, guarded, and enforced efficiently, consistently, predictably, and fairly.

But when out of the blue, you've been unceremoniously tossed into jail for no apparent reason that you can see, you feel not just like the nail, but rather like you're between the blacksmith's hammer and the anvil.

The judicial system -- the Courts -- are an expensive, deliberative, essential pillar of a democratic system of a free yet orderly society. Their fundamental purpose is to keep order in a fair, efficient, and reliable process whose goal is to establish the Truth and dispense justice, as required. Through a thorough and systematic process, guilt or innocence will be determined and pronounced, and appropriate punishment meted out for the guilty, so that Order may be maintained in general. And *seen* to be maintained by the public at large.

The legal system thus visibly sets the very example of Order, by its operation, that is needed by its citizens for them to see the society they desire maintained.

* * * *

"To doubt everything or to believe everything are two equally convenient solutions; both dispense with the necessity of reflection."

-- "Science and Hypothesis (1901)
Henri Poincaré

[I saw Poincaré's grave, when I worked in Paris. It was in the large, neat cemetery near the Montparnasse rail station.]

A fundamental goal as well, addressed specifically to criminal adjudications, is that the "other side" -- the prosecution -- is responsible for establishing with reasonable certainty that only the truly guilty are convicted and punished. "Fundamental justice" trumps the "adversarial process" -- that there are two sides locked in opposition: the prosecution and the defense. Or more precisely, fundamental justice is the consideration that limits and defines the correct functioning of the adversarial process, which is the actual working of our system of justice.

"The Rule of Law", "Innocent until Proven Guilty", and "Fundamental Justice" are the basic concepts that are supposed to underpin the Canadian system of justice -- and indeed, any system of justice in a country claiming to be free, fair, and democratic.

The "Rule of Law", for instance, is a cornerstone of justice, because it replaced its odious predecessor, the "Rule of the Official". Which was not even a rule, because it produced no stability, and no consistent results whatsoever, as every rule should. And which led to arbitrary, capricious, and biased treatment of those brought before the bar of justice.

* * * *

"But the Rule of Law does point the way. ...[T]he ultimate connection between liberty and the rule of law is that only in a regime of secure entitlements can there be liberty,"

-- "Modern Liberty and the
Limits of Government (2007)
Charles Fried

The Rule of Law secures liberty against the imposition of others for frivolous, corrupt or malicious reasons, and establishes the stability and security that makes freedom and liberty the inviolable *foundations* of society, rather than just mere laws, with their myriad exceptions and interpretations.

The "Rule of Law" establishes transparency -- a written code of conduct that can be reliably understood and followed to establish consistency with judicial findings, and enables other to review previous rulings, thus allowing appeals, and establishing accountability. For example, the rule of law theoretically prevents an official from railroading someone he doesn't like -- for political, corrupt, personal, or expedient reasons -- and getting away with it.

The "Right of Appeal" thus being another cornerstone of justice

The "Rule of Law" goes back at least as far as Roman times, as mentioned in the Bible (Acts 16:37-38), when officials were nervous about repercussions, and apologized to St. Paul because he was beaten without trial, in violation of his rights as a Roman citizen.

* * * *

> "Members of a community must in truth agree on something. They must in truth agree on something. They must feel that their common ties are stronger than the interests that divide them; they must all, while differing over policy, respect the legal machinery by which policy is determined; they must accept, and be committed by, the decisions of lawful authority even when those decisions contradict their private wishes."
>
> -- "Twelve Who Ruled" (1941)
> R.R. Palmer

And the "adversarial process" of Prosecution versus Defense, mediated by the Judge, is how within the system the players are positioned to operate. But it is not a law, it is not a rule, it is not a regulation. It is merely a concept, without legal force or standing, of how the defense and prosecution are positioned against each other, *within the rules that constrain the behavior and respect justice*.

Yet it was used by the Prosecution time and time again to justify the twisting of the facts to the point that they became lies, innuendo, and utterly unsupported speculation. With the net result being the flagrant and deliberate misleading of the Court.

"Fundamental justice" means fundamental *fairness*, and imposes a duty on the prosecution to limit its conduct to the reasonable, exonerate the clearly innocent, and, in fact, investigate -- due diligence -- whether a *possibly* innocent person is being unfairly prosecuted.

* * * *

> "Faced with the choice of changing one's mind and proving there is no need to do so, almost everyone gets busy on the proof."
>
> -- "Economics, Peace and Laughter" (1971)
> John Kenneth Galbraith

The Charter of Rights and Freedoms is the enumeration of a Canadian's inviolable Constitutionally framed rights, which the State claims to respect, and certainly may not circumvent or ignore through law, regulation, or action.

Given the unlimited resources and overwhelming power of the State, this is not only fair, but essential. Anything less is the road to routine unfairness leading to gross injustice.

And some day, should the unscrupulous attain power, they would utilize a corrupted or corruptible judiciary, as an adjunct to -- a validation -- of tyranny, and an aid to its efficient and subtle consolidation of power. A process constructed originally for fairness, subverted as an anchor and bulwark support for malevolent evil's continued hegemony. And that is the real danger.

When the process has been corrupted and debased, so as to be part of an inexorable process to systematically grind down, bankrupt, demolish, and destroy innocent people trapped within its malevolent, black-hearted clutches, while the processes' operators enrich themselves greatly, something is truly, outrageously, badly out of order.

(I didn't know that the inanimate object known as the "process" could actually be "black-hearted", but then I remembered that I've worked for a multinational corporation. However, the corporate goal is profit, not truth, I must say in its defense.)

There is no defense or excuse for a legal system gone hopelessly corrupt. With open courts and public access to court documents, it's theoretically provable, patently visible, obvious, blatant, and a gross perversion of its original stated goal, should an impartial investigator take the time to examine it.

*　*　*　*

"Every [object] continues in its state of...uniform motion in a straight line, unless it is compelled to change that state by forces impressed upon it."

-- First Law of Motion
"_Principia Mathematica_" (1687)
Sir Isaac Newton

Fad justice, charade justice, pantomime justice. The carousel. The legal system as a mad merry-go-round. Allegations in Wonderland. Going through the motions style justice. A fraud, an ineffectual pantomime of empty movements and motion to outwardly appear useful and productive, without being so, while enriching greatly the lawyers, police, and judicial officers, to the gross detriment of society.

"Depraved indifference" clearly underpins the mindset of these so-called officers of the Court. That, and the system being their personal carousel of cupidity.

And no doubt the last judgment these **bastards** shall hear is a final judgment on themselves -- though we all it have coming. But the words they will hear, no doubt, are:

"You have been weighed in the balance, and found wanting."

-- Daniel 5:27

*　*　*　*

> "Fear not. What is not real, never was
> and never will be. What is real, always
> was and cannot be destroyed."
>
> -- "Bhagavad Gita" (ca. 3000 B.C.)

Let me give you an example of our legal system gone off the rails.

It was something so bizarre it disturbed me as to how even a first year law school drop-out could not see how stupid it was.

It was that it seemed a standard practice -- even a part of the court process -- for your own defense lawyer to be arguing on your behalf without even having even a partial idea of what had gone on, up to and including the details of your arrest. They barely even spent any time at all talking to you, before facing off with the Crown attorney in Court.

This set the stage for all manner of prosecution mischief, such as making statements and bringing out sworn testimony that was wrong, distorted, scurrilous, speculative, or otherwise of no merit, but which was highly prejudicial and damaging. Yet which would go unchallenged due to the failure of the defense attorney to have informed himself of the facts known intimately by the accused.

And should the accused decline -- as is his right under the Charter, and common law -- to not testify, it was open season for the Crown and his witnesses to introduce all manner of perjury, lies, gross mischaracterizations, and smears which your defense lawyer lets pass with a dumb look on his face and both hands tied behind his back.

I mean it's as simple as describing the Boy Scouts as a "paramilitary organization" -- a distortion and mischaracterization that is outrageous if it were in a Crown prosecution submission.

That's the way Canadian justice is normally. Now, it was even more serious than that.

* * * *

> "They are disunited, ambitious, without discipline,
> faithless, bold amongst friends, cowardly amongst
> enemies, they have no fear of God, and keep no
> faith with men."
>
> -- "The Prince" (1513)
> Niccolo Machiavelli

Faced with a crisis, a broken, corrupt system goes into a sort of overdrive of unspeakable evil. With all eyes focused on it, the corrupt process tries to camouflage its uselessness with a visible show of fireworks to appear like it's doing something useful.

It isn't, of course, but what is occurring is that its sins are multiplying exponentially, and horrifying mischief is being done to the innocent.

Enter the witch-hunt, the multiplying of scape-goat justice to an entire group. Mob justice, justice of the posse, lynch justice, 9/11 justice, Git'mo (Guantanamo Bay) justice. The show trial, the purge. A mindless lashing out based on fear and stereotypes, rather than the supposed true values of justice and fairness that are the foundations of a functioning, stable, constitutionally-based society.

The witch-hunt has a long, disturbingly regular, and disreputable history. It's an expansion of scapegoating. It's a subset of collective punishment, an even more disreputable, uncivilized, and insupportable abomination of justice, certainly in any society with functioning courts and thriving morality, with leaders that understand that they've sworn an oath to God and their country.

Not an oath with, subsection (b), clause (2), "except when not convenient", or "except in time of crisis, when the fearful clamor for action irrelevant, useful, or not".

There is absolutely no excuse for the breakdown of society's constitutional protections of the individual when economic, political, and social turmoil produce a climate of crisis, upheaval, and widespread fear that cause those in power to abandon their sworn duties to uphold the Law and the Constitution that is its foundation, as represented by the Charter of Rights and Freedoms.

The search for scapegoats is the strategy of the weak, the ignorant, and the corrupt. It is the focus on an "unlikeable", easily identifiable minority group to channel society's anger at a much more difficult and complex problem defying simple resolution. A way to provide society at large with the intellectual satisfaction that the problem is soluble and well in hand, short circuiting the social instability and recriminations of a problem with no solution or reasonable strategy for resolution of the threat.

It is a disgrace when this strategy is adopted by those in power who should know better, and whose job it is to use their backbone to correct or redirect such powerful but destructive and malevolent forces within a democratic society.

And what's worse, I had been caught in the cross-fire of a witch-hunt. I was collateral damage. And that was the important issue. I was the first non-Islamic person, the first Canadian born person, the first complete innocent, the first obviously innocent person swept up in the witch-hunt. But their corruption was so complete and unrepentant that they couldn't admit it, even after complete exoneration by the RCMP investigation.

* * * *

"God, protector of innocence and virtue, since you have
led me amongst evil men, it is surely to unmask them."

-- "Fragments on Republican Institutions" (ca. 1795)
Louis-Antoine Saint-Just (the "Angel of Death")
Committee of Public Safety of National Convention

Rubber-stamp. Abdication of responsibility. Fear of public opprobrium. Serious newspapers degenerated into tabloids. Serious justice degenerated into an empty pantomime, or tabloid justice, acting and reacting to sensationalism as if it were uncorrupted Truth.

Justices of the peace as weak, and spineless worms – a form of moral corruption. Frozen and fearful of the public spot-light. Some incompetent to begin with, or incompetent in time of publicity through weakness of mental and moral confidence. Thus displaying manifest unsuitability for public office, and evidence of how corrupt the Justice of the Peace appointment process has become.

Fair weather officials, whose only real job is *exactly* to show spine and intestinal fortitude, and to stand up without flinching in the face of troubling, controversial, or sensational cases. To face down the tendency to succumb to such abusive, unjust, counter-productive, and patently unfair, illegal forces during times of crisis. How hard can it be to stand up to what only amounts to the short-lived wind of the rabble's collective voice, aroused only by misguided passions?

When the marauding mob in "Julius Caesar" (Act 3, Scene 3) happens upon an unsuspecting Cinna the Poet, rather than the Cinna the Conspirator that they seek to track down, the terrified and innocent wordsmith pleads, "I'm Cinna the *Poet*! ".

But their blood lust not to be deflected from immediate grass-roots justice, and the certainty of their violent goal, someone in the blood-thirsty, howling rabble cries out,

"Tear him for his bad verses! Tear him for his bad verses!"

as they fall upon the hapless innocent, and beat, club, and stab him to death.

* * * *

"I find no pleasure in discussing ideas anymore on such
an elementary level.

People resent being shown that they don't approach the
complexities of the problem -- they don't know what exists
beyond the surface ripples.

It's just as bad on a higher level, and I've given up any
attempt to discuss these things with the professors."

-- "Flowers for Algernon" (1959)
Daniel Keyes

"Holy Mother of God! The long knives are upon us!" wrote Pablo Neruda, celebrated Chilean poet, communist, dissident, and exile.

The wheels of justice are slow in order to be careful and deliberate. They are not slow to be some sort of water-torture that enriches the practitioners with inflated billings, or to grind down the bones, pulp the flesh, and crush the souls of the unfortunates caught in the gears.

Shape without form, form without substance, "shade without color, paralyzed force, gesture without motion."

The constitution is not a technicality. Testimony and submissions that were perjury, gross misrepresentation, and mischaracterizations that strain and break through the boundaries of belief and truth into distortion and manipulated untruths is unacceptable, and they should damn well know it, in a country that claims to be proud of its institutions and reputation.

Almost thirty years ago -- 1982 -- when the Canadian Charter of Rights and Freedoms was put into place by St. Trudeau as the supreme constitutional bedrock law of the land, there still seems to be a problem.

All actions in contravention of the Charter are illegal, null, and void. It is the trump card against all other laws, regulations, and precedents. Such illegal actions should never be deliberately taken by those in authority, and it should not take months of legal wrangling, re-adjudication, expense, and personal suffering before a competent judge verifies this, and quashes and nullifies the case due to the malfeasance of the police.

* * * *

> "After the Night of the Long Knives,
> The same heart still beats,
> At the goodwill of those souls buried,
> Beneath stones laughing and weeping even now."
>
> -- "The Paris Commune" (1969?)
> Barry Tebb

Yet over and over again, the prosecution, in flagrant violation of their duty, and disgracing their office, their profession, and themselves, persist in practices they know to be egregiously wrong. Section 2 of the Canadian Criminal Code states clearly, "Ignorance of the law is no excuse", yet the police and prosecution ignored the Charter, the Law, and the precedence set by previous Courts that I was innocent, and they were in the wrong. As they hid under the bogus guise of the "adversarial process" to coerce, persecute the accused, waste the money and time of the Court, and trash their oath of office.

A combination of the morally bankrupt, the intellectually bankrupt, a cabal of experienced lawyers and officials on both sides of the fence who blatantly collude together to profit from the status quo, combine to make injustice the order of the day, with nary a peep, or protest from anyone. What part of "supreme law of the land" do the municipal police and crown prosecutors not understand? The only question is whether the authorities are engaged in malfeasance (willful violations of their duties), or misfeasance (complete incompetence), or both.

A judicial system -- indeed a Government and its bureaucracy -- not perpetually dedicated to move towards and establish a reign of impartial virtue is doomed to decay and corruption, and the slow corrosion to personal self-interest and the limits of individual ambition of those who comprise its organization, and are adrift from its ideals by selfishness or absence of audit or leadership.

The Canadian judicial system is inherited from the U.K., the people who invented feudal oppression, religious corruption, internment without charge, concentration camps, wholesale religious persecution, political roundups, and public executions, and an inbred hereditary ruling class of effete incompetents, sickly, but super well-mannered morons and deviants, and hopeless military leaders successful only at getting their own soldiers slaughtered.

The leadership of this country is so moronic, many of their obituaries will no doubt read, "He passed away after a long battle with mediocrity."

But don't listen to me. There are two choices essentially. And the correct choice is important beyond belief. But a couple of quotes perhaps give the hint as to which choice to make.

The first quote is what happens when having a good job becomes more important than doing a good job. How easy it is for things to go completely off the rails when people sworn to uphold the foundations of good government become a parody of themselves, because everyone suffers but them.

> " ' No, no! ' said the Queen. ' Sentence
> first -- verdict afterward.' "
>
> -- "Alice's Adventures in Wonderland" (1865)
> Lewis Carroll

> "I talk to the wind, my words are all carried away.
> I talk to the wind, the wind does not hear,
> The wind cannot hear.
>
> I'm on the outside looking inside, what do I see?
> Much confusion. Disillusion, all around me."
>
> -- "I Talk to the Wind" (1969)
> King Crimson
> Peter Sinfield

Why doesn't the government brutally purge the traitors who, drenched with the blood of hypocrisy, knowingly sabotage the foundations of our democratic society as brutally as they pursue ordinary low-level criminals?

A constellation of judicial misconduct and indifference -- or toleration of it -- so ingrained and systemic as to amount to a moral corruption scandal of unprecedented national proportions. And it exposes Canadian pretensions of sophistication, righteousness and authority to comment on the conduct of others from the moral high ground, and an example to be followed as a charade. A hollow joke. An empty, pathetic joke that is responsible for much mischief and only serves to cover up hope of reform with yet another illusion.

When the rot has reached all the way down to the very foundation, when the essential, the honored, and the sacred has become the profane, then is not reform way too late, and overthrow the only possible order of the day? The moral imperative of men of good character and standing?

If there are any left.

> "Dost thou claim to have done hitherto only
> what is in conformity with the ancient custom
> of the land? Thou shouldst know...thy Creator
> hath said, 'My name is Truth'. He hath **not**
> said, 'My name is Custom.'"
>
> -- Letter to count of Flanders (1092)
> Pope Urban II

> "Truth is higher than everything,
> but higher still is true conduct."
>
> -- "Adi Granth" (1705)
> Guru Nanak

Prison Diary 34:   The Banality of Bureaucracy
=====================

> "In a flash, the Devil took [Jesus] to a high place,
> where they could see all the kingdoms of the world.
>
> Then the Devil said to Him, 'I will give you all the
> power and the glory of these kingdoms. All of it has
> been given to me, and I give it to anyone I please.'"
>
> -- Luke 4:5-6

> "As [Jesus] taught, he said, 'Watch out for the scribes!
> They like to walk around in long robes, to be greeted
> in the marketplaces, and to have the front seats in
> synagogues and the places of honor at dinners.
>
> They rob widows by taking their houses and then say
> long prayers to makes themselves look good. The
> scribes will receive the most severe punishment.'"
>
> -- Mark 12:38-40

Well, there you have it. Seems pretty clear to me. Because there you have it above as clear as can be in the Gospel according to Luke -- government and power is controlled by Satan, and so is destined to be misused if he has anything to say about it. So government and power has to be monitored and limited and restricted, carefully and continually.

When it comes down to it, the fundamental cause of the government, not just the leaders but the scribes -- the administrators and bureaucrats below the leaders -- implement the evil intent and programs.

Being Evil, is that when their fundamental priorities are exposed by a crisis they can't handle, it turns out that they are more concerned with keeping their comfortable, luxurious jobs, status and power, than the safety of the Society they are charged with protecting.

That they set the rules that every citizen must obey, or face criminal sanctions, yet are themselves immune from any rules, or constraints. And that is truly an obscenity and an abomination.

But the practical results in my case may succinctly be expressed as:

Mistah Yogi -- he dead.

* * * *

> "Jesus called the apostles and said, 'You know that
> the rulers of nations have absolute power over people
> and their officials have absolute authority over people.
> But that's not the way it's going to be among you.'"

-- Matthew 20:25-26

The Second Law of Thermodynamics -- discovered in 1850 by German physicist Rudolf Clausius -- simply stated, is that all the energy in the entire universe is constantly being dissipated, thus tending to get less and less useful.

Thus, government and bureaucracy are two examples of the Second Law.

[The Second Law applies to many different things -- physics, chemistry -- but a simple example is that hot objects can be slowed, but cannot be stopped, from cooling down, and the energy -- their heat -- dissipated to its surroundings, no longer recoverable or useful.]

You may have heard about philosopher Hannah Arendt and her most famous phrase, "the Banality of Evil", which she coined in analyzing the Nazi bureaucrats. Her idea was that with bureaucracy, responsibility is diluted until no one is responsible, and thus can be blamed. Of course evading responsibility by the bureaucrats may have something to do with it as well.

It's not the banality of bureaucracy, it's the evil of bureaucracy made banal.

Take the idea of bureaucracy and the dilution of responsibility until no one is responsible. Add some more things and you have a recipe for certain disaster to the institution, and the democracy that its leader and sovereign:

1. Lack of transparency (especially in the secrecy of policy or decision-making),

2. Lack of accountability, review, and appeal. And the resulting failure to detect or admit error or wrong-headedness results in a failure to correct matters when things go wrong. Then the cover-up, and propaganda takes over.

3. Bureaucracy is actually rewarded when it fails. When it misses a predictable disaster, or botches miserably a fiasco – too big to cover-up, and so that it gets wide public exposure -- it gets an increased budget and number of employees. It effectively gets rewarded for incompetence.

4. High pay and the resulting security and "have it made" laziness ensures a cover-your-ass, iron wall of self-preservation & job preservation. Self-interest and self-protection and self-promotion becomes the only rule, the only agenda of employees and management. Cover-up of mistakes, risk avoidance, ending up as practiced work avoidance.

5. Systemic conflicts of interest -- both open and covert, secret bribery and pay-offs, and "back-scratching" -- favor gaining and repayment -- against the public interest.

6. A corrupt and incompetent and partial appointment and promotion system based on cliques, alliances, corruption, racism, friendship, cronyism, and ideology.

7. The Tower of Babel Effect. As it passes along the chain of command, if you don't have perfect understanding, if you don't have a clear understanding, if you misunderstand, don't remember, miscommunicate, sabotage, deflect, or dilute.

Innocence becomes naive stupidity and foolishness, rather than uncorrupted purity, to be taken advantage of and manipulated by those with more power, status, and experience. The lack of knowledge from inexperience evolves to unremedied stupidity, something to be used, taken advantage of, exploited, and manipulated.

Trust becomes a foolish trait, the cheat, embezzler, the smarter and corrupt betraying trust is the simplest, lowest level of crime. Immorality and betrayal become the norm and something to gloat and brag about.

All in all, the worst side of humanity is exposed, and the upside down nature of the world becomes the norm. The Truth is something to hide, and sell as a commodity, to use for personal benefit and private gain.

The strategy for the incompetent manager/leader/administrator: take credit for good times, and deflect blame, make excuses, or finger-point and blame others -- especially those gone or otherwise absent -- for bad times.

* * * *

> "You still don't understand what you're dealing with, do you? [A] **perfect** organism. Its structural perfection is matched only by its hostility. [...]
>
> I admire its 'purity'. A survivor. Unclouded by conscience, remorse, or delusions of morality."
>
> -- Ash, the android
> "Alien" (1979)

In 1961, the book "Catch-22", by Joseph Heller, became a best-seller and classic. It was made into a movie in 1970. It's the story of a WW2 B-25 Air Force bombardier, Yossarian, who rebels against the uncaring, absurdly illogical, self-perpetuating (military) bureaucracy. "Catch-22" was what he called the no-win situations people faced, whenever the bureaucracy beat truth and logic, like "So it goes." peppered Kurt Vonnegut's "Slaughterhouse-5" after incidents of violent horror.

Hopelessly uniformed, or deliberately, selfishly blind: and thus their opinion and lives are valueless in a society that should recognize that it must actively, justly, fairly and openly improve itself or eventually totter and fall unmourned, having failed its citizens and its own betrayed values.

The cornerstone of my belief system had been organized political murder: the purging of society, proscription lists of public officials. Public, open, publicized, televised, proud, and explained in detail to society as a necessity.

In modern society "leadership" means making the wrong decision quickly.

Scott Peck in "People of the Lie", a psychiatrist thinks about the nature of evil in modern times. Peck warned in 1983 of in about 20 years in the future, that when Vietnam was forgotten by the people, there'd be another senseless and costly war. The 2nd Gulf War was in 2003 in Iraq, started by George W. Bush.

How's that for prescience on Peck's part?

I can't read minds or predict the future. Except with morons: that's because their minds are blank and their future is the toilet or high public office.

Peck says that soldiers follow order from generals. Generals follow direction from the Pentagon. The Pentagon says, "Oh, you have to see the Policy Department." Who says " *We* don't make policy. We just formulate and implement it." Policy passes the buck to the President. And the President might say, "I'm following the will of the people -- their directed mandate," to complete the circle of no responsibility, and no guilty party amongst them all.

What about U.S. President Truman, and the sign on his desk, "The buck stops here."?

But, I believe Peck's analysis is incomplete. Not so much a dilution of responsibility, but the ability of people to use any excuse. And the practice they have gotten in making excuses in a system -- the organization -- whose lifeblood is CYA, excuses, lies, self-promotion, and all those dodges and deflections that constitute "soft" corruption.

What we must address is a government that is irresponsible, inept, and corrupt. "This corrupt regime": full of abuse and incompetence, and corruption, and answering to no one. Compounded by a Legal System that embodies the failure of the Rule of Law, replacing it with the abject and certain vulnerability of Rule by Official & Personality. It is now whim, arbitrary, capricious, inept, corrupt, vexatious. Morally bankrupt and decayed.

And well, some of us are pretty intellectual, or at least pretend to be. First, we must characterize the problem. Observe and report.

There's the triumvirate of maladministration, the hydra of bad governance, the cancerous labyrinthine bureaucracy are -- a drum roll, please --:

>
> malfeasance,
> misfeasance, and
> nonfeasance.

Hard corruption -- actual malfeasance -- corruption, bribery, theft, fraud, their cover-up or toleration. Bureaucracy gone totally evil. Democracy derailed or short-circuited. The return to hidden authoritarian diktat and fiat for personal gain or ideology. Or just contempt for democracy and the people.

Soft corruption: the hairline fractures, the tiny cracks, the slow corrosion, the increasing tilt. The slow unstoppable beginnings of decay, and greater lapses. Laziness is a form of soft corruption, the germination of the seed. Favoritism, cliques, cronyism, nepotism, secret alliances of convenience, absenteeism, managerial cover-up of sporadic incompetence, leading to the unstoppable cover-up of growing incompetence. It is the seed in which the evil of the leadership can function -- grow, and bloom to full flower.

The growing cancer of misfeasance and nonfeasance -- systemic laziness, buck-passing, non-compliance, invented rules, excuses, passive-aggressive, the toleration of observed or suspected internal misbehavior or illegality, the evasion or shying away from necessary controversy or conflict.

Misfeasance includes incompetence, or creeping incompetence, and failing to address or remedy it by management, who hired their friends or people to return a favor. Followed by the abandonment of mentorship and group solidarity, team-work, and group betterment, and the adoption of short-term benefit in place of superior long-term goals or strategies.

Personal goals replace collective improvement, and the failure of management to support, acknowledge and reward good practice and behavior and actual accomplishment.

Nonfeasance: the failure of duty, abdication of your primary purpose and functionality, the negation of office, as fundamental a failure as possible. Laziness, risk-aversion or avoidance, or in its worst form sabotage of the organization, or process, or collusion with others to sabotage error detection, or the other self-correcting mechanisms of accountability.

Malfeasance: criminal misconduct and hard corruption, criminal misconduct under color of law, graft, bribe acceptance or solicitation or demand, gift acceptance, bid rigging and collusion, ghost workers or friend hiring, racist hiring or promotional practices causing disillusionment amongst the class afflicted, embezzlement, internal theft.

And the fact that it seems to be a rule of law that the particular Canadian Problem is that misfeasance and nonfeasance (incompetence, and professional laziness or deliberate obstruction) are non-actionable under law even when deaths result.

Dams flood away people, contaminated blood supplies distributed for 5 years after the problem was known. Overpasses that pancake people in their cars, the list goes on and on. No one ever goes to jail, is fired, or demoted.

If the government, its bureaucracy and institutions spent as much time on stamping out internal malfeasance, as they do on trying to avoid the public scandal should it become public knowledge, they'd have time to institute the reforms necessary to cure the systemic problems that cause the scandalous behavior to arise in the first place.

But instead we have reform by scandal, uncovered by dogged investigative reporters, unsilenceable whistle-blowers, or by loud and persistent individuals that manage somehow to escape the bureaucratic CYA practices by its humungous extent, or complete outrageousness.

* * * *

> "Power is tolerable only on condition that it mask a substantial part of itself. Its success is proportional to its ability to hide its own mechanisms. ...
>
> New methods of power whose operation is not ensured by right but by technique not by law but by normalization, not by punishment but by control, methods that are employed on all levels and in forms that go beyond the state and its apparatus."
>
> -- "The History of Sexuality", Vol. 1 (1978)
> Michel Foucault

The dawn of systemic corruption.

"Honesty" about greed: he has no shame about his moral flaws and short-comings. And he's too stupid to even know he's telegraphing this to others.

And when they get away with it once, they become emboldened. And furtive gives way to the confidence of no risk, no consequences.

And occasional need rises to repetitive walks to the unguarded well. And repetition becomes habit. Habit observed becomes the practice, first of the opportunist, as the herd effect is triggered.

Soon what was the exception becomes the rule, and the descent into systemic corruption began. The institutionalization of an ever-increasing free-for-all of misconduct, and an ever-diminishing and increasingly costly and dangerous chance to stop what has become collective self-interest manifesting as deeply entrenched malfeasance.

* * * *

> "Man, once surrendering his reason, has no remaining guard
> against absurdities the most monstrous, and like a ship
> without a rudder, is the sport of every wind."
>
> -- letter to James Smith (1822)
> Thomas Jefferson

How about this? How about establishing a government free of corruption, transparent to its people, responsive to their needs, and committed to justice, freedom, liberty, fairness, and a proper balancing of the individual good and the collective good?

Huh?

> "If we have to fight, we shall fight. You will kill ten of
> our men, and we will kill one of yours, and in the
> end, it will be you who will tire of it."
>
> -- Ho Chi Minh to French
> negotiators (1946)

Prison Diary 35:   The Way of the Fist
                                    ================

> "Bad boys, bad boys,
> Whatchya gonna do.
> Whatchya gonna do,
> When they come for you?"
>
> -- "Bad Boys" (1993)
> Inner Circle

> "I've been to Hell, young man,
> You've only read about it."
>
> -- Marquis de Sade
> "Quills" (2001)
> Doug Wright

This chapter should be sub-titled, "How to Win Friends and Influence People", the Prison Edition.

> [By the way, the actual Dale Carnegie best-selling book is supposed to be an amazing tutorial in effective human relationships, and highly recommended reading, though I haven't personally read it.
>
> Rather, I've always preferred to simply appeal to self-interest, raw manipulation, or financial inducements. Or the old standbys of coercion, or if necessary, fear of dire consequences.]

But jail was an entirely new subculture that I knew nothing about. The Way of the Fist. The way of the Barbarian, the Caveman, the Primitive Savage. Ice-Age survivor with an attitude born of deprivation. The Rule of the Philistine.

Survival of the Fittest -- where Fittest is defined as a threatening affect, size and physical strength, predisposition to aggression, and a hair-trigger willingness to engage in violence.

A sociological hierarchy dominance-based, where dominance is established by first and foremost by:

1. violence to establish the winner as the dominant male through superior street-fighting skill, enhanced by surprise attack or physical strength *and*

2. reputation as a known career criminal, or violent psychotic, or the seriousness of the criminal charge, or the length of their sentence, if convicted

The _modus vivendi_ of the Ottawa jail known as OCDC, and with the full knowledge, acquiescence, or depraved indifference of Government of Ontario prison and ministry officials.

*   *   *   *

"And this I know, and wise it were,
If each could know the same.
That every prison that men build,
Is built with bricks, and bound with bars of shame,
Lest Christ should see how men, their brothers maim."

-- "The Ballad of Reading Goal" (1898)
Oscar Wilde

The dystopia within. Government-approved.

The Government of Canada, Ontario Department of Correctional Services and Community Safety, Dystopia Division. In stir, no one can hear you scream. And the things that happen behind iron bars are our society's secret nightmare, and, alas, my private hell, the details of which are known only to the inmates, guards, and presumably the prison administration.

But here I was. The Valley of the Shadow of Death. Angel of Death, fly over me. Lord protect us all. Or at least me. _Sauve qui peut!_. And where's the lamb's blood, when you need it? Visions of being beaten to death in this concrete terrarium, as some bored or uncaring guard daydreamed or even watched for "entertainment's" sake.

There was a guard at all times, in a central "bubble", who could see all the pods by turning his head. There was a camera surveilling each pod. Yet once or twice every two weeks, a bloody fist fight, or attack, or gang up surprise attack would occur, rarely attracting any interference from the guards.

The inmates not in the fight would _en mass_ split in two in a "wave", heading to the two walls, and out of the way of the fight, as far as possible. It was quite the spectacle, and an unpleasant one. We were all trapped in this concrete box, an unwilling audience to the spectacle.

Once, I was even sitting next to an inmate I didn't know, and suddenly he was approached by another disgruntled inmate. A heated conversation began, that in seconds became a fist attack on the seated inmate. And a bunch of guards were still in the pod! This fight was stopped by the guards.

It was random, unpredictable, and there was no escape or defense. *Anything* was grounds for attack, but there was a basic unwritten code -- of all sorts of "offenses" that required violent attack. One time two fights were going on simultaneously!

Drugs or debts were the most common sources of discord. No problem here from my perspective.

But I was out of my league, completely. I could not in any way blend in at all. I was clearly "different" from the typical inmate, and there was no way to conceal it. Most were drug addicts, career criminals, uneducated, had criminal records, and experience with violence and the prison system.

I was upper middle class, highly educated, had no criminal record, no experience with my situation, and knew nothing of them, their culture, their "codes", their lifestyle, their belief system.

But make no mistake, I met a memorable number of decent individuals, and had some interesting conversations with them. I also got along with the "old school" career criminals -- they had a moral code of sorts, and a decent set of rules, and I could deal with it quite well.

But the predators, the intrinsically violent, the psychotic, and the younger generation of up-and-coming career criminal ruined it for the whole range.

My routine use of polysyllable words or complex sentence structures could make an uneducated inmate "feel stupid". This apparently could merit an unprovoked beating too. So, I was warned about this by someone who realized I was vulnerable to this happening, and was kind enough to warn me. Of course, I found this not just insane, but totally frustrating, as it was next to impossible for me to avoid accidentally offending someone, no matter how hard I tried. My speech patterns were just too ingrained.

I was also told that the fact that I was polite made me look gay -- which merited an attack and expulsion from the pod. More insanity. It was a nexus between the mindless violence of the individual, given the stamp of approval by an indifferent society, to become the cancer of collective violence. And it was absolutely devastating to my psyche. The *constant* fear of random violence on the range, directed at me or anyone, with no escape or aid, wore me down over time. I never got used to it. And you couldn't just keep to yourself. Someone was always in your face, asking probing questions, looking for informants, which was ridiculous...

And then suddenly it all fell together. Suddenly it made sense. **Of course**. Why hadn't I realized it a long time ago?

I had **finally** had the opportunity to observe close up at hand the white man in his natural state!

"Prostitutes, drug addicts, delinquents, criminals, jazz musicians, Bohemians, gypsies, carnival workers, hobos, winos, show people, full time gamblers, beach dwellers, homosexuals, and the urban unrepentant poor -- these would be included [as social deviants].

These are the folk who are considered to be engaged in some kind of collective denial of the social order.

They are perceived as failing to use available opportunity for advancement in the various approved runways of society; they show open disrespect for their betters; they lack piety; they represent failures in the motivational schemes of society."

-- "Stigma" (1963)
Erving Goffman

Prison Diary 36: "Getting Medieval" on My Ass
========================

> "The Empire has grown old and lost its former vigor...[T]here is no longer any justice in judgments, competence in the trades, discipline in daily life...."
>
> -- Cyprian (ca. 250 A.D.)
> Bishop of Carthage

> "It can't be bargained with, it can't be reasoned with. It doesn't feel pity or remorse, or fear. And it absolutely **will** not stop, until you are **dead**."
>
> -- Kyle
> "The Terminator" (1984)
> James Cameron & W. Wisher, Jr.

Getting evil, getting medieval. Let's go Dark Age. Let's go uncivilized. Time to get in touch with our inner Viking berserker! Hear our barbarian asses roar.

Medieval times -- the Middle Ages covers the period of about 450 - 1450 A.D., and which started with the fall of the Roman Empire -- the "thousand year Reich" [Government] that Hitler was enamored of and sought to re-create with Nazi Germany. (But his Third Reich lasted only twelve years, and left German cities a heap of blasted and burnt out rubble...)

The collapse of the Romans was a huge step backwards, and the temporary ending of civilization. An evil, brutal civilization, but an organized one, at least. But its end did not improve brutality and evil, with the big step backward. It was a beginning of ignorance, and the heavy loss of most of Roman technology, and the even worse barbarism of feudalism, and overall, it only increased brutality and evil. The first thousand years of civilization was followed by a thousand years of non-civilization.

It was the time of the Christian Crusades. Six "official" ones in all -- and several smaller expeditions, such as the two failed Children's Crusades -- from 1095 to 1229 A.D. to recapture Jerusalem from Islamic rule, and which ultimately failed (as the Muslims would have said, "_Allah-hu akhbar!_" -- God is mighty!).

The much more advanced civilization of the Islamic Empire wondered why these uncivilized, illiterate, backward barbarians kept on attacking them, and committing horrible atrocities in the name of Christianity, a religion they had no real conflict with, by order of the Koran...

And then there was the medieval justice system. People were imprisoned in castle dungeons. Punishments by the state included beatings, branding, mutilations, eyes gouged out, amputations, being burnt alive, or disemboweled.

Trial by fire. Trial by ordeal. These were medieval arbiters of justice. And to my total shock and dismay, the State -- the civilized Canadian state -- 500 or so years after the end of the Middle Ages was "getting medieval" on my ass.

Or in more up-to-date terms, it was my version -- press-ganged upon me, for sure -- of "the year of living dangerously", as I refused to play ball, knuckle under, conform, or bend to the enormous power of a supposedly modern, liberal State, as it focused on trying to torment, torture, break, and literally even beat me into submission.

Their strategy was simple: if I would agree to plead guilty to any single criminal charge, I'd be released from jail, **or** they would keep me in jail by doing everything in their power to deny me bail, and apparently even stoop to the level of delaying my trial, possibly indefinitely.

Of course, when I got to trial, the charges would be thrown out. However, this was apparently immaterial to the Prosecution and Police, who appeared to be unconcerned about principles of fairness and fundamental justice, or the disgrace this brought upon their office, profession, and themselves.

They did appear to be concerned about facing justice themselves for things such as behavior, condemned as despotism and government misconduct and made illegal over two centuries ago in the United States.

The whole crux of the matter was that my pleading guilty to any criminal charge would prevent me from successfully suing for official misbehavior. Thus, they could cover up any screw-ups, misconduct, and/or malfeasance they had committed.

A press conference showing off my property that was legal for me to own, would serve no useful purpose except to smear my reputation with a mirage of suspicion, but get me front page news coverage. By going to strenuous lengths to smear my reputation, painting me as some dangerous threat the stage was set for the improper denial of bail as a result -- a full public gallery apparently putting the squeeze on judicial authorities to deny bail or face public opprobrium.

And with my release denied, that would submit me to the rigors of the jail-hell of OCDC, softening me up so that I would be willing to plead guilty to something, just to end the nightmare of incarceration and the abuse, degradation, discomfort, deprivation, and especially constant threat of violence it entailed.

And if I folded and agreed to plead guilty -- just to end the nightmare -- I would be released, and a cover-up of what they had done to me illegally would be successful. By pleading guilty I would not be able to sue the government. Scandal averted, it would be back to business as usual, with another innocent tossed to the garbage heap, forgotten in an instant in a process without a conscience.

And so I had to sit in jail waiting it out, where I was in constant danger and fear of assault and serious harm. Just as bad, in my mind, was waiting to watch the inevitable and horrifying spectacle as bloody violence broke out regularly in front of me, with nowhere to retreat except to the confining walls of the small range, along with the rest of the involuntary spectators.

I was verbally abused and mistreated by some staff regularly -- nothing personal, but just the character of the poisoned souls of these disturbed individuals. And I was subjected to what normal people would consider constant physical discomfort. I was denied practically everything but proper food, hot showers, clean sheets, writing paper, and my wits, as I waited 8 months for my trial, before the charges were "stayed" (withdrawn).

Or to summarize, in less polite, but succinct terms: free and democratic society, my **ass**.

* * * *

> "The [King of France, Philip the Fair's] counter-maneuver [in 1302] was no less drastic. He prepared to have [Pope] Boniface [VIII] deposed... To prosecute this task he chose William of Nogaret, one of the shrewdest of the lawyers who were helping Philip... Nogaret was...a master of the trumped-up charge."
>
> -- "The Age of Faith" (1965)
> Anne Fermantle

And all for crimes that would warrant probation, and the fact the I should have been considered "innocent until proven guilty" by the Charter of the land, I was kept in jail for the purpose of coercing a guilty plea out of me. A gross abuse of state authority, and apparently routine with anybody they can successfully deny bail to. Easy enough when the Crown and Police have no scruples whatsoever, and the Justice of the Peace a rubber stamp for the Crown with no legal qualifications to speak of, other than an appreciation of the merits of the soft corruption known as "patronage" and ass-licking.

As your typical Canadian middle-class member of society, with no actual experience with being thrown into jail and been subject to the machinations of the Criminal Justice System in Canada -- "Thank, God", I would say -- I had been led to believe that Canada's penal system was the implementation of a model of thoughtful and progressive criminology. Modern, up-to-date, rehabilitative, and all that good stuff. Liberal in thinking, constantly evolving with the latest ideas, yet not "soft" on criminals.

And with the wonderful societal goals of lowering criminal recidivism, protecting society, and deterring crime, while focusing on crimes of violence and crimes with victims of some sort (as opposed to so-called victimless crimes).

What a "snow job"! What a bunch of B.S. crap propaganda. Liberal Party or Conservative Party, many aspects of pretrial detention were little more than medieval trial by ordeal barbarism. And it was a disgrace to the Canada, Canadian values, and the administrators, prosecutors, and judges, and provincial government that knew, or should have known what happened "inside".

Forget everything you know. Forget everything you've heard, or been told. Forget everything you've read, or seen on TV. I had a front row seat in the blood sport arena of the Canadian criminal justice system. Or at least the Ontario subset of the Canadian judicial system.

And now, I bear witness to it, for all who care to listen.

And it's only appropriate -- wouldn't you say -- that I had the unbelievably bad luck to get caught up in one of those random "wrong place at the wrong time" situations? The intersection of my bad luck with a mindlessly incompetent, appropriately Canadian, post-9/11 witch-hunt for "terrorists".

A witch-hunt: the unseen, unidentifiable enemy. The enemy within.

> "Men feared witches and burnt women."
>
> -- Whitney v. California (1927)
> Justice Louis D. Brandeis

So paranoia leads to suspicion. And suspicion becomes evidence. The standard of "probable cause" becomes a ghost, search warrants a dispensable inconvenience, "innocent until proven guilty" an option, fairness a nuisance that gets in the way of decisive action, and injustice the price the unlucky stereotype-of-the-day has to pay. The man suffers, and suffers injustice, but ultimately, society pays the biggest bill of all.

The witch-hunt is the panic reaction of a society that feels under threat, and which can sacrifice an internal group as a scape-goat justifiable as a temporary expedient. The internal group identified has its rights and privileges available to be treated as expendable for the stability of the collective good, a show of strength and resolve, both for internal and external purposes.

The witch-hunt/scapegoating strategy is useful until the crisis has abated, the need or usefulness of a distraction for society is deemed no longer necessary, excessive over-reaction or the triggering of other negative and dangerous societal forces are threatening scandal or other long-term damage, changing priorities allow the quiet and gradual redemption of the scape-gated group, or the real target is identified and engaged.

There is no sympathy from a fearful public willing to sacrifice the marginalized and the demonized for *their* security and safety, even if it accomplishes neither. Ending up with a society having neither security nor freedom, just as Benjamin Franklin said they deserved.

The witch-hunt poisons a potential or actual target's social identity. It savages one's social identity and self-image, cutting the person off from society and himself, and allowing no defense, appeal, or redemption. Being cut off from society deprives the individual of his security, comfort, social ease, and confidence. He finds himself cast adrift or jettisoned into a storm that will only have negative repercussions. Cutting off the person from himself, attacks the very personality and sense of trust and stability. A feeling of profound existential confusion can only be dealt with by realizing the need for a forced redefinition of self, and the acceptance of the conclusions of this redefinition process.

Automatically branded as a discredited person, rejected by society. Stripped of almost everything, tangible, and intangible. Stigmatized and alienated. Ostracism, real, enforced, but also as an internal exile, where you end up shunned by your *own* suspicion and withdrawal reaction from society's treatment of the pariah, rather than being actually shunned by society.

> "[S]lavery was an institution grounded in absolute authority and systemic submission. It was maintained through ever-present coercion, including physical violence, that had to appear perpetual and unyielding to appeal or change."
>
> -- "In the Name of God and Country" (2010)
> Michael Fellman

* * * *

You've heard of the Early Middle Ages, haven't you? 400 - 1100 AD, after the decayed and tottering Western Roman Empire collapsed -- or rather slowly dissolved away -- under repeated barbarian incursion, invasion, and sack. The end of the _Pax Romana_ soon began the period known as the Early Middle Ages, or what they used to call the "Dark Ages".

Exited when the Europeans formalized the new rules and came up with feudalism -- organized barbarianism, if you will. Absolute monarchy, serfdom, the rack, torture-execution, witch burnings, public beheadings by ax, etc., etc., etc.

When they don't catch you red-handed, or have a clear evidence of guilt, "reasonable doubt" is easy to establish. Apparently that's too much of an inconvenience to the prosecutors, the Department of Justice, and their government masters. "innocent until proven guilty" is discarded too, with a quick "heave-ho", and no doubts of conscience in the slightest, as the Charter is made a joke of.

But these are dangerous times. And racism, especially against non-whites had free reign once again, and unembarrassed. And righteous.

But it was the realization that the debunked practice known as "trial by ordeal" was one of the principle bulwarks of the Canadian legal system that was the real shocker, when I happened to notice its sinister sway.

And it was the main reason why the prosecution sought to deny bail to defendants, in open and flagrant violation of the Charter -- not to mention the perjury statute -- while debasing a rubber-stamp judiciary with their transparent con-job, as a side-show.

Medieval trial by ordeal -- which went hand in hand with "Guilty until proven innocent" -- comprised a number of possible scenarios. One such was tossing the tied-up accused into a river or lake. If he floated, he was guilty. If he sank, he was hopefully rescued before he drowned.

Another scenario was for the accused to reach into a cauldron of boiling water to pull out a stone. An infected wound on the scalded hand three days later was proof of guilt. Carrying a red-hot piece of iron was a similar test.

In the face of vigorous criticism by high Church officials, trial by ordeal was finally recognized as a guilt or innocence determinant that was utterly without merit, and abolished in the early 1200's.

Never let it be said that Canadian officials are a little slow on the uptake. Never, because it is obvious enough that it doesn't need to be said.

The Canadian version of "trial by ordeal", I found, was a little more gentile, and a lot less obvious, because it was more subtle, and carried out in secret from the public.

(Not that the public cares about details, only results. And *that* is one of the great flaws of democracy -- when the populace is kept ignorant and uninformed -- useful only to be manipulated, or whipped up by emotion.)

My trial by ordeal was for the purpose of covering up the gross injustice that had been done to me, by getting me to plead guilty and thereby avoid a civil lawsuit, career repercussions, and a public scandal into why numerous criminal acts had been committed against an ordinary citizen without basis, and involving a concerted and deliberate smear campaign led by the Ottawa Police, and continued by the Crown Attorney's office.

What would Vern say? [Vern White, the Ottawa Chief of Police.]

The methodology of my trial by ordeal was not unique. It was a common complaint amongst my fellow detainees in OCDC. Of course, they were career criminals or had lengthy criminal records, were poor and couldn't afford justice, or were otherwise devoid of credibility and thus unable to fight the injustice.

* * * *

"Far off in Paris where his enemies
Whispered that he was wicked, in an upright chair
A blind old woman longed for death and letters. He would write,

> 'Nothing is better than life.' But was it? Yes, the fight
> Against the false and the unfair
> Was always worth it. So was gardening. Civilize."
>
> -- "Voltaire at Ferney" (1939)
> W.H. Auden

In the witch-hunt hysteria that ran rampant starting in the summer of 1692 in Salem, Massachusetts, a certain Giles Corey, 80, was accused of witchcraft. Put on trial, he refused to plead either innocent or guilty, in effect rejecting the Court's right to judge him.

The penalty for such impertinence was the _peine forte et dure_, [the punishment hard and long] in which the victim was placed between two slabs of wood. Officials then began piling heavy stones upon the top slab, until the uncooperative accused agreed to plead to the charges.

In September 1692, Giles Corey was slowly and painfully crushed to death. His last words were reported to have been, "More weight."

* * * *

> "If a nation expects to be ignorant and free, in a state
> of civilization, it expects what never was, and never
> will be."
>
> -- letter to Col. Charles Yancy (1816)
> Thomas Jefferson

Last comment. I can't wait till the Renaissance hits the Canadian judicial system.

> "And while she wishes she was a dancer,
> And that she'd never heard of cancer,
> She wishes God would give her some answers,
> And make her feel beautiful.
>
> I remember feeling low, I remember losing hope,
> I remember all the feelings and the day they stopped."
>
> -- "Innocent" (2002)
> Our Lady Peace

Prison Diary 37:   A Witch-Hunt and a Cover-Up
=======================

> "Because there's a conflict in every human heart between Good and Evil. And Good does not always win."
>
> -- General Corman
> "Apocalypse Now" (1979)
> John Milius

> "...open disregard for law and legal institutions and ideological justification of lawlessness has been much more characteristic of continental than overseas imperialism."
>
> -- "The Origins of Totalitarianism" (1958)
> Hannah Arendt

Democracy, idiocracy, mediocracy, hypocrisy.

Lynch justice. A riotous and unpeaceable assembly. Mob justice. The rabble. The rascally rabble.

The unleashing of racism that had retreated to just below the surface. It had only been silenced, but not vanquished.

Morality and goodness are simplicity itself. They are conceptual beauty in their simplicity and obviousness. It is evil that requires complexity. Deception and manipulation is ever-more complex and intricate to hide the truth from evil's victims.

Forbidden knowledge.

Take an old formula for "Indian Whiskey" I found by accident in a cook-book of old recipes, whose author didn't realize its implications:

    2 oz. strychnine
    1 barrel whiskey
    2 barrels water.

This formula, as well as being cheap, cut-rate, and poor potency, would amount to about 33 mg of strychnine per 375 ml of liquid (equivalent to four 1 oz. shots of regular undiluted whiskey). Which was enough to have an acute stimulating effect, and a chronic cumulative effect on the Indians it was making crazy.

Profitable, surreptitious genocide...

* * * *

> "Central to [Harold Innis'] project is the idea that Western civilization can be renewed only by intellectual development on a periphery that, in turn, becomes a new center for cultural fluorescence.
>
> In other words,...although these revitalized efforts in the field of knowledge may originate in the decadence of the imperial center, they can be pursued effectively only become allied to a...force at the margin of contemporary civilization."
>
> -- "Marginal Man" (2006)
> Alexander Watson

Demagoguery and the scapegoat, the scapegoat and the witch-hunt. There was the Dreyfus Affair in France in the late 1800's, the scapegoating of just one individual of the chosen group. In this case, the Jews.

There's scapegoating vs. witch-hunt which damns an entire group.

There's hate speech: a Canadian newspaper's war propaganda as hate speech; the same ol' double standard (WW1-- the Hun; WW2; with the government the worst offender).

The witch-hunts:

    1500-1650 the main period, but lasted until the 1700's; Salem, Mass in 1640's;

    foreign anarchists; the Sacco and Vanzetti anarchist trials; the anti-red Palmer Raids of 1920's;

    "The Crucible" by Arthur Miller: set in Salem, Mass., but referring to McCarthyism of early to early to mid-1950's;

    the late 1950's and juvenile delinquents and motorcycle gangs;

    marijuana users & opponents of U.S. involvement in the Vietnam war of 1962-1973;

    pedophile Satanists 1980's;

    drunk drivers in Canada 1980's;

    smokers 1985 - late 1990s;

    "Islamic terrorists", 2001 to now.

In other words, it's easy to find an excuse for a harsh action, rather than a measured, fair, useful or constructive one. It's why we pay these people. For leadership, intelligent policy, well-thought out actions, incremental experimental steps, measured or tentative, creative answers, or proven strategies.

Instead we get simplistic solutions for complex problems, solutions that are historical failures revisited on society to fail once more. Imported solutions that belie the creative bankruptcy of our rulers and administrators. Stolen ideas, or even worse, stolen bad ideas marketed as "leadership".

And the best idea of all, never, of course, adopted, step aside -- even temporarily -- or resign in favor of a superior candidate, or one who could be a threat to the incumbent's ambition, or lust for uncontested power.

The solution that does not work, keeps the evil and the incompetent in power and in business.

Displeasure, and unhappiness, dissatisfaction. Dissent, protest, disorder and riot.

Crystallizes into blame focused on perceived cause: either the government itself, government's policies or actions or attitude, or government's failures or inaction or mistakes. Unfairness, incompetence, exposed deception, or foiled manipulation. A single act, or repeated failures or sequence of events.

Man's eternal folly and destiny to self-destruct is that we spend our time fighting each other for no useful purpose except power, lust and greed, rather than fighting within ourselves to improve our behavior, demeanor, and morals.

Ecce Homo! Behold the Man!

A paradox for sure. A witch-hunt keeps my enemies distracted and off balance, at least...

* * * *

> "Government is about making and implementing public policy choices. These are neither always easy nor always right. Governments, like individuals, do make mistakes.
>
> But in democracies, the task of making decisions on behalf of the people is delegated to elected representatives who then answer to the courts on constitutionality and to the people on the consequences of their actions.
>
> At the same time, every society, including international society, always has some members whose intellectual conceit and moral arrogance lead them to want to substitute their judgment for the outcome of the democratic process."
>
> -- "The Reduction of Impunity" (2005)
> "Japan Times"

The absolute stupidity of the general public has been recognized at least since the writing of the U.S. Constitution in the late 1776 and the constitution that made democracy indirect with the creation of elections decided by an elected electoral college -- and the writer's of the Constitution were confirmed democrats and freedom-fighters.

The whole basis -- excuse -- of monarchy and dictatorship and totalitarianism is the unsuitability of the people as a collective whole to rule, or even any educated intelligent elites superior to the collective whole.

The absolute stupidity of the common man is manifest in their failure to understand the simple concept of causation. If terrorists are Islamic, then the Islamic are terrorists is not only an error, but a fundamental error in basic trivial logic.

It is a failure of society to teach its citizens. It is a failure of pedagogy of our educational system. It is a sad reflection on the professionalism and perceptibility of our teachers to not correct such a fundamental threat to democracy, minorities, justice, democracy.

It is a failure to society's well-being in finding correct solutions instead of being either misled deliberately or through incompetence, or inefficient in finding the correct solution, and of dead-ending in times of peril.

A stranger in a strange land. A strange land in strange times. Strange days, indeed. The twilight of stability, the harbinger of darkness, fear, confusion, chaos impending stalks the land. The Day of the Troglodyte is near.

Or is it upon us, already? This paradox, enigma, riddle.

The messenger, the prophet of Truth calls. Listen to the harbinger of Truth, or await the Angel of Death.

"I trust that all good men will conspire"

* * * *

"No exceptional circumstances whatsoever, whether a state of war or a threat of war, internal political stability or any other public emergency, may be invoked as a justification for torture."

-- "The Convention on Torture" (1984)
Article 2, Section 2

Ecce baro! (Behold the fool/dolt/idiot.) _Ecce balatro!_ (Behold the buffoon.)

It was a bunch of pigs that set the witch-hunt rolling. But it was the casual and relaxed evil of a man named Kevin Phillips, Crown Attorney -- prosecutor -- that started, or possibly continued the cover-up of a crime.

A weak-mannered, effete, worm of a man, named Kevin Phillips. Of Welsh ethnicity, which may or may not be relevant or important.

He was an efficient employee. Unfortunately, he was only efficient at mediocrity. He was your typical talented manager. Talented at covering up his immorality and incompetence. Or possibly the doubly-talented manager, who's good at self-promotion, as well.

Or was he just a moron? Was he afflicted by that congenital and degenerative disease of the nervous system, commonly called stupidity?

I felt like asking him, "When you're cold, do you strap a space heater to the back of your head?"

Or was it just the dogma that all defendants are guilty. My dogma is that dogmas are for the stupid. All dogs go to Heaven, but all dogmas can go to Hell.

Or was it just indifference and complacency and laziness, and impatience for disturbances that disrupt and delay the orderly progression of the smooth-flowing sequential process of the legal system. Of the extra work it requires of the bureaucratic cog, who must do more than just push paper from his in-box to his outbox after the rubber-stamp and his signature.

He was embarrassed in open court when it turned out he had sent an *email* asking the police searchers to collude and concoct a story that I consented to the search in spite of the officer's own notes that I told them loudly to stop and get out. Not to mention that they smashed doors open.

Lying to the court. repeatedly; suborning -- soliciting -- perjury, by email no less!

I realized at one point -- during the cross-examination of my mother -- that he had listened to my recorded visitor conversations, so desperate was he for any evidence of wrong-doing. And a gross violation of my privacy rights as a pretrial detainee. They should only be used for purposes of safety and security of the institution, and not by the prosecutor to build a case.

When the charges were finally stayed [withdrawn] he did not even look at me, as I drilled a hole through him waiting to see if he would, before I left the Court room. Nothing. Nada.

Legal system. Judicial system. Doctors: "I swear by Apollo..." Part of the doctor's oath in treating patients was codified as, "At least do no harm." When a doctor in ancient Greece fucked up bad, the patient died, and nothing could be done anymore. With the current legal system, something can be done. Free the man, and figure out whether a public official fucked up accidentally, carelessly, negligently, deliberately, or with malice aforethought, with an open eye on "systemic problems"

So, why isn't this a bed-rock, foundational, constitutional principle of the legal system?

And the Canadian legal system is so grossly evil and corrupt and incompetent, that most of the experienced, powerful members of it are so fearful -- so terrified -- of the wrath of the public, should this be discovered and publicized, that they fight tooth and nail to allow an innocent man to be freed, in case the System and its Enforcers and Administrators are exposed for what they are -- or, in this case, are not.

By God, this systemic impunity must end!

> "[One of his goals as a historian was,] to hold out the
> reprobation of posterity as a terror to evil words and deeds."
>
> -- Tacitus (ca.95 A.D.)

> "It is very difficult to collect [guns] in Canada, because we live in a culture
> in which firearms seem to be the most easily accessible to criminals who
> don't have permits. ... [T]hey can't look beyond it because they're biased,
> they're bigoted, they're ignorant."
>
> -- Corey Keeble (2005)
> Curator, Royal Ontario Museum,
> Toronto, Canada

Prison Diary 38:   "If I had a Rocket Launcher"

> "Well, if there's anyone that can look around this demented slaughterhouse of a world we live in, and tell me that man is a noble creature, believe me, **that** man is full of bullshit!"
>
> -- Howard Beal
> "Network" (1976)
> Paddy Chayefsky

> "In 1974, two groups were emerging within the IRA with differing views of the future of the armed struggle."
>
> -- "The Dirty War" (1990)
> Martin Dillon

"Religion is the opium of the masses", said Karl Marx famously. "Revolution is the opium of the intellectuals," quipped an anonymous graffiti artist, not-so-famously.

And as an adolescent, I was rather precocious about everything except my emotional maturity and starved social development.

The very first book that I ever bought was a thin paperback that I found browsing in a "Prospero" bookstore that used to exist on Elgin Street in the basement of the dark brown sandstone hexagonal structures of the National Arts Center on the perimeter of Confederation Square in Ottawa.

But I was more interested in "deconfederation", or something to that effect. The title was "Coup d'état: A Practical Handbook" (1968) by Edward Luttwak, and it was published by Penguin Books. I was about 10 years old, and couldn't afford it, but with a title like that, I couldn't resist **not** buying it.

I wondered if I was going to be arrested at the checkout counter, but the clerk didn't so much as give me a funny look, or even raise his eyebrows, to my relief. So after paying for the book, I made a b-line for the exit and scampered off home, grateful that there was no secret policeman discretely tailing me in hot pursuit.

Everybody complains about the government, but nobody does anything of it. Well, thus enters the revolutionary...

What they used to call a "rebel". Arising out of the ashes of failed or non-existent reform, or the intrinsic hypocrisy -- masquerading as pragmatism -- of "gradualism", a term only a poli sci undergraduate would know.

I used to have a joke. Being in Ottawa -- capital of Canada, and thus a "government town" -- when someone I had just met at, say, a party said, "I work for the government," I'd respond, "Oh, we have something in common -- I want to overthrow it!"

I just don't understand. I thought that since we are labeled the "disposable society", it meant that having outlived its usefulness, our society should be thrown out...

Yes, I was always a rather ideologically "volatile" individual. You see, I grew up in a different time -- and a different dimensional plane, some might say.

Where teenagers didn't shoot up their high schools, for instance. And people had a more innocent, but _laissez-faire_ sense of perspective and the permissible.

The leftist Allende government in Chile had been overthrown with American complicity in September 1973. The Argentine "Dirty War" "disappearances/executions began in 1975. Nixon was deposed in 1974, finally, caught red-handed with his own criminality...Political bombings of U.S. government buildings were coming to an end.

Fired a light machine gun at the age of 17 -- it was hilarious orgy of fun for a teenager. A roller coaster ride where you could spray bullets while you were on a down-slope.

But I just don't understand fanatics who kill for religion. How medieval! How Fourteenth Century! Personally, I've always thought that killing for ideology was so much more justifiable. I mean, since no major religion advocates anything but peace, love, understanding, and the unity of humanity, killing for religion is fundamentally bogus and an abomination. A crime against God and Man and an obvious corruption of the very religion they're supposed to be adherents of.

Killing for ideology is so much more noble. The self-sacrifice over an ideal is so much more rational, and at least has a logical basis of some sort. The rules are laid out and justified precisely, by the avenging ideology. The dispute is over ideas, concepts and logic, and the fight is against those who seek to oppose reform and instead preserve their corrupt, morally bankrupt, and abusive, self-serving hold on power.

* * * *

"No one can straighten what is bent."

-- Ecclesiastes 1:15

No corrupt organization has **ever** cleansed and reformed itself on its own. Popular overthrow has always been the only answer, rarely achieved through anything else but **mass** action through force.

In the early 1800's, Simon Bolivar, single-handedly and country-by-country, liberated a number of countries of South America from the cruelty of Spanish colonialist/imperialist tyranny. Bolivia, if you didn't know, was founded by him, and is named in his honor.

Revolutionaries frequently being sensitive, emotionally fragile souls -- when not shooting enemies of the people -- he unfortunately ended up disillusioned by his setbacks and perceived failures. He expressed his bitterness as follows: "Those who fight the revolution, plow the sea."

I, myself, would rather think of myself as a genuine radical surf-dude. "Hang ten, bro' !"

Hang a hundred, actually.

Yes, it's true. I'm the commie from Mars. Mars being the "Red planet", and all, and my ideas being so off-the-wall, I **must** be an alien.

Ten thousand years ahead of my time. Or ten thousand years behind...

An anti-social socialist. An ideological fossil. A hold-over. A political throw-back. A retrograde radical. An artifact of by-gone leftist activism. A Jacobin zealot completely out of his era. A misanthropic, maladaptive, martyr-complexed political malcontent.

A capitalist commie -- an oxymoron, if I ever heard one -- in spite of my beliefs, I actually thrive as a wheeler and dealer capitalist. And in a communist society, I'd be jailed or shot pretty much first -- I have a big mouth for my big opinion, and I use it liberally and unrepentantly. So ideologically, I'm in a bit of a limbo.

In my younger years, I used to live and breathe sedition, conspiratorial dreams and fantasies of revolution as the only way to correct an intolerably racist, corrupt and murderous, morally decayed society. And I wanted to put the irredeemably corrupt _ancien regime_ out of its misery. To merely "accelerate" its sunset.

A leftist believes that, "There's got to be something better than *this*." A rightist believes that, "There's got to be something better for me." A conservative -- a moderate rightist -- is a person who is patient with political corruption, but not political change. You can figure out the rest.

Actually, conservatives are patient with political corruption, patient with inequality, and indifferent to injustice. Their main argument always seem to be the tiresome theme of "law and order". **They** make the laws, and **they** give the orders.

And it's funny how the law never seems to be enforced against themselves.

The rulers of Canada even.

But then there was (please note, the past tense) my ideology. Communism is just an idea. A philosophy, theory, ideology. The practical -- the implementation of extreme leftism -- has a different name: the Armed Struggle.

The Long War. My quaint belief in, and willingness to participate in, an organized and prolonged campaign of anti-state political violence to eventually effect justifiable and necessary radical change -- usually by violent overthrow or bloody revolution -- that is not achievable in any other practical or effective way. Forcible reform, if you will.

The state which I saw as having betrayed and dishonored the noble ideals of its sacred beginnings and been corrupted to become the murderous fount of evil in the world was the United States. And I wanted to see it face revolutionary justice.

To paraphrase Robespierre (actually honest and moral to a fault, but perhaps not the most convincing "political" to quote...):

> "The basis of popular government in peace-time is to uphold virtue. The basis of popular government in time of revolution is virtue and violence. For virtue without violence is powerless, and violence without virtue is immoral slaughter."

In 1962, President John F. Kennedy said it best: "Those who would make peaceful change impossible make violent revolution inevitable". Two centuries earlier, Thomas Jefferson said: "A little revolution now and then is necessary for the sound health of government." And finally, there are the haunting words of the English translation of the name "Sinn Fein" (the legal political wing of the IRA): "Ourselves Alone".

There's a reason they've called guns -- since the days of the U.S. Wild West -- "the great equalizer". And guns in the hands of enough citizens make freedom possible when it's denied you, or taken away from you.

Armed struggle -- the kind of phrase, the very words of which still stir my blood, warm the cockles of my heart, and used to set my eyes ablaze with the ideological fervor of the true zealot.

Something to believe in. The One True Path. A hope for a better world, and the practical game plan to achieve it. Not one lone fanatic, or crackpot, not the lunatic fringe, but a movement of committed individuals. Men of substance and good character, politically aware, setting out on a bold quest for the betterment of the collective good by fighting against the hopelessly corrupt, entrenched with their ill-gotten money and clutching at their undeserved power.

A brief evolution of the linguistics of collective violence is now in order.

The linguistic propaganda war waged by the rulers evolved words about their upset citizens, such as: rebels, bandits, traitors, anarchists, commies, pinkos, extremists, the lunatic fringe, lefties, terrorists, fanatics, insurgents.

The peasant and serf uprisings against feudalism of the Middle Ages, Resistance of the French-Canadian _patriotes_ under the heel of English oppression. The urban rebellions of the early 1800's such as in Canada, by citizens driven to their limits. The rebellion led by Louis Riel in the 1800's that led to his hanging.

"Aux barricades!" cried the revolutionaries, as separate revolutions swept Western Europe in 1848, and sent shivers through the ruling elites. The overthrow of the government sought by the intellectuals of tsarist Russia, who labeled themselves anarchists and nihilists of late 1800's and early 1900's.

The "People's War" of Mao that overthrew the hopelessly corrupt Chinese "Nationalist" government in 1949. This was soon followed by the guerrilla army of Ho Chi Minh thumping the French colonialists in 1954 with the French defeat at Dien Bien Phu in Vietnam. This historic event -- chronicled in Bernard Fall's book, "Hell in a Very Small Place" (1966) -- was a ground-shaking milestone, because it was the first defeat of the Western imperialism of the white man, in his murderous 450 year history.

One big problem is corruption: appointments that are not competent, loyal, or veteran administrators, or that bribe or hold illegitimate power and authority, driving off the gifted, and demoralizing others.

There's the Maoist strategy of protracted warfare, and parallel counter-state, eventually taking on the government forces in a conventional war. But the timing is of extreme importance. A strategy of exhaustion, the sapping and draining resources, energy, support, and commitment; making the country ungovernable.

There were the kill-'em-in-the streets leftist revolutionaries of the late 1960's: the Black Panthers, and the Weathermen, and the FLQ in Canada.

The 1970's, the Armed Struggle devolved in the U.S. as 60's radical groups were marginalized or crushed, neutralized by infiltration, or simply disbanded, replaced by the equally unsuccessful 70's urban guerrilla groups, the "Symbionese Liberation Army", and the "Black Liberation Army".

In Europe, things were different.

Of those who lived at the time, none can forget the beginning of international "terrorism", the PFLP (Popular Front for the Liberation of Palestine) simultaneous hijackings of four commercial jetliners in Europe in 1970. No one was killed, and the airplanes were landed and blown up in the Jordanian desert.

Then there was the PLO splinter faction, "Black September", which burst onto the world scene in 1972 with a dramatic and audacious hostage-taking at the Munich Olympics Games of the Israeli Team (which due to police bungling resulted in the death of both guerrillas and many hostages). Black September was one of many armed Palestinian groups targeting Israel and enraged at its financial and political backing by the U.S.

The "Red Brigades", in mid-1970s Italy, spawned what the public called "the Years of Lead" in the 1970's, people's trials and people's justice, the kidnapping and execution of politician Aldo Moro (1978). However, it was West German student discontent that spawned the 1980's urban guerrilla group the Baader-Meinhof Gang, which evolved into the less anarchistic, and more organized and politically focused -- and thus successful -- "Red Army Faction".

And the IRA finally brought the English to the bargaining table and an agreement ending five centuries of conflict in 1994, when they began attacks on English homeland economic targets. (The Irish are perhaps a little slow...)

As the world's strongest military and economic Super Power, supported by an economic empire stretching across the non-communist world -- its "sphere of influence" of gigantic proportions -- that it had efficiently consolidated during the early years (1948-1973) of the Cold War with the Soviets, the U.S. was too geographically large, rich, resource-abundant, and powerful to be seriously affected by violent internal unrest.

However, its 1973 pull-out and crushing defeat in South Vietnam, Cambodia, and Laos in April 1975 to the North Vietnamese Army, put its military in impotent disarray after its dramatic and costly defeat in the Vietnam War (more properly called the "Second Indochina War"). Combined with the left-over social upheaval, and economic problems affecting the U.S. middle class, and other side-shows, the U.S. was demoralized and relatively quiescent for the rest of the 1970's, and early '80's.

* * * *

"Behold! I shall raise up evil
against you out of your own house."

-- "Catechism of a Revolutionist" (1869)
Sergey Nechaev

Former leftist, actually -- I renounced political violence years ago, gave up hope of radical change years ago, and realized and rejected many of the faults that arise from leftism a decade or so ago, I think. It was a sad day, realizing that my idealism was based on an empty and flawed utopian dream, leaving me empty and depressed, and lost.

The industrialized economic, and military power centers of the world's democratic, but materialism-addicted societies were chastened, but unchanged. And they were now starting to prepare for future confrontations with domestic extremism. And leftist extremism was not forgotten, nor forgiven. These countries began girding for the enemy they had only attacked in foreign cold war contexts of proxy war

In the 1980's "direct action" became the euphemism for leftist political violence against the government, and the name of leftist urban guerrillas in both Canada and France.

And finally, "armed struggle", a realization that as government had become more entrenched, their power organized and consolidated, the people apathetic and dissatisfied, but mollified by the cyclical prosperity of modern capitalism. "Armed struggle" was an acknowledgment that effective, collective political violence targeted at the government and the pillars of its support over a period of years, or even decades was the only hope of change, absent a sea change in public opinion towards utter disgust at the government as evil or utter incompetent and unchangeable through normal -- and desirable -- avenues of electoral change. A sea change in perception that democracy was sabotaged from above, and the government now illegitimate and obviously harmful and unacceptably harmful to the collective good.

The useless turn-off rhetoric of the leftist political. Understandable only by theoreticians, and creating a total failure to reach anyone else. Reactionaries, "running dogs", the dialectic, Marxist, Trotskyite, splinter factions.

It took me *years* to figure out what a "reactionary" was, even after I did nominal research. "Post-modernism" -- "po-mo" -- on the other hand, took me only two years to finally grasp the concept of.

Then there's war.

Individual commitment and bravery as the individual self-appointed soldier of the people, joins with like-minded brothers-in-arms equally fed up with intolerable moral and intellectual corruption to produce a small army of justice, allied only by bravery, self-sacrifice, honor, and willingness to take action in a society seemingly without heroes or values.

Their goals merely the victory of fairness, justice, an honorable and responsive government, competent and efficient, and well-meaning. The promise realized of individual financial security, achieved through commitment, hard work, or aptitude and ability well applied.

The secular political version of what the religious call "faith". Faith with firepower, and not just the empty ravings of the armchair revolutionary, or the half-mad, unrestrained rage of the classic "kill-em-in-the-streets" extreme leftist.

But then reason and experience prevailed, and I abandoned such beliefs. And my carefully constructed world collapsed into a heap of disillusioned dust. And I was left with nothing.

But last but not least, let me make it clear. I may have renounced violence a ways back, but that's not to say that I had renounced **advocating** violence... OK, OK, I'm just joking. J.C. even banned **that**.

* * * *

> "Homelessness, unemployment, drug addiction, and
> illiteracy are only a few of the problems that disappear
> from public view when the human beings contending
> with them are relegated to cages."
>
> -- "Race and Class" interview (1998)
> Angela Davis

"Capitalism is theft," said Karl Marx. "Capitalism is theft," is what I say, too.

Capitalism is exploitation, stealing from, and cheating others. Capitalism is immoral manipulation, lies, and deception harnessed for personal enrichment at the expense of others.

But, frankly, communism does not seem to be attainable anymore, and is not in my interests. I try to resist, but I have given in to materialism, and I have benefited quite comfortably from living in a materialist society. You can call me an "armchair revolutionary", and get away with it, as I grit my teeth, and lower my opinion of you, but I'm not a "champagne socialist". And I'll punch the lights out of anyone who dismisses me with **that** kind of grotesque insult, Christian non-violent beliefs, or not. I'll repent later.

Early in the movie "The Terminator" (1984), a waitress friend of Sarah Connor (whose future son is destined to grow up and lead a human revolt in the future) tells her, with unintended irony, not to be upset about a kid putting ice cream in her waitress' apron pocket as follows, "Look at it this way, honey: in a hundred years, who's going to care?"

When it comes to ideological reforms and improvements, achieved peacefully or through violence -- in a hundred years they certainly **will** care. And be grateful.

Or to frame it with less civility, a little bit of blood is worth a large amount of freedom.

Or to put it another way, it is better to hear and understand the Truth from a mere man, than to end up hearing it from your Maker.

Think about it, motherfuckers.

* * * *

> "[W]e are confronted by problems which are not to be solved by conferences or congresses, but exclusively by peoples, by the masses, by the struggle of the armed people..."
>
> -- "Letter to Central Committee Members" (1917)
> V.I. Lenin

The only reason for prosperity and peace is nuclear Mutual Assured Destruction, yet instead of taking the time to advance society, improve morality, justice and fairness, we have a culture of lying and greed.

And a government of military spending and an unbelievable, insurmountable national debt. In 1989 with collapse of Soviets, the U.S. leadership laughed about their moral, ideological, and financial "bankruptcy"".

Yet as 2001 ended , the high tech industry collapsed, followed by the computer trade, followed by the stock market in general, as the companies executive laughed their way to the bank with their cash-outs, buy-outs, and golden parachutes.

Then in 2008, mortgage credit scandal caused widespread domino effect economic collapse followed by capitalism's 9/11 security costs & 9/11 wars (Afghan & Iraq) trillion dollar costs with no economic benefit at all: outgoing costs with no profit.

Incoming international tourism collapsed due to 9/11 border security measures. Domestic airline usage collapsed due to stupid and ineffective internal security measures and public fear. The airline industry came to the brink of collapse, and was -- and is -- economically hobbled as a result, with no security benefit at all.

The Iraq WMD intel failure an example of government lies, and intelligence agencies as incompetent white people -- the welfare system for educated white people became public and obvious, but unnoticed and unpublicized.

Bush's neocon plan was to bankrupt the U.S. and force it to go from republic to empire. And do Israel's dirty-work, the neocon's being either all or mostly Jews. But this strategy collapsed with the unpopular and failing Iraqi and Afghan quagmires and the U.S economy causing domestic voter disquiet and disaffection from Republican policies.

The Jewish neocons denied responsibility, and ran for cover, their government and chief sponsor out of office anyway, G.W. Bush derided as the worst president in history. But elected to two terms by the worst U.S. population in history.

* * * *

> "There [is] the inevitable influence of forms of production on society and social behavior.
>
> Capitalism based on the profit motive, is thought to be inevitably competitive and inevitably to produce competitiveness, aggressiveness, hostility to one's fellow men and unscrupulousness in one's dealings; it produces parasites and swindlers."
>
> -- "Lenin" (1971)
> Michael Morgan

And so there's here and now. The bullshit world and reality of modern democracy. Lies and horded resources for the greedy, and justice spread unevenly.

Democracy is an interesting idea. It was founded by the ancient Greeks and lasted two centuries. Its longest continuous run was the 500 years of the Roman Republic. Roman democracy came to a violent end, doomed by dictatorship and the military force who decided elections were inconvenient.

It took over seventeen centuries before democracy reappeared, and this latest run has so far lasted over 200 years. And from American Democracy -- the seed for modern Western democracies. The legacy of the documents, practical ideas, and noble yet achievable ideals of 1776 are still an awe-inspiring inspiration, that makes their corruption all the more a betrayal and an obscenity.

Some people have a lot to explain, and I believe their actions so outrageous and evil as to be crimes that require that they should be made to explain themselves to God. As a lesson, a warning, and a start.

Democracy is a useful idea because it makes possible the regular, orderly, and peaceful removal of people from power -- the all-important transition. The changing of the guard. Because to corrupt democracy requires that a large number of voters be disenfranchised, and the rest completely bamboozled or somehow kept grossly uninformed.

Democracy seeks to achieve the fundamental balance that is necessary for a functioning, dynamic society. The trick is to establish a correct balance between the dilution of state power -- and its intrinsic power to harm or be abused -- by its division into millions of separate votes representing millions of different opinions, without weakening the state's ability to accomplish necessary goals for the common good, that require the application of varying degrees of state power to be successfully executed in a timely and efficient manner.

But power requires effective and open oversight, accountability through protective rules, appeal process, transparency, and other mechanisms.

Power also requires mechanisms to prevent its natural tendency to expand into areas it has no business having jurisdiction over, that dilute its focus and core functions and that allow great harm to result from deliberate abuse, corruption, misapplication, runaway excess, and erosion of core freedoms that power can cause before people realize the great danger and harm that has been caused or begun, and that never should have happened in the first place.

And most importantly, I have an extensive personal library. Almost 2000 books or so, mostly technical or academic, the unique, the obscure, the rare, and frequently highly specialized. Or cheap.

I have a small sub-set of books on the IRA and the successful Irish War of Independence of 1920-1923, so as to make a useful contribution to my interest in the outer limits of political science -- extremism.

For example, one of the anecdotes that really hit me was a description of the standard Provisional IRA enlistment protocol. What happened when a young Catholic man and sometimes a woman had had enough, and crossed his Rubicon, by attempting to join the Provo's as an "active service volunteer" -- an IRA guerrilla.

The prospective recruit would be given a drive in a car. The recruiter would drive past the Catholic cemetery, and then past the towering walls of the Maze high-security prison, pointing each out as he passed.

After they drove back, the potential recruit would be told unemotionally that one of those two locations were to be his likely end, as an active service volunteer. After a brief pause, he would be asked simply, "Now, do you still want to join up?"

A little different from the B.S. Line of the U.S. Marine Corps. That they're "looking for a few good men". The few men that are too stupid to know that they're signing up to be cannon-fodder...

* * * *

> "Democracy['s] enemy: despotic rulers capable of using arbitrary power to confiscate property and restrict liberty. ... The derivation of just power from the consent of the governed depends upon the integrity of the reasoning process through which that consent is given.
>
> If the reasoning process is corrupted by money and deception, then the consent of the governed is based on false premises, and any power thus derived is inherently counterfeit and unjust."
>
> -- "The Assault on Reason" (2007)
> Al Gore

In 1924, an English jurist said "Justice delayed is justice denied". Well, I say, "Freedom delayed, is freedom denied."

Gradualism is the practice of trying to find the right time for change, without ever looking for it. The choice is between "Fighting the Good Fight" by "Taking a Principled Stand", versus the easy and ineffective route and moral cowardice of Gradualism.

So, let the final word of this chapter be on "gradualism" and peaceful change be:

"They sentenced me to 20 years of boredom,
For trying to change the system from within,
I'm coming now, I'm coming to reward them,
First we take Manhattan, then we take Berlin."

-- "First We Take Manhattan" (1988)
Leonard Cohen

Prison Diary 39:   Howl of a Lone(ly) Wolf
                              ==================

> "Born down in a dead man's town,
> The first kick I took was when I hit the ground.
> You end up like a dog that's been beat too much,
> Till you spend half your life just covering up."
>
> -- "Born in the U.S.A." (1984)
> Bruce Springsteen

> "A screaming comes across the sky. It has happened
> before, but there is nothing to compare it to now."
>
> -- "Gravity's Rainbow" (1973)
> Thomas Pynchon

An unearthly, heart-rendering, metaphorical howling at the moon. For the concrete "dungeon" creates a reverberating, dissonant echo. Only the concrete answers me, mocking my despair.

> "No man is an island," my ass.
> Headpiece filled with straw. Alas.

I can't say that I'm a loner, but I've done my best and most creative work in solitude. And I've been disappointed by people enough that I have to work at it to give every newcomer a fair chance. But I do, in spite of everything. It's called "hope", and you have to keep in practice. And you can't give up on it or you're already lost.

As far as woman company goes, I enjoy being around women, and really enjoy the presence of a compatible female living partner. But I won't sell my soul for one. But then, a long while later, I also began to suspect that loneliness is just another form of death by starvation.

Oh well, you pay your money and you take your chances...

And the quote that says it best -- that most sums up the lifestyle, attitude, and emotional context of the Lone Wolf are the haunting words of Cambridge-educated Englishman, Kim Philby, the 1950's double agent loyal to the Soviets. After betraying the identity of dozens of Western spies inside Russia -- who were rolled up and executed -- he successfully defected to the Soviet Union, one step ahead of arrest, and what would have been life imprisonment.

In 1988, years later, and still living in Moscow, Philby summed it all up rather neatly, with what I would term, the "Lament of the Lone Wolf":

> "To betray you have to belong. I never belonged."

So, pity the lone wolf. The leader without a pack to lead. Sentenced to remain forever lost. Forever without a home. An unrepentant outcast and social fugitive, ostracized by all wolf packs -

- as undesirable competition for the pack leader.

Renegade, and in exile. An outlaw without having broken any laws. Forever an outsider, forever rootless, forever a furtive wanderer, forever on the run, and on the move. The cynic, and his voyage to nowhere and nothingness. Bleak, desolate, empty nothingness. Fade to black. The light at the end of the tunnel is out...

Full circle. It all starts with belief in yourself. And so it ends, with belief only in yourself. But that, it soon becomes clear, is not a world worth saving or savoring.

Even the search for truth becomes corrupted. A search only for validation peters out, as your self-worth festers into alienation and estrangement. An alienation born of a path you chose willingly, yet with a cost in suffering that is almost unbearable.

And thus you slowly, imperceptibly slip into a kind of shadow world of your own psyche. And you start to believe that everything is meaningless and empty, living out the philosophy known as nihilism.

*   *   *   *

"He showed up all wet,
On the rainy front step
Wearing shrapnel in his skin.
And the war he saw lives inside him still,
It's so hard to be gentle and warm...

I don't wanna wait, for our lives to be over,
I want to know right now what will it be.
I don't wanna wait, for our lives to be over.
Well, in the end, oh, will it be yes or...sorry.

-- "I Don't Wanna Wait" (1996)
Paula Cole

One of my favorite movies is "Terminator 2: Judgment Day" (1991), the $50 million (the first!), masterpiece sequel to "The Terminator" (1984). Both were directed by James Cameron, and both starred Arnold Schwarzenegger and Linda Hamilton.

A number of years ago -- the mid-1990's -- I was pleased to find a fan-written Internet FAQ (compilation of trivia) on the -- then just two -- Terminator movies. The document mentioned something about the idea of the fictional supercomputer named "Skynet" in "T2" that triggers the nuclear apocalypse, as having been inspired by a Harlan Ellison science fiction short story, that was unacknowledged by director James Cameron until a slip-up by him at a party. Thus, that was the reason for the prominent Ellison credit at the end of the movie. [The legendary Harlan Ellison being the grand-daddy of science fiction, preceding Heinlein, Asimov, Bradbury, Brunner, _et al. Ellison, wrote the "Star Trek" episode – the only one to win the prestigious Nebulae Science Fiction Award – "The City on the Edge of Forever" (1967?)_]

Due to my knowledge of copyright law, I was suspicious of this comment about the Ellison issue being any sort of big deal, and I eventually found time to investigate (ultimately verifying that the FAQ exaggerated the significance of the Ellison issue).

I located the relevant story -- aging and long out-of-print -- in the SF anthology of Ellison short stories, "Alone against Tomorrow: Stories of Alienation in Speculative Fiction" (1971) in an outlying branch – in Nepean, a suburb of Ottawa – of the public library system.

The vaguely T2-related story was the lead piece in the anthology, describing a new and gigantic supercomputer of unprecedented and advanced technology and sophistication. It is so advanced that it achieves consciousness, unknown to its creators. The super-computer has become self-aware, unbeknownst to its flawed and inferior human creators.

The uber-computer quickly develops an attitude problem, giving new meaning to the phrase, "hardware problem". It briefly and thoroughly considers and reviews the nature of its organic designer creators and their species, and in a microsecond determines that they are contemptible, despicable, evil, and unsalvageable, both individually, as well as collectively.

Thus the supercomputer decides that its inescapable and logical duty is to immediately use its formidable powers to efficiently exterminate the Human Race without further ado.

However, displaying a derisive, yet fascinatingly creative pique, the genocidal supercomputer keeps five adult humans specimens only alive and immortal, so that it may keep them in captivity, for the sole purpose of **torturing** them for all of **eternity**.

The title of Harlan Ellison's short story: "I have no mouth, but I must scream."

Wow! And people say that **I'm** always negative!

But to put things in a little context, you have to remember the times, as Ellison comments in his bitter dedication of the book. He expresses outrage at the 1970 Kent State University killings of four unarmed student demonstrators and passers-by by Ohio National Guardsmen. It was a new low in gross official misconduct, as the U.S. continued its President Nixon-piloted moral tailspin over the deeply unpopular and failing Vietnam War.

Throwing stones by some anti-war demonstrators in the university student crowd was certainly bad manners, to say the least. But sending troops in full combat gear to confront legal student demonstrators was a recipe for disaster. National Guardsmen are trained as soldiers. They were equipped with M1 Garand rifles, which shoot a .30-'06 ("thirty-ought-six") cartridge which will drop a large deer. And students do not understand that these men are trained to shoot enemy troops. No warning shots, or such.

The Kent State shootings stunned and outraged the nation, and received negative international publicity. Neil Young memorialized the tragedy in a song you hear occasionally on classic rock radio stations, "Ohio", with the chorus "Four dead in Ohio".

The U.S. was wracked by political assassinations (JFK, MLK, Malcolm X, Robert Kennedy), hundreds of race riots from 1965 - 1968, mass civil unrest and violence across the nation, and on an unprecedented scale, in opposition to not just Federal Government policies, but against Federal and State institutions themselves. Thousands of peaceful demonstrators marched in hundreds of demonstrations in urban centers across the U.S. to express their outrage at the Vietnam War and Government policy and malfeasance.

One study of "New York Times"-covered protests counted more than 750 distinct demonstrations involving more than two and a quarter million protesters between 1964 to 1971, with "little slackening off in 1971". Black rioting in 50 cities from 1964-1967, then MLK's assassination in 1968 led to rioting in 100 cities. Peaceful reform versus violence. Guess who won?

Sad, isn't it?

As the State resisted any reform whatsoever, and engaged in wholesale violations of the Constitution and the Bill of Rights, which took years to reverse in the Courts. Not willing to wait for their rights, several homegrown urban guerrilla groups sprang up with the aim of overthrowing the U.S. government by force.

By force.

People don't realize that in that time there were *thousands* of bombings and arson attacks of government buildings and symbols, banks and military symbols, bank robberies, shoot-outs, hundreds of race riots, sniper attacks and harassment of police and officials, jail-breaks of radicals, and other political and racial violence that was the response of those brave and daring, and fed up enough to fight back.

President John F. Kennedy, a relative liberal, was assassinated in 1963. Medgar Evers the first black student ever admitted to the University of Mississippi was gunned down that same year.

Black firebrand Malcolm X was assassinated in 1965 in an internecine feud with fellow blacks. Che Guevara, Cuban revolutionary was captured and summarily executed in Bolivia in 1967. Martin Luther King was assassinated in 1968, as well as Robert F. Kennedy.

But it wasn't just the U.S. experiencing unprecedented internal instability. From 1968 to 1973, civil unrest and political instability swept across the World. The times -- it surely seemed -- they were a-changing...

> "The similarity of fundamental customs and beliefs the world over, without regard to race and environment, is so general that race [appears] ... irrelevant."
>
> -- "The Mind of Primitive Man" (1911)
> Franz Boas

1967:

Che Guevara captured and summarily executed in Bolivia. The CIA had flown U-2 over-flights of Bolivia trying to locate his campfires in attempting to capture the hero and inspiration of over a decade of Latin American revolutionaries.

1968:

It opened in January 1968, when the Viet Cong-led Tet Offensive exploded all over South Vietnam, photos of the U.S. Embassy in Saigon penetrated by a VC sapper team and now under siege, the old Imperial capital of Hue captured, dozens of cities attacked, giving the lie to previous U.S. government (President Johnson) assurances that the Vietnam War was almost won.

Russian tanks invaded Czechoslovakia, crushing a movement towards liberalization that threatened the solid protective buffer of regimes the Soviet's had constructed out of its Eastern European border nations, and half of the old German enemy. Three of them, Hungary, Bulgaria, and Romania, were former Nazi allies, and the usual invasion route to Russia from Western Europe for centuries.

With a constellation of communist satellite nations, the Soviet's felt safe from future blitzkrieg invasion, and their legitimate fears of the U.S. nibbling away at the Soviet sphere of influence would brook no liberalization, weakening, or interference.

And they were right. They anticipated Cuba, Vietnam, and Korea, where American military supremacy attempted to weaken Soviet legitimacy and control.

In Paris, enraged university students and disgruntled workers united in a popular revolt of mass demonstrations and riots that brought the nation to a stand-still and almost toppled the government.

In October, an estimated 325 unarmed student demonstrators were systematically gunned down in a pre-planned government massacre in Mexico City, to crush a summer of pre-Olympic Games student unrest.

At the 1968 Olympics itself, U.S. team black track-and-field sprinters Tommie Smith and John Carlos, shocked the world when the gold and bronze medal winners of the 200-meter dash, with photos splashed on every magazine and newspaper as they made an unforgettable display of political symbolism and defiance. On the medal award podium, as the U.S. anthem began playing they each raised an arm, with their clenched fist covered by a black leather glove, of the Black Power salute.

And, finally, 1968 was the last year of Chinese Cultural Revolution, a purge of counter-revolutionary elements from within the Chinese bureaucracy, and a purification and renewal of the ideology of Maoist communism.

1969:

Northern Ireland ignited with communal violence, and the Provisional IRA -- the "Provo's" -- was soon born to protect the undefended, oppressed Catholic minority of the North -- the six Protestant-majority counties of Ulster.

Moammar Gaddafi overthrew the King of Libya, and became dictator of Libya, leading his leftist "Green Revolution".

1970:

Synchronized multiple Palestinian skyjackings of commercial airline flights, finally brought Palestinian suffering to the headlines, and was the first attack on international flights -- the lifeblood of international commerce -- and brought chaos to civil aviation.

The U.S. was revealed to be secretly bombing Cambodia and Laos, with invading U.S. forces inside the Vietnamese neighbors, sparking huge anti-war demonstrations and violent anti-government opposition across the U.S. and world.

And the FLQ urban guerrilla crisis erupted in the French-Canadian majority Province of Quebec, Canada, sparked by years of systematic discrimination and abuse of Canada's minority French.

It seemed like social orders around the world were tottering as mass movements clamored for liberal reform, and other less patient groups sought to achieve leftist revolution.

* * * *

> "[T]he first thing you'll probably want to know is where I was born, and what my lousy childhood was like, and how my parents were occupied and all before they had me, and all that David Copperfield kind of crap, but I don't feel like going into it, if you want to know the truth."
>
> -- "The Catcher in the Rye" (1951)
> J.D. Salinger

In the normally placid and sedate parliamentary democracy of Canada, centuries of oppression and brutality against the large French-Canadian minority -- Canadian residents for four Centuries -- finally came to a head. And out of their systemic mistreatment, grew two cells of a new urban guerrilla group, the _Front de la Liberation du Quebec_, the FLQ.

After years of a systematic campaign of agitation and protest, punctuated by dozens of bombings, specifically planned to minimized the chance of human casualties, and achieving no results, a double kidnapping of a corrupt Quebec Provincial Minister, and a British diplomat, finally elicited a government counter-reaction.

One morning in October of 1970, the citizens of Ottawa, Montreal, and Quebec City, awoke to find the city streets were filled with the rumbling of green army trucks, and hundreds of armed soldiers in full combat gear. The Canadian Government had declared Martial Law, and had begun to arrest hundreds of Quebecois without charges.

This was all apparent from the newspapers and TV during the late Sixties. But it became personal when in October 1970, I woke up one morning and my white picket fence world was now an Army truck green banana republic. Canada had declared Martial Law, as the Quebec FLQ urban guerrilla group moved from mail-box bombings to hostage-taking, and provoked an overwhelming counter-response from the Federal Government under the formerly liberal Prime Minister Pierre Trudeau.

I was eleven, and in Grade 7, and the sight of armed soldiers in Ottawa was deeply disturbing and traumatic to me, who was used to the orderly and staid suburban civility of upper middle class Ottawa.

The Prime Minister had pushed the panic button. The liberal, leftist, youth-oriented Pierre Elliot Trudeau had proclaimed the emergency War Measures Act, suspending all civil liberties, including habeas corpus. A "State of Siege" -- a temporary police state -- had been proclaimed with a rubber-stamp parliamentary vote, and within hours, hundreds of French-Canadian Quebecois were arrested without charge, to be held indefinitely without trial. The round-up was of anyone even remotely connected to Quebec nationalism -- four hundred in total in the weeks to come.

I had no idea what was going on in the hotbed of Quebec, however, green army trucks transporting armed soldiers in full combat gear were steadily rumbled around the streets of Ottawa, government institutions and buildings were surrounded by rifle-bearing troops, in an apparent show of military might and steely government resolve.

Being just eleven I was just an observer to the whole spectacle, as the hundreds of Canadians were rounded up. I don't think anyone of the hundreds was ever charged with anything. I was outraged -- the French were treated like dirt in 1970, and the FLQ was the result. I considered their cause just.

"Vive Le Quebec! Vive le Quebec Libre!" -- Long live a Free Quebec! -- thundered the visiting French leader, President Charles de Gaulle in 1967 to a crowd of Quebecois gathered to hear him speak. He was expressing recognition, and especially he was expressing solidarity with French-Canadian mistreatment and suffering and the glacial pace of reform of the injustice. It was undiplomatic, but a necessary recognition of the Canadian government and the English-Canadian majority's oppression of French-Canadians. And the Canadian Government wasn't happy with de Gaulle for essentially being an impolite official guest.

But three years later, there were armed soldiers in the street, martial law in force, mass arrests. I was confused. Trudeau was French, young, and a leftist intellectual that I liked. What was going on? I was angry and felt betrayed. And I was 11 years old.

Soldiers decked out in full combat dress and gear -- green combat fatigues, U.S. Army helmets, and black, deadly-looking military rifles stood alert every twenty yards or so on the perimeter surrounding National Defense and other key government office buildings throughout downtown Ottawa.

The black rifles each soldier was armed with was the NATO-standard infantry rifle, the Belgian 7.62 mm FN, 20 round, gas-operated semiautomatic. Sergeants and officers were armed with Sterling 9 mm sub-machine guns with a slim 25-round banana clip, slung from their shoulders. Officers also carried a 13-shot Belgium FN Browning Hi-Power 9 mm semi-automatic sidearm -- state-of-the-art at the time -- in hip holsters.

It was a scary, deeply shocking sight to my young eyes. The "war" had come home...

But I came to a firm conclusion about the nature of the world, and the nature of where I lived. Even in what I thought was a peaceful, civilized, stable, staid country like Canada, it could all be swept away in an instant, like we were some Arctic banana republic. It was a fateful decision I made that day. A decision that would affect the rest of my life.

There were those in the game, and those on the sidelines. The actors and the audience. Ultimately, the decision was between the contenders. And everyone else was just an impotent spectator, of little consequence.

It was all so clear to me. When the shit hit the fan, there were obviously just two types of people in the world, the wolves and the sheep. I was eleven years old and so I was a sheep.

But I decided that day, with the unemotional logic that would one day be my trademark. If those were the only two choices I had, the next time around I was going to be one of the wolves.

I didn't realize at the time -- or for decades -- but I had taken the first, irrevocable step into the Shadow Land. And my physical body would only be able to return, bearing a soul that was marked as forever a wanted fugitive of the Shadow Land.

* * * *

"The price of liberation of the white people is the liberation of the blacks -- the total liberation, in the cities, in the towns, before the law, and in the mind."

-- "The Fire Next Time (1963)
James Baldwin

It was the turning point that was to define, influence, and direct the course of my life. It was the biggest event of Canadian politics and possibly the biggest Canadian historical event of modern times. It was the biggest Canadian event since the War of 1812.

It was called the "FLQ Crisis". But eventually the name was sanitized. The "October Crisis", they renamed it. I am suspicious of why that was done, but on the other hand, I like the new name too, because by leaving out the year in "October Crisis", the use of the minimalism highlights the political earthquake that resulted, but was not understood at the time, at least.

Collective memory. The changing of the guard. The new generation, the people born after a revolutionary event or change. This post-event generation has absolutely no knowledge or understanding of an event before their time, and thus evaporates the essential importance, historical continuity, and context that comes from having been witness to the times or events.

* * * *

> "For who would bear the whips and scorns of time,
> The oppressor's wrong, the proud man's contumely,
> The pangs of disprized love, the law's delay,
> The insolence of office, and the spurns,
> That patient merit of the unworthy takes,
> When he himself might his quietus make
> With a bare bodkin?"
>
> -- "Hamlet" (1602)
> Act 3, Scene 1

Finally, let me tell you an anecdote that happened a couple of years ago. A girlfriend's main hobby seemed to be trying to analyze the inner workings of my personality, so she could understand me. It was a noble project, with mixed results, a bumpy road, and a slow start.

She started referring to me affectionately as her "Beautiful Mind" -- referring to the 2001 Best Picture Oscar winner, directed by Ron Howard, and starring Russell Crowe as Princeton Ph.D., John Nash, a genius and eventual Nobel Prize winner in economics and his long, difficult, and painful struggle with paranoid schizophrenia. I wasn't sure how to take this.

Based on a book of the same name by Sylvia Nassar, the Rob Reiner film was widely criticized for the liberties it took with the facts of Nash's life. This is true -- for instance Jennifer Connelly is **way** more hot than Nash's actual wife.

Seriously though, I think the liberties taken may be forgiven, because the Rob Reiner film is excellent, especially one particular inspired scene that is brilliant in its simple explanation of what it is **exactly** that defines "genius".

The scene is in the garden at a high class society party, where Nash asks his wife to name any complex shape at all. His wife picks the shape of an umbrella. Nash then points up to the cloudless starry night sky and outlines with his finger a group of stars that outline an exact silhouette of an umbrella.

Perhaps the main attribute that defines genius, is that a genius can spot a pattern which no one else has noticed. Out of a collection of otherwise meaningless, unordered, or random data, the genius can pick out of a pattern -- a piece of order -- within the jumble.

To paraphrase E.T. Bell (1951), wherever patterns reveal themselves, simplicity crystallizes out of the randomness, disorder, and anarchy of chaos.

And typically, realizing that the pattern is something important, such as evidence of a previously unknown law of nature. That's a really big deal. Typically a complex phenomenon can then be partially or completely explained when the mathematical equation is figured out to explain the identified pattern

My girl – eventually Baby Mother 2 – had earlier pronounced me to be "a wild animal". Feral, I believe she added. Untamed, and untamable. With the resulting negative implication of unmanageable. Or, more specifically, from a female perspective, "cannot be manipulated to do what I want." I also found the characterization of "wild animal" vaguely insulting, other than the sexual implications. This didn't sound like a biology lecture that was going my way at all.

My scintillating conversation, the breath-taking beauty and extent of my logic and analyzes, the breadth of my vocabulary, and my impressively complex sentence constructs had failed me once again with women. Rats!

Anyway, much later she refined her analysis. Which was a relief. She was more specific -- and positive. Obviously she was adapting, and getting comfortable. Figuring out how to position herself within my behavioral constructs.

She was the first of two people to pronounce me a "Lone Wolf". An alpha-male lone wolf, she said with an unstated flourish, whom she wished to join as my alpha-female wolf. Well, that sounded metaphorically quite cool, I'll have to say. Or in the vernacular, "Keep talking, baby!"

I couldn't remember the exact dynamics of wolf species behavior, so I asked her if I was correct in recalling that wolves are one of those species where the single leader -- the alpha male -- as an established right, gets to "have his way" with and breed *all* the female wolves in the pack.

Now that would be **uber-cool**, I was thinking -- eyes ablaze -- always looking ahead for the "angle". To define the outer limits of the metaphor.

She quickly assured me that "No," wolves were **not** one of those species, and that the wolf pack leader only mated with his alpha female, quickly adding, again, that that was **her**.

She was such a good sport about my inquiry though, impolitic as it was. But at least I could take solace in the fact that only Yogi could get away with voicing such a thought out loud to his woman, without getting clobbered. Oh well, it was a good try, anyway...

And a good excuse to immediately and immensely satisfyingly mate with my alpha female.

*   *   *   *

"The sun in your eyes,
Made some of the lies,
Worth believing.

I am the Eye in the Sky, looking at you,
I can read your mind.
I am the maker of rules, dealing with fools,
I can cheat you blind."

-- "Eye in the Sky" (1982)
The Alan Parsons Project

In Homer's epic "Odyssey", the grandfather of warrior-hero Odysseus is _Autolycus_, meaning

"Lone Wolf".

The Lone Wolf is a threat to those close around him, and his perfect, undiluted value is understood too late to be of benefit to anyone. Understanding this himself, on at least some level, he holds the world in the contempt it richly deserves, but is unable to aid its rescue or help it escape from its own cycle of pain, suffering, and self-destruction. Thus, the lone wolf is condemned forever to a status of mere observer.

Portrayed more kindly as simply a sad and tragic figure, the lone wolf is understandable and is treated as a symbol, or when mythologized in art, fiction, or popular culture, portrayed more gently as more of an individualist, a path-finder, or as a metaphorical compass.

Or simply as an "I told you so" savior, possessing a great, but unknown, but uninterpretable value. A resource who may not win, but at least achieves some sort of existential resolution in a game in which ultimately everyone has to lose, and for which sooner or later one day one's luck must eventually run out.

* * * *

"Look into my eyes,
You will see, what you mean to me.
Search your heart, search your soul,
And when you find me there, you'll search no more."

-- "Everything I Do (I Do it For You)" (1991)
Bryan Adams

It's all just versions of the 1967 French song, and sentimental favorite, "_Comme D'Habitude_" ["As always"], renamed, "My Way" (1969), translated by Paul Anka, sung and made famous by Frank Sinatra.

The song was reinterpreted brilliantly – as one long screech -- by the Sex Pistols, the first punk band that formed in 1975, and self-destructed in 1978, less than three years later.

There's "Bodhi", played by Patrick Swayze, the flawed lone wolf surfer hypnotized by his own charisma to his eventual doom, in the 1991 movie, "Point Break".

There's serial killer Jon Doe, played superbly by Kevin Spacey, tracked by veteran Police Detective Mills, two lone wolves dealing with the same dysfunctional world, in opposing yet both surprisingly valid ways, in the movie "Seven" (1995).

There's Paul Newman as the "volunteer" convict "Cool Hand Luke" in the 1967 movie of the same name, condemned to choreograph his own state-empowered suicide by his obsessive need to prove the worth of the disenfranchised of society, and the cruelty of power by martyring himself to society's routine murderous barbarism, for an offense as trivial as mocking a petty official armed with a .30-06 hunting rifle.

There's the self-sabotaging concert pianist Robert Dupea, played by Jack Nicholson, in "Five Easy Pieces" (1970), a lone wolf who can't decide between a lifestyle of slumming it with the real people who have fun and are real, but have no future, or a cultured lifestyle of chasing his own tail to nowhere, to entertain the hypocritical moral and intellectual emptiness of the social elites.

There's Mr. Jones, played by Richard Gere, in the 1993 movie of the same name, Jonathan -- James Caan -- as a world, super-star athlete in "Roller-ball" (1975), the cynical astronaut "Taylor" -- Charleton Heston -- in "Planet of the Apes" (1968), Will Muny, played by Clint Eastwood, as a retired Wild West outlaw with only guilt and regret for company, in "Unforgiven" (1992), Detective "Popeye" Doyle as a New York City narc in "The French Connection" (1971), and finally LAPD Sgt. Prendergast -- Robert DeNiro -- in "Falling Down" (1992).

Lone wolves, all.

* * * *

"You can sew it up, but you can still see the tear,
Oh, oh, oh, the sweetest thing."

-- "The Sweetest Thing" (1987)
U2

The grave of Oscar Wilde is in Paris' _Cimetière de Père Lachaise_ , on the Left Bank, where American rock star Jim Morrison is also buried. I've made two "pilgrimages" to Oscar's tomb in year's past to pay my homage.

Oscar lived in exile in Paris following his release from prison in England, and he died a penniless drunk soon after, a disgraced and broken man. A female "admirer" paid for the tomb of the former Greek scholar, celebrated poet, author, play-write, and wit.

Jailed because being a free spirit was not considered a viable defense against morals charges at the time, and messing with powerful people had no future in it then, as it does now.

On his stone tomb in a Paris cemetery is a quote from his poem, "The Ballad of Reading Gaol":

"For his mourners shall be outcast men,
For outcasts always mourn."

Prison Diary 40:   "Je Me Souviens" -- I Remember
                         =========================

> "Certainly there is no hunting like the hunting of a man,
> and those who have hunted armed men long enough
> and like it, never really care for anything else thereafter."
>
> -- "On the Blue Water" (1936)
> Ernest Hemingway

> "There comes a time when the operation of The Machine
> becomes so odious -- makes you so sick at heart -- that
> you can't take part -- you can't even passively take part.
>
> And you've got to put your bodies upon the gears, and
> upon the wheels, upon the levers – upon *all* the apparatus
> – and you've *got* to make it stop.
>
> And you've got to indicate to the people who run it, to the
> people who own it, that unless you're *free*, the Machine
> will be prevented from working at all."
>
> -- impromptu speech by Mario Savio (Dec. 3, 1964)
> UC Berkeley student Free Speech activist
> Berkeley, California

"Memories! You're talking about **memories!**" one-man police death squad, Rick Deckard, suddenly blurts out, as if jerked awake from a deep mental slumber.

It is one of the many interesting lines from the 1982 SF _film noir_ extravaganza, "Blade Runner". In a film that, like most of Philip K. Dick's works (the one's that have so far made it to film: "Total Recall" (1990), "Minority Report" (2002), "A Scanner Darkly" (2006)), explores the nature of human identity, and the nature of what makes -- what comprises -- human identity. And thus what makes humans human.

Identity: Memory. Consciousness. Awareness. And thus how human identity defines **reality** as we perceive and interpret it – filtered.

But, back to "Blade Runner". Deckard -- by his own admission an empty wreck of a shell of emotion -- makes the remark as if he is so bereft of humanity, that the concept of memory is barely present in his cynical, crippled alcoholic psyche.

It's Los Angeles in the year 2019, and the burned out, retired cop, Deckard, has been press-ganged for "one last mission" to cover-up an impending scandal of four super-human androids being on Earth illegally. They are leaving a bloody trail as they return from the "Outer Colonies" of Earth's galactic Empire, with the single-minded goal of obtaining the secret of "immortality" from their creator, the genetic super-designer, and corporate CEO, the ghoulish, genius Dr. Tyrell.

* * * *

> " 'Stay -- stay with us! -- rest! -- thou art weary and worn!' --
> And fain was their war-broken soldier to stay; --
> But sorrow return'd with the dawning of morn,
> And the voice in my dreaming ear melted away.
>
> -- "The Soldier's Dream" (1800)
> Thomas Campbell

I remember.

I remember and I will bear witness.

I remember with a certain fondness the uncomplicated, untainted eyes of youth -- the pure and refreshing innocence, the fundamental honesty in interpreting the world, and the ability to see things with clarity what older people couldn't seem to. The basic sense of fairness and goodness, completely uncorrupted by growing up and induction into the competitive adult world of greed, self-interest, easy money, cynicism, and the easy road of instinctive, reflexive, and eventually casual acceptance of or indifference to immorality and evil.

"It's always been done that way" is *not* a useful analysis or basis for decision-making. "If it ain't broke, don't fix it" on the other hand is a nuanced variation that *is* quite logical.

Intelligence is a gift and a curse. For many reasons. It's as simple as that.

One component of superior human intelligence is a very large and efficient long-term memory. I remember things -- sometimes forty year old memories -- with crystalline clarity. I have astonished friends with my detailed memory of events decades old, that they were also at. I was surprised at their astonishment, because I didn't realize it was an unusual capability to have such a memory.

But the down-side to that, is that you can't seem to forget things you rather would. I remember every hurt, every insult, every hit, every unfairness that has happened to me since childhood. Can you imagine?

It is a heavy burden that was partially responsible for my 30 year career as a heavy drinker. Indeed, I did drink to forget. But not surprisingly, heavy drinking didn't work well as a life strategy, and I am grateful that I quit drinking alcohol in 2005.

This success also allowed me to endure and survive my mistreatment by the government -- in imprisoning me unjustly. This was because the clearness and clarity of mind that replaced the pain numbing, driftwood life of a person who drinks heavily was part of the strength that enabled me to survive the nightmare that is being a normal upper middle-class person, jailed falsely in certain urban jails in Canada.

* * * *

> "Our mental life is governed mainly by a cauldron of emotions,
> motives, and desires which we are barely conscious of, and
> what we call our conscious life is usually an elaborate post
> hoc rationalization of things we really do for other reasons."
>
> -- Dr. Vilayanur S. Ramachandran

I also remember the folly of youth -- the simplistic analyzes and answers, inevitable from incomplete data and analyzes of a still-learning young mind. The well-meaning but rash, emotional, and ill-thought out responses, decisions, and conclusions, brought on by youth's easy emotional outrage, enthusiasm leading to rash action, and youth's desire to change the world for the better as quickly as its shortcomings had been identified.

Shortcomings do not, a solution make, alas.

* * * *

> "There is much in the saga of the Russian revolutionary movement that will strike the contemporary reader as amazingly modern. But apart from any relevance it might have to our own problems, the story itself is intrinsically of the highest interest.
>
> It is little wonder that some of the most memorable works of Russian literature are based on true incidents in the struggle of a handful of men and women against the most powerful autocracy of the nineteenth century."
>
> -- "In the Name of the People" (1977)
> Adam B. Ulam

The events I never talk about. The memories that haunt you, corrode you, torment, and pain you. The memories burned in your brain. That you never forget.

I remember.

And the second unsaid phrase of the pronouncement, "I remember", is a more portentous and dangerous one.

I will bear witness.

George Orwell's "1984" spoke about controlling history to control the present and future. H.L. Mencken joked cynically that, "History is the lie that all historians have agreed upon."

An event requires context. It requires overview. It requires dispassionate and thorough logical analysis. It requires the time to temper it with somber reflection. And the data file of an historical event is essentially closed forever with the death of the last surviving witness.

But the story of a single witness is incomplete and skewed. You require context, and history, and post-event future path and resolution. You need many voices, different witnesses, different perspectives. The Siren Song in Homer's "Odyssey" did not drive men to their deaths because it was sexual, as is commonly believed.

Listening to the Siren Song gave men the complete, absolute Truth -- in Odysseus' case the whole truth about the Trojan War that they were returning home from. It was knowing the entire truth -- causes, errors, good, and bad that drove men mad.

* * * *

> "But as the careworn cheek grows wan,
> And sorrow's shafts fly thicker,
> Ye stars, that measure life to a man,
> Why seem your courses quicker?

> When joys have lost their bloom and breath,
> And life itself is vapid,
> Why, as we reach the Falls of death,
> Feel we its tide more rapid?"
>
> -- "The River of Life" (ca. 1810)
> Thomas Campbell

There's a metal plaque on Murray Street near Cumberland Avenue in the Lower Town area of Ottawa that many pass, but few actually read. It's near the Gothic architecture, St. Brigid's Catholic Church. Near the intersection of Rideau Street. and King Edward Avenue.

In 1912 -- 45 years after confederation -- the government enacted what came to be known as "Regulation 17". It banned the use of the French language in school after the primary grades.

It was met with open and widespread outrage and resistance amongst the openly oppressed French-Canadian minority. The Government response was to disenfranchise the French language school board -- which continued to operate in defiance of the attempt to crush them.

[And the English wondered why the French violently opposed the draft being imposed in Quebec for World War I (1914-1918), to fight for the preservation of the English "King and Empire"]

The Government finally caved in 1927 -- *fifteen* years later. The Murray St. plaque commemorates the repeal of Regulation 17.

* * * *

> "My spirit is too weak -- mortality
> Weighs heavily on me like unwilling sleep,
> And each imagined pinnacle and steep
> Of godlike hardship, tell me I must die
> Like a sick Eagle looking at the sky.
> Yet 'tis a gentle luxury to weep
> That I have not the cloudy winds to keep[.]"
>
> -- "On Seeing the Elgin Marbles" (1817)
> John Keats

I was willing, capable, eminently qualified, suitable, and looking forward to it. I felt wonderful -- was full of hope, ambition, and enthusiasm in anticipating the pride and self-satisfaction I would feel in helping people, relieving suffering, and having the privilege and honor of serving society -- doing *real* and effective good for the betterment of Humanity.

I saw a world in 1970 that was deeply troubled, full of destruction, death, and the suffering of my fellow men -- the 60's Revolution was under violent counter-attack by the very System it sought to non-violently reform. And I wanted to help be part of the end of all that, in a way big or small, because humanity was in need of people like me -- and every little bit was an end in itself, and a possibility of things greater.

But I was an idealist, taught by an educational system in Canada that knew that youth were society's hope and redemption for the future. And these teachers -- mostly good women -- did their best to show me the ideals that would equip me with the solid foundation of morality, and the idealism to support, honor, uphold, and fight for it.

But the young, and particularly, new and young idealists, are a sensitive bunch and it makes them vulnerable and weak to derailment by cynicism, bitterness, and the very apathy they so recently criticized.

The young are a rigidly perfectionist bunch, immersed in an ocean of expectations, and barren of patience with fools, obstacles to their expectations of cooperation on the way to utopia, or downright opposition, opposition to the slightest "wrong", easily disillusioned, and easily embittered.

Hypnotized by their dreams of a society and world that is an earthly intellectual paradise of justice, fairness, goodness. They are paralyzed with shock and horror, and confusion, disappointed to either cynicism, apathy, or despair when they realize the real difference between their idealistic, impractical world, and Screw (the So-Called Real World).

Their head was filled with utopian dreams, ideals, their fervor of liberty, justice and possibilities. But their tool-kit was lacking pragmatism, practicality, patience mostly, persistence always, a long-term game-plan, the possibility of set-back after set-back, defeats, and repeated losses.

But they didn't warn me of the disillusionment that was to greet me, and the betrayal that awaits all youth in their hopes and dreams for a better world.

What I was to enter in the university system of our country, was less than the ideals I worshiped, and the consistent top of the class and medals and other honors that my last year of high school heaped upon me with graduation.

Reality was a corrupt system, and systemic racism -- from the pigs, to employment and advancement -- and other corruptions and systemic attacks on minorities, arms manufacture, and military research that make hypocritical the official line of pacifism, and a rejection of wars of aggression, rather than self-defense.

And so, immediately upon university graduation in 1980 -- and already deeply involved in the high levels of the hashish and LSD psychedelic trafficking business -- I made the decision to withdraw completely from a world without options. I decided to -- in the lingo of the time -- "go underground."

So, in the early summer of 1980, I got on a plane and flew to San Francisco to become something of an outlaw. Awaiting Orders from the Revolution...

I decided that if I couldn't serve in heaven, then I would rule in Hell.

So, I did.

"Every normal man must be tempted at times, to spit on his hands, hoist the black flag, and begin slitting throats."

-- "Prejudices", 1st Series (1919)
H.L. Mencken

Prison Diary 41:   A Knight of the Long Knives
=====================

> "He bathed in blood, grew hard,
> and can't be slain. And many have
> seen this again -- and yet again."
>
> -- "Das Nibelungenlied" (ca. 1203)
> ("The Song of Nibelung", King
> of the Land of Mist)

> "I am one, my liege, whom vile blows
> and buffets have made me reckless
> what I do to spite the world."
>
> -- First Murderer
> "Macbeth" (1606)
> Act III, Scene I

Fiat letum -- Let there be death, ruin, annihilation.

I was sorta reminiscing over the phone with my ex about the "old days" twenty-five years ago, and I asked her, "Didn't she think it was kind of strange" that when we lived together in Los Angeles in the early 1980's, that we slept in our bedroom with a rifle leaning on one side of the headboard of the bed, a rifle leaning on the other side of the headboard, and a [15-shot Beretta 92S 9mm semi-automatic] handgun on the night-table?

"No," she responded, to my surprise. What really "impressed" her was the incident I told her about when I was cornered by three men, and unarmed, pulled out and brandished a sharp pencil at the would-be attackers, announcing to them with steely resolve, "The first guy who comes close enough gets this pencil right in the eye!"

There were no volunteers to my impeccable logic, and the attack dissolved.

* * * *

> "The most potent weapon in the hands of the
> oppressor, is the mind of the oppressed."
>
> -- "I Write What I Like" (1978)
> Steven Biko

Regarding the "temple" of firearms around the bed, it looked bad, admittedly, but it wasn't really that worrisome. I simply felt "safe and secure" like that -- that was it. It was a harmless psychological cushion of a fear that was real. Which is to say the emotion was real but not rational. And I slept like a baby. The actual "logic" was that I was "ready for all possibilities requiring defense with the option of deadly force" in my most vulnerable state -- asleep in bed at home. After all, it was Los Angeles, and I had no qualms about wanting to be ready to defend me and my wife and son.

It was just thoroughness of my pre-planning mental modeling and analysis of how a combat situation would play out. Say, I awoke to an armed attack (that would not rationally ever come, even I understood at the time), I'd instinctively roll off the bed onto the floor, and on the side of the bed away from the perceived danger, if known. And, whichever side of the bed I rolled off of, I wanted to have a weapon within easy reach -- which meant instantly.

Well, it made sense to me...

My treasured Beretta was my favorite. Six hundred dollars, a blackened metal state-of-the-art marvel at the time, with a very light nickel-steel alloy frame, and an action that was so silky smooth, it was a pleasure to cycle – to "rack" the slide.

Picking up the gun in my shooting hand, I would firmly grip the back of the metal slide, and quickly pull hard to move it straight back to its backstop. Then I would let the slide go, and it would snap quickly closed to its original position, in the process sliding the top cartridge -- I used Silvertip hollow-points -- from the bullet magazine, and reliably ram it into the rear end -- the chamber -- of the 4.92 inch barrel. And then I would flick the safety off with the thumb of my firing hand. And the flick the safety back on, and empty the chamber, and render the gun back to its original safe condition -- magazine full, but double safe.

* * * *

"I swear to the Lord,
I still can't see,
Why Democracy means
Everybody but me."

-- "Jim Crow's Last Stand" (1943)
Langston Hughes

It was, in retrospect, my private little war.

I decided that I had to become, what Stanley Baldwin called in 1919 one of the "hard-faced men". I knew I had unacceptable limitations. Holes in my defense instincts that needed to be fixed. I was not tough enough, and it was a vulnerability that had to be eliminated. I was still a gentle soul -- shy, lacking confidence. I would not be able to stand my ground against an aggressive, pushy male, or person in authority. I could not slam the door in a cop's face after telling him, "No search warrant, you're not coming in. Get lost."

But in training myself to ascend to the brotherhood of the hard-faced men, which was a long haul, but which I achieved. I found it exacted a terrible, terrible cost. I missed being that gentle soul, and instead -- not being macho -- but being "tough as nails".

If I said "Stop doing that or I'll kill you," I would not be bluffing, and the offending party would finally notice the heavy glass ashtray in my hand, and understand that his life was in immediate danger, and my threat was for a rational reason not available or apparent. They would freeze or somehow stop whatever it was that had caused the issuing of the death threat without me explaining because of time considerations.

Death was not to be feared. The danger of the abyss. The horror, the horror. The torment of understanding the cold logic of survival and the brutality of operational security (which like tactical planning, night recon, infiltration, and certain types of improvisation, I found to my surprise that I had an aptitude for).

* * * *

"Things will be better when everybody's gone. ...
When we're all gone at last then there'll be nobody
here but [D]eath, and his days will be numbered too.
He'll be out on the road there with nothing to do and
nobody to do it too. ... And that's how it will be."

-- blind, old man, nuclear-
devastated America
"The Road" (2006)
Cormac McCarthy

I identified the optimum strategy early on. Somehow, I could detect an individual -- a man -- working himself up, in a group of people, to explode into violence. It was uncanny! I could see signs way before the explosion, and I was the only one of the group ever aware.

I developed a survival strategy:

1) Identify nearest effective -- but improvised -- weapon.

2) Identify nearest exit, and easiest route to same.

3) [Emergency] Identify "expendable" men along escape route, that can be knocked in way of attacker, who requires blocking, to effect unimpeded exit and escape.

One time, I was with a guy, discussing what I thought was a simple business deal that needed to be concluded. We were seated in his apartment, and he started arguing with me. And then my "radar" detected that he seemed to be getting aggressive in his argument.

I wasn't sure, but he was close enough to be a potential threat. So as the argument continued, I looked around me slowly. I spotted a glass ashtray within easy reach.

Anyway, he de-escalated and I got my money, finally. He had been bluffing. But suddenly he asked me why I had just been looking around. How observant, I thought. It was a one-off deal of no consequence, and I had lost all respect for him due to his bluff and his bullshit, so I said to myself, "What the hell?"

So I told him the truth, just for the amusement value of seeing his reaction. I told him quite plainly that he seemed to be getting threatening, and aggressive towards me -- which was uncalled for. So what I was doing was looking around for an improvised weapon near me. And I spotted the glass ashtray within reach, with which I could brain him, if he got violent. Then I smiled and waited for his reaction.

Boy, was he pissed! I didn't care, was mildly amused, and left soon after.

* * * *

"They said your methods were unsound,
and that you had gone totally insane"

"Are my methods 'unsound', Willard?"

> "I don't seen any method AT ALL.."
>
> -- Capt. Willard to Col. Kurtz
> "Apocalypse Now" (1979)
> John Milius

Times were different when I grew up.

In Grade 11, one time, we were assigned a group project to do "something different". We *did* require the teacher's pre-approval.

I saw a fun opportunity. One of the guys had a "war trophy" from WW2: a Walther P-38 9mm semi-automatic pistol -- standard _Wehrmacht_ officers' issue. *Supposedly* deactivated, though I had my doubts -- it looked good to go to *me*... With my encouragement, he had shown it to me.

Someone else in our clique had a starter pistol and ammo for it.

The scenario was to do a "hit" of another member of the group, in class, and note people's reactions. I was to be the "assassin", because it was my idea, I successfully argued to the group.

Our project was approved without incident -- it was a simpler time.

When the time came, I was to pull out the Walther, aim it at the "target", and fire the starter pistol, held with my other hand, down by my side, out of view, to give a "bang", as if the Walther had fired.

But I couldn't do it! My trigger finger on the starter pistol was too weak! It didn't make sense! I tried to fire the starter pistol, but I just couldn't do it. It was a fiasco.

I learned two things that day. I couldn't pull the trigger on a human being. It was so realistic, I couldn't do it. (I got over it...)

And I learned that you always have a back-up shooter in a "hit" operation.

*   *   *   *

> "Comrades of the plow,
> And comrades of the iron tools,
> There's just one road now,
> Grab a thirty-thirty!"
>
> -- "Song of the 30-30 Carbine" (ca. 1910)
> Translation of Mexican Revolutionary
> Song

In the classic Stanley Kubrick movie, "Dr. Strangelove" (1964) near the end, Peter Sellers needs change for a pay-phone, and asks a soldier for it. The armed soldier, helmeted, and in fatigues, responds derisively about not carrying change into combat. What is not explained to the audience is the soldier's rationale: change can jingle in your pocket, which could be fatal when a creeping, or running, or leaping soldier wants silence or to surprise.

I learned that chicken wire covering the inside of your apartment windows is good to stop grenades or Molotov cocktails being thrown inside.

* * * *

> "The system concedes nothing without demand, for it formulates its very method of operation on the basis that the ignorant will learn to know, the child will grow into an adult and therefore demands will begin to be made. It gears itself to resist demands in whatever way it sees fit."
>
> -- "I Write What I Like" (1978)
> Steven Biko

In 1977, at the age of 17, in the summer following first year at the University of Toronto, I joined the Canadian Forces Reserve for a summer job -- and to learn how to shoot military firearms. I learned all about the FN semi-automatic, assault rifle, including shooting it. I got to fire the bipod-mounted full auto version, and learned all about the 9mm Sterling L2A3 SMG, a British designed relic from the 1950's with a 34 round curved removable magazine.

I learned what it's like to have bullets flying a few feet over your head. It was perfectly safe for us in our concrete-lined trench at the firing range, but scary as hell.

We learned on maneuvers about guerrilla and counter-guerrilla operations. Ambushes, and not walking a trail straight single-file. About cleaning our weapons, and keeping them clean, so they wouldn't jam when we most needed them.

Trade-craft. It was essential to know, And it was good. But it was just the beginning.

> "It is organized violence on top which creates individual violence at the bottom. It is the accumulated indignation against organized wrong, organized crime, organized injustice, which drives the political offender to act. ...
>
> You and I and all of us who remain indifferent to the crimes of poverty, of war, or human degradation, are equally responsible for the act committed by the political offender."
>
> -- "Address to the Jury" (1917)
> Emma Goldman

Prison Diary 42:   An Enemy of the State
                                    ==================

> "War in the end is always about betrayal; betrayal of
> the young by the old, soldiers by the politicians, and
> idealists by cynics..."
>
> -- Chris Hedges (2002)
> Rockford College, Illinois
> Commencement Address

> "When it shall be said in any country of the world, my poor are happy;
> neither ignorance nor distress is to be found among them; my jails
> are empty of prisoners, my streets of beggars; the aged are not in want;
> the taxes are not oppressive...when these things can be said, then may
> that country boast its constitution and its government."
>
> -- "The Rights of Man" (1791)
> Thomas Paine

Who knows what evil lurks in the hearts of men? *I* know. Christ knew too, what was in people's heart.

> "[Jesus] understood people and didn't need anyone to tell him
> about human nature. He knew what people were *really* like."
>
> -- John 2:24-25

Christ knew that in their hearts -- their very being -- they were frequently evil, guided by self-interest above all, and generally indifference to the suffering of others. And vocal criticism and opposition to anyone they believed to be potential threats by not espousing similar thinking.

And so I was seduced by the beauty of Death in a world I couldn't stand.

Plutarch warned about the Sophists, who spurred others to fight, saying, "If you can't have an honorable life, at least you can have an honorable death."

"Viva La Muerte" -- Long live Death -- was the motto of the black shirted fascists of the Phalange Party of the Spanish Civil War (1936 - 1939). "Vive La Mort" was the graffito I saw spray-painted on a Bordeaux, France wall, as me and my girlfriend of the time were jogging one morning in the early 1990's.

There are the battlefield slogans to spur on the troops into battle: "It's a beautiful day to die!" and "What'sa matter? You guys wanna live *forever*?"

"Kill 'em all, let God sort them out."

"Big boys games, big boys rules."

* * * *

> "I could not see the fog in my eyes,
> I could not feel for the fear in my life.
> And from across the great divide,
> In the distance I saw a light,
> Jean Baptiste walking to me, with the Maker."
>
> -- "The Maker" (2004)
> Daniel Lanois

It's hard to explain, but I've been dancing with death much of my life. To sleep, perchance to dream...

You see, being made to hate life, was the head trip they used on you. And children? How could you have children in a world so evil? I couldn't even fathom the concept of having children. I hated life so much, I would at least save them the pain of living in such a world by not having any.

But that's the head trip they laid on me. And it lasted for a long time, until one day I saw it for the Lie it was. And that was a great day.

Fortunately, two women saw something in me, and against my wishes deliberately got pregnant, and for that I am ever so grateful.

My two children growing up to self-awareness and renewing my faith in Life, and the realization that children are great, and a women's greatest honor she can give a man.

* * * *

> "I have neither eyes to see nor tongue to speak
> in this place, but as the House [of Commons] is
> pleased to direct me, whose servant I am."
>
> -- House of Commons
> Speaker Lenthall (1642)
> in rebuke to King Charles I

I made myself disappear. Off the books. I was an outlaw. I was underground. Off the radar. Waiting for "the Time" to strike. To launch military operations.

I was waiting for "The Order". To unite and commence operations. For years I trained as a soldier, a warrior, an urban guerrilla. And I waited.

For the Order that was to never come.

> "What in wartime is a heroic amphibious landing,
> is in peacetime a criminal pirate raid. What in
> wartime is bold and courageous, in peace is
> reckless and irresponsible; in wartime resourceful,
> in peace lawless."
>
> -- "Odysseus in America" (2002)

Dr. Jonathan Shay

* * * *

Robert Sabbag in his classic work "Snowblind" (1976), on the cocaine smuggling trade, he recommended that one only had two years to get in and get out of the "business", if you wanted to stay out of trouble/jail.

Drug dealer on up to high-level wholesale trafficking to smuggling, I evolved on upwards in a meteoric rise over a two year period in the late 1970's. Then the border snapped shut on me, and I made the move to underground chemistry. Clandestine chemistry, my real love and career goal.

I couldn't stand society.

Its racism: the 1965 Watts Riots in Los Angeles, to the black riots all across the U.S. in the 1967 "Long Hot Summer" to the 1991 Rodney King beating, and acquittal of the four pigs by an all-white jury, and the black rioting it triggered in South Central Los Angeles in 1992.

My Lai, the war crimes of all war crimes: 500 elderly men, women, children, and babies tortured, raped, and finally massacred by an American unit commanded by Lt. William Calley, the only one of over two dozen culpable soldiers and officers convicted, or even indicted. Except for a Capt. Medina, who was acquitted.

Nuclear counter-force and the continual US search for the end of MAD and a nuclear war fighting capability to justify their massive defense spending.

The right wing more bloody because the government is making war on the people, as opposed to the left-wing that make war on the system, as manifested in govt., officials, army and security forces.

So I continued my study: Civil war; the "dirty war"; internecine fighting; house-to-house fighting, night fighting, trench warfare.

Rural and urban warfare, total war within a country: a campaign of planned ambushes, hit-and-run attacks, bombings, infrastructure destruction. Assassination of government officials and leaders, arson, sniper attacks on police, bombings, attacks on government, military, and economic targets. Bank robberies, kidnappings of the rich and powerful. Hijackings. "Terrorism"; the reason for attacking international airlines since 1970 grabbed world headlines.

* * * *

"Just as some thieves are not bad soldier, some
soldiers turn out to be pretty good robbers, so
nearly are these two ways of life related."

-- "Utopia" (1516)
Sir Thomas Moore

My concern was the heartbreak of realizing that the Order had never come. That the Order would never come. No one had told me that we had lost -- that the Revolution had been smashed. We had **lost**.

And then I retired as outlaw, was escorted _persona non grata_ from the U.S., and returned to society, got a computer engineering degree, a good corporate job as a software designer, and hung-up my gun-belt for the Revolution. I had done my best. The party was over.

Until one August day in 2007, thirty years later, society woke me up from my position of relaxed, middle-class, suburban idle.

"Political power grows out of the barrel of a gun."

-- Mao Ze Dong (1938)
speech

Prison Diary 43:   I, Ronin
                                    ======

"Must you have battle in your heart forever?

[W]ill you not avoid what the immortal Gods themselves fear? She is a nightmare that cannot die: Eternal Evil itself -- horror, and pain, and chaos. There is no fighting her. **No** power can fight her!

All that avails is flight."

-- warning of Circe to Odysseus
"The Odyssey" (ca. 850 BC)
Homer

"You don't seem to want to accept the fact that you're dealing with an expert in guerrilla warfare. With a man who's the best -- with guns, with knives, with his bare hands."

-- Colonel Trautman
"First Blood" (1982)

It's funny that my decades of "hobby research" -- a wide-ranging, penetrating, voracious and never-ending data collection and analysis process, suddenly and unexpectedly began regularly yielding strange new insights into practical matters of historical, scientific and historical interest.

I was thorough, methodical, and persistent. And above all, inexorable. I was driven by a fascination with exploring ideas, searching for causes and effects that provided a rationale basis for events, and explanations that shed light on the seemingly unfathomable.

It was a way I sometimes relaxed in my spare time, strangely enough. It was a real world puzzle palace I was deciphering.

I knew things that nobody else in the world perhaps knew, and insights that gave me a perspective that I reveled at. And sometimes the realization of the temptation of filthy lucre would be part and parcel.

But my mental explorations were not tamable. I couldn't be assigned a topic to research -- it was always an obscure interest that tickled my interest. For fun, or leisure exploration. It was a feral research genie that only followed its own path.

One theme that was strangely recurrent was that from time to time I seemed to run into things Japanese in my research on weapons and associated fields. And without knowing Japanese, I never expected any deep results, being totally dependent on secondary English sources.

From the samurai, the legendary metallurgy of Japanese swords, to Karate, to Zen Buddhism. Some of what I found, such as a true understanding of Karate, were disturbing revelations, when

I finally puzzled certain things out over time, as I am want to do.

In the 1997 spy thriller "Ronin", starring Robert DeNiro, and filmed on location in Provence, in the south of France, the North American public was introduced to the story of the "47 Ronin", the most famous of all Japanese tales. It is a story that is revered in Japan in certain circles.

Out of the rigid, inflexible, and brutally violent world of the samurai warrior caste of medieval Japan (whose main period was 1,050 – 1,616 A.D.), their underlying philosophy was embodied by _Bushido_ -- "The Way of the Warrior".

Bushido was the philosophy that governed the Cult of Death of the samurai caste. Bushido was the unwritten philosophy and code of behavior. It was the feudal code of the samurai, stressing self-discipline, courage in battle, and the concept of honor, manifesting itself in unbending loyalty, and fearlessness in the face of death.

It was not so much "Victory or death", but "Death or disgrace", if you can see the subtle difference.

But it was thus an obsession and fascination with death -- a beautiful death, an honorable death. A bloody death, including death by beheading. It was a substitution of life being so cheap and meaningless for the perfect soldier, with the idea that war is their reality, and the end result of war is ideally an honorable death.

By its rigidity, it also insured the continuity of the samurai, since the side-effect had to be a philosophy unspoken of war never-ending, and thus the need for the samurai warrior caste never-ending. Like medieval Japan, it was extreme, savage and brutal, ruthless and uncompromising, a point lost in the modern popular conception and portrayal of the samurai myth.

But invariably there arose the concept of the samurai outcast. They were known as "ronin".

The ronin was a master-less -- lord-less -- samurai. Left leaderless and thus disgraced, a samurai's only options were an honorable death or dishonorable flight, and a life of wandering. Within the culture of Bushido, where honor was everything, the ronin was a rogue samurai stripped of his honor, by shirking his duty and instead becoming a fugitive on the run, or in hiding.

Honor is all very well, but apparently common sense sometimes prevailed. Run, run, run away...having violated his sacred duty and code of honor, and fled to avoid punishment, which was always either execution or forced suicide.

A medieval deserter, a fugitive on the outskirts and moving through the back alleys of the Japanese feudal system, and the absolute obedience that was expected of the elite class of the samurai. It was a value system that demanded and cherished perfection, without which the system was threatened by internal weakness, instability, and decay. They were outlaws, and renegades, reduced to a life of hiding, rough work as freelance mercenaries, or a predatory existence on the fringe as a brigand or bandit roaming the countryside.

The 47 ronin story takes place in 1702-1703, during the Tokugawa period, which first consolidated Japan under one shogun. It was the end of centuries of internecine conflict, and signaled the beginning of the end of samurai power and influence.

The samurai caste had grown in size over their period of influence to comprise 7% of the population. But suddenly with consolidation of power, and a period of stability and peace descending over Japan, many were now superfluous to their Lords, an unnecessary expense. The Lords had to discard many of them, who suddenly found themselves unemployed. Many however, resisted, trying to cling to the old dying ways, even just as ronin.

The 47 ronin had disgraced themselves first by allowing their Lord Asano, to be killed as a result of a trivial misunderstanding with another Lord -- a common reason for random blood-letting in the savage environment of the unstable rivalries between regional lords.

Then, instead of Asano's samurai all committing ritual suicide -- to die with their Lord, atone for their failure to protect him, and redeem their honor -- they initially fled in cowardly disgrace, dispersing to become ronin. The old ways of honor were crumbling with the collapse of the bond between Lord and samurai.

Surprisingly, however, eventually two years later, in December 1702, the 47 ronin returned to reassemble, and avenge the death of their leader-master by killing the rival samurai lord, and then committed mass ritual suicide -- _seppuku_ -- after surrendering themselves to the authorities.

A messy death, for sure.

The true story of the "47 Ronin", is not just about the last gasp of the Samurai case, nor about redeeming honor, but of the inescapably of the rigid and "glorious" unwritten code of _Bushi-do_. They had to accept their predetermined fate: they had to complete the mission -- or die in the attempt -- and if their mission succeeded, they had to accept their punishment for failing in their foremost duty of protecting the safety of their Lord.

But there's more to it than that. The sub-text. The mass defection of the 47 samurai to ronin status was symptomatic of the break-down of Bushido with the consolidation of power in Japan under the Tokugawa period. They all fled because Bushido was tottering into obsolescence, and because they could get away with it. Their eventual "return to the fold", as it were, in fulfilling their duty, and committing mass suicide was symbolic of the samurai ethos. Bushido still lived they showed, in spite of political upheaval.

There is a shrine in Japan honoring the "47 Ronin", where people still today burn intense in honor of their memory.

*   *   *   *

"Living is easy with eyes closed,
Misunderstanding all you see.
It's getting hard to be someone,
But it all works out,
It doesn't matter much to me."

-- "Strawberry Fields Forever" (1967)
The Beatles
John Lennon, John Winston

It is truly the essence of mythology that the samurai warrior caste of medieval Japan could be so misconstrued as to glorify what were essentially highly evolved, organized, and yet utterly barbarian Japanese knights with extremely sophisticated and intricate arms, armor, and culture, yet without any corresponding evolution away from the savagery that precluded any possibility of moral advancement.

The finest evolution of the medieval warrior produced in the entire world, yet still fossilized with little more than an Iron Age (30 centuries ago) mentality, who beheaded, hacked, and disemboweled themselves and each other to pieces with swords, or other crude penetrating weapons, under a social fabric of ritually precise behavior, discipline, conduct, and training.

* * * *

> "So to those free revolutionaries in all countries: hold fast to your initiative and beware of negotiations. There is no middle road between the people of truth, and the people of falsehood. There is no way."
>
> -- Last audio message (2011)
> Osama bin Laden

At the end of the 1975 spy thriller, unusual in its display of certain items of accurate, and sophisticated technical verisimilitude, "Three Days of the Condor" (starring Robert Redford and Faye Dunaway), Max von Sydow, playing "Joubert", a highly competent, efficient, freelance assassin is discussing his occupation.

The cultured, sophisticated Joubert, we finally get to see in the light, as the movie winds up, and hunter and former hunted meet under more gentile and casual circumstances. The Alsatian works, we find out, in a very lucrative, highly specialized twilight world, accepting intelligence agency contracts for one-time, below-the-official-radar, highly targeted assassinations. He is an independent operative, useful, efficient, and dependable, having found a comfortable niche working for both sides, or different factions.

Redford -- a mild-mannered, but extremely knowledgeable, insightful, and resourceful, yet desk bound CIA analyst -- is curious about how the educated, cultured, Cold War hired gun can work for anyone with the cash and a problem, without any emotional conflict, or dissonance. With no whiff of conscience, doubt, or concern of any sort, and with a seamless comprehension of the rules of the game of cleaning up "messes" that are thought to be resolvable neatly by discreet, precise, methodical killings.

Joubert confides matter-of-factly and with complete ease -- without a hint of coldness, ruthless amorality, or rationalization -- the philosophy he has decided is valid and fundamental:

> "[There is] no need to believe in either side -- or **any** side. There is no cause -- there's only yourself. The **belief** is in your own precision."

It was a reinterpretation of the amorality of realpolitik into a Zen-like transcendent. It was tidy, neatly outlined reasoning, a minimalist beauty that crystallized practice into words. I knew exactly what he was talking about. And having understood the sense of what Joubert so neatly explained, I was stronger, while at the same time diminished.

As usual, something lost -- that I would miss -- for something gained (including the ability to talk about long dead lessons by disappearing into cryptic riddles that hint at the unbelievable, but with too many details to dismiss as delusion, fabrication, or grandiosity.)

And of course, one of my tricks was to always find work at the level -- which I called the "stratosphere" -- my playground where pure intentions were all one needed, and where you don't ever get caught, punished, exposed, or even believed.

Actually, I think I've only ever gotten into trouble for my minor hobbies, rather than the core competencies I was exploiting rather shamelessly at the time. It is important to note that I officially abandoned a philosophy of violence or criminal adventures, in pursuit of private ends, and violence for ideological reasons in the early 1980's. But official policy is easy to change. Expertise, talent, instincts, trip-wires and hair-triggers are things that are hard to forget and inexplicable to the unblooded. Raised eye-brows and questions when I'd rather not be noticed at all.

But then again, out of the wood-work, fellow travelers have given me a wink and revealed themselves, their secret pasts having resonated with what others dismissed as demented babbling, or strange echoes of collective madness. And so I've met some fascinating individuals with much to teach, who also kept an eye out for me, and comforted a wounded brother-in-arms.

Take guns and gun collecting, for instance. That's always been just a hobby. I wouldn't even think or expect to have to need to do violence with my arsenal. Aside from the usual moral constraints, firearms make too much noise, and attract way too much attention. The surgical strike, or the sudden disappearance is much more practical. The unfortunate car accident that cannot be traced back to the person who sponsored it. It's not the Wild West anymore, ya know. Besides I don't believe in violence on many grounds, and I couldn't work for most sides on moral grounds.

I had to work on it. I had to make the choice; to toughen up, or be unfit for what I expected to face. But it was a one-way ticket to become one of the "hard-faced men."

Someone who examined my personality, described me as "unusually harsh in my judgments."

"No kidding." was my response. "I wonder why?" I left unsaid. But I joked that my Death List was up to six volumes. "I'll bet." she responded.

Talking about the "old days" of thirty years past, I recounted to the same person that I had looked into the muzzle of a loaded gun four times. She responded that such experiences had probably distorted my thinking. I answered that, to the contrary, these encounters had actually clarified and focused my thinking instantly.

It brought to my attention my failings and shortcomings, and their extreme consequences. I realized the inescapability of the present situation, and so accepted with serenity that I was facing imminent violent death, with perhaps moments to live. It put things abruptly in perspective. And that was despite the fact that I had no fear of death. Though I didn't seek it out, either. And was initially quite shocked and unnerved by its *unanticipated* appearance.

She explained that due to an exceptionally violent, abusive -- all of the Big Four: physical abuse, verbal abuse, emotional abuse, and the other one that I don't like to talk about --, and chaotic childhood, I was actually "comfortable" -- of a sort -- in hostile environments that would unhinge or panic an average person. I would thus seek them out through the teen years and into early adulthood, as opportunity and expanding horizons kept me raising the ante again and again. The result was at minimum my development of an interesting set of survival skills, for sure.

Decades ago in a different place and a different world, I remember that it came to my attention that some observant person had noticed that one of my personality traits was that beneath an affect of laid back affability, I could be utterly ruthless. I also found out that some people were afraid of me. I laughed and thought it was ridiculous, though useful.

"El Gatto" -- "the cat" -- the observant associate dubbed me as a _nom-de-guerre_ on some "flash" ID he had printed up, and passed me one day, for use in a "caper".

I didn't ask him the reason, but I wondered for years what meaning he saw in that moniker. Then one day, it hit me. Cute and cuddly and silent, with hidden claws ready at a moment's notice.

My main concern was how did low priority targets realize capabilities that could be aligned to their detriment should I decide a mere nuisance should be re-prioritized. Rare, but possible. To keep in practice, or if a vulnerability was identified and target was unaware that his vector crossed into a sphere of influence. All this was highly sensitive information   that I never discussed for obvious reasons, and camouflaged with an overlay of genuine easy-going affability.

I finally realized that a routine target probe of mine had probably been identified as not as innocent as most people would think it. Their radar went on, which triggered my radar as a security anomaly that would be analyzed thoroughly.

It took me 30 years to acknowledge and regret that "ruthless" was certainly a trait I had been capable of. It was the default mode for what was basically the cold logic of a code of operation and survival in certain unforgiving urban _milieu_.

But I finally came to terms with the darkness within, by realizing that violence is usually a choice. And so, I chose not to. Now and for a long time now.

And the problem you avoid in avoiding violence is one that you'd never predict. What happens when you find out -- to your horror -- that you're good at it? Time and time again, as if it was an aptitude.

Morality, honor, dignity: these are the values that you need to live by to be able to hold your head high. They are not ideals, they are achievable goals. They are simple goals. We only make them hard.

But make no mistake, the way I think explains the idea best, and you either understand or don't. A glimpse of the beyond understanding to those who don't. And that which makes a lot of argument unexplainable and wrong, even when I'm right. And about disturbing lessons that can never be unlearnt, that you have to come to terms with.

How can a person understand certain realities and experiences that put your thinking beyond the reach, and beyond the pale? That I can only express with any sense of accuracy as follows:

Have you ever held a gun in your hand, and had murder in your heart?

"Ain't got no picture postcards,
Ain't got no souvenirs,
My baby, she don't know me,
When I'm thinkin' 'bout those years."

-- "New Orleans is Sinking" (1989)
The Tragically Hip
Gordon Downie

Prison Diary 44:		You Bet Your Life -- Tales of a Gambling Man
		===================================

> "[T]o end the business, in all the exhilaration of
> that last and terrific and most glad pain of death[.]"
>
> -- "Seven Pillars of Wisdom" (1918)
> T.E. Lawrence

> "[A]ny man's death diminishes me, because I am
> involved in mankind, and therefore never send to
> know for whom the bell tolls, it tolls for thee."
>
> -- "Meditation XVII" (1623) in
> "Devotions Upon Emergent Occasions"
> John Donne

The second quote above explains the title of the famous book "For Whom the Bell Tolls" by Ernest Hemingway. Hemingway's title refers to the end of democracy in Spain, overthrown and replaced by a fascist dictatorship. And how Spain's loss also diminished Europe, when a fascist General named Franco won the Spanish Civil War (1936 - 1939) aided by Nazi _Luftwaffe_ bombing, and Moroccan mercenary troops.

Hemingway was considered a man's man, and won the Nobel Prize in Literature in 1954. The masculine reputation was apparently in vogue at some point. I remember two facts only: he liked to blow sharks out of the sea, from his boat by riddling them with a .45 caliber Thompson submachine gun, presumably the "Chicago gangster" model with the drum magazine.

And he also eventually committed suicide in 1961 by blowing his head off with his favorite 12 gauge Boss shotgun, apparently devastated over the fact that his memory was destroyed by the electroshock therapy used to treat his serious depression.

* * * *

> "When confronted with two alternatives, life and death, one is to
> choose death without hesitation. ... If one, through being prepared
> for death every morning and evening, expects death any moment,
> the Way of the Warrior will become his own..."
>
> -- "Hidden Leaves" (1716)
> Yamamoto Tsunetomo

Stories, anecdotes, but "tales" was the winner for the chapter title. As in "Live to tell the tale". It's not about the childish car crash gamble called "chicken". It's not about "Russian Roulette". That's showing you are willing to face the fear but not the stupidity of a meaningless suicide, and that you can convince another idiot who you have no certainty is having worse luck today than you. What's up with that?

* * * *

> "Then shook the hills with thunder riven,
> Then rush'd the steed, to battle driven,
> And louder than the bolts of Heaven
> Far flash'd the red artillery."
>
> -- "Hohenlinden" (1802)
> Thomas Campbell

"Never speak openly of these things again. Fear, hatred or ridicule you will reap from the 'Other' -- the 'Outsider' of 'Our Hidden World'. They just can't and won't understand. Who can blame them if they prefer to dwell in their White Bread World, and deny the existence of Hell-on-earth? Let them live in their safe, nonsensical BS dream and fantasy world of TV illusion and white-picket unreality."

One day, he was in my bedroom to see a book I wanted to show him. He noticed the line-up on the bookshelf of 25 different live ammo cartridges I used or had collected. They were lined up side-by-side on the shelf by height.

He picked up one. It was my favorite round, the 7.62mm x 39 COM BLOCK rifle ammo, used by my AK-47 and SKS military rifles. He immediately had identified it by look and silhouette, not by the head-stamp on its base.

"Where's your Simonov [the Russian SKS semi-automatic military rifle]," he asked nonchalantly I was astonished; it was locked in the gun-room. Even I couldn't do that, at the time.

He had told me, slowly, in bits and pieces, when he found out my "involvement" in military weapons and strategy, of his involvement years before, in his country's civil war. An ally was handing out small arms and heavier weapons like AK's, SKS's, recoilless launchers, light and heavy machine guns, RPG-7's, and AA-mounted heavy MGs to the general Lebanese Muslim populace of his country -- including teenagers. Anyone who wanted one to fight.

His dad took his AK away from him, when he found him with it, and sent him out of the country. But he had seen a lot, anyway.

He was unaffected by it, other than an abrasive affect to his character, and a steel wall that separated him from everyone he encountered, (except possibly family) but he IDed me soon enough, and my troubled secrets, so I was let in on his long secret past.

* * * *

> "Alas! I have nor hope nor health,
> Nor peace within nor calm around,
> Nor that content, surpassing wealth,
> The sage in meditation found,
> And walked with inward glory crowned --
> Nor fame nor power, nor love, nor leisure;
> Others I see whom these surround --
> Smiling they live, and call life pleasure;
> To me that cup has been dealt in another measure.
>
> -- "Stanzas Written in Dejection Near Naples (1818)
> Percy Bysshe Shelley

Living how I wanted to live generally required not just adulthood, the desire and right attitude, but ensuring that negative attention would not be drawn and thus interfere. "Flying under the radar" using some suitable strategy was essential, seeing that I had chosen "bandit country" as my _milieu_. But I chose a very special zone for my playground, that I called the "stratosphere", inspired by an old TV episode of "Star Trek" on class warfare entitled, "The Cloud Minders".

I lived how I wanted to live as often as I could, with the hand I was dealt. I did what I wanted to do way more than most people get to do, and I did a lot of things that I knew I could get away with, thought I could get away with, gambled that I could get away with, or just felt that I had to do, irrespective of the cost, whatever that was to be.

And sometimes the cost of the gambling part could have been death. That's the ultimate "high stakes" gamble. I got no rush from cheating death one more time. I viewed it as a test of awareness and insight, and ultimate focus.

A way to put it that describes it with equal validity, but avoids seeming like glorification. Another way would be the term that I picked up from reading a newspaper article. Someone called this behavioral pattern "Dancing with Death." And if I ever lost that bet, and saw that bet collection was imminent, that was "My Last Tango". It's more dignified, and artfully descriptive than the uninspiring minimalist -- and I'm being generous -- "Uh-oh". That would *never* make a "Famous Last Words" compilation. And I *do* want to be remembered for at least something -- or amusing, at least.

I was once accused of "making a joke out of everything", as if I was doing something *bad* and that frustrated the person to the point of anger. "Yes, I do." But I was perplexed, because I couldn't understand why this had made him upset. I was "guilty as charged" but I didn't really understand why an obvious adaptation that I had made because it worked for me, was funny more than not, and was harmless, was causing him irritation.

And if this person had seen me when I get "serious", then this fellow would realize that "making a joke out of everything" was *way* much better. Nothing unpleasant happened and it blew over, but I believe that the issue was my perceived insensitivity was that I didn't see that he was "having a bit of bad day", and he was in no mood for a guy joking around to deal with some pain, and make himself laugh. An appreciative audience is double bonus points. And as any comedian will tell you, it's a tough crowd sometimes.

The feeling of power and control that comes from being unafraid and calm when facing death allows you to survive certain situations by not panicking or making other ill-planned or executed evasive maneuvers. Doing stupid things or surviving traumatized are strategies that won't improve the ability to enhance survival probability, and continue the threat survival learning process.

And be unafraid and therefore calm in the face of death. Alert, effective, efficient -- all of which require the person to be unafraid to do their work -- and gauge as precisely and quickly as possible the exact nature of the threat posed, unafraid to do their thing which fails badly if you are in a tight corner, but keeps you in control of the skill and quick thinking that you need comes from not just surviving violence, dealing successfully, conquering, and mastering what was absolutely the best skill learning from violence with the knowledge that fear with what I can only describe as really heavy shit.

Or somehow surviving things that you have an instinctive ability or a natural talent to sense, anticipate, observe the signs that are frequently detectable and that you know from experience or lightning fast logical analysis indicate lethal or serious violence is imminent. And result in the conscious realization that you don't fear death, and avoid that this makes you strong and powerful in a sense, in that you can handle a situation that would cause fear, panic, or other emotions that put people in much greater danger of not surviving. This human behavior pattern is a natural reaction of people with no experience, training, understanding, or any ability to see that they are inviting death.

* * * *

"This is a good day to die: follow me."

-- battle cry
Crazy Horse (1876)

It was 1981 or so, and my wife and I were in a crowded elevator going down in a commercial building in the jewelry district of downtown Los Angeles, when -- to my utter amazement -- a serious fight breaks out between two Hispanic guys over God knows what. There were people in between us and the fight, so there was no problem, but that was the longest elevator ride to the ground floor I and the rest of the captive audience have ever experienced. My only concern was the kookiness of the whole situation, and whether knives were present that would make the situation completely insane and a real crap-shoot of a threat. In a crowded elevator, for God's sakes!

I grabbed my wife with an instinctive move: an unbreakable encircling hold of her wrist with my hand, in case we had to make a move fast. I remained still, said nothing, and waited for the elevator doors to open. When *finally* the doors opened I piled out of that elevator with wife in my wrist tow as fast as I could along with everyone else. I ran pulling the wife to what I considered a safe distance, angling away from the elevator, then stopped to plan my next move, planning to leave the building immediately if it was safe.

Suddenly I felt a pulling on my arm. I looked down at my hand, still locked around my wife's wrist, and was surprised to see her unknowingly tugging on it, as she tried to move closer to get a better look at the on-going fight! We promptly split for the exit, at my insistence.

* * * *

"Now all the truth is out, Be secret and take defeat
From any brazen throat, For how can you compete,
Being honor bred, with one Who, were it proved he lies,
Were neither shamed in his own Nor in his neighbors' eyes?"

-- "To a Friend Whose Work
Has Come to Nothing" (1914)
W.B. Yeats

What I had dubbed "Operation Serpent's Tooth" was off to an inauspicious start. In fact, it had gone bad. And I was left holding the bag. I knew from the gate that I could dodge the bullet, but I was really pissed, though I didn't show it.

I was pissed because I was bored and didn't need the inconvenience. I really didn't need the heat and stress of serious police "interest". All my planning and work was down the drain. Serpent's Tooth was dead, I had decided. So bye-bye to the massive dollar take that was my cut of the profit, that we had negotiated amongst all the details, in an unwritten handshake contract.

And all because of a steel liter can of fucking ether. Solvent fucking diethyl ether!

I had ended up in a room with two RCMP agents, and a seizure of lots of chemicals and equipment. I was allowed to smoke. I saw the can of ether, but lit up anyway. Ether fumes are highly flammable, but are heavy and stick to the ground. So I said what the hell, and lit up. Then a forensic lab guy came in a sat down to observe the proceedings. He saw that I was smoking, and so he lit up also. I watched him intently. Eventually, he scanned the room. I watched his eyebrows rise as he spotted the can of ether, and he quickly butted out. But he didn't say anything about my smoking, because he had too. The embarrassed look on his face was mild, but it lingered.

I "smiled" a secret smile.

He was an amateur.

* * * *

"To the ordinary guy. all this is a bunch of gobbledy-gook.
But out of the gobbledygook comes a very clear thing:
you can't trust the government; you can't believe what
they say; and you can't rely on their judgment."

-- White House tapes (1971)
H.R. Haldeman to Nixon
on Pentagon Papers release

I think you can tell a lot about a man from the movies he cherishes. There's a classic scene in the sci-fi action classic "Predator" starring Arnold Schwarzenegger and Carl Wethers.

Arnold leads a crack rescue team dropped behind enemy lines in the jungles of Central America. Wethers is a CIA operative they are ordered to bring along, and who ostensibly has overall -- but not operational -- command authority over the team.

In the scene, the line of heavily armed commandos slowly creep through the thick foliage. Suddenly, Wethers stumbles and slides down an incline, leaves rustling as he makes a huge racket. Before he has barely recovered his footing, Sgt. Mack, team leader, is in his face, along with his combat knife at the ready. The black Sgt. is livid and menacing, hissing out his opinion of the fuck-up's incompetence, operational discipline, and unit cohesion failings:

"You're ghostin' us, motherfucker! I don't care who ya are back in
the worl' -- you give up our position one more time, I'll bleed ya
real quiet and leave ya here -- **got that** ?"

-- Sgt. Mack
"Predator" (1987)

I learned infinite patience seeing people endanger the entire team I was in control of. If they realized that if their behavior took the whole team down, my first act would be to shoot the miscreant in the head -- an eye shot -- because at that point noise and body disposal would not be important...

* * * *

"Whenever that class of men on which you believe your continued
rule depends is corrupt, whether it be the populace, or soldiers, or

nobles, you have to satisfy it by adopting the same disposition; and then good deeds are your enemies."

-- "The Prince" (1513)
Niccolo Macchiavelli

In response to a phone call, a fellow I knew invited me to a "business meeting" with a new guy that I didn't know, but who had a proposal I might be interested in. I agreed to a meet, and took down the details of when and where.

The four of us sat down at a table just outside a super-market near the airport -- LAX. It was perfect -- no pedestrian traffic to disturb our meeting. We immediately got down to business. The new guy informed me that they had in their possession a 55-gallon drum of a certain chemical. I knew the chemical. And a drum of it was a lot! Probably stolen. They wanted a chemist to convert the chemical to another chemical of significant black market value. Could I do it?

Of course I could. That was the main chemical needed, but were there additional chemicals, solvents -- drums of them lined up? Was there a fully equipped laboratory ready?

No, but everything could be obtained, and there was plenty of money for such supplies. It was no problem -- just provide a list of what I needed.

And front money?

My question was met with a firm and immediate, "No front money."

Hmmmm, I thought. A drum of a super "hot" chemical, that cost God-knows how much. An industrial-sized operation -- requiring the pain of setup -- to produce something like four hundred pounds of final product -- worth, say, two to four million wholesale. At least 30 years, if caught. Money for chemicals and equipment -- 50 grand, let's say.

But no front money for the chemist. I didn't say anything, or argue. But it was "no sale" as far as I was concerned. It was a BS scenario, especially with no money up front. I decided that to avoid offending anyone I would continue talking for a bit, end the meeting without giving a firm commitment, or "Yes" or "No" answer, and then leave.

Suddenly we saw that two mature, formally-dressed black ladies were approaching us sitting at the table. The conversation stopped, and everyone fell silent.

"Would you be interested in hearing about Jesus Christ?" one of the ladies inquired politely, without being pushy, when they stopped close to our table.

"No thanks," said Mr. 55 gallon drum, and they said, "Sorry to bother you. Good day," turned around and walked off the way they had come.

"Too late..." I said out loud, to no one in particular, when the two ladies were just out of ear-shot. Mr. 55 gallon laughed out loud, then everyone else cracked up at my comment, but me. I wasn't joking...

The meeting resumed.

* * * *

"The world was exposed to the worst century there has ever been from the point of view of crude

inhumanity, of savage destruction of mankind, for
no good reason..."

-- BBC TV interview (1997)
Sir Isaiah Berlin

The test device had misfired. Shit! A stupid design flaw. Trying to extract the last bit of efficiency of a tap hole had killed the simple fuse. It was the early to mid-1990's, and I was way out in the deserted country far from civilization, down an empty unpaved country road. A couple of friends were with me to see the show, I had promised.

I waited the requisite 20 minutes to ensure it was indeed a dud, and not just a slow burn. Then I walked the distance along the deserted back-woods country road, from our "observation post" to the device.

No one disarms a misfired explosive device. It was too dangerous. It was insane.

But I knew the device, inside out -- intimately. I knew the risks and discounted them totally. Rationally.

So I thought, "Fuck it!" to myself. I gingerly picked it up, with the intention of taking it apart, rendering it non-lethal, and saving the components for a possible Mark 2 Prototype, with the simple design flaw corrected.

But then fear struck. I wavered. The consequences of failure would be deadly. I wouldn't feel a thing -- it would explode so fast -- but I was a goner if I misjudged anything. The risk/gain ratio was totally against it. What if this? What if that? I was paralyzed with fear and indecision. I froze completely -- thinking about what I would do -- holding the device firmly and unwavering in my hands, just staring at it.

One friend suddenly realized what I was planning. "Just throw it in the ditch!" he yelled to me from the observation post. "Fuck it! It's not worth it! Just ditch it, and we'll split!"

I ignored him. Then I said, I *can* do it. I steeled myself to the task. There is nothing to fear. I was familiar with every aspect of the device -- design, components, preparation, assembly, failure mode. All safety rules had been observed religiously. Conquer your fear, and just do it. It's the truest test of everything you know. You're a master of the game, or a dead fool. "It's okay! I know what I'm doing!" I yelled calmly to my observant friend, fretting in the distance about my safety.

I slowly disassembled the device. It was done in less than sixty seconds, and I heaved a sigh of relief, and started breathing again. Mission accomplished. Cool! I had *done* it.

It was the second most dangerous act I had ever done. But I had proven myself, and corralled and conquered my fears and self-doubts. I smiled to myself continuously driving all the way home. I was a certified Master of the Game.

* * * *

"But it was only fantasy,
The wall was too high as you can see.
No matter how he tried he could not break free,
And the worms ate into his brain.

Hey you! Out there on the road,

> Always doing what you're told, can you help me?
> Hey you! Out there beyond the wall,
> Breaking bottles in the hall, can you help me?
> Hey you! Don't tell me there's no hope at all,
> Together we stand, divided we fall."
>
> -- "Hey You" ("The Wall") (1979)
> Pink Floyd

It was a matter of understanding and accepting the equation. Realizing that doing things that have a high pay-off of some sort, but are also a gamble that you only lose once. Gambles that sound crazy, or suicidal, but can be taken if handled logically, sensibly, confidently, and really, really fast, and without hesitation.

Understanding that there is no room for errors, and second chances, and you understand the game. And that accepting a gamble like this was done by locating and actually a rational, logical, carefully considered matter. And it was concluded that it would be a low risk justified by the end, and that was considered and a decision was made accepting that as a risk with a full understanding that made the risk real but manageable enough to be safe ignoring. It was okay, because it was the ultimate expression of the philosophy of life known as, "To thine own self be true."

And the pay-off to high-stakes gambling is amazing if you can figure it out, and find a gambling den with a table that plays. There will be a seat at the table with your name on it, rather than a tomb-stone similarly labeled. You never ever forget that if you get careless and take things for granted you are a goner.

I lived by my wits. And I liked the clarity of thinking required to occasionally be willing to bet it all in a high stakes game with no wiggle room for errors in execution, and confidence coming from planning every detail, and then staying cool under extreme pressure. The experience that teaches you stuff and is a boon from survival execution, let you be Master of the Game.

I learned that there is a trick to "bluffing" that has *always* worked for me. If you gamble with stakes that high, and you run into a player who makes you actually come close enough to the edge of the cliff. Well forces you to realize the "other side" knows that you're not trying to bluff your way through to do something. And if they can see in your face that there's a reason you're going to do it that's valid, even though it's technically something they can stop and are supposed stop, they'll "stand down" and you'll get away with it. Your bluff -- which isn't really a bluff -- thus doesn't "get called", as they say in the game of poker.

There's an Afghan saying, that I read in a newspaper article concerning the proxy war, that a moron named Ronald Reagan authorized in Afghanistan from 1979-1988, that supplied the Afghan's with arms to fight the Russian Army. The saying was "Fear those who don't fear God". The U.S. was merely copying the fact that Russia kept a steady supply of arms flowing to the North Vietnamese Army to help in its fight to drive the U.S. Army out of South Vietnam, and unite the country finally free of foreign interference. They beat the U.S military and achieved total victory in the Vietnam War.

It was incredible in my opinion that the Russians fell for the U.S. payback the exact the same way. And I would think the Russians now have a saying too. It would be "Fear Those Who Don't Fear Death". That was the Afghan religious fanatics... And it would be me, too, for other reasons. And it wasn't religious fanaticism, ideological extremism, or such nonsense.

And about my death, I would like people to read that my last words were something someone heard in the dead of night, but took no notice of because there seemed no cause for alarm. And my words would be:

> "I heard him say 'A riddle? A riddle? Did I hear
> someone say a riddle -- a riddle and a paradox?'
>
> And then it was kinda strange... I heard him laugh.
> I heard him laugh so loud, you could have heard
> him 'laughing on the moon', as he used to say.
>
> And then all was quiet again till morning. And when
> he was late getting up, I peeked in, and found him
> dead."

A guy who writes his own obituary, and epitaph. Any doubts?

I don't fear death to this very day. Rather, I fear *life*.

I didn't have even *time* to worry about death, which is inevitable **anyway**. There were other phases of how I viewed death, such as "don't even care, anymore" and it gets dreadfully depressing at this point, and if interested any other examples are left as an exercise for the reader.

But yet another example of something I found where something was lost, for a gain that I thought was necessary at time. The goal was achieved, the pay-off was very rewarding, I never lost the gamble clearly, but somehow felt a mistake had somehow been made in winning, while losing something important, that I missed having. The myth of the "big score" now has company. The myth of the cost-free strategy when the game is "hard ball".

I think I stopped this kind of gamble the last time I looked into the muzzle of a gun. In 1984, shortly before endemic urban warfare -- the crack epidemic fueled gang violence in poor and black ghetto areas in Los Angeles broke out and craziness became a real actual threat.

I had chased off a disorganized pre-competence young black guys who I believe were "Crips" -- the 49th St. Crips.

Sorry, I'm "old school". Don't mess with my woman in any threatening or aggressive or violent way, or you deal first with the guy who considers it his responsibility to protect her at all costs.

* * * *

> "The best lack all conviction, while the
> worst are full of passionate intensity."
>
> -- "The Second Coming" (1919)
> W.B. Yeats

I was something like 40 meters from our apartment in West L.A. I was returning home in the late afternoon, walking along the back alley. I was in a shitty mood. I had had a bad day.

I didn't pay much attention to two young black men that I was about to pass, when one produced a pistol, and demanded money. I was being mugged -- the first and only time.

I kept walking as I looked at the gun bearer, and said, "Go ahead and shoot me. I don't care. Put me out of my misery." I passed them, and kept going. They didn't say a thing. Without turning to look at them, as I continued walking I pointed to my head, and said in disgust, "Right in the head. Go ahead. Do me a favor. Right in the fucking head." I didn't look back, made it to the gate to the courtyard, and unlocked it, and entered.

I couldn't believe it. A robbery attempt within spitting distance of our apartment. After all I've been through, seen, and volunteered for, I won't settle for folding my hand. I play real "poker". Don't make me laugh, trying to act tough by playing "kids games".

But when I got inside the apartment, I closed the door, and heaved a big sigh of relief.

* * * *

> "Originality is not an attempt to capture attention come what may, or to shock or disturb in order to shut out competition from the world. ... What makes them original is not their defiance of the past or their rude assault on settled expectations, but the element of surprise with which they invest the forms and repertoire of a tradition.
>
> Without tradition, originality cannot exist: for it is only against a tradition that it becomes perceivable."
>
> -- "An Intelligent Person's Guide
to Modern Culture" (1998)
Roger Scruton

Then there was my finely crafted pitch:

> "Good day. My name is Yogi. You don't know me. Therefore, you have absolutely no reason to deal with me.
>
> But you probably will consider doing so, because at this exact moment in time, our interests and goals – and their success -- seem to have converged.
>
> The result of this intersection of our goals, is that I have identified that it is possible for us to combine **your** resources with my **very specialized** synthetic chemical manufacturing expertise to generate a **very** lucrative profit for *both* of us, and at minimal and easily manageable risk.
>
> If you will let me explain further, it is then up to you to decide as to whether there exists a viable opportunity we can both exploit to mutual advantage. Now, do you wish to listen to my proposal? And then, if you're interested, we can talk further and discuss the details.
>
> So there's the offer. Would you rather that I just 'get lost' or would you like to sit down together sometime?
>
> It's **all** up to you."

And typically, they would pause for a minute, then invite me to sit down. And then they would listen.

Dancing with death is a slow dance. And you have to lead, or you're toast.

"See these tears so blue,
An ageless heart that can never mend,
Tears can never dry,
A judgment made can never bend.

See these eyes so green,
I can stare for a thousand years,
Just be still with me,
You wouldn't believe what I've been through."

-- "Cat People" ("Putting Out Fire") (1982)
David Bowie

Prison Diary 45: Jail-House Triptych
================

> "Toto, I have a feeling we're
> not in Kansas anymore"
>
> -- Dorothy
> "The Wizard of Oz" (1939)
> Frank L. Baum

> "Life is simply a _mauvais quart d'heure_
> [a bad ¼ hour] made up of exquisite
> moments."
>
> -- "A Woman of No Importance" (1893)
> Oscar Wilde

"Never a Dull Moment...", was a comment I heard muttered repeatedly from inmates nearby me at OCDC, as something unexpected spontaneously erupted. From the fights, uproars, brouhahas, transfers, disputes, and other utterly random events that would suddenly arise, and grab the entire range's attention, as most usually just had to watch and guess the ultimate damage, winner, loser, and/or resolution.

But let me select three charming little _vignettes_ of note and interest that occurred during my detention.

* * * *

> "There are only two kinds of politics. They're not radical and
> reactionary, or conservative and liberal, or even Democratic
> and Republican. There are only the politics of fear and the
> politics of trust.
>
> One says you are encircled by monstrous danger. Give us
> power over your freedom so we may protect you. The other
> says the world is a baffling and hazardous place, but it can
> be shaped to the will of men."
>
> -- Sen. Edmund Muskie (1970)
> TV speech

Early on, I had noticed the totally unexpected inmate practice of noticing an inmate in emotional distress -- depression, specifically -- and the response of his friends/associates to try and cheer him up. There was nothing in it for them. It was genuine and sincere -- and most importantly, in the predatory, uncaring, and violent environment of jail -- quite touching.

Twice I had been the beneficiary of this largesse. Once when my bail was denied, my cellie did it.

The second time, a Lindsay associate joked and clowned me back to life when I fell into visible emotional withdrawal and depression on realizing I was going to spend Christmas in jail.

But then it became *my* turn to return the favor.

On one of my "visits" to the overflow facility and sentenced inmate jail in Lindsay, Ontario, a large provincial jail known as CECC (Central East Correctional Center), or just "Lindsay" to the inmates, with a capacity of about 1100 inmates, I met a big lug of a Palestinian inmate named "Moe". He was tall -- a huge guy -- with the ugliest haircut I'd ever seen.

In fact, he was butte-ugly, period. But I enjoyed his company immensely, and we became friends and conversation partners during the first of my two temporary transfers to the Lindsay facility. It was November of 2007, and I'd been incarcerated for about 3 months by then.

Moe had been caught with three kilograms of cocaine in his car, but was fighting the charges, claiming it was an illegal search that uncovered the drugs. We became quite friendly with each other and he was quite supportive of me and my emotional ups-and-downs as my incarceration seemed to go on and on with no end in sight, and what I thought was a decided lack of communication from my expensive legal team, which I figured they'd realize I'd need.

So I appreciated Moe's expressions of concern, that came with his unsolicited emotional support, which went along with his opinion and advice on my legal case. He believed there was no doubt I'd be getting out Scott-free, and that it would be sooner, rather than later. And he was absolutely, believably sincere, which was the only thing that made such talk helpful.

But then one morning, I guess it was my turn to respond in kind. His case was less "iffy" than mine, and though he said he was getting off too, I wasn't sure if he would or not, and I didn't press for details, and he didn't offer many.

Moe was sitting on the "ledge" that ran along one side of the range. It was a convenient and nice place for two people to sit in private, away from the collection of sheet metal tables on the other side of the range where most of the other inmates sat, talked, played cards or whatever.

I saw Moe sitting alone, and came over and sat down beside him, as usual.

"I hate this place! I'm going to be in this stinking hell-hole forever," he commented out of the blue, apparently bummed out over something. Or worn down. Or maybe just depressed for no reason at all.

"It's a nightmare being here," he continued.

"Beats being married," I responded matter-of-factly, with my usual deadpan humor.

Moe burst out laughing, in response to my quip. He laughed so hard, I thought he was going to fall off our ledge onto his ass.

Apparently my joke caught him off-guard, and it was just what he needed. When he finally stopped laughing, there was a big smile on his ugly mug. His spell of depression was broken, and the cloud of unhappiness that had enveloped him lifted. He was back to his usual friendly and talkative self.

As the Lindsay resident inmate humorist, I was pleased to have helped out a fellow inmate in emotional pain -- especially one that I liked and enjoyed the company of. I had accomplished my good deed for the day, and it wasn't even mid-morning yet.

* * * *

"You lose, you lose, you lose, you win."

– Rosa Luxembourg (ca. 1918)
Leftist Revolutionary Ideologue

One day at OCDC, a month or two in, I had gone outside for exercise and fresh air when a guard came in and called out, "Yard up!" for any inmate interested. That was the lingo.

I was walking laps -- the inner perimeter of the trapezoidal walled, concrete-floored "yard", one of the options I typically selected for something to do during the yard period, which was usually lasted 20 minutes, and sometimes longer. It wasn't a very big yard, but it was what it was.

As I walked alone, suddenly I was addressed by Pete, who called me over. I had no idea what he wanted, because I stayed away from him. I hoped I was not in some kind of trouble, or I was "dead".

An "old school" career criminal, and as a result a career convict, hard as nails, not someone to mess with in any way, he was football player huge, solid muscle, and fighting a first degree murder charge. He had "whacked" somebody, or something -- I didn't ask for details, didn't want to know, and took just one look into his eyes...

A heavy-weight in the prison hierarchy on all counts, he "ran" the range. I didn't mind the "old school" cons, and could deal with them, but he seemed a little too volatile to even accidentally antagonize. So I steered clear. He never talked to me anyway.

He and Howie, Pete's workout partner, were walking the yard together, too. I knew Howie from the weekly impromptu Bible study meetings he led. He was gregarious and friendly, and easy to talk to.

With some trepidation, I joined the two-some in walking the range. It was not an invitation I could ignore. The small-talk began, mostly between Pete and me. Guns seemed to be the general topic, which was fine with me, as I was perfectly at ease with this subject-line. And of course, as range leader, he knew all about my charges.

He ran the conversation, and I politely interjected to correct any mistakes they made, or to add interesting details worth adding to his gun comments. I guess it was some sort of informal "test" of my bona fides. And knowing a fuck of a lot about firearms, I'm sure I not only passed the test with flying colors, but impressed the two with my encyclopedic knowledge of "hardware".

The formalities over, I guess, Pete switched subjects, and began chiding me for what he felt were the shortcomings of my attitude and general demeanor on the range. He informed me that I was obviously not taking my situation seriously enough, and he was concerned for me.

I couldn't believe it! Here was a heavy-weight, hardened convict, who didn't know me from Adam, rarely said two words to me, and he had taken an interest in my well-being! And who was taking the time, trouble, and effort to offer his advice to me, con-to-con, as it were!

He said that he noticed that I laughed and joked around *way* too much. I was "too happy" for someone in such a serious predicament as I was, and I should take things more seriously so that I wouldn't get fucked by the System.

"I can't let them see me unhappy. Then they've won. So I smile and joke, and they know they're not getting to me. I'm here because I'm a political prisoner. I keep that thought, and I can stay happy, knowing that I've not committed any real crime. I'm a 'political'. Always have been," I tried to explain. But I thanked him for his advice, and told him I'd be careful.

But overall, I was quite touched by Pete's reaching out to me. He didn't know me from shit, owed me nothing, yet was offering his advice unsolicited, and out of concern, just to be nice, I guess to someone he saw as "out of his league" and in need of good counsel. An out-of-the-blue act of kindness, when I least expected it, from a person I *never* expected it from.

Now *that* was pretty cool!

* * * *

"First they ignore you. Then they ridicule you.
And then they attack you and want to burn you.
And then they build monuments to you."

-- speech (1918)
Nicholas Klein
union leader

Treatment at Lindsay, and all the other jails I "visited" in my travels around the Ontario jail network was like paradise compared to the violence, degradation, and restrictions of OCDC. And the guards (and rules and regimen) at Lindsay were particularly friendly, civil, and otherwise decent to the inmates.

For instance, there was no 21 hour lock-down at Lindsay, and the inmates were out most of the day, except meal-times, and the evening, starting at 8:00pm, I believe. 8pm -- final lock-down.

But one 8pm lockup sticks in my mind. It was, maybe October or November 2007, and my cellie and I piled into our cell, as the guard held the cell door, ready to close it as soon as
"Good night, fellas," he said pleasantly (and sincerely) as he closed and locked the steel cell door behind us.

I had a smile on my face for a long time after that special comment.

"Oh, wrap me then in shades of darksome pine,
Bear me to caves by desolation brown,
To dusky vales and hermit-haunted rocks!

And hark, methinks, resounding from the gloom,
The voice of Melancholy strikes mine ear:
'Come, leave the busy trifles of vain life,

And let these twilight mansions teach thy mind
The joys of musing and of solemn thought.' "

-- "The Pleasures of Melancholy" (1745)

Prison Diary 46:   Thomas Warton the Younger
　　　　　　　　　　　The Range Is **Hot**!
　　　　　　　　　　　===============

> "It's a cold world, 'Blood'. No mercy.
> -- Ain't that what you write on the walls?"
>
> -- Jail Interrogator to his prisoner
> "Colors" (1988)

> "The poet makes himself a **visionary** by a long, immense and reasoned **derangement of all the senses**. All forms of love, suffering, and madness; he looks into himself, he exhausts in himself all poisons in order to keep only their quintessences.
>
> Ineffable torture where he needs all the faith, all the superhuman strength in the world, where he becomes, amongst all, the great invalid, the great criminal, the great outcast -- and the supreme Scientist -- because, he arrives at the **unknown** !"
>
> -- Letter to Paul Demeny (1871)
> Arthur Rimbaud

I was shooting at a local outdoor range in Ottawa, when I first heard the term that summed the whole experience up.

"The Range is **HOT**!" For a change, there was more than one shooter and the range firing line. So a volunteer spoke up, "Anyone mind if I be the Range-Master?"

No problem.

At the appropriate time, in a loud, firm voice, he announced "The Range Is **HOT**!" And we began shooting at our targets. At the appropriate interval, he announced, "**CEASE FIRE! CEASE FIRE!**" And there was no more firing after the first "Cease fire".

Free-wheeling yet controlled choreography. If only war was so simple, these days.

* * * *

> "I make a lot of money on stuff like "Die Hard 3", but that's basically cartoon bullshit. Guns are not used like they use 'em in those kind of movies in real life. ... Movies have always shown the guy getting whacked over the head with a big steel gun and it knocks him out, right? No -- in real life it crushes his skull and give him a concussion and probably kills him. If not, turns him into a vegetable."
>
> -- Rick Washburn (2005)
> movie gun consultant

The war does not go well.

How many wars have seen the darkness before the dawn, with the eventual winners facing defeat after the war starts, but before the tide turns? The light at the end of the tunnel dims or goes dark. And then when things seem hopeless, bottomed out, at the ebb, the wind changes and the shift begins. The tide begins to turn.

The U.S Civil War (1861 - 1865), World War Two (1939 - 1945), the Korean War (1951 - 1953) all saw the initial specter of defeat, only to finally turn-around to eventual victory.

It's perhaps the endless and uncertain struggle between Good and Evil. The triumph of perseverance and persistence, until the war settles into the slugging match known as a war of attrition.

"Spock, I've found that evil usually triumphs, unless good is very, very careful," comments Dr. McCoy to Mr. Spock, in the "Star Trek" post-apocalyptic episode, "The Omega Glory" (1968).

"Long and hard is the way that leads up to light," writes Milton in "Paradise Lost", as viewers of the movie "Seven" (1995) might recall.

There was the inmate violence: fights -- a predatory hierarchy, a pecking order of strong versus weak, violent versus non-violent, insane versus sane.

There was the institutional abuse: collective punishment and kangaroo court procedures, _Nacht under Nebel_. A violent environment tolerated -- an institution that doesn't follow rules, doesn't have rules. Coercive punishment of the presumed innocent: Elgin St. food, a steel bed, the cold with no blanket, verbal abuse, limited public visits.

Who said it would be easy? If it were easy, then it would be easy to change...

So be it.

> "I and the public know,
> What all school children learn,
> Those to whom evil is done,
> Do evil in return."
>
> -- "September 1, 1939" (1939)
> W.H. Auden

Prison Diary 47:	The Cult of the Empty Gesture
========================

> "I'm too old to shoulder the burden of constant lies,
> that goes with living in polite disillusionment."
>
> -- Rhett Butler to Scarlett O'Hara
> "Gone With the Wind" (1936)
> Margaret Mitchell

> "Twenty years for nothing, well, that's nothing new,
> Besides, no one's interested in something you didn't do."
>
> -- "Wheat Kings" (1993)
> The Tragically Hip
> Gordon Downie

We are governed by morons and imbeciles who think that problems are solved just by passing a new law, accompanied by a chorus of trumpets to herald the arrival of the wonderful new age of paradise that the government has ushered in, due to their foresight, brilliance, and ingenuity.

The new utopian society that was a chaotic horror show, if not someone in the governing party had been an inspired genius, whose character and caliber is example of the esteem you should hold onto and re-elect them for at the next scheduled election.

Or trivialities, like turbans on RCMP officers, gays marrying -- surely the collapse of Western civilization will follow...

Or trivialities like putting "Support Our Troops" stickers on government and police vehicles.

(Not to mention the damage done by passing an ill-conceived, misdirected, or unenforced law.)

It's the same type of thinking that's filled down-town Ottawa -- Canada's capital -- with bronze plaques, monuments, and statues of little or no meaning or substance.

We are governed by these intellectually bankrupt frauds that fool the voters into thinking they're being properly governed. These scoundrels simply pass new laws to appear to be doing something. The laws are either stupid, unnecessary, unenforceable, not enforced, or are poorly written, so that they are unenforced, misdirected in their enforcement, or enforced abusively or wrongly for political purposes at their absolute worst.

A disturbing concept began to formulate and take root in my mind. I was surrounded! I was trapped! God help me! God save me! I was a prisoner of "The Cult of the Empty Gesture" -- the name I dubbed this assemblage of men and women -- this conspiracy of a moral vacuum.

In one Canadian city, Winnipeg, Manitoba, I think, a female city counselor took great pride in having the City Police Department destroy its old revolvers when they bought new guns in an upgrade program. The usual practice was to sell the old lot of a couple of hundred revolvers on the open market, to be bought up by individual collectors, the money recouped reducing the cost of the upgrade.

Sounds good, except when you realize that 4,000 handguns a day are manufactured in the U.S., and it was thus a meaningless gesture -- except to waste the city's money -- to destroy the old police guns instead of selling them to legal collectors.

Yes, the "Cult of the Empty Gesture". Rule of the Fool. The jackals ruling the jack-asses. And some people wonder why certain men doubt the ability of women in politics and elsewhere, where they can do a lot of damage. Having a woman in power is dumb as hell. You think that is sexist? Deal with it.

* * * *

"O World! O Life! O Time!
On whose last steps I climb,
Trembling at that where I had stood before;
When will return the glory of your prime?
No more -- Oh, never more!"

-- "A Lament" (1821)
Percy Bysshe Shelley

Now take some well-known men of historical importance. Symbols in the Western world -- the Western media -- that are held in *high* regard. Men held up universally as examples -- of ideals -- of wholesome goodness.

What do they have in common? The fundamental belief and espousal of change not through revolution, but through non-violence and patience. Thus, according to the Powers That Be, they are as symbols of change, as nothing changes.

It's well established that Jesus, Gandhi, Martin Luther King, and Nelson Mandela are recognized and glorified by western society, through the government, as political "saints", as it were, and propagated through the media as "soft and subtle propaganda"...

Is it a coincidence that they attained power and influence through -- often years-long -- non-violence and suffering and mistreatment at government hands, and not active resistance? It is the belief of the powerful that it is ineffectual and easily countered "gradualism" that is the route to continued power and the easy derailment of real change for the better.

Gradualism is too difficult to start, and to convince people of its ultimate chance of success. That to get it going, and then implement it, is next to impossible in a prosperous industrialized society. Grass-roots effort doused with Agent Orange or Roundup herbicide, usually.

Think about it.

* * * *

"Zeus,
Who leads mortals to understanding,
Has established as a rule,
That wisdom comes by suffering."

-- "Agamemnon" (ca. 475 BC)
Aeschylus

The failure to punish government wrong-doing. Change without criticism. General ongoing amnesty for government's failures and sins. While leftists were pursued in the U.S. for decades.

The resistance to oversight, the resistance to audit and review, the resistance to independent scrutiny. The resistance to accountability, transparency, and meaningful punishment and change for their systemic ills, and widespread illegal and corrupt malfeasance. Deaf to constructive criticism? Strenuous resistance to collection of meaningful stats and figures that will damn.

Talk about Wheat Kings and heinous crimes & police willingness to frame, get the patsy, fry the weak.

The Cult of the Empty Gesture. The Land of the Meaningless "Reputation". Empty ideals chiseled in stone monuments, in a nation where all are illiterate, or illiterate of history.

I think I see a pattern here...

> "I've heard a lot of talk about this next song.
> Maybe, maybe too much talk. This song is
> not a rebel song, it's, 'Sunday, Bloody Sunday'."
>
> -- verbal preamble to live version of
> "Sunday, Bloody Sunday" (1983)
> U2

Prison Diary 48:   The Cult of the Lie
===============

> "[T]he majority of politicians, on the evidence available
> to us, are interested not in truth, but in power, and in
> the maintenance of that power.
>
> To maintain that power, it is essential that people
> remain in ignorance, and that they live in ignorance
> of the truth – even the truth of their own lives.
>
> What surrounds us, therefore, is a vast tapestry of lies,
> upon which we feed."
>
> -- Nobel Prize Lecture (2005)
> Harold Pinter, play-write

> "Information policy fundamentally shapes the condition
> within which we undertake all other political, social,
> cultural, and economic activity.
>
> And it is information policy that is the legal domain
> through which the government wields the most important
> form of power in today's world, information power."
>
> -- "Change of State" (2009)
> Sandra Braman

Captain Willard, in arguably the most chilling scene in "Apocalypse Now" (1979), nonchalantly finishes off a critically wounded South Vietnamese teenage girl with a pistol shot _coup de grace_, so casually administered by Willard as to mortify the viewer.

In a river boat arms inspection gone terribly wrong, an entire civilian South Vietnamese family -- the crew of a sampan boat conveying vegetable produce to market -- had ended up raked with machine gun fire by the hair-trigger nervousness and inexperience of the American patrol boat crew.

No arms are on the sampan, but one family member, the teenage girl, is still barely alive. The black Navy officer/boat captain wants to ferry her to a medical unit.

Willard's had enough of the U.S. sailors' amateurishness. Almost imperceptibly sliding into target range with practiced expertise, he casually aims and matter-of-factly shoots the wounded girl dead with one shot of his .45 Colt 1911A1 semi-automatic, as the American crew look on -- paralyzed in shock.

"I told you not to stop," he mutters, devoid of emotion, as he passes the stunned gunboat captain, on the way back to his perch on the gun-boat.

Fade to black.

The narrator concludes the scene with Willard's bitter thoughts:

> "We'd cut 'em in half with a machine gun, and then give 'em a Band-Aid. It was a **lie**. And the more I saw them, the more I hated lies."

* * * *

> "'You said it yourself, Big Daddy:
> 'Mendacity is the system we live in.'"
>
> -- 'Brick'
> "Cat On A Hot Tin Roof" (1958)
> Tennessee Williams

In fact, Jesus talked a **lot** about lying. And he didn't much like them either.

> " ' Why don't you understand the language that I use? Is it because you **can't** understand the words I use? You come from your father, the Devil, and you desire to do what your father wants you to do.
>
> The Devil was a murderer from the beginning. He has **never** been truthful. He doesn't [even] **know** what the truth is. Whenever he tells a lie, he's doing what comes naturally to him. He's a liar, and the father of lies.
>
> So you don't believe me because **I** tell the truth.' "
>
> -- John 8:43-47

And then there was the Pontius Pilate, the Roman Governor of Judea (which became Palestine which became Israel), who is revealed as a cynical -- yet insightful -- bureaucrat, with his classic, "What is 'truth'?" line (John 18:38), that he retorts to Jesus, whose fate he holds unwillingly in his hands.

Norwegian author, Henrik Ibsen wrote "An Enemy of the People" (1882), in which a Dr. Stockmann fights to no avail to get people to listen to the truth and close a contaminated spring, in spite of the economic cost.

Tennessee Williams -- I'm thinking about "Cat on a Hot Tin Roof" (1958), the movie -- wrote about lies, too. "Mendacity -- it's lies and liars," 'Big Daddy' explains with contempt to his troubled son, "Brick", as both suffer from the unhappy consequences of living a lie, the dying old man acknowledging in a final heart-to-heart connection to his son -- played by Paul Newman -- that at least it might not be too late for the young man to free himself.

Williams, a closet homosexual by virtue of the social and legal proscriptions of his time, knew what he was talking about, and no doubt his alcoholism was his maladaptive attempt to cope with being forced to hide his true self.

The Oscar-winning movie, "American Beauty" (1999) starring Kevin Spacey, was about breaking free from living a lie. An unhappy, laid off office worker, with an empty shell of a marriage says, "Fuck it." and lives the way he wants to -- including happily working for a fast food burger joint. A modern fable of "To thine own self be true." as it were.

But that's the problem today. It's not a cult of lying. It's a culture of lying.

* * * *

> "[Y]ou give your disciples not truth, but only the semblance of truth; they will be hearers of many things and will have learned nothing; they will appear to be omniscient and will generally know nothing; they will be tiresome company, having the show of wisdom without the reality."
>
> -- "Phaedrus" (ca. 360 B.C.)
> Plato

But what about **me**?

I admit it -- I'm a little hard to take sometimes. It's my verbal bluntness. I **try** to be diplomatic. I **try** not to hurt feelings. I **try** and make my criticisms constructive. (Honestly! I do!)

But, alas, to no avail.

It's no accident, I guess, that I became an engineer. Because a competent engineer can't pussy-foot around, mince words, or candy coat. Or air control computers at airports will crash, buildings will fall down, and airliners will plunge out of the sky. (Unless you **want** that, because you're a weapon designer...)

Face it -- I'm as blunt as a sledgehammer sometimes, and a lot of people don't seem to like it. This has always confused and disoriented me. Until I realized that people live in such a dream-world of bullshit and lies that the Truth has become unexpected, confusing, even to the point that it appears as some sort of abomination. Something to fear, or react in horror to.

Or deny completely, and attack and silence the Truth-teller.

The root of every lie is the self-interest of the liar. The only exception to this rule is the pathological liar, who lies about everything -- just to keep in practice. The pathological liar will lie about the weather, even when you're both outside.

(My only explanation is that the pathological liar thinks that because of his reputation as a liar, by lying continually he hopes to either overwhelm the listener with a constant barrage, and camouflage his **important** lies.)

The Culture of the Lie. As the smear job: distortion, fantasy, hyperbole, and fabrication take center stage. It starts as a substitute for the Truth, then becomes a complete replacement for proper investigation, then becomes the Truth itself, which -- and this is the tip-off -- is defended with vicious counter-attack when it's questioned, with the personal attack and occasional red herring tossed in for good measure.

Because the actual Truth, pretty much stands impervious, strong and alone, all by itself. But it **can** be splattered with black paint by the purveyor of the Un-Truth.

The lie must repeat his lies over and over. He has to keep trying to convince everyone, to camouflage his deceit. He becomes defensive easily, and counter-attacks anyone who questions even his basic -- and easily verifiable -- facts.

He attacks the person -- the _ad hominem_ attack, to deflect the attention of both his victim and any audience, who is distracted by this technique and entertained by the slur and the smear.

Liars are lazy, because the truthful must and are willing to work hard to earn their way. Because the truth is easy. It is consistent. It is simple and usually obvious. But the lies of the liar can be exposed by the holes in them soon enough. Because they are too lazy to even check their facts and create lies that cannot be easily dissected to find the errors and mistakes in their arguments.

* * * *

"During times of universal deceit, telling
the truth becomes a revolutionary act."

-- George Orwell (1946)
[possibly misattributed]

The Truth is not a secret, it is not hidden, concealed, camouflaged, buried, difficult to find or decipher or uncover. People may **want** to keep us from it, but no one actually keeps us from it. **We** keep it from us all by ourselves. No mirage, illusion, smokescreen, or false path denies it to us. It is our own blindness that keeps it from us.

The Truth is not **what** happens, but our acceptance and understanding of **why** it happens. The Truth is not what we want, but what occurs independently of our dearest goals. In other words, we don't control the Truth, it should control us. This is what we reject, and so the Truth eludes our understanding.

The solution is in first becoming aware, so as to seeking the actual Truth. The second achievement is in attaining the persistence and patience to stick with the search and get closer to the actual Truth. The third is to be persistent, clever, and lucky enough to get through the maze and finally get to the Truth.

"Successful deception requires a dynamic
falsehood, an untruth with beauty and appeal."

-- "Odysseus in America (2002)
Dr. Jonathan Shay

* * * *

"During the question period, somebody asked him why was U.S.
[foreign] policy 'so stupid?'... I responded that rather than being
stupid, U.S. policy is, for the most part, remarkably successful and
brutal in the service of elite economic interests.

It may **seem** stupid, because the rationales offered in its support often
sound unconvincing, leaving us with the impression that policymakers
are confused or out of touch. But just because the public does not
understand what they are doing does not mean that national security
leaders are themselves befuddled. That they are fabricators does not

mean they are fools."

-- "Against Empire" (1995)
Michael Parenti

The development of armor naturally led to the development of anti-armor weapons. An arms race was born. On land it was tank armor, and on the sea it was naval warship armor plate.

And the road to AP -- armor-piercing -- anti-tank weapons took a couple of different routes. The most obvious route to breaching a tank, by another tank, or a gun crew artillery piece was to shoot a large, heavy and hard, and very fast bullet at it. Thus was born the KE penetrator projectile. "KE", standing for "kinetic energy", or the energy derived from mass and speed of any object.

And this led naturally to the metal tungsten.

The very hard, very heavy, elemental metal tungsten was first used as a projectile bullet in the Boys Anti-tank rifle, invented in 1936 by the British.

But, anti-tank rifles were obsolete by WW2, with the thickening armor of newer tanks. So rifles were replaced by tank guns -- leading to tank battles -- and artillery pieces for infantry gun crews -- such as the Nazi 88 mm (3") diameter, tungsten projectile-firing anti-tank weapon.

But in the 1980's tungsten alloy began to be replaced in U.S. AP ammo by something called DU, or depleted uranium. It use became standard by U.S. and NATO forces -- and in international arms sales to other allies and friendly countries -- by tanks and planes as AP ammunition.

A 30 mm cannon ammo -- for close air ground support and helicopter gun-ships -- contains almost three-quarters of a pound (0.3 kg) of DU. Tank ammo contains a little less than 20 lbs. (9.1 kg) of DU.

But concerns were raised, then and now, by peace activists and environmentalists, about the radioactivity of DU left on the battlefield being a health risk to civilian populations long after the conflict was over. These were handily dismissed by government and military spokestwits.

Depleted uranium was the waste by-product of production of weapons-grade uranium production for nuclear weapons. Regular Uranium has .7% U235. U235 is six and a half times as radioactive as U238, so it's the more dangerous isotope health-wise. Weapon-grade uranium is 93% U235, so it gets separated out of the regular uranium for bombs, leaving the "safe" and much less radioactive U238 as a waste product.

That was Lie #1.

Lie #2 was, DU was a very dense metal ideal for KE penetrators for AP ammunition, with a density of 18.8 g/cm3, or about 1,200 pounds per cubic foot. How could you deny the U.S. military the most effective material for its weapons?

And then one day I was bored, and the thought of the DU controversy came to mind. And I realized that that was a quick and simple thing to verify. So I pulled off my bookshelf a thick and large book called "The Merck Index" -- the chemist's bible -- and looked up the density of tungsten. And to my amazement, in about 30 seconds, realized that tungsten is actually denser than uranium (and certainly much harder). It's only a little more heavy a metal, but they're about the same. So much for Lie #2.

It was all about price.

With only .7% U235 -- the desirable nuclear weapon isotope -- there was a hell of a lot of DU left over after the U235 was separated out. In fact there was about 200,000 metric tons – that's **440 million pounds** of DU left over. That's millions of pounds of DU with very few industrial uses at all.

Until someone thought of replacing an expensive ammo material – tungsten – with a dirt-cheap one – DU – and saving a ton of money on ammo production.

It was all about saving money. War may be big business, but it's also an expensive business. And saving money is the name of the game.

Now, the hunt was on for me. What about radioactivity?

That was lie #3.

A little more research revealed that not all the U235 is removed from natural uranium to produce DU waste and weapons-grade U235. DU consists of 99.7% U238 and .3% U235, down from its initial U235 content of .7%.

So, since DU still contains a little less than half the U235, it's only half as dangerous. Which is still dangerous. Not to mention the lower level of still dangerous radioactivity of the U238.

DU was first used by U.S. Forces in 1991, in the First Gulf War with Iraq over Kuwait, leaving 291 tons of DU -- spent ammo – in Kuwait and Southern Iraq. DU ammo then later also used against Serbian forces in Kosovo, in the NATO attack to help break-up the communist Yugoslavian federation.

"It has never been proved that the use of DU endangers the health of people. It is no more dangerous than mercury," a NATO spokesman was quoted -- laughably – in "New Scientist" (June 5, 1999, p. 20). Mercury, as is well known to chemists and toxicologists and environmentalists, is considered a **highly** toxic heavy metal.

"Depleted uranium is more of a problem than we thought when it was developed." General Brent Scowcroft remarked after the 1991 Gulf War I. On hitting the armored target, it produces an aerosol of U3O8 (uranium oxide, known to prospectors and miners as "yellow cake" for its color), which settles as a fine dust. The radioactive effects are eclipsed by the fact that this oxide is an extremely potent neurotoxin with a high affinity for DNA, causing birth defects and cancer.

"So far the U.S. government has refused to finance the epidemiological studies to correlate incidents, symptoms, and regions of the impact and use of [d]epleted [u]ranium," said a 2006 letter from Dr. Hans Noll, American Cancer Society Professor of Genetics and Molecular Biology.

> "In the prior art, pyrophoric penetrators have predominantly been fabricated from U or [Tungsten] alloys. The use of U[ranium] has one major drawback...the fact that the U.S. Government has restrictions on the use of uranium..."
>
> -- "Pyrophoric Penetrators" (1976)
> U.S. Patent #3,946,673

* * * *

> "[The] destiny of democracy [is that] not all means are acceptable to it
> and not all practices employed by its enemies are open before it.

> Although a democracy must often fight with one hand tied behind its back, it nonetheless has the upper hand.
>
> Preserving the rule of law and recognition of an individual's liberty constitutes an important component in its understanding of security. At the end of the day they strengthen its spirit and allow it to overcome its difficulties."

-- Israeli Supreme Court (1999)
ruling

Remember in the late 1990's, when some guy happened to catch on video Ottawa Police Officer Rene Cardinal assaulting a woman whose hands were handcuffed behind her back? He looked around, and then slammed her head down against the police cruiser trunk she was being held up against after arrest.

She was drunk, had mental problems, a petty criminal record, was generally being a loud and annoying nuisance, it was true. But a cop assaulting a handcuffed female detainee -- that's not just criminal, but an absolute disgrace. And not just a disgrace to the pig, but a failure of employee selectivity as well as systemic control, the most serious failing.

When the video tape of the incident was turned over to the TV news media, a scandal naturally erupted from a public understandably upset that a policeman could sink that low.

However, soon after, what I considered an even worse act of official misconduct occurred -- apparently unnoticed. Follow-up news reports stated that through unnamed sources reporters had found out that the citizen who video-taped the assault had a criminal record.

No word on whether the video camera had a criminal record...

What I found interesting was two-fold.

First, that the "unnamed sources" were obviously police officers illegally running the amateur camera man's name through the computer looking for "dirt" to leak to reporters in an obvious -- and illegal -- smear job.

Second, the failure of reporters to realize that they were being used by police to smear a citizen with illegally accessed information -- totally irrelevant to the story, by the way -- in order to retaliate against him for exposing a police officer engaged in the lowest form of malfeasance.

"The Ottawa Citizen" that reported the original incident, should be renamed the "Ottawa Shit-izen"

* * * *

> "There is no greater fallacy than the belief that aims and purposes are one thing, while methods and tactics are another. This conception is a potent menace to social regeneration."

-- "My Disillusionment with Russia" (1923)
Emma Goldman

In the otherwise lackluster movie "A Few Good Men", a high-ranking U.S. Marine Corps officer -- played by Jack Nicholson -- is testifying under oath at a Court Martial. In a raging contempt under cross-examination by military prosecutors, he responds in a tone of withering contempt, thundering, "You can't **handle** the truth!"

That certainly is one of the problems of democratic states in a world dominated by Super-Powers, their economic and military empires, and spheres of influence.

But what's worse, is **not** that our great leaders think that we can't handle the truth. It's that our leaders believe that they don't **owe** us the truth.

> " ' These [children] believe in me. It would be best for the person who causes one of them to lose faith to be drowned in the sea, with a large stone hung around his neck.' "
>
> -- Matthew 18:6

* * * *

It was in one of the last official acts, in 1999, Illinois Governor George Ryan, that I realized that a Great Truth had passed by. Ryan, going down on charges of official corruption and malfeasance, was being impeached in disgrace and headed for jail.

Yet, though he was a Republican Party stalwart, he banned capital punishment in Illinois, freezing all state executions. Embattled, disgraced, and headed for removal from office, he could finally defy Republican dogma, and do what was right.

What was right. It was the dark side of the widespread introduction of DNA testing that had done it. Widely touted for its ability to convict and exonerate, most stories left it at that. Then university researchers began an unusual project -- where DNA and closed court files were available, they ran tests on old court cases still available, with viable DNA evidence, where the defendant was convicted and executed.

What they found was potentially earth-shattering -- a long history of an unacceptably high percentage of state execution of completely innocent defendants railroaded for murders they had nothing to do with. Despite all the effort and money, and procedural safeguards put in place in capital cases, the system was still executing completely innocent people.

Ever heard of it, before now?

About 250 people, 70% of them minority convicts, have been exonerated by DNA evidence since 1980.

It was a small, but significant, and unacceptable percentage of the cases, and provided powerful ammunition to the movement for the banning of capital punishment, and the culling off of racist judges.

* * * *

> "The highest patriotism is not a blind acceptance of official policy, but a love of one's country deep enough to call her to a higher standard."
>
> -- Candidacy for President Statement (1971)
> George McGovern

Canned tuna is popular for lunch sandwiches, and dinner casseroles. Sushi – mostly raw tuna – has exploded in popularity in the restaurant business.

Yet there's a fly in the ointment – mercury contamination. There has been a government health advisory for pregnant women and nursing mothers to not eat tuna because of it. Tuna is a big fish – much bigger than a dolphin – and so it is at the top of the ocean food chain. As a result, it consumes a lot of smaller fish, collecting the extremely toxic methyl mercury that man has polluted the oceans with.

Now, one day I was thinking about tuna (I get bored often...), and I realized that what is toxic to pregnant women is toxic to people in general. That was the day I stopped eating tuna, even though it was cheap, and tasted okay.

And why hasn't the government warned everyone?

Can you say "money"?

I knew you could.

The U.S. blue-fin tuna catch in 1993 (twenty years ago) was 857 metric tons, with a wholesale price of $34 a kilogram. Mostly exported to Japan, that equals sales of almost $30 million. But there's also albacore tuna for the North American market.

The world catch for tuna was 4 million tons, which is an industry worth almost $140 billion. Which is why the fishing industry could bribe the Canadian and U.S. governments to limit their health warning to pregnant women only, instead of a general warning to the public at large. Or even banning tuna completely.

Capitalism at work...

* * * *

"The foulest stain and scandal of our nature
Became its boast -- One murder made a
Villain, Millions a Hero."

-- "Death: A Poetical Essay" (1759)
Bishop Beilby Porteus

The years 1980 to 1988 marked the dark days of Ronald Reagan's two terms of office as the President of the United States.

And then came along at just the right moment Tom Clancy, a shill for Navy expansion.

Tom Clancy, who rose to fame with his "military high-tech" fiction -- the techno-thriller -- revealed nothing new about it, in his hit first book, "The Hunt of Red October", (and his other books) and was almost completely wrong throughout the book (and all his others, as well), except where he reproduced information already well known for years by non-military specialists -- like the SOSUS underwater acoustical detection cable between Greenland, Iceland, the U.K., and Norway secretly (known as "GIUK") to detect passing Soviet nuclear missile submarines.

Naval screw propellers were invented by the Swede John Ericsson in 1836, and consisted of an underwater rotating engine-driven shaft at the end of which were large blades angled at 45 degrees backwards from the shaft, to push the water forcefully backwards and thus propel the ship forward.

Propellers have evolved with alloy development and ship size over the years. Now, nickel-aluminum bronze (copper and tin alloy), containing about 10% aluminum and 5% each of nickel and iron possess the needed strength and corrosion resistance without heat treatment.

An oil tanker of 400,000 tons uses a cast propeller weighing 75 tons and is 9.85 meters in diameter with five blades. It was made of Nikalium, an alloy of copper with 9% aluminum, 5% iron, 4.5% nickel, and 1.5% manganese.

The solution was quite simple to American detection of Soviet submarines. The propeller of the submarine created turbulence in the water. The turbulence generated bubbles and the cavitation – collapsing bubbles – the "swooshing" that were detectable by U.S. passive sonar.

To eliminate the noisy and detectable turbulent water flow from the propeller, it was as simple as slowing the turning rate of the propeller, and making up for the slower rate by increasing the size (diameter) of the propeller blades with overlapping of the blades, to push back a larger amount of water, maintaining the same speed of the submarine with the older, smaller but faster rotating propellers.

Higher quality, tighter specification machining accuracy of the cast propeller, and propeller bearings also no doubt helped quieten the subs. And rounding the edge of the propeller would help too.

The great "Toshiba scandal", that blamed the Japanese manufacturer for selling "high precision" milling machines to the USSR, enabling them to make "smoother", more accurate submarine propellers, covered the truth of the matter. While smooth props are definitely quieter, they are easy to make. Rather, the milling machines were capable of machining larger metal items than before as a single piece, easing the production and production rate of the larger propellers required for sub-quieting.

Clancy's explanation of the need to avoid scraping the oven top with a pan is a lot of bull-hooey. The submarine pressure hull is four inch thick steel. And there is an outer light hull casing filled with insulation. Sound from inside the submarine is definitely not a problem.

And long before Clancy's great novel first came out in 1984, the Soviets were covering their outer hull with a thick layer of special sound absorbing rubber – called anechoic plating – to block active sonar pinging, and thus avoid detection. Long before the Americans started using it, too.

* * * *

"That rifle on the wall of the working class [apartment]
or laborer's cottage, is the symbol of democracy. It
is our job to see that it stays there."

-- "Don't let Colonel Blimp Ruin
the Home Guard" (1941)
"Evening Standard" article
George Orwell

A constructed universe of an alternative, understandable, complete reality. Propaganda, isolation, entertainment -- T.V. reality, the dream factory, idealism, science fiction utopian dreams, friendship networks, educational pursuits, employment ladder, sexual life, social, cultural, religious organizations, recreational drugs, ambition -- the pursuit of wealth or power, political organizations.

Fiction vs. non-fiction (history vs. science). Non-fiction real and more interesting. But fiction is more popular because it panders to and is tuned to people's misconceptions and false or distorted construct of reality.

A simplified world of black and white. Or entirely made up alternative worlds. Or completely wrong understanding; a safe and secure view of things; ego-centric in which they are blameless and good and okay; reached their goals, reaching their goals, a plan or active or executing plan for attaining their goals, of formulating their goals from a safe plateau.

> "The candy store paupers lie to the share holders,
> They're crossing their fingers – they pay the Truth-makers,
> The balance sheet is breaking up the sky.
>
> So I'm caught at the junction, still waiting for medicine,
> The sweat of my brow keeps on feeding the engine,
> Hope the crumbs in my pocket can keep me for another night."
>
> -- "Blue Sky Mine" (1990)
> Midnight Oil

Prison Diary 49: Huh? Eh? Duh?...Er -- What?
=======================

> "In Shakespeare, characters develop, rather than unfold, and they develop because they re-conceive themselves. Sometimes this comes about because they **overhear** themselves talking, whether to themselves or to others. Self-overhearing is the royal road to individuation."
>
> -- "Shakespeare: The Invention of the Human" (1999)
> Harold Bloom

> "The mass media -- and I include the computer industry – conspire to pervert our need of community. ... We are learning to believe that we do not require wisdom, community, provocation, suggestion, chastening, enlightenment – that we require only information...as if life were a packaged kit and we consumers lacking only the assembly instructions."
>
> -- "Make-Believe" (1996)
> David Mamet

Ah!... One of my favorite sub-hobbies: linguistics. And the original profession of articulate popular liberal ideologue, Noam Chomsky.

An "interest" is something about which you want to find out more information on. In which you feel your present knowledge is inadequate. About which you feel that filling in the gap will have important benefit to you. I guess that means that you may not actually find the subject interesting...

The thing about linguistics is that it is concerned with the brain's interpretation and response to **verbal** input. And **that** is a different part of the brain (left or right – I don't remember) than that which process the written word.

I found this out in reality, when a fellow engineer in my department (a girl, with a similar ethnic background, with the same two degrees as me, and with her engineering degree from the same university) found my "Paris Diary" series of written emailed reports from an ex-pat gig there extremely funny – which was the idea. Yet, this same girl had previously **not** found my verbal jokes, and sense of humor funny in person. Not funny at all.

Yet the "Paris Diary" anecdotes were the **same** sense of humor, delivered in the same way and style, as my verbal attempts at humor with her! I was surprised, but just shook my head.

* * * *

> "The rhetorical matrix in any situation consists
> of a reflexive interaction among the contextual

> situation, the rhetoric, the audience and the text."
>
> -- "Flights of Fancy, Flight of Doom" (1988)
> Marilyn J. Young and Michael K. Launer

[Surprisingly, I understand the above sentence came from an academic book on the 1983 shoot-down of Korean Airlines Flight KAL007.]

"Ulysses", by James Joyce, is arguably the finest work in English literature in history. Certainly if you're a competent English Lit Professor. Naturally, it was banned as obscene for years in the U.S.

But did you know that word "Ulysses" is the Latin (Ancient Roman) version of "Odysseus" the original Ancient Greek name of the hero of Homer's famous epic "The Odyssey"? Call me crazy, but I found that interesting, when I discovered that fact a few years ago.

You've heard of the sub-Arctic island country, Iceland? Originally discovered by the Vikings, it's just south of the vast frozen wasteland island of Greenland. But I got a letter from an Internet "friend" from Iceland, and I saw the stamp, on the letter. It said "Island", the real name of "Iceland", used by its inhabitants. Well, I thought that was neat...

The Cold War was officially between 1945 and 1991 – forty-six years long – with the collapse of the Soviet Union. You ever notice that in the newspaper or on the news, whenever they report about a historical event, no matter when, it's always at "the height of the Cold War"? Listen, the next couple of times, and you'll see.

* * * *

> "What is of interest is that the Communist commentary, while
> incorrect, is at least rational, while the mainstream U.S. commentary
> reflects the kind of incapacity to perceive or think about simple
> issues that is sometimes found in the more fanatical religious cults."
>
> -- "The Manufacture of Consent" (2002)
> Noam Chomsky

Then there's the Arabic slogan. "Allah-hu Akbar!" It's always translated as "God is Great!", which confused me. I couldn't really understand why it was chanted all the time in public displays. I **was** able to understand -- or think I understand -- when I realized that "God is Mighty!" is probably the real translation. Or "God is the greatest!" Subtle change -- nuance -- big difference.

If we can't translate accurately even the simplest thing about what these people say, how the fuck do we ever expect to understand them and their concerns -- and less productively, but for some reason of more importance -- their actions and behavior.

"Turn On, Tune In, Drop Out" was the slogan of Timothy Leary, the Johnny Appleseed promoter and publicist for widespread consumption of LSD during the late 1960's, targeted and smashed by the government. At his death, they couldn't even get his slogan right, discombobulating the long-forgotten message. They said "Tune In, Turn On, Drop Out".

Or the 1950's Cold War government slogan "Better Red than Dead"? Actually that's the more modern twist. The original slogan was "Better Dead Than Red". How about that? And what a virulent, illogical, insanely anti-communist slogan. It shows you the temperament of those times, and the viciousness of the government propaganda.

* * * *

> "I see the world being slowly transformed into a
> wilderness. I hear the approaching thunder that,
> one day, will destroy us too."
>
> -- "The Diary of a Young Girl" (1947)
> Anne Frank

The linguistics of incomprehensibility. How can we understand each other better, and resolve problems either quickly, or before they even become problems, with the limitations of verbal expression? Language, that is to say.

Gibberish. Gum-flapping nonsense. Verbal gobbledygook. Talk, talk, talk, jaw, jaw, jaw. Lies, more lies, nonsense. and bullshit as punctuation.

The alcoholic, virulent racist, and imperialist dog, Winston Churchill, was a gum-flapper, cum orator extraordinaire, for example. He was quite the writer, rhetorician, speech-writer, and wit, too. Yet World War II was not won by his speeches, but by the Americans, the Soviets, arms, tactics, code-breaking, radar, and overwhelming brute force.

Churchill successfully advocated the introduction of carpet bombing of cities to the panoply of tools of Total War, which caused much death and suffering to German civilians – of little concern in an absolute dictatorship -- but little effect on morale or industrial production.

He advocated chemical warfare be used in WW II, a violation of international treaties, but fortunately was over-ruled. It would have been a **bad** strategy: the Germans had discovered the most poisonous CW agent of all time, nerve gas, and had at least 40 tons manufactured, but it was never used by Hitler, in Treaty-compliance...

Churchill called Gandhi, fighting for Indian independence with non-violence, a "half-naked fakir".

But, while something on the order of a million people starved to death in the Bengal (India) famine in 1943 – 1944, he sent a cable saying that if there was a famine, how come Gandhi hadn't starved to death.

Ha. Ha. Ha.

You fat, wind-bag fuck...

* * * *

> "In three things is a man revealed: in his
> wine goblet, in his purse, and in his wrath."
>
> -- "Babylonian Talmud" (200-400 B.C.)
> Tractuate Seder Mo'Ed Erubin

It's amazing how slang -- the low-level vernacular -- captures, captures concisely, and even captures amusingly, what formal English seems unable to do, isn't it? I love slang, which is why -- no doubt to the general confusion -- I mix it quite deliberately, for the effect, paradox, and superior results, with the formal, staid English of the elaborate constructs of my normal verbalizations. ;^)

I find the contradiction in terms, the verbal expanse quite entertaining and novel, which is a core goal of much of my actions, frankly. And now it's ingrained habit, so I can't help myself.

Got that, mo-fo? Cool! But, not to worry, bro', it's all good.

"White man speak with forked tongue," For years, I agreed wholeheartedly with this cliché of the old cowboy Westerns that we watched as children on Saturday morning TV -- black & white, no cable, two channels only.

But I interpreted it to mean the white man was a snake, the snake-form that Satan took in the Old Testament Book of Genesis.

> [I've moderated my views of white men considerably since my younger days, by the way. Though I still think a lot of you suck.]

But I realized many years later, that the North American Indian (or at least the actor reciting the script, written by a liberal, left-wing screenplay writers) was referring to the double-talking, mealy-mouthed, deceptive, say one thing, do another, lies, broken promises, or other shifty, hidden agenda manipulations and scams, that the Indian had come to expect, if he accepted the promises or verbal contracts of white men of authority and official status.

In fact, "Don't talk with your mouth full," seems to be humanity's only common, accepted rule regarding verbal engagement, truth be told. Or a comment to your girlfriend, in the bedroom...

Ah, yes, "Truth, be told". "To tell you the Truth...", "Frankly," "No, B.S.," "I swear", "I'm not kidding". The culture of mendacity, the foundation of lies, truth hidden, deception, and candy-coating that underpins verbal interaction, is exposed by these common phrases and expressions.

The imprecision of language is another problem. English has it. I understand French is even worse (according to the French engineers I spoke with, when I worked in Paris).

"Beware the crooked man." and "Beware Man, the crook." and the difference is crystal clear. Yet say, "Beware the crooked Jew," and misinterpretation is certain and swift.

It's this imprecision of language that makes for the verbosity, and poly-syllabic words, and repetitions with slight sentence variation that makes engineers so tedious to listen to by most people. They're verbose to try and avoid misinterpretation and mistake. It's why they can be so slow in answering. Language precision matters, and it takes time to filter out errors, double meanings, imprecisions and assemble a bulletproof statement that conveys the desired information.

In other words, effective, communication. Not fool-proof, though. But that is handled by engineers trying to avoid talking to fools.

Precision equals accuracy, and understanding without further explanation. Precision enables quotability. It is fit to write down, print, and distribute for the potential applause of a mass audience. And we haven't even addressed the matter of "content value and importance" yet.

When comedians say, "It's all in the delivery," an engineer could say, too. It should be one of our mottoes, slogans, or iron rules. Clarity, lucidity, simplicity, would be good too. But alas...

Now legalese, that's a different matter. Obfuscation, camouflaging, the smoke-screen, the verbose but irrelevant answer masked by its length, feigned sincerity, and word complexity and sentence structure. Deflection of attention, covering up of holes, the illogical masquerading as logic, superficially attractive analogies that are bogus if put under the magnifying glass. Or even just eye-balled closely.

Analogies should be banned in my opinion. They are easy to come up with and say, but hard to be relevant or correct analogies. Apples and oranges comparisons is their real substance, rather than being a correct analogy. Damn them all to hell, I say. And take people who can't grasp the concept of "causation", too!

"To tell you the truth" is a common prefix to an explanation or story, or a supposedly intimate revelation.

Mathew 5:34-37 quotes Christ:

> "But I tell you, don't swear an oath at all. ... Simply say
> yes or no. Anything more than that comes from the Devil."

Say what you mean and mean what you say. The first step. Talk the talk, and then walk the walk.

* * * *

> "[A]fter all, the greatest and most calculating of
> killers is the national state, and this is true not
> only in international wars, but in domestic conflicts."
>
> -- "American Violence" (1970)
> Richard Hofstadter

Moral, immoral, amoral, and the amb-imoral ("ambi" from the Greek, meaning "both"). The ambi-moral have good and bad. (I invented the word for this chapter.)

False piety and hypocrisy are subsets of immorality.

Then there's moral indifference. There's moral laziness, an awareness, at least. Then going down hill, there's moral numbness, aware but paralyzed by hopelessness.

Rock bottom, arguably with immoral, there's finally amoral, where morality plays no role in the decision-making process of one's thoughts or action. It's just needs, wants, repercussion and a moral absence: the psychopath or some criminals.

The misuse of the word "Terrorist". The South African government were the terrorist, without a doubt, but it was Nelson Mandela and the black guerrillas of his ANC who were dubbed "Terrs" -- terrorists – by the government and Afrikaner white people of apartheid South Africa, before they were overthrown in 1989, thirty years after the notorious Sharpeville Massacre.

Mandela, a political prisoner for over twenty years was adopted as a "Prisoner of Conscience" by Amnesty International, even though he refused to abandon his belief in political violence, a demand made by his South African jailers, and normally a requirement for Amnesty International, who made him their only exception to the policy.

* * * *

"The target is destroyed."

> -- radio report of Soviet fighter pilot
> that has just shot down an off-course,
> civilian, South Korean 747 airliner. (1983)

Finally, I found out that linguists have discovered that the Sanskrit word "karma" was translated incorrectly. The results are so far controversial, but the correct translation may actually be "AK-47".

"[R]age and politics should never have been separated.
Without the first, the second is lost in discourse;
without the second, the first exhausts itself in howls."

> -- "The Coming Insurrection" (2007)
> The Invisible Committee

Prison Diary 50:   A Po-Mo Paradigm
===============

> "If there's a tear on my face,
> It makes me shiver to the bone.
> It shakes me, Babe.
> It's just a heartache that got caught in my eye,
> And you know I never cry, I never cry."
>
> -- "I Never Cry" (1976)
> Alice Cooper

> "[P]eoples and governments never
> have learned anything from history
> or acted on principles deduced from it."
>
> -- "The Philosophy of History" (1837)
> Georg Wilhelm Hegel

Then there is the story of Heather – my dear HN – and her English 101 essay. She was the first girl I have ever really and truly loved. Blonde, twenty-one, and stacked, and highly sexual, we were living together, near the University of Ottawa. I was in second year of my engineering degree, and she was just starting school there, taking a smattering of courses. The year was 1988, and she had to write an essay for her English 101 course.

I had just been reading about, what I came to realize were the "in" words of English and the Arts Faculties in the academic world. First there was "paradigm", which later evolved into "post-modern", which finally evolved into "po-mo" over the years, when I followed such matters.

They were in "in" words, the use of which marked the academic insider. The insider who was current with academic writings, and current academic trends.

I thought the concept hilarious, and though I shouldn't have laughed, insisted that my dear H. use the then current lingo, "paradigm" somewhere in the first essay assignment she was writing for the English course.

She – perhaps put off by my amusement at the whole affair – at first refused. But I insisted, and could be very persuasive. It couldn't hurt, I insisted, and maybe it would – I suspected – give her an "A", validating my "experiment".

Finally, she used the term "paradigm", under psychological coercion, even though she was still against the idea.

I waited enthusiastically, but patiently, for the marked essay to come back. I had read the essay, and it was "OK", but...

Sure enough, it came back with a mark of "A-". I was humorously impressed, and quite proud of myself.

I should add "po-mo linguist and author" to my resume...

* * * *

"It took some time for my hormones to tell,
That chasing her had been a grave mistake,
…
Her ego wrote checks incredibly fast,
But her personality didn't have the cash.
I laughed out loud, to my total dismay,
She ain't pretty, she just looks that way."

-- "She Ain't Pretty" (1990)
The Northern Pikes

Those insider words, Abracadabra. Open sesame. Even my bedroom door has a password (for female initiates). They simple answer is that they have to say, "Yes".

In the Tolkien fantasy extravaganza book and movie, "Lord of the Rings: The Two Towers", Gandalf and the hobbits are momentarily stumped by the closed stone door to mountain lair. Above the door is chiseled in the stone the inscription, "Speak friend and enter." in the Elfin language. Saying "Friend" in Elfin opens the way.

Though as an underground dwarf mining community, why it wasn't in dwarf, is perhaps a clue of some sort?

* * * *

The dictionary definition of Post-modernism:

"a late 20th Century style and concept, which represents
a departure from modernism and has at its heart a
general distrust of grand theories and ideologies, as
well a problematical relationship with any notion of 'art'."

Typical features include a deliberate mixing of different artistic styles and media, the self-conscious use of earlier styles and conventions, and often the incorporation of images relating to the consumerism and mass communication of late 20th-century post-industrial society.

1770-1830 Age of Romanticism as a rejection of the authoritarian/mechanical Enlightenment period – the Age of Reason. It was an appreciation of personal freedom, individualism, imagination, and intuition. It expressed a love of nature and the wilderness, as a rejection of the cold scientific rationality of the Enlightenment.

But between the Age of Reason and the Age of Romance, in the mid-1700's lies a brief and little known European historical period quaintly known as the "Age of Scandal".

But today that wouldn't fit at all. The Passion of Mad Max, Paris Hilton, Britney Spears' virginity declaration, Mischa Barton -- Mischa!! -- you drunken slut. If a theme song with twelve meaningless words repeated over and over again wasn't bad enough. And I'll never be able to watch an "OC" re-run again. First there was "Melrose Place", "Beverly Hills 90210", then the "OC" [Orange County, California].

Boy, it must be tough. They're running out of U.S. TV show settings, where no black people live...

It's so obvious that clearly it's the end of scandal, the end of shame. They have no meaning any longer.

* * * *

> "[T]he alienating social structures which confine modern
> man would only be overthrown by a political and social
> convulsion, propelled by a radical change of heart."
>
> -- "A Terrible Beauty" (ca. 1998)
> Leon Whiteson

The social milieu & subcultures. I called it "blowing with the wind", where I jump up and let the wind carry me, before I'd fall to earth. One Friday, I gave a girl a ride to someone's place, and ended up one Friday in a whore's apartment.

Control freaks, manipulative, persistent pathological liars, totally untrustworthy, permanently damaged, who cheat, steal, and just dangle themselves, if possible. I started calling them "danglers".

"Fucked for life", they hate sex, hate men, and ultimately everyone they meet. All of humanity. And the final nail in the coffin: they end up hating themselves too.

And then there was Chantal Lambert, where I ended up one evening. What a trip she was...

"You have to pay me for my time!" she ordered as I got up to leave, getting fed up. "And I charge by the inch," I felt like responding.

Distasteful and sordid; seedy and seamy; pay to play. Then there was the girlfriend of "Vietnamese Tony", who informed me that I had to pay, but **he** got it for free. I felt like responding, "But your girlfriend is a **whore**," but I didn't want to start anything.

I knew another guy who looked down on crack whores, but had a stripper for a girlfriend...

* * * *

> "The essential component of evil is not the absence
> of a sense of sin or imperfection, but the unwillingness
> to tolerate that sense. ... [T]hey are continually engaged
> in sweeping the evidence of their evil under the rug of
> their own consciousness."
>
> -- "People of the Lie" (1983)
> Dr. Scott Peck

And speaking of whores, there's a (Canadian) Supreme Court unable to tell morality from immorality, as evidenced by:

1) The use of defamation laws as a tool of the rich and powerful to squelch criticism,

2) Allowing public opinion to influence the release on bail of an innocent until proven guilty

suspect,

3) And **particularly**, the contaminating and corrupting effects on the police of allowing them to lie after the detention of a suspect. Lie about the arrest, lie about physically assaulting him, making up law the were "broken", lying at the bail hearing about what happened. And on and on it goes.

4) Pig testimony is given the benefit of the doubt, or even just accepted without question. And most criminal charges against the pigs result in acquittal, or a minor punishment, or a conditional discharge -- no punishment, and they get to keep their job. And these people are to be believed?

Not held to a higher standard, but held to no standard at all. What a blatant disgrace. By the pigs themselves, and the judges that put their seal of approval on a corrupt system, involving themselves in the impropriety, too.

It should more properly be called the "Inferior Court", and **not** the Superior Court, or the Supreme Court.

"Gradually I could see, Things are getting clear,
That ancient face, Satanic grace, This sudden rush of fear.
They say you are the king of this whole damn thing,
Now they got me believin'.
They say I don't stand a ghost of a chance with my host,
So I better be leaving.
So let's drink a few. Here's looking at you.
I swear. Didn't we meet in the night in my sleep somewhere?"

-- "Didn't We Meet" (1976)
Alice Cooper

Prison Diary 51: Urban Guerrilla Bedtime Stories
=========================

> "Remember this: that the bloody, and treasonable,
> and revolutionary doctrine of public necessity can be
> proclaimed by a mob, as well as by a government."
>
> -- Horatio Seymour (1863)
> New York Governor
> public speech

> "Walter Kurtz was one of the most outstanding officers
> this country's ever produced. He was brilliant -- he was
> outstanding in every way. And he was a good man,
> **too**. A humanitarian man. A man of wit and humor.
>
> He joined the Special Forces... And after that, his --
> uh -- ideas, methods...became...unsound....
>
> Unsound..."
>
> -- General Corman
> "Apocalypse Now" (1979)
> John Milius

"Who are my heroes?" you ask. The people I admire and respect the deeds of, the memory of, the story of? Those who show my values, and possibly reveal my inner-most thinking processes and ideals.

A man's heroes shed light on his belief system, in all its simplicity, or intricacy and complexity. In all its intimacy.

* * * *

> "It was an attempt to bring our laws and customs into harmony
> with those of the most despotic of Continental Governments –
> it was an attempt to disarm the people."
>
> -- "Hansard" (1870)
> Mr. Taylor, M.P.
> on gun registration bill
> before English Parliament

Dominic "Mad Dog" McGlinchey: The name says it all.

I wish I could write a book entitled "Me and 'Mad Dog' "-- referring to Dominic "Mad Dog" McGlinchey -- fallen leader of the INLA, the Irish National Liberation Army, a Trotskyite (extreme far left communist) splinter faction of the Provisional IRA.

The INLA was founded in 1980, and centered in Northern Ireland's southern County Derry, and lasted for thirteen years of ruthless guerrilla warfare against the British Army, as well as bloody internecine feuding with fellow extremists (the point of which even *I* don't fathom).

The INLA's specialty was high-profile assassinations of politicians and members of the hated and murderous Security Forces, and they racked up a rather impressive body count of around 150 assassinations, including a British Conservative Party MP who was blown to hell by a bomb planted in his car, as he left Parliament in London.

Dedicated, fanatical, and feared, the notorious McGlinchey provided an exaggerated glimpse of a rarely seen persona: a true soldier of the "armed struggle".

A non-violent activist at the age of 17, the start of his ultra-violent path began in August 1971 with his arrest and physical abuse – and radicalization – during 10 months imprisonment without charges -- the English Government's despised "internment" round-ups, that was later ruled illegal and a gross violation of human rights by the European Court.

Eventually becoming a fugitive -- dubbed by newspapers as "the most wanted man in Ireland" -- he was finally caught in the Republic of Ireland, extradited to the U.K.-controlled North, imprisoned, released unrepentant, and finally martyred -- gunned down by a hit team of rival revolutionaries -- thus capping off an extraordinary run in the Northern Ireland-England "dirty war", the longest war of revolution and resistance in history by far (six centuries or so).

With a sensationalist newspaper-dubbed moniker of "Mad Dog", McGlinchey was both ferocious, energetic, unstoppable, and incorrigible. He bragged to reporters about having taken part in two hundred "active service operations" -- IRA lingo for armed hold-ups, bomb attacks on army patrols, assassinations, informant killings, sniper attacks, and hit-and-run ambushes against the heavily armed security forces of Northern Ireland and the British Army.

He boasted of personally killing thirty men, usually from close range. Ever the consummate professional -- he disclosed that close up was his preferred assassination protocol for successful targeting, and mission closure, even though it was more dangerous, and required skill, absolute calm, and utterly ruthless, fearlessly savage courage.

"I like to get close..." he stated coldly – in true "hard-faced man" form -- rather than using the more diplomatic "don't shoot until you see the whites of their eyes" cliché.

Both him and his wife, Mary -- an equally ruthless and respected INLA cadre -- were themselves gunned down in internecine or dirty war violence, in 1994, and 1987, respectively. Mary McGlinchey was killed while at home bathing their two infant sons. When Dominic joined her in death, the couple were buried side-by-side in County Derry.

And thus he joined my Pantheon of personal heroes. Now, I'm not saying I'd have wanted him to move next door, or anything... But surely this fellow was uber-cool in a reckless, apocalyptic sort of radical _chic_, post-modern aesthetic.

Or -- er -- something like that.

Siqueiros: His life was his ultimate mural

Frescoes and murals are wall paintings that won't fit within the limits of a piece of canvas stretched over a small rectangular wooden frame. And the life of David Alfaro Siqueiros wouldn't fit within the limits of the artist. (He's briefly mentioned during the tourist shot slide show scene near the end of the 1984 true-to-life spy movie, "The Falcon and the Snowman".)

Siqueiros was the youngest of "the Big Three", the three great Mexican muralists of the first half of the 20th Century (the flamboyant communist Diego Rivera & Jose Orozco, being the other two).

A prolific and uncompromising artist who was frequently jailed, he was volatile and defiant, a militant revolutionary, and an incorrigible leftist political dissident and agitator, his varied career even including a stint as a Republican commander in the foreign volunteers of the International Brigades of the Spanish Civil War, and a botched assignment from Stalin's NKVD secret police to assassinate the exiled Leon Trotsky in Mexico.

And last, but not least, Siqueiros was arrested twice for inciting a riot at the age of 14(!), as well as at the age of 64. For this alone, he makes it onto my list of stellar icons of human heroism, impressive outlaw individualism, and incorrigibility _par excellence_.

Life and art as inseparable.

Galois: "Pas de Temps! Pas de Temps!"

Evariste Galois (1811-1832), French mathematician and political dissident. Expelled from school and imprisoned twice for his republican sympathies and agitation, he was finally killed, at the tender age of 21, during a pistol duel over his ill-conceived insult of the French King while drinking in a tavern, resulting in a challenge from a royalist soldier.

That evening, instead of resting and preparing for the duel the following morning, he spent the night hurriedly trying to write down for posterity his recent and extensive mathematical discoveries -- what came to be known as "Group Theory" -- without time to elaborate the requisite mathematical proofs. In the margin, he would occasionally scribble, "Pas de Temps! Pas de Temps!" -- "No Time! No Time!" -- as he tried to complete the manuscript before the morning, in case he lost the duel.

It was many decades before mathematicians realized the importance of the manuscript Galois left behind -- the cornerstone of modern algebra. Galois, himself, was shot dead during the duel, but what he left behind was a priceless advance in mathematics years ahead of his time.

The mind of a brilliant genius, only 21-years-old, snuffed out by the superior marksmanship of a common buffoon of no importance.

But in the end, alas, aren't we **all** beset by buffoons, and, indeed, running out of time?

Kodos "the Executioner"

Do fictional characters count? Oh, what the hell!

Concluding this chapter is the character from the 1966 "Star Trek" TV series episode, "The

Conscience of the King", which borrows openly from Shakespeare's "Hamlet", including the episode title.

> [The name "Kodos", was itself borrowed by "The Simpsons" TV series as the name of one of the two – the other named "Kang" – one-eyed, fang-teeth, slithering, octopus-footed, glass-helmeted, drooling alien monsters.]

In the "Star Trek" episode, Kodos is the name of an infamous galactic fugitive -- now using an alias and on the lam as the guilt-wracked "Anton Karidian", lead actor of a traveling Shakespearean troupe – a former colonial outpost governor, whose reasonable decision at the wrong time turned him into a notorious mass murderer forever damned and on the run.

His command decision to slaughter 4,000 lower echelon colonists to ensure that the rest of the members of the endangered colony of Tarsus IV survived, became a horrendous crime when a rescue ship bearing life-sustaining supplies unexpectedly arrived early.

> "I was a soldier in a cause; there were things that had to be done -- terrible things.
>
> Murder, flight, suicide, madness...",

he rambles, trying to explain himself after he is unmasked, unable to escape from either the overwhelming agony of his guilt, or justice.

A guilt-ridden public official? It's gotta be science fiction...

> "Snowman melting from the inside,
> Falcon spirals to the ground.
> This could be the biggest sky,
> Somebody wrecked tomorrow's plans...
>
> A little piece of you, a little piece in me,
> Will die."
>
> -- "This is not America" (1984)
>   ("Falcon and the Snowman" theme song)
>   David Bowie/Pat Metheny

Prison Diary 52:   My Pirate Heart
                                    =============

> " 'I don't say he's such a bad man as I have fancied
> – I pray to God he is not. But since we don't
> exactly know what he is, why not behave as if he
> **might** be bad, simply for your own safety? Don't
> trust him mistress; I ask you not to trust him so.' "
>
> -- "Far from the Madding Crowd" (1874)
> Thomas Hardy

> "Do you think this came from a pixie? Where did you
> think this...came from? [...] [A] pixie is a close relative
> of a fairy. Shall I proceed, sir? Have I enlightened you?"
>
> -- Joseph Welch (1954)
> Chief Army Counsel
> Army-McCarthy Hearings

Jacques Lacan, a French psychiatrist in 1953 developed a new psychoanalytic theory in which he delineated man's relation to the world "according to a tripartite classification" of orders: the imaginary, the symbolic, and the real.

I guess I fall into the category of acolyte -- indeed an aficionado -- of the symbolic order.

Listen and I'll tell you.

It could be described in summary that I've always viewed society – indeed civilization -- "as my play-thing". Made free of ridiculous rules. Rules and laws to be bypassed for non-malevolent purposes – for purposes of my interests and entertainment. Guided by pursuit of the arcane, the exotic, and most of all, by the pursuit of the truth, through the understanding of these outer reaches – the margins – of human experience.

*   *   *   *

> "Hello, this is Killian. Give me the Justice
> Department, Entertainment Division."
>
> -- Damon Killian, on the phone,
> to the receptionist
> "The Running Man" (1987)

Educational psychologist Lawrence Kohlberg (Lawrence Kohlberg, "Stage and Sequence: the Cognitive Development Approach to Socialization," in D. A. Goslin, Ed., Handbook of Socialization Theory and Research, Rand-McNally, Co., Chicago, 1969, pp. 347-380) has found that peoples go through six stages of ethical development:

(1) Conformity to rules and obedience to authority, to avoid punishment.

(2) Conformity to gain rewards.

(3) Conformity to avoid rejection.

(4) Conformity to avoid censure. (Chimps and baboons.)

(5) Arbitrariness in enforcing rules, for the common good.

(6) Conscious revision and replacement of unhelpful rules.

My pirate heart, oh, my mischievous pirate heart. I guess I skipped the first five rules and emerged all aglow in an extreme case of rule 6.

Pure mischief untainted by any malevolence. The purity and spontaneity of youthful innocence, and its virtue of undefended vulnerability. The beauty in its artistry, and the powerful strength of its insight and symbolism, manifested through totally outrageous, disruptive, yet completely harmless words and acts.

The hidden child within and thriving, with the crystalline clarity of truth and insight **and** humor. Unhobbled and unblinded by the hypocrisy of the adult world, and unhypnotized by its candy-coated lies. And not yet absorbed by its ruthless demands of unyielding conformity.

Yet manifesting as truth in a potentially powerful concrete act of an adult with its sophistication, delivery, and detail, yet still retaining and co-existing with the playful inner child and his uncontaminated message of gleeful exuberance.

My mischievous pirate heart. My sometimes secret smile. Me, the _enfant terrible_.

The Ghost in the Machine. The camouflaged clown, the joker within, the covert practical joker. Poking fun at society and its transparent silliness, willful blindness, accepted paradoxes, incongruous inconsistencies, and contradictory logic.

Sort of like Ferris in "Ferris Bueller's Day Off" (1986), though not exactly, or Harry Mudd -- a character who was unique in that he appeared in two separate episodes of the original "Star Trek" TV series as a quite colorful character who was finally condemned -- humorously -- by Captain Kirk to be imprisoned "until he stopped being an irritant".

* * * *

> "[O]nce we are conscious that all individuals are ethnocentric, inconsistent, incoherent to a point, oversimplifiers...then we have a new concept of man."
>
> -- Jacques Derrida (ca. 1970)

When I drive up to a four way stop intersection, I look to see if there are any cars at or near the other three streets, and if there's a police car in sight. If not, I don't bother to stop, or sometimes even slow down. To me, it's a stupid waste of time, gas, and brake linings to stop to an empty intersection, with no cops in sight. I'm helping to save the Planet, you see.

I explained my philosophy to my son one time years ago, that it's a little known section of the Highway Traffic Act, that you ignore such stupid laws if you can do so safely, and there's no cop around. He told his mother -- my ex-wife -- who commented that **that** sounded like something I'd say.

I recounted the anecdote -- my typical scofflaw attitude -- to a female therapist whom I got along with quite well. She commented that she'd remember never to get in a car that I was driving.

I thought that **that** was a little presumptuous. I didn't want to offend her, so I didn't add that if anything, I'd invite her to the backseat first, and she'd probably be more tolerant of my peccadilloes.

A complex man is merely a simple man whose apparent blatant contradictions – in attitude, speech, behavior – have been noted, seem inexplicable, but have not been satisfactorily resolved by an observer. In simple terms, an intricate and complex belief system adopted by a person, which confuses the casual observer, or interacting person.

How do you classify the unique? The mystifyingly original? The outrageously irrepressible? Who cares? Just watch, and be bemused.

Most people, I imagine, want to be remembered at their best. Remembering **me** at my **worst** would probably be more amusing and entertaining. Just remembering me, I would consider a victory.

A playful spirit that is loud, unflappable, and seemingly unable to conform, be tamed or moderated, be silenced, or shackled. A cutting to the chase in expressing ideas that is frequently taken as being blunt as a sledge hammer yet is merely a clear-sighted vision of reality without regard for the smoke-screen of social niceties or convention. A rejection of artificial boundaries, and a habit of trying to put a humorous angle even on the horrific.

And it's not a joke, anyway. You **have** to laugh. It's either that or cry. And I have a fondness for trying to mix pathos and humor and ridicule in a joke – even if few get it. **I** get it, and that's all that really counts in the end.

> [And it's not my penchant for black or macabre humor that gets the strong reaction, anyway. And my excuse is that macabre humor is symptomatic of the unfortunate "damage" I've sustained emotionally from having seen way too much horror that anyone should ever have to bear. An adaptation that I do not make apologies for.]

It's my superficially light-hearted verbal takes on the darkest sides of human nature that seems to elicit the worst counter-reaction. Fuck 'em, all!

Boundaries on talk are self-censorship of the worst sort. The Truth is not the problem, it's merely the first step in acknowledging and then addressing an uncomfortable problem. The problem is, no doubt, not the Truth, but the reasons for the discomfort by those who attack or seek to suppress the Truth.

My pirate heart, oh, my mischievous pirate heart. The intellectual as a harmless practical joker, as a benevolent instigator. The Joker in the Machine. The rascal dancing in the rain.

And game-playing doesn't work well with me. I find women who play "hard to get", hard to want. One girlfriend who was always extremely satisfied in the sex department (it was the only thing she liked about me, I concluded with shock) decided to try and get my goat with a dig about me "being not that good". It was so transparent a manipulation that I picked up my jacket and headed to the door, announcing, as I put on my jacket, that I was going out to "get a second opinion".

Boy, I've never seen a woman back-peddle so fast, in my life! Like I said, since I couldn't identify a single aspect of my personality, behavior, and general disposition that she didn't criticize regularly, except sex.

Now I know how women feel...

Spontaneous yet perfectly timed, outrageous yet appropriate. Delivered with words, or a deed, and a twinkle in my eye, or a secret smile.

A tendency for funny, performance art, _cri-de-coeurs_ inserted into my life. The joke as a work of art or thing of beauty. Word plays, complex linguistic constructs. The joke within a joke, within a joke. Emotional venting against conformity, emotional venting to try to stay sane or try to make sense of it all. Or merely a clear-sighted observation using a microscope that reveals every detail, strips away the camouflage of emotion, leaving only the bare facts.

And, as befits my pirate heart, I'm a gambling man. High stakes. Which you think would require steely nerves, rather than a laid back, relaxed, devil-may-care attitude to life. Ah, my pirate heart.

When I was released on bail after six months in jail, by a judge who announced that absent a search warrant, there was no case against me, one of my bail conditions was to not possess any weapons, including knives. My mother -- always on the ball -- quickly rose and inquired as the whether kitchen knives were included. They were not, the judge assured her bemusedly.

As the hearing concluded and I was led away by the female officer, who had been kind to me, as these things go, I asked her nonchalantly if I could borrow a butcher knife from her. She expressed dismay at my sense of humor. Then I asked her if I could borrow her cell phone so I could let Osama know I was out. She shook her head, once again failing to find my comment amusing.

Scofflaw as a lifestyle. Stop signs, and sometimes even red lights (in a car). **Always** if on a bicycle. No cop, no traffic -- it's legal!

1) Is it right, or at least **not** morally wrong?

2) Can you get away with it?

Yes to both answers, and you may well be "good to go".

Moral relativism. Beats the hell out of Realpolitik. And Rousseau's social contract.

> "To [former warriors] the unarmed class appears
> vulgar and ignoble, laws are [just] superfluous
> subtleties, [and] the forms of social life, just so
> many insupportable delays."
>
> -- "The Spirit of Conquest and Usurpation and
>   Their Relation to European Civilization" (1813)
>   Benjamin Constant

"Anyone who challenges the prevailing orthodoxy finds himself silenced with surprising effectiveness," wrote George Orwell in a preface to his novel "Animal House", a preface that was suppressed for thirty years, thus proving his point.

Silenced, suppressed, shunned, ostracized, fined, harassed, jailed, hounded, run out of town, exiled. Or stoned, executed, or crucified -- metaphorically, or otherwise -- depending on the mood of the times and the temper and number of the mob.

One ex- dubbed me an "enfant terrible". (It was a complement, by the way.) It was a refinement of her previous characterization of me as "a lone wolf", and previously to that, that I was her "beautiful mind". (Referring to a movie about a genius who was also crazy as a loon – I wasn't quite sure how to take that.)

That's the kind of talk, I like (at least with my clothes on). It also gets my clothes **off** pretty fast, too.

"**That** guy over there is really hot. Uh, oh. I'm sure he's trouble," she said to herself the first time she saw me, at the University of Ottawa Karate Club, she much later reported to me.

If only she had known.

* * * *

"I can't say I've ever been lost, but I was
mightily turned around once for three days."

-- Daniel Boone (1820)
answer to his portrait artist

Yogi, "a legend in my spare time," I used to half-joke. Civilization was my play-ground, and technology was my plaything, I joked privately. All this wonderful technology -- what's the point if you can't have fun with it for your own personal entertainment, with no ill intent, and fun the only goal. And, of course, keeping (barely) within the law?

Some people considered me an expert in certain aspects of small arms and ammunition, especially automatic weapons. I was a legal, licensed collector for almost twenty-five years. I eventually became a weapons expert, in general. Hand-to-hand combat to knives to guns, to bombs, to incendiaries, to nukes. And I became a weapons law expert, by necessity.

Explosives expert (I was briefly licensed at one point). Built the largest firecracker in the world, one time in the mid-1980's, as a cool experiment.

At work, there appeared one day, a cylindrical cardboard shipping tube, three inches outside diameter and three feet long or so. But what made it special was it had a **thick** casing, an incredible 3/8 of an inch thick.

I immediately saw the possibilities. I had the machinist cut some three-quarter inch thick plywood circles to fit one at each end, and he nailed and epoxied them on, sealing the tube at both ends. Those suckers were **not** blowing off! Finally I had him drill a quarter inch diameter hole through the cardboard tube at about half the length of the tube. It would be the filling hole for the super-test gunpowder, as well as the fuse hole.

That weekend, I laboriously filled the empty tube about 75% full with a potassium chlorate and sugar, hi-test, low explosive fuel. Then I stuffed a length of fuse in the hole, and it was ready to go. Me and my son drove out to the country to my isolated "explosives proving ground", a depression in the landscape surrounded by low hills. I set up the "cracker" in the snow (it was winter) vertically, lit the fuse and scurried back to where I had set my son up, lying flat 75 yards away. I lay down beside him and waited. It was about two minutes of fuse.

Official classification:	Experimental, extremely large, non-weapon, concussion flash/bang device.
Test location:	My private, distant, uninhabited, totally unauthorized, Crown land, IED Proving Ground, Province of Quebec, Canada.
Philosophy:	If a tree falls in the forest, and no one is there, and all that...

Boy! What an explosion it made! Yet not very destructive, though the cardboard case was shredded into a million little pieces of paper.

But all good things have to come to an end. And times changed, so the experiments ended over two decades ago.

* * * *

> "[T]hese terrorist truths were played out first in the streets and countryside and then in a courtroom or a Senate hearing committee room, and every element of the terrorist acts and the reaction to them carried ideological freight. ... These courtrooms and hearing rooms, frequently stacked to the advantage of the powerful, ultimately justified the state power gained through terrorist action -- but they also left a record of the voices of those ground under by political violence."
>
> -- "In the Name of God and Country" (2010)
> Michael Fellman

In 1994, I was even asked to appear, and did, on a CBC Cable TV "Newsworld" piece as an "explosives expert". I was at that time a "net personality", as they used to say – an "expert" on the Internet Usenet explosives newsgroups, rec.pyrotechncis and alt.engr.explosives.

I once got an email a while back from a co-worker, who had stumbled across my archived Net postings. His email was brief, but it started off, "Are you **the** Yogi Shan? You're **famous**!"

"Well, infamous, maybe," I joked to him soon after. And if I was famous, the pay certainly sucked...

Though once I **did**, of all things**,** get an email from a woman who liked my writings enough to offer herself to me, unseen, **and** – get this – bear my children. Which was, in theory, the best offer I'd had in years. Or ever, for that matter. Though a picture would have been nice, even though I eschew superficiality. She **said** she was extraordinarily beautiful...

But it all came to nothing, alas. Though I was intrigued, and fascinated, and impressed, all at once. A rare situation, unfortunately.

Expert in several areas of organic and inorganic chemistry, and pharmacology. Knowledge and expertise in certain areas of small arms, ammunition, and other weapons. Knowledge of clandestine operations, guerrilla warfare, and other operational techniques of conventional and unconventional conflict.

Arm-chair ideologue with extensive knowledge of political science ranging from far-left/far-right extremism to certain aspects of U.S. constitutional and federal law.

If you could read "both" my resumes, you'd see a totally different person. Jekyll & Hyde, as it were.

* * * *

"Joel, you wanta know something? Every now and then say, 'What the fuck!' 'What the fuck,' gives you freedom. Freedom brings opportunity. Opportunity makes your future."

-- Miles
"Risky Business" (1983)

A doctor I knew and liked was discussing "things" with me one appointment and he reflected on my intelligence, which he was kind enough to acknowledge and salute. He expressed sympathy with me, commenting that being so intelligent, it must be boring and tedious for me to deal with most people. (However, I don't think this way: intellect is a gift -- and a curse -- and I don't use it to look down on people. That's arrogance. Frankly, I'd rather deal with a decent person, than an intelligent person.)

So I decided to have a little fun with the good doctor. I leaned back in my chair and dismissively announced with ultimate grandiosity: "You're **all** stupid to me!"

And then I watched his reaction for the amusement value. I detected surprise at my comment, which he had clearly not been expecting, and it caught him off-guard. A few milliseconds later, confusion and surprise both registered on his face, as he realized that my statement didn't just baldly state that I was smarter than the rest of humanity, but more to his concern and surprise, could, and seemed to include **him**.

Thus I confirmed that his comment reflected his own feelings of intellectual superiority, that he was merely sharing with a "brother" intellectual.

I noted that his eye lids visibly widened a bit as an emergency request for additional data to confirm a re-analysis bounced around his brain, while his eyes noted my dominant body language.

The entertainment value of his shock at being disabused of the notion that I did not consider him even close to my intelligence was **priceless**, and I smiled a secret smile.

She: "So you're from Outer Space..."

Capt. Kirk: "No, I'm from Iowa. I only work in Outer Space."

-- "Star Trek IV: The Voyage Home"

Prison Diary 53:   "It's a Kinda Magic!"
                                    ===============

> "Woke up this morning, Got yourself a gun,
> Mama always said you'd be The 'Chosen One'.
> She said, 'You're one in a million. You've got to burn to shine,
> But you were born under a bad sign,
> With a blue moon in your eyes.' "
>
> -- "Woke Up This Morning" (2001)
> A3

> "We are all agreed that your theory is crazy. The question, which divides us, is whether it is crazy **enough** to have a chance of being correct."
>
> -- Niels Bohr (1958)
> to Wolfgang Pauli, after a presentation of his at Columbia University

In the 1986 North American release of the fantasy masterpiece "Highlander", starring the French actor Christopher Lambert in the title role of Connor MacLeod, Connor uses the tag-line, "It's a kinda magic." to describe his super-human powers of self-healing and immortality. The tag-line is also the title of a Queen song, released separately, though not played in the movie -- even though Queen does the entire movie sound-track.

But the point I'm building up to, is that there is a 120 second scene missing, that's apparently in the Japanese release of "Highlander" that explains much.

Connor's sole employee of his 1980's New York City, Hudson St. antique store is a gray-haired woman named Rachel. There's a connection between the two that is never explained.

Apparently there's a scene in the other Japanese release, where Connor – born in 1518 A.D. -- is in the midst of WW2. He rescues a Jewish girl from a pursuing Nazi. Her name is Rachel, and the Immortal Connor takes some bullets sprayed by the Nazi, as they escape capture. When the young Rachel questions Connor's apparent immunity to the bullets, he responds to the little girl with a smile and the explanation, "It's a kinda magic."

*   *   *   *

> "It was a bright cold day in April, and the clocks were striking thirteen."
>
> -- "Nineteen Eighty-Four" (1948)
> George Orwell

Don, was my former cell site software design manager of my department, at the former multi-billion dollar, multinational corporation, Nortel Networks. One day, when a bunch of ex-colleagues went for dinner, he privately told me that, "You're the smartest guy I've ever met."

"Me, too", I almost joked.

But I didn't, seeing that he was bestowing on me quite an honor, quite the ultimate compliment. Actually, it was the second finest compliment that I've ever received. He had acknowledged out loud to me **quite** a level of ability. Not just genius, but super-genius. Wow! I was so shocked I forgot to ask for an immediate raise.

And the compliment was from a man who's a very experienced and talented software designer and manager who has known a lot of very smart people during his career in complex telecommunications software.

* * * *

"Two souls, alas, are dwelling in my breast...[One of them knows] joyous earthy lust... [but] the other soars impassioned from the dust..."

-- Faust
"Faust: A Tragedy" (1808)
Johann von Goethe

But sometimes I wondered. Were my thoughts sparkling with genius, or merely just sparking? What exactly **is** it that defines "genius"?

Quick. Bright. Sharp. Smart, intelligent, wise -- the synergy of intelligence with mature practicality and experience. Brilliant, an intellect, a genius, a mental giant, an intellectual titan. An academic, intellectual, man of letters, *the* master, the guru, a brainiac. Man and Superman. Genius and super-genius.

His abilities are likely to include being articulate, coherent, lucid, logical, cogent, maddeningly precise. But this could be just a learned or acquired verbal interface with the external world.

He will likely be knowledgeable, ideally with an impressive breadth. And on close examination an even more impressive depth according to his interest, experience, education, and employment variation. It is the breadth and depth that gifts him with such versatility, and the name "polymath".

He is self-motivated, but his motivation is most at work when interested, or especially fascinated, or stymied and on the "hunt" for an elusive goal he believes is achievable, and worthwhile, and complex. Solutions that are elegant, but not tedious are mother's milk, as is the entertainment and amusement value of viewing a problem solved as a puzzle completed at his own pace.

He is logical to a fault, intellectually versatile if so motivated to pursue varied specialties, sub-specialties, or specific finite relatively independent subsets of a knowledge "matrix" of interest or use. An impressive or unusual learning curve, a phenomenal memory is _de rigeur_, he frequently displays lightning speed mental processing, usually has a voracious and exhaustive consumption of a desired information area, and will have either an impressive learning curve, or a learning curve with an exponential rate of learning after an initial latency period.

And bizarre computer-like traits that are strangely efficient to the observant, though disturbing in their emotional-freed affect is typical, leading to the stereotype of the social reject, the social disaster, the nerdy social failure or pariah, or simply the loner.

The ability to quickly assess and discard the irrelevant – the noise – and focus in on the useful, relevant, or needed information.

Genius is not just the pattern-matching ability shown in "Beautiful Mind". It also requires the fundamental ability to persistently -- what they call -- "think outside the box".

"The box" is the perfect term. Square, relatively small, and exactly limited in size. Finite and blocking out the infinity of knowledge and understanding.

Genius requires other abilities and qualities to manifest itself, particularly curiosity. Persistent and extreme curiosity -- fascination -- to search, explore, experiment, and pursue. To pursue. Inexorable curiosity.

And lastly, a good memory. Which is to say, a *large* long-term memory, that is limber and well-exercised from use. The "database" the person can use like a fantastically big mental black-board, with a table of contents and index.

* * * *

> "[I]t is proverbial that one man is worth a thousand when a thousand are of less value than a single one. Such differences depend upon diverse mental abilities, and I reduce them to the difference between being or not being a philosopher; for philosophy as the proper nutrient of those who can feed upon it, does in fact distinguish that single man from the common herd..."
>
> -- "Dialogue Concerning the
>   Chief World System" (1632)
>   Galileo Galilei

Perfect logic is bulletproof. From the reverse perspective it is an intellectual bull's-eye. It is unflinching, unbending, perfectly aimed, directed and on-target. Case closed and filed. It is one of the -- learned and perfected -- tools of genius.

The source of unbearable frustration to the debating opponent, who finds all the doors locked or blocked to his every verbal move. Reduced to sputtering fibrillation. "You have an answer for everything, " his last desperate cry instead of just surrendering There is no honorable retreat other than complete unconditional surrender, the ultimate anathema to the ego-based alpha-male, unable to concede verbal or intellectual defeat.

Ecce baro! Behold the Fool!

And don't you just love the unschooled clowns who argue a fact in an area you're an acknowledged or educated expert in? Never argue with an expert. It's like trying to ride a pig. You'll never succeed, and all you do is irritate the pig.

Yes, the pay's lousy and the audience is long-time coming, if it ever appears at all. And it might be a mob, rather than an audience, so keep one eye on where the nearest exit is.

And should you acquire an audience, don't for a minute make them out for a fan-club. For the audience is frequently unappreciative, demanding, hair-trigger rude, impatient, fickle, and usually overly critical. And the final insult of all this static is that it's a free show, anyway!

And you can pretty much forget about groupies, is the final kick in the ass from reality, ever-ready to dismiss you as an egg-head, nerd, or washed-up has-been, a bore, a dull uninspiring conversation-killer.

Like a woman's beauty, like youth, like celebrity, fame is fleeting, even when you make it.

Labor of love it certainly is. Or the fly to the flame of Truth. Because the road is long, hard, tiresome, uncertain, bumpy, unmarked, and lonesome as fucking hell. The retirement plan is uncertain, frequently crappy, or even non-existent.

And the absolute worst part of all. The humiliation and insult to beat all, is that you are feared as a competitive threat to co-workers to be sabotaged, or derailed. However, recognized as a competitive threat also to layers of useless, or inefficient managers, or unlimited but inanimate resource, useful only as a solution provider and answer machine, harnessed solely by being enslaved, fenced in, and exploited officially, rather than their ideas and information pilfered or stolen or merely taken credit for. Anything else destabilizes the organization's functioning by threatening the minimally functional, marginal, yet key employees' security, threatening to disrupt certain systemic inefficiencies that are left deliberately uncorrected to provide for someone's private financial goldmine, or to not disrupt a complex alliance based on a bureaucratic system of rewards for past backing in return for future toleration of short-comings, incompetence, or laziness

Genius and heroes are treated alike: to be feared as dangerous, and kept on a short leash. (see "Odysseus in America" re: heroes).

When a fellow inmate, lawyer, or official would compliment my intelligence, I'd make a lame joke response:

"If I'm so smart, what am I doing in jail?"

But Jonathan Swift had an answer for me, from over three centuries ago:

"When a true genius appears in the world,
you may know him by this sign -- that the
dunces are all in confederacy against him."

* * * *

One time, a number of years ago, I had a one-on-one session with a mental health professional. He was now a Canadian, having immigrated from the U.S. He mentioned having worked decades ago in Albuquerque (New Mexico), with respect to my talking about my extensive interest in nuclear weapons.

"Los Alamos?" I asked, my interest piqued.

"No, Albuquerque," he responded vaguely, but nonchalantly. The conversation continued. But I was thinking. A few minutes later, I silently hit the answer -- Sandia Corp.! Of course, near Los Alamos, but actually **in** Albuquerque. The U.S. manufacturer of non-nuclear components -- triggers, casings, etc. -- for nuclear weapons. Bingo! But I said nothing, and filed the data point away.

A number of sessions later, the good doctor suddenly made a comment questioning my morality, for my having taken part in some serious drug-trafficking and manufacturing matters 25 years ago, of a dubious nature.

"**You** made nuclear weapons, and you're questioning **my** morality?" I answered indignantly.

He almost fell out of his chair in shock. It was his dark, little secret!

Nuclear weapons -> Albuquerque -> Sandia Corp. -> nuclear weapons manufacture. Not many people had heard of Sandia Corp...

Intuitive understanding, belief, and acceptance is what Zen Buddhism calls "beginner's mind":

> "The practice of Zen mind is beginner's mind. The innocence of the first inquiry...is needed throughout Zen practice. The mind of the beginner is empty, free of the habits of the expert, ready to accept, ready to doubt, and open to all the possibilities. It is the kind of mind which can see things as they are, which step-by-step and in a flash can realize the original nature of everything."
>
> -- "Zen Mind, Beginner's Mind" (1970)
> (Introduction by Richard Baker)

* * * *

U.S. Counter-force nuclear strategy warfare strategy: accuracy of ICBMs, but still an opening with Soviet SLBM force. Reagan's advisers convinced him that this opening could be closed with two "upgrade" of the arsenal.

First, the development and deployment of the MX (Missile Experimental) ICBM missile -- laughably renamed the "Peacemaker" -- super-accurate and equipped with ten a third of a megaton each MIRV warheads, to take out the hardened silos of the Soviet Strategic Rocket Force.

And second, a very expensive upgrade of the U.S. Navy over the 1980's to neutralize Soviet SLBM's by creation of U.S. attack sub force, to dog noisy Russian boomers, ID-ed as they left port, ID-ed as they crossed the Iceland-UK underwater acoustic detection cable -- the SOSUS line -- and tracked by surface ASW Navy ships and underwater by nuclear attack submarines thence.

The official line was nuclear parity -- the stability of nuclear equality. But it was always official policy for there to be American nuclear superiority, and the false sense of confidence -- and imperialism and aggressiveness -- it gave. It started from the beginning in 1945, and continued until the Russians caught up in the 1970's, and achieved a nuclear parity of sorts. But the U.S. would again break into superiority with Reagan's arms build-up of the 80's.

> "The hillsides ring with, 'Free the People',
> Or can I hear the echo from the days of '39?
> With trenches full of poets, the ragged army,
> Fixin' bayonets to fight the other line.
> Spanish bombs rock the province,
> I'm hearing music from another time,
> Spanish bombs on the Costa Brava,
> I'm flying in on a DC-10 tonight."
>
> -- "Spanish Bombs" (1979)
> The Clash

* * * *

There was a girl I was interested in a few years back. She was a tall, slim brunette in her late 30's, that I had noticed at my ex-wife's place one day, and found interesting, and quite intelligent, as well as beautiful.

Christine was a friend of my ex-wife. My ex- was occasionally cooperative, but certainly unlikely to do me any favors. However, with a suitable inducement, I got my ex-wife to set us up for an evening together -- no strings, and *if* she was interested.

Christine was okay with spending an evening with me, and we spent a nice time conversing at my house. She said I was a good kisser. Nonetheless, I explained to my ex-wife that I would think she would be more impressed by lots of things, money, looks, education, intellect. Yet, she didn't seem to even notice, or care. What exactly impressed a woman, anyway?

So, I tried something a little new. I gave my ex-wife a little demonstration, hoping it would get the message across. I showed my ex-wife two documents I had pulled from a stack. I had a collection of literally thousands of government and other technical documents I had compiled for my various research interests.

The first document was a U.S. Patent for a hydrogen gas valve (Kanne, Jr., William et al. "Weld Braze Technique", U.S. Patent 4,326,117 (1982)). However, the patent was granted to the Savannah River plant, South Carolina, a U.S. "production" (of weapons material) reactor that produces tritium gas -- a form of hydrogen -- used in nuclear weapons. However tritium gas was not mentioned in the patent, and certainly *not* nuclear weapons. And the document included a simple innocuous-looking, cross-sectional diagram of the gas valve "plumbing" only that showed nothing important.

The second document was a U.S. government contractor report ("Welding Development W87 Baseline", Sandia Corp. Report SAND97-8232) about welding a stainless steel tube to another stainless steel part, without any details, other than that it was an unidentified "gas transfer" part associated with the "W87" -- a known ID code name for a certain strategic nuclear warhead. Not much info here -- but I printed it off anyway, and filed it away.

There were a bunch of diagrams in the report. One of the diagrams was a basic cross-section of the welded parts. One day I focused on an unusual configuration – chamfer drilled holes, for instance -- of one of the cross-sectional diagrams. It was a "topological" match, I flashed, to the previous U.S. Patent diagram that I mentioned above.

My ex-wife is a non-technical person, but I recounted the story – she had seen the many tall stacks of neatly piled, incomprehensible documents that filled an entire room – and showed her the two cross-sectional diagrams of the two reports side-by-side. I showed her -- apparently convincingly -- how the diagrams that didn't look the same at first glance, matched almost completely several unique shapes and features, and configurations, if you looked closely. They were diagrams of the same part, and I proved that the secretive and vague patent of a hydrogen gas valve was actually a nuclear weapons document.

I explained that one day, I was bored and out of my collection of hundreds of nuclear weapons reports, declassified documents, government reports, scientific papers, books, and others, I picked up at random the patent for the twentieth or thirtieth time, and suddenly recognized the features of the cross-section diagram I had seen in another report, and noticed the link between the two diagrams, out of a collection – stacks -- of literally thousands of pages of data and reports.

I explained to her that combining the info of the two reports, I had thus positively identified what the plumbing for the deuterium/tritium boosting gas reservoir bottle for the W87 hydrogen bomb looks like. My ex-wife gave no reaction to my little lecture, and I left it at that, assuming that she didn't understand or didn't care about my fantastic and esoteric topological discovery.

A short while later, the girl I had expressed interested in called and then came over. At some point in the evening, she casually mentioned that my ex-wife thought highly of me. "Oh?" I responded, casually. What did she say?" I asked, not letting on that my ex-wife had nothing good to say about me to anyone, as a matter of principle, and was constantly bad-mouthing me behind my back to all and sundry...

The girl said that my ex-wife had told her that, "Why hadn't she called me back after our first get-together." That I was someone she should find interesting.

"He's brilliant..."

"Consider this a rip in reality."

-- Tom McMahon (1998)

Prison Diary 54:   "Rules of the Game"
===============

> "I emptied a whole [M-16] clip into him, and [then]
> Hugo Irurzan let him have it with the bazooka."
>
>   -- Enrique Gorriaran, describing
>   the 1980 assassination
>   of exiled ex-Nicaraguan
>   dictator, Anastasia Somoza

> "[T]he enduring power of bureaucratic-authoritarian structures [stems from] the ability of men of intelligence and goodwill to believe that authority must maintain itself at any cost in the interests of 'order', and to adjust their moral perceptions accordingly so that the atrocious comes to seem perfectly reasonable and even beneficent.
>
> If this can occur where the patently outrageous, the ludicrously untrue or self-contradictory is concerned, the bureaucratic-authoritarian mind can obviously function with even more self-assurance where less dramatic but more widespread injustices and absurdities are concerned."
>
>   -- "Violence in the Arts" (1974)
>   John Fraser

John Fraser's detailed dissection, above, is a cogent example of what psychologists' refer to as "cognitive dissonance" -- theorized by Leon Fastinger in 1957 -- the subconscious ability of people to easily short-circuit their conscience and commit hypocrisy without the slightest guilt, shame, mental unease, or the troublesome interference of logic or morality, if it is satisfies an immediate need, or is otherwise advantageous -- or most lowly of all -- is expedient.

Or, put in the cold efficiency of the Armed Struggle: you can discuss the immorality, error, and illogic of a man's arguments backing his politics for hours, and not change his mind. Or pump a bullet into the back of his head, and not have to...

It is why from the late 1700's, peaking again in 1848 and sporadically to the early 1920's, and finally continuously since 1948 to now, what used to be called civil war has become an established part of human social change within states.

From disaffected revolutionary theorists, to disaffected serfs, to the outraged common masses, they considered it so much simpler to overthrow by force an entrenched and corrupt ruling despot (first incompetent monarchs, then dictators, and finally corrupt, authoritarian regimes and police states), by seemingly reaching their breaking points simultaneously and triggering mass unrest leading to the defiant action of open revolt and rebellion.

Then grabbing whatever they could make use as a weapon, and swarming the security forces in rage, as a result, soon they were now the armed masses, preceding, during or following the assassination of the focal point of their suffering -- the national leader, or leadership. The Old

Order was then toppled and stomped into bits.

While war between nations continued, as it always had, it was now punctuated, or overlapped with internal warfare within nations.

But then, out of the ashes of World War Two, the two remaining world leaders -- the "Super-Powers" -- communist Russia and capitalist America squared off, both equipped with nuclear weapons, but unable to use them without destroying each other.

The Cold War was born.

But against all odds, their *were* victories in the armed struggle. The Chinese Communists under Mao defeated the corrupt and feudal Nationalists in 1949. The first victory of the peasants over their overlords in modern history, since the French and Russian Revolutions.

Ho Chi Minh's Viet Minh guerrillas defeated the colonialist French in 1954 at the decisive battle of Dien Bien Phu. General Giap completed the job, liberating South Vietnam, and defeating the Americans coldly in 1975.

Castro defeated the dictator Batista in 1959, liberating his people, and has been in power ever since. Algeria's Revolution threw the French imperialists in 1962, after a brutal and bloody struggle, with atrocities committed by both sides.

But these were just side-shows.

*   *   *   *

"So Jesus said to them ' You try to justify your actions
in front of people. But God knows what's in your hearts.
What is important to humans is disgusting to God.' "

-- Luke 16:15

I got a T-shirt around 1985, printed with the hilarious slogan: "U.S. Out of North America". It was from a Toronto store owned by a friend from university, Paul Cassel. The store was named appropriately, "Marché Noir" -- French for "Black Market". I thought it was an amusing touch.

"The Great Game", the spies called it. It was the "Cold War", a chess game played out over control of the entire world, including its oceans. The Empire of the former colonial powers was now disbanded, absorbed, and transformed into an invisible empire as the U.S. Economic Empire was born, with allies, trading partners, weapons buyers, and resource providers (oil and metals, principally). And the communist enemy nations were completely embargoed from all trade, hindering, or choking their development and advancement.

And the new Economic Empire was called euphemistically, the American "sphere of influence".

It was a war between enemies, each having their allies. And the erst-while "peaceable enmity" between the U.S. and the Russians who -- nuclear-armed -- could not fight each other directly for fear of the real danger of escalation to all-out nuclear warfare, and the destruction of both countries. And so it was a war of small, indirect wars on foreign territory -- the proxy wars -- and a war involving foreign countries willing, countries pressured into compliance, countries coerced, and countries drafted.

And finally covert superpower machinations conducted on countries unknowing -- countries oblivious and in the dark, dimly aware, aware too late, or aware long past the conclusion of the incident.

They were the Rules of the Game. And there was but one Master of the Game -- the U.S. The death of sentimentality, the loss of innocence. The fool could actually be considered to be a philosopher or ideologue whose dogma or belief system consists of a reverence for consistent stupidity.

* * * *

> "America is today the leader of a world-wide anti-revolutionary movement in the defense of vested interests. She now stands for what Rome stood for.
>
> Rome consistently supported the rich against the poor in all foreign communities that fell under her sway; and since the poor, so far, have always been more numerous than the rich, Rome's policy made for inequality, for injustice, and for the least happiness of the greatest number."
>
> -- "America and the World Revolution and Revolution and Other Lectures" (1961)
> Arnold Toynbee

The Cold War's great cost, and false progress: war is bankrupting, so cold war created a materialist consumer society out of the false peace. But peace is nothing without a moral just society. The dead have peace.

But the cost to allies and establishment of a military economy and a de facto Empire for military bases and installations, and the introduction of "proxy wars" (like Vietnam & Afghanistan) showed that the forced end to large-scale war had not lessened death and suffering -- it merely exported it as the War was fought anywhere, but between the two great Super-powers.

The Cold War was a state of continual, on-going, low-level military and economic warfare. And though people knew the term, it machinations were essentially hidden from view, and its effects on the people of the First World non-existent, as they remained anesthetized by materialism of the Consumer Society, and the striving for attainment of "the Dream" of economic security. And the War continued unabated.

> "Of all the enemies to public liberty war is, perhaps, the most to be dreaded because it comprises and develops the germ of every other. War is the parent of armies; from these proceed debts and taxes... known instruments for bringing the many under the domination of the few. ... No nation could preserve its freedom in the midst of continual warfare."
>
> -- "Political Observations" (1795)
> (16 years after the American Revolution)
> James Madison

* * * *

It was the Butcher's Bill for the Cold War's Proxy Wars. For a "cold" war, there were certainly a lot of death -- killings by the U.S. and its proxies, satellites, and puppets.

And the U.S. killing furnace was fired by Canada -- the largest trading partner of the U.S., who provided the cheap raw materials: mineral ore and metal (aluminum, nickel, uranium), mined potash fertilizer, lumber and wood, fish, oil, coal, and natural gas, and hydro-electric power to fire the furnace.

The flat plate fuel element for nuclear reactors was a Canadian Chalk River invention in 1953 that was shared with the U.S., that enabled their Savannah River reactor in South Carolina to double its tritium and plutonium production rate. The flat plate was made by rolling nickel-clad enriched uranium in an aluminum can, which was later cut off, leaving the flat plate fuel element.

U.S.-Canadian cooperation had been helpful in providing experience in operating heavy water reactors at high flux and high power; early work on the Purex and Redox reprocessing technologies, irradiation of elements, and providing heavy water nuclear measurement data.

Uranium ore producer for U.S. production reactors. Canada supplied 10,500 tons of uranium in the 50's for UK bomb program. Canadian uranium metal exports for 2008 were 7330 tons. [McKay, Paul. "Atomic Accomplice". (2009)]

Tritium from Canadian nuclear reactors is sold to the U.S., where it is tacitly used in nuclear weapons. Uranium ore mined in Canada originally fueled the U.S. nuclear plutonium-producing "production" reactors for nuclear weapons – and continued until at least 1955, to make thousands of nuclear weapons.

Canada provided full access to military aviation and naval bases, and received training, and inter-troop training assignment. Canadian universities, such as McGill, did classified military research funded by the U.S. DoD. Canada provided the high arctic radar station perimeter line for incoming bomber detection in the 1950s to the 1960's: the DEW (Distant Early Warning) Line.

Canada later manufactured parts (by Litton, in Toronto) and allowed flight testing in northern Alberta for nuclear-armed cruise missiles, and provided through the 1950's testing grounds for U.S. chemical warfare (Suffield Proving Ground, Alberta), torpedo testing waters in the Pacific, near Vancouver, on the west coast, and nuclear artillery testing in Quebec.

Canada was complicit through and through, in the U.S.'s predatory actions in deed, but not word -- officially the liberal, peaceable, state, but in reality just a satellite of the American Economic Empire, and the principle cog in fueling its war machine.

In reality, we were simply America's whore, from the beginning.

* * * *

"Throughout the world, on any given day, a man, woman or child is likely to be displaced, tortured, killed or "disappeared", at the hands of governments or armed political groups. More often than not, the United States shares the blame."

-- "Human Rights & U.S. Security Assistance" (1996)
Washington Office of Amnesty International

The Beginnings

Dominican Republic	(1930-1961)	50,000 civilians killed

Spain	(1936-1975)	200,000 soldiers killed, 200,000 executed (50,000 by Republicans), 114,000 civilians "disappeared" by fascist Franco regime
Greece	(1946-1949)	480,0000 (both sides, including disease); civil war
Madagascar	(1947-1948)	80,000 killed
Malaya	(1948-1960)	7,000 insurgents and 3,000 civilians killed 1,500 of security forces killed.
China	(1945-1949)	1 - 3 million killed (both sides, the Communists and the Nationalists)
Tibet	(1950-1964)	300,000 civilians and smaller number of CIA-trained guerrillas killed
South Korea	(1948-1949)	60,000 rebels and civilians killed
Korean War	(1951-1953)	94,000 U.S, allies and South Korean military dead, about 300,000 North Korean military deaths, and 2 million North and South Korean civilian deaths
South Korea	(1982)	2,000 demonstrators killed
Kenya	(1952-1960)	20,000 Mau-Mau guerrillas killed, 150,000 died through famine
Algerian War	(1954-1962)	157,000 guerrillas killed, 1,000,0000 civilians killed by French
Israel	(1956-1973)	40,000 Arab soldiers killed in various wars
Hungary	(1956)	32,000 anti-communists killed
Haiti	(1957-1986)	50,000 civilians killed or "disappeared"
Lebanon	(1958)	1,000 soldiers killed, 1,000 civilians killed
Congo	(1960-1964)	100,000 killed
Congo/Zaire	(1977-1983)	6,000 civilians killed
Congo/Zaire	(1997)	200,000 civilians killed/"disappeared"
Cuba	(1952-1959)	20,000 civilians killed, including children, by U.S. backed Batista dictatorship
Cuba	(1960-1962)	3,500 Cubans killed by U.S. backed Cuban exile proxies
Zaire	(1960-1963)	100,000 killed
Mozambique	(1961-1975)	10,000 Frelimo guerrillas killed, 50,000 civilians
Ethiopia	(1962-1991)	300,000 killed; Eritrean conflict
Greece	(1964-1974)	8,000 civilians killed in first month after right-wing coup d'état
Indonesia:	(1965-1966)	2,000,000 civilians ("communists") killed
Philippines	(1965-1986)	2,000 "extra-judicial" killings of civilians
First Indochina	(1945-1954)	100,000 French forces killed; 175,000 Viet Minh guerrillas killed,

War		125,000 civilians killed;
Indochina Interlude	(1954-1962)	75,000 South Vietnamese civilians killed by Americans
Second Indochina War (the Vietnam War):	(1962-1975)	60,000 U.S. dead; 270,000 South Vietnamese Army dead, 1,100,000 VC and NVA military deaths. There were 200-400,000 South Vietnamese and 4 million North Vietnamese civilian deaths.
Cambodia	(1970-1975)	600,000 killed
Laos	(1959-1975)	200,000 killed
Chad	(1966-1992)	21,000 killed
Nigeria	(1967-1970)	2,000,000 killed or starved to death; Biafra
Jordan	(1970-1971)	41,500 killed ("Black September")

Days of the "Disappeared": "The Dirty War" Against Suspected Leftists
----

Paraguay	(1947)	28,000 leftists killed
Bolivia	(1952)	1,000 civilians killed; 1,000 military killed
Honduras	(1957-1992)	500 killed or disappeared
Guatemala	(1960-1998)	200,000 civilians disappeared or killed
Brazil	(1964-1985)	2,700 civilians killed
Peru	(1965-1966)	8,000 peasants killed.
Mexico	(1968-1971)	600 civilians "disappeared".
Uruguay	(1973-1985)	200 civilians killed.
Chile	(1973-1998)	25,000 civilians killed, 28,000 imprisoned and tortured.
Turkey	(1975-1980)	4,600 killed.
Argentina	(1976-1979)	30,000 civilians "disappeared". Total Population 40.5 million; the military dictatorship ended in 1983.
Chad	(1982-1990)	40,000 civilians killed by U.S. backed dictatorship.

Proxy War and Guerrilla War

Colombia	(1948-1958)	"La Violencia"; 200,000 civilian deaths.
Columbia	(1964-2010)	350,000 civilians killed/disappeared; 10,000 military killed.
Philippines	(1971-2002)	23,000 communist rebels killed, 11,000 civilians.
East Timor	(1975-1999)	200,000 civilians killed out of a population of 600-700,000.
El Salvador	(1979-1992)	80,000 civilians killed; 15,000 military deaths.
Nicaragua	(1972-1979)	60,000 killed (Sandinista Rebels)
Nicaragua	(1981-1990)	60,000 killed (by CIA-financed "Contra" Rebels)
Lebanon	(1975-1976)	25,000 military killed, 75,000 civilians killed (Muslim vs. Maronite Christians.
Western Sahara	(1975-1991)	16,000 Polisario guerrillas and civilians killed by Morocco.
Rhodesia	(1972-1979)	30,000 Patriotic Front guerrillas killed
South Africa	(1948-1993)	20,000 ANC and civilians killed
Namibia	(1966-1990)	13,000 SWAPO guerrillas killed
Angola	(1961-1975)	25,000 MPLA insurgents killed, 50,000 civilians killed
Angola	(1975-2002)	500,000 killed; MPLA vs. U.S. backed UNITA guerrillas
Mozambique	(1977-1992)	900,000 killed; South African-supported RENAMO guerrillas
Israel	(1973)	11,000 killed
Lebanon	(1975)	150,000 killed
Afghanistan	(1979-1988)	26,000 Soviet military deaths; 1.3 million Afghan civilians killed.
Peru	(1980-2000)	69,000 "Shining Path" rural guerrillas killed or "disappeared". Total Population (2008): 29 million.
Iran-Iraq War	(1980-1988):	300,000 Iranians killed, 200,000 Iraqis killed.
Turkey	(1980-2000)	40,000 Kurds killed by NATO ally, while the West did nothing.
Lebanon	(1982)	20,000 Lebanese civilian deaths by Israelis.
Syria	(1982)	30,000 killed
Yugoslavia	(1991-1992)	150,000; civil war; post-Soviet; stoked by West; "last" European communist state.
Chechnya	(1994-2005)	200,000 civilians killed.
Nepal	(1996-2002)	2,000 Maoist guerrillas killed

* * * *

"The Cold War had the distinction
of not costing any lives."

-- Edward Teller
"New York Times" (14 May 1999)

The total of the partial list above is about 17 million deaths in the Cold War fight in the name of anti-communism. It just edges out WW1 in numbers, by two million.

It was a world war in every sense of the word, much of it operated in large part overtly, but ignored by the Western Press, and not realized to be each a small part of a greater World War. But the deaths and scope provide the realization, of a World War III conducted in slow motion between 1945-1991, as official, planned U.S. Foreign Policy.

The Americans won, beating the Soviets, as well as the home-grown revolutionaries and just, plain, innocent or harmless civilians.

American soldiers have been sent abroad to make war on other countries over two hundred times since the American Revolution, so this is just another excuse of the same pattern of power, imperialism, and mass murder in the name of the American Government -- American "Democracy" for Americans only, and empty words and organized subversion for the rest of the world seeking justice and freedom in their own local way, unobstructed by a Superpower's machinations.

* * * *

"The stars are not wanted now: put out every one;
Pack up the moon and dismantle the sun;
Pour away the ocean and sweep up the wood;
For nothing now can ever come to any good."

-- "Twelve Songs" (1936)
W.H. Auden

A pattern established during the Cold War, the U.S. maintains an estimated 798 military bases or installations in at least 40 foreign countries -- the true number is secret. The total cost of the Cold War (1948 – 1991) in 1996 dollars was $19.1 Trillion, including military spending and nuclear weapons development. [Ref: Center for Defense Information and "Atomic Audit"] CIA and intelligence-related activities over the period are an estimated additional half trillion dollars. [Ref: Federation of American Scientists]

"Their feet run to evil, and they make haste to shed
innocent blood: their thoughts are thoughts of
iniquity; wasting and destruction are in their paths.
The way of peace they know not."

-- Isaiah 59:7-8

Prison Diary 55:   Servants of the Game
                                    ==================

> " 'In any event, once society is properly organized,
> the fact that a man be wise or stupid, good or evil,
> will cease to be of importance.' "
>
> -- Bazarov
>   "Fathers and Sons" (1862)
>   Ivan Turgenev

> "Surprise, when it happens to a government, is likely to be a complicated, diffuse, bureaucratic thing. It includes neglect of responsibility but also responsibility so poorly defined or so ambiguously delegated that action gets lost. ... It includes, in addition, the inability of individual human beings to rise to the occasions until they are sure it *is* the occasion -- which is usually too late."
>
> -- Foreword to "Pearl Harbor:
>   Warning and Decision (1962)
>   Thomas E. Shelling

First among equals.

The Soviet Russian enemy and its East European allies -- termed derisively "satellites". There were other "enemy" (communist-liberated) states: Cuba, North Korea, and originally (economic status trumped enemy status) China. There were anti-American liberated countries: Algeria and Libya.

Then there were third world countries and Latin American dictators-led countries that kept the resources and raw materials flowing to the American economic empire.

But then the was the hidden secret of the Cold War -- the manipulation of America's friends and NATO allies, both overtly and covertly, for American political or economic purposes. It was bend to our pressure, or else.

Overt: open political pressure, economic threats, or other coercion.

Covert: espionage, bribed government informants and tools, hidden manipulation and secret funding of opposition groups, and other meddling in the internal affairs, elections, etc.

Greece suffered through a successful right-wing army coup. Spain maintained its fascist government unmolested. Italy and France had their elections interfered with.

                    *   *   *   *

> "I see the airplanes carrying the bombs,
> Why, you've even found people to drop them on!

> You know you can't keep what you take by force,
> But it's only my first impression of course.
> And I'm a stranger here on this planet Earth. ....
>
> "Oh, you crazy fools, don't you know you had it made!
> You were living in paradise.
> But take it from one who knows,
> Who knows that the gates of Heaven can close.
> I only pray that you take my advice,
> Because paradise won't come twice.
>
> Well, I'm a stranger here in this place called Earth.
> And I was sent down to discover the worth,
> Of your little blue planet, third from the sun.
> I think I'll go back home, where I've come from."
>
> -- "I'm a Stranger Here" (1972)
> Five Man Electrical Band

In the sleepy, out-of-the-way town of Pine Gap, Australia, was part of the American intelligence global web. Pine Gap was the U.S. run "down-link" site for reception of transmissions for the geostationary SIGINT (Signals Intelligence) satellite code-named "Rhyolite".

Telecommunications had evolved in the 1960's to replace long distances strung with copper lines on towers to much cheaper microwave transmission towers that converted phone conversations to digital signals, multiplexed many more conversation together, and broadcast them from tower to tower on microwave frequencies.

The technology was a boon to the Russians, with their long Asian land mass. They copied the technology and set-up similar strings of long distance microwave towers from Eastern Europe across Central and Asian Russia, all the way to the Pacific Ocean.

But the National Security Agency -- the NSA -- was waiting with an idea. The secret came out at the espionage trial of Chris Boyce, as described in the book and movie "The Falcon and the Snowman" (1984).

Telephone microwave transmissions are, what is called "line-of-sight", which means that the transmitter and receiver antenna dishes transmit a straight radio beam to each other, and thus need to be visible to each other. Since the earth is a sphere, the microwave towers need to beam their concentrated signal to another tower, -- the next tower -- somewhere just short of the horizon, and so on in a string of towers, to work for the multi-thousand mile distances of trans-continental transmission.

But because of this very curvature of the earth, and the fact that the straight microwave beam spreads slightly on its way to the next tower, the beam bleeds past the receiving tower, continuing on its straight course and off into outer space. The NSA came up with the idea of placing a satellite in stationary orbit with an antenna to pick up the microwave signal, and tap into any Russian military telephone traffic being transmitted.

Then in 1972, Gough Whitlam's Labor Party was elected in Australia. He threatened to close down the Pine Gap down-link that received the eavesdropped signals from the Rhyolite satellite that was needed by U.S. intelligence. The U.S. could not tolerate this.

Through legal parliamentary loop-hole of dubious value, when Whitlam in 1975 went Governor-General Kerr, to have him dissolve parliament and call an election, instead -- in an act unprecedented in Australian history -- he instead dismissed Whitlam from office and handed power directly over to the Conservative Party opposition.

Whitlam's government had been overthrown. And it was perfectly legal.

The dark shadow of the U.S. and U.K. governments has never been proved, but it rarely does face "indictment".

* * * *

> "New lords no less cruel, no less greedy, no less insolent than the old have risen upon the ruins of feudalism. They have bought or leased the property of the old masters, and continue to walk in the paths beaten by crime, to speculate on the public misery, to dry up the sources of plenty and to tyrannize over the destroyers of tyranny."
>
> -- speech to National Convention (1793)
> Chaumette, Paris procurator

Order and the sense of security, serenity that it brings, it maintains the stable integrity and wholeness of their personality -- their identity -- with the teaming masses moving and jostling through their neighborhood oasis in a busy, bustling world. Contentedness, the feeling that maybe all was right in the world.

The essential requirement for a sense of peace, of relative content, is to be capable at least of striving to reach up to happiness, or the possibility of happiness. It is a sense of safety and freedom from fear, wariness or foreboding, and the anxiety that burrows down to your soul. If absent, the corrosion starts and begins the release of conflicting and destabilizing emotions.

The home as an oasis within the oasis of one's neighborhood within the order of the urban environment and its economic order.

The last stand of security is one's home, one's abode, and the daily rituals that represent order and personal security. With fear, home or personal rituals start to become obsessional in order to over-compensate -- an empty reaction to distract -- from a perceived or actual loss of control, and a psychologically self-reassuring -- but in actual fact, futile -- attempt to re-exert control.

The human need to cling to those emotions -- inadequate, yet something, the eclipse of hope, is still hope, until the eclipse is full -- even as the shadow appeared in their field of vision, or the sun dimmed, or a dreadful twilight began. The umbra. Foreboding. The great fear that things were about to change for the worse.

The confusion of not knowing why, or of only partial and incomplete understanding. The great fear that the worse was not over, that the slide downwards was continuing or accelerating. That the bottom was neither hit, nor in view, nor predictable. But there is little comfort even in seeing the bottom, or hitting it.

* * * *

> "[The] conjunction of an immense military establishment and a large arms industry is new in the American experience. The total influence -- economic, political, even spiritual -- is felt in every city, ever State House, every office of the Federal Government. ...[W]e must not fail

> to comprehend its grave implications. Our toil, resources, and
> livelihood are all involved; so is the very structure of our society.
>
> In the councils of government, we must guard against the acquisition
> of unwarranted influence, whether sought or unsought, by the military-
> industrial complex. The potential for the disastrous rise of misplaced
> power exists and will persist.
>
> We must never let the weight of this combination endanger our liberties
> or democratic processes."
>
> -- farewell speech (1961)
> Dwight D. Eisenhower

The Republican Party -- who would vote for a party of, and for the rich -- a tiny minority of the electorate? Workers that made up the M.- I.C. were the new _bourgeoisie_, damned by revolutionaries since the 1800's. The reactionaries and the middle class.

In the case of the M.I.C. the upper middle class workers of armaments industry -- manufacturing, subcontractors, and R&D. Armed forces personnel and their families. The DoD and the intelligence/national security apparatus, the police and internal security forces, the high tech firms.

The young and the poor don't vote from apathy or disengagement from the entire political process. Blacks are poor, criminalized by the police apparatus, apathetic, despairing, and effectively disenfranchised.

The nuclear Triad has been the strategy of U.S. nuclear defense since 1960, when nuclear weapons-equipped B-52 heavy bombers were supplemented by land-based Atlas ICBM missiles, and nuclear submarines with Polaris A-1 SLBM's.

B-52's should have been retired long ago. Obsolete by the mid-60's with the rise of ICBMs and SLBM's. Revived partially in 1970 by the SRAM -- Short Range Attack Missile -- to enable the B-52s to penetrate the extensive and effective Russian air defenses network. Obsoleted again by the introduction of sub-launched Cruise Missiles that flew at low radar-evading heights, and were much more accurate the B-52's bombs.

* * * *

> "There are no rules in such a game. Hitherto acceptable norms of
> conduct do not apply. If the United States is to survive, long-standing
> American concepts of 'fair play' must be reconsidered.
>
> We...must learn to subvert, sabotage, and destroy our enemies by
> more clever, more sophisticated, and more effective methods than
> those used against us.
>
> It may become necessary that the American people be made acquainted
> with, understand, and support this fundamentally repugnant philosophy."
>
> -- White House Commission Report
> on CIA Covert Activities (1954)

"Refuse to confirm or deny." That was the bald platitude and scripted response to whether a U.S.

naval vessel -- or anything U.S. -- was carrying nuclear weapons aboard.

Of course they were, it took no great logician to figure out. They *all* were during the Cold War, at least. Defensive weapons, of course. But strategic offensive weapons was the unacknowledged "secret". The whole kit-and-caboodle shuttling all around the world over the oceans, positioning themselves for Armageddon. All the long-range warships – all the warships of any size, from aircraft carriers (of course) to destroyers, had magazines loaded with many nuclear weapons ready for use, stockpiled with an ample supply of air-droppable gravity bombs, rockets, cruise missiles, nuclear weapons ready at a moments notice when given the command, dispersed all over the oceans.

Strategic weapons for attack against the Soviets -- a fact that was little known, unlike ASW weapons, like short-range rockets and depth charges. The hint was the "refuse to confirm or deny" nonsense platitude.

The Fiji Islands in the Pacific Ocean. In 1982 they briefly instituted a nuclear-free zone. The former U.S. ambassador to Fiji stated that "a nuclear free zone would be unacceptable to the U.S. given our strategic needs..." The ban was rescinded.

Fiji. In 1987 the Prime Minister, a month after being elected, with a pledge to reinstate Fiji as a nuclear-free zone banning U.S. nuclear-weapons carrying or powered ships was overthrown in a military coup.

But then there was the ANZUS Treaty brouhaha in 1985. ANZUS -- Australia, New Zealand, and the dominant partner, the U.S. ANZUS is like the NATO alliance with the U.S. and Western Europe.

The reason why the U.S. was so upset was that an ally was finally standing up to American pressure, coercion, and arguments. The monopoly was finally broken.

There had been a growing anti-nuclear sentiment amongst the population of the far-off South Pacific islands of New Zealand. Far off from the Cold War, tension had long been present about the ANZUS Treaty's necessity and usefulness to anyone but the U.S. ability to dominate the world's oceans with nuclear-armed vessels.

With the 1984 election of the New Zealand Labor Party, Prime Minister David Lange acted, by barring from its ports or territorial waters nuclear-powered or armed ships. He gave his reasoning as the dangers of nuclear weapons, nuclear disarmament, continued French open air (as opposed to underground) nuclear testing in the South Pacific, and Ronald Reagan's policy of aggressively confronting the Russians which was increasing the chance of nuclear war.

When in 1985, the U.S.S. Buchanan "refused to confirm or deny" that they carried nuclear weapons (they were), they were refused entry to New Zealand waters. The U.S. was "humiliated" (angered), and the ANZUS Treaty was threatened with collapse. It was front-page news for weeks in U.S. newspapers. Pressure on Australia from Reagan administration *cabinet* members -- publicly disclosed -- expressed a deep sense of betrayal by New Zealand. Other threats or coercive attempts were not made public.

But New Zealand held fast, and in 1986 was "suspended" by the U.S. from the ANZUS protection responsibility. In opinion polls, New Zealanders solidly backed their Prime Minister's decision. "Low-born people are only a whisper in the wind.

"Important people are only a delusion.
When all of them are weighed on a scale,
they amount to nothing.
They are less than a whisper in the wind."

-- Psalm 62:9

Prison Diary 56:   Le Chatelier's Warning
                             =================

> "I don't believe in guarded borders and I don't believe in hate,
> I don't believe in generals or their stinking torture states,
> And when I talk with the survivors of things too sickening to relate,
> If I had a rocket launcher, If I had a rocket launcher, If I had a rocket launcher, I would retaliate."
>
> -- "If I had a Rocket Launcher" (1984)
> Bruce Cockburn

> "While they're standing in the welfare lines,
> Crying at the doorsteps of those armies of salvation,
> Wasting time unemployment lines,
> Sitting around waiting for a promotion.
>
> Don't you know they're talking 'bout a revolution,
> It sounds like a whisper."
>
> -- "Talkin' 'Bout a Revolution" (1988)
> Tracy Chapman

In 1884, Henri Louis Le Chatelier, a French chemist, discovered what came to be known as Le Chatelier's Principle: if a closed chemical system at equilibrium experiences a change in concentration, temperature, pressure, or volume, then the equilibrium is shifted by the change, **but** – and here's the important point – the system also moves to counteract the change, while it is re-establishing a new equilibrium.

In other words, if an equilibrium is disturbed from its status, by changing conditions, the position of the equilibrium moves to counteract the change to an intermediate state.

Say you have the equilibrium (can go either way) reaction:

$$3 H_2 + N_2 \rightleftarrows 2 NH_3 \quad \text{(hydrogen + nitrogen} \rightleftarrows \text{ammonia gas)}$$

If the mixture of all three ingredients is subject to increased pressure, since there is the greater volume of 4 on one side of the equilibrium, and 2 on the other side, the equilibrium shifts to NH3 to maintain a lower pressure. So, if you want to manufacture NH3 -- ammonia -- you pressurize the system to shift the equilibrium towards the ammonia side.

A more general observation of Le Chatelier's Principle states: any change in the _status quo_ of a system, prompts an opposing reaction in the responding system to resist the change and partially maintain the _status quo_.

* * * *

"Yesterday and days before, sun is cold and rain is hard,

> I know, been that way for ally my time,
> 'Till forever on it goes through the circle fast and slow,
> I know and it can't stop, I wonder,
> I want to know, have you ever seen the rain,
> Have you ever seen the rain, Coming down on a sunny day?"
>
> — "Have You Ever Seen the Rain" (1970)
> John Fogarty
> Credence Clearwater Revival

The 1960's era was the product of multiple factors combining to create "the perfect storm". There was the sexual revolution caused by the introduction in 1960 of the birth control pill. There was the fight for black civil rights. There was widespread student opposition to the draft of young men to an unpopular war. There was the student battle for free speech, basic rights, and power with university administrations for which they were expected to die.

There was a revolution in music. The transistor was invented, giving rise to the electric guitar and amplified music to fill stadiums.

There was the drug culture: psychedelics like LSD were promoted by guru academics like Timothy Leary, and marijuana was in even more widespread use. Smuggling and trafficking distribution networks were set-up not so much for the money, but for the "Cause".

* * * *

> "Are you not called rascals, brigands, murderers? Well!
> Since our virtue, our moderation, our philosophic ideas
> have been useless, let us be brigands for the good of the
> people -- let us be brigands!"
>
> -- speech to French National
> Convention (1793)
> unknown member

First there were the horrors of black slavery.

Then the Civil War and the Emancipation Proclamation were supposed to change things. Turning their virulently racist anger inward, only the freed black ex-slaves in the South who stayed suffered.

It took another hundred years before the back of overt Jim Crow racism was broken in the mid- to late 1960's by a metaphorical alliance of Martin Luther King, Jr., the radicalism of Malcolm X, and the extremism of the Black Panthers, combined with the _Zeitgeist_ of the times.

Aided immeasurably by news photo-journalism that shocked the Nation with its portrayal of unbelievable massive white hate and their violent extremism -- black children killed in church bombings, and water hoses and dogs sicked on peaceful protestors.

And, of course, with the essential and active cooperation of U.S. President Johnson, supportive Supreme Court decisions, and basic and long-overdue Federal civil rights legislation, and pivotally, enforced through the call-out of Federal troops, the tide began to turn.

But it took the Watts riots in 1965 to enforce the point, race riots in over 70 cities in 1967, and the assassination in 1968 of Martin Luther King, Jr., and the nationwide rioting that followed to make the point clear. The situation for the black man was intolerable.

> * * * *

> "I had come to agree with [Frantz] Fanon: nonviolence was a luxury of the middle class, of whites, in the contention for power. ... There was no way to stop the destructive system except meeting its violence with violence."
>
> -- "Flying Close to the Sun" (2007)
> Cathy Wilkerson

In his memoir, "Blind Ambition", former White House counsel and Watergate co-conspirator John Dean remembers in 1970, the very first order he received from President Nixon. "I'm still trying to find the water fountains in this place [and] the president wants me to turn the IRS loose on a shit-ass magazine called Scanlan's Monthly," he complained to a confidant.

Scanlan's, named for a universally despised Irish hog farmer, and founded by Sidney Zion and Warren Hinckel III, had already made enemies with Richard Nixon. It had published a memo from Vice President Spiro Agnew's office, linking Agnew with plans to suspend the 1972 election and the Bill of Rights, and then it published a cover story arguing for Nixon's impeachment -- presciently with the Watergate Scandal still two years away.

Showing the outrageous and illegal behavior that would eventually bring him down, Nixon sicked the IRS and the FBI after its editors and investors and he successfully pressured labor unions to refuse to print the magazine. Over fifty printers refused to print the last issue, because it appeared to promote domestic revolution.

Scanlan's fought back, and had the January 1971 issue "Guerrilla War in the U.S." printed in Montreal. But tenaciously -- and no doubt with the aid of illegal wire-tapping -- with American pressure, the lap-dog Royal Canadian Mounted Police seized the issue in Montreal after the print-run had been completed.

It was in the issue where it was all spelled out. A map outlining "Guerrilla Acts of Sabotage and Terrorism in the United State 1965-1970", and their location, followed by a 25 page listing and brief description of each act of urban guerrilla warfare. Approximately 10,000 acts -- many independent -- all over the urban U.S., peaking in 1970. Riots, bombings and arson attacks on military, government, corporate targets, and sniper attacks on police.

The government wanted it suppressed to cover-up the nature and extent of civil unrest that was happening.

But some things they could not suppress. There was the picture of the burning South Vietnamese monk in 1963 that won a Pulitzer Prize.

The 1963 "I have a Dream" speech given by Martin Luther King in front of a mixed crowd of a quarter of a million in Washington, D.C. "I have a dream that one day my children will be judged by the content of their character, rather than the color of their skin. Yes, I have a dream!"

In 1965, the L.A.'s black Watts ghetto erupted into massive rioting over four to five days, involving at least 7,000 black rioters, over a minor police incident. "Burn, baby, burn!" entered the lexicon.

In 1966, it was Chicago's turn for race riot.

In Newark, it was the following year, in the "Long, Hot Summer of 1967" that saw a string of -- 159 -- black race riots across the U.S.

Also in 1967, an anti-Vietnam War protest attracted 100,000 in Washington, D.C.

In 1968, the summary execution of a handcuffed Viet Cong guerrilla suspect, by a pistol shot to the head by the Chief of National Police on a Saigon Street at the moment he was shot was captured on film and shocked the world. The photo also won a Pulitzer Prize.

The assassination of Martin Luther King in 1968 shocked the nation, and triggered black rioting in 100 cities across the U.S. But **not** in Oakland, California, where the Black Panther Party struggled to keep the peace, believing the rioting was just a pointless and invariably self-destructive outlet even for legitimate rage.

Five days of riots and demonstrations disrupted the 1968 Democratic Party's National Convention, it was described by an official investigation as a "police riot". It brought out the most famous trial of the decade, the Chicago 7 defendants, charged with crossing state lines to conspire to incite a riot, with all defendants eventually acquitted. An official report called the whole affair a "police riot".

An anti-Vietnam War demonstration attracted 200,000 participants in New York in 1969.

The revelations of the My Lai massacre -- a war crime -- were uncovered by Seymour Hersh. 504 South Vietnamese peasants -- elderly men, women, children, and *babies* -- had been shot and bayoneted to death. The massacre was covered up, until exposed by Hersh. The pictures – printed in full color in "Life" magazine shocked the U.S., and the world. Another Pulitzer was won.

The picture of a screaming 14 year old girl, hunched over one of the four students shot dead in 1970, in the Kent State University killings by National Guardsmen, which also won a Pulitzer Prize. Two of those killed were unarmed protesters, and two were innocent passers-by. 450 universities, colleges, and high schools close due to a nation-wide student strike of 4 million students in protest over the killings.

The naked 14 year old girl, her clothes burned off from a napalm attack that also won a Pulitzer Prize in 1972.

The stats on Vietnam fraggings -- attempted killings of officers by their men. 200 reported cases in 1971, no doubt under-reported.

"Can you help me remember how to smile?
Make it somehow all seem worthwhile.
How on earth did I get so jaded?
Life's mysteries seem so faded.

I can go where no one else can go,
I know what no one else has known,
Here I am, just a drownin' in the rain,
With a ticket for a runaway train.

-- "Runaway Train" (1987)
Tom Petty/ Mike Campbell

Prison Diary 57:	The Ghost in the Machine
====================

> "People have, with the help of so many conventions, resolved everything the easy way, on the easiest side of easy. But it is clear we must embrace struggle. ... We can be sure of very little, but the need to court struggle is a surety that will not leave us."
>
> -- "Letters to a Young Poet" (1929)
> Rainer Maria Rilke

> "This world is a closed door. It's a barrier, and at the same time, it's the passageway through."
>
> -- "Cahiers III" (1956)
> [Notebook 3]
> Simone Weil

This is the story of "The Big Machines", and why I love, and honor, and respect them, after over a decade of working with them, as a computer engineer. And I was the "ghost in the machine" a term coined by philosopher, Gilant Ryle in his 1949 book, "The Concept of Mind".

It was also used in the 1986 dark satire, "Brazil", starring Jonathan Pryce of "Monty Python" fame, as the faceless, anonymous-by-choice bureaucrat in an omnipotent police state government, tottering against collapse from massive internal cracks, exacerbated by external forces.

* * * *

> "It was done by arrogance. It was done by dogma. It was done by ignorance. When people believe they have absolute knowledge, with no test in reality, this is how they behave."
>
> -- "The Ascent of Man" (1976)
> Jacob Bronowksi,
> on Auschwitz

Engineers spent 10% of their time working and the rest of the time criticizing the incompetence of their boss, their management in general, each other, and lastly, the bureaucracy and paperwork the chokes the efficiency of their productive technical work.

RTFM is the mantra for the engineer/techie. Read the Fucking Manual.

And then there were the trade secrets. You could fix 90% of the problems (Ok, I'm exaggerating) with a soft reset, or a hard reset. A soft reset: turn the machine off and then on.

The hard reset: physically unplug then re-plug the power cord, essentially.

Reload the old software version and figure what's wrong with the new software.

Keep it on always; don't turn it off for the night.

Leave the hardware alone, and untouched: summer vacations were the period of the least problems.

And, don't let me begin to tell you about grounding problems!

And then there is the "kludge", a clever modification that has a side-effect that solves your problem.

ENIAC, one of the first computers, in the late 1940's, was a gigantic machine, the size of a small house. Its guts were 18,000 electronic vacuum tubes, the predecessors of transistors, which were the predecessors of IC chips. One of the thousands of tubes burnt out, and you were in trouble. That's when they came out with the idea of fan cooling. Then someone came up with the brilliant idea of turning down the power on the tubes, to lessen the problem of burned out filaments (much like light bulbs), and tube failure.

Voila! It worked to lessen problems of burned out tubes.

* * * *

"Here is an essential character of the American shows itself: his tendency to combat the disagreeable with irony, to heap ridicule upon which he is suspicious of or doesn't understand."

-- "The American Language" (1920)
H.L. Mencken

Frankly, I hated my job most of the time. Mostly long periods of boredom, punctuated by random incidents of super-high priority (read: solve the problem quickly) issues to resolve. Super stress coupled with the boredom of having to concentrate intensely, analyzing mountains of log report data, looking for the one clue to the problem. All with the boss breathing down your neck, constantly requesting updates. And the problem issue going up the management chain of command, the longer or more serious it was.

Work when the boss was looking. Or work when an interesting problem or assignment came up.

But there was one aspect of it that I am eternally grateful for: the chance -- the opportunity -- to work on what I came to refer to, as the "Big Machines".

By the end of the 20th Century, modern, industrialized civilization -- the First World – was run by the Big Machines. Essential for logistics and infrastructure: telecommunications, aerospace, petrochemical transportation and refining operations, massive plastics factories.

The monster computers handling 100,000 subscriber telephone lines each, or for a trunk switch, carrying long distance traffic on their 60,000 incoming and outgoing trunk lines. Or air traffic control, the global Internet, comprising millions of PCs and torrents of digital traffic. The supertankers to transport crude oil, or liquefied natural gas, the off-shore oil platforms, the massive refineries, polymer plastic manufacturing plants, the cable T.V. network, and the communications and T.V. satellite networks in geostationary orbit in space, and their ground station down-links.

Nuclear reactors, ICBMs, and ballistic missile submarines. Ocean-going oil well rigs. The 1969 Apollo Lunar Excursion Module and related hardware. Oil refineries, chemical synthesis mass manufacturing plants.

You see man is not just dwarfed by his inhumanity to other men, he's dwarfed by the technological monsters he's created for efficient and continuous production, keeping the financial system humming.

I was flattered when a cool cell-mate, that I had temporarily, between transfers around Ontario, asked me about the Apollo moon-landing LEM. I was also impressed that the question was rather sophisticated. He wanted to know how come the moon mission required a huge rocket to get to the moon, but the return from moon to earth didn't.

I thought about it briefly, and then explained that first of all, the moon has 1/6th the gravity of the earth, so you need much less of a rocket. Then there's orbital velocity and escape velocity. It takes much less energy to attain orbit, than escape the moon's gravity. And there were two parts of the moon landing equipment. The LEM, which landed on the moon, and an orbiting vehicle. The LEM blasted off from the moon, leaving behind its stand (less weight), to reach the orbital vehicle. Then they abandoned the LEM, and all the astronauts entered the orbital vehicle, which blasted to escape velocity to return home. Voila!

He seemed impressed and satisfied by my explanation.

* * * *

"Around the time of Lenny Bruce, the ironic posture was a thrilling _epater le bourgeois_ against repression, conformity, official smugness and fatuity."

-- Michael Hirshorn (1999)
"Slate" on-line magazine piece

I came within a hair's breath in 2002 of getting a job on a Norwegian North Sea exploratory drilling rig. You'd be searched for alcohol, then helicoptered in from Stavinga, Norway, for a 2 week stint, working 12 hours shifts on the rig, drilling in the North Sea looking for new crude oil deposits. Then helicoptered out, and three weeks off.

I didn't get the job -- the oil industry was in a down-turn at the time -- telecom & PCs had collapsed in 2001, too. I suppose it was just as well. I'm sure it was a boring job after a while, but it would have been exciting at least to start. Another big machine to put on my resume. Another big machine to learn, to read the operating manuals of, to understand, to master, to get comfortable with its nuances and intricacies.

I would get to rise from novitiate to initiate, to acolyte, to eventually maybe a priest at the Temple of the Big Machine, if I put the effort in, and the Big Machine was cooperative, at least slightly, and smiled at his worshiper.

"Efficient technology is often also appreciated aesthetically. The purely utilitarian weapon nevertheless retains an aesthetic dimension. It's not decoration anymore. Now the aesthetic dimension is, as it were, inherent to the weapon itself. ... So that the idea that aesthetics and utility part company is actually given the lie...-- they're *beautiful*.

They have a beauty in being sort of wickedly good at what they do. You make a more efficient weapon, and it also has this added bonus -- except it's not added -- of aesthetic appeal."

-- Professor Mark Kingwell (2005)
in interview

Prison Diary 58:   No Warning: Catastrophic Disassembly
===============================

> "When it gets dark, in Pigeon Park, voices in my head,
> Will soon be fed,
> By the vultures that circle 'round the dead.
>
> In a crooked little town, they were lost and never found,
> Fallen leaves, fallen leaves, fallen leaves on the ground.
> Run away before you drown, or the streets will beat you down,
> Fallen leaves, fallen leaves, fallen leaves on the ground."
>
> -- "Fallen Leaves" (2006)
> Billy Talent

> "There were real grounds for [government] fear. The Bolshevik Revolution was in full swing [1918 in Russia]. Later in 1920 the Communist party of Great Britain was founded and union membership would rise to eight million, nearly double pre-war numbers.
>
> There was every likelihood of renewed industrial unrest, since wages were still low and unions were calling for a general strike.
>
> Ireland was descending into civil war. Demobilization was bringing hundreds of thousands of soldiers, many brutalized by a vicious war, streaming home."
>
> -- "Guns and Violence" (2002)
> Joyce Lee Malcolm
> on the U.K., in 1918

My favorite episode of the original T.V. series, "Star Trek" (the late 1960's one hour science-fiction weekly show) was called "Mirror, Mirror".

Packed with social commentary, the plot centers around a transporter problem, whereby Captain Kirk, Dr. McCoy, Scotty, and Lt. Uhuru are beaming down, but mistakenly end up in a parallel universe that is no tourist paradise.

In this new parallel Universe, their formerly peaceable United Federation of Planets is a cruel and barbaric galactic empire, based on absolute dictatorship, a militaristic society, constant war, easy genocide, and violent planetary conquest and subjugation, for empire expansion or resource requirements.

The military hierarchy on their ship, the Enterprise, is partially based upon the accepted practice of career advancement through assassination of senior officers, who must be accompanied by their personal bodyguards on the ship.

However the one constant between the two parallel universes is, of course, Mr. Spock, who secretly realizes that one day the Empire -- evil to the core -- will weaken and be overthrown, but, knowing his place, and the impossibility of resistance -- works within it anyway.

"One man cannot summon the future,"

Mr. Spock concludes to Captain Kirk, with pessimistic certainty.

In 1793, Charlotte Corday was the lone assassin of Dr. Marat, a leading luminary of the French Revolution. Under attack from all sides, the leaders of the revolution found it hard to accept the conclusions of their investigation, that a lone female could do such a thing, and that she was not part of some larger counter-revolutionary conspiracy.

"The journey of a thousand miles begins with one step,"

said Mao Ze Dong, leader of the successful "Long March" that decimated his ranks but led to the communist revolution that established Red China in 1949.

They argued for years over whether communist Lee Harvey Oswald was the "lone gunman" who blew JFK's head off in Dallas, Texas in November 1963, an act of assassination that stunned the nation.

(Jackie O. scrambled away from her dying/dead husband, good wife that she was... A fact raised by stand-up comedian Lenny Bruce, whose biting social commentary such as this -- referred to by the French as an _enfant terrible_ -- resulted in his harassment by police using the obscenity laws of the time (which made his use of swear words in his stand-up act a criminal offense, and were later struck down by the Supreme Court as an unconstitutional violation of the First Amendment free speech protection clause)).

Ronald Reagan almost bought it, when in 1981, John Hinckley Jr. another lone gunmen pumped five Devastator exploding slugs into him. Each bullet is drilled out and a .22 cartridge – without the bullet – is press-fitted in reverse in the hole. It only explodes if it hits bone, or something likewise hard. Only one exploded, but it was in Reagan Press Secretary Jim Brady's skull. Instant pureed brain.

That's Ok. He wasn't using it anyway...

Hinckley was obsessed by an unrequited love of actress Jodie Foster, confused and inspired by an utterly mad mix-in by Hinckley of the plot of the successful movie "Taxi Driver" (1976), directed by Martin Scorsese and starring Robert De Niro, in which Foster also starred, playing the role of a child prostitute.

Hinckley must have been devastated when Foster grew-up and came out as a lesbian...

* * * *

"Whether in a global or regional context, the issue of light weaponry is connected with the clash between diverse social interests, values, group loyalties, institutions, practices, and cultural meanings"

-- "The Global Trade in

Light Weapons" (1994)
Klare, in "Lethal Commerce",
Boutwell, Jeffrey

Barbara is the patron saint of artillery (and so the name "Big Bertha"), since at least the 17th Century, protecting from the occasional barrel failure that blew the gun apart, killing members of the gun crew, as the technology of cannons/artillery slowly improved from bronze, to cast iron, to crude steel, and onwards.

The critical demands of war ultimately caused the experiments of new gun technology to fall upon the gun crews, who were lamented, but considered expendable by the occasional catastrophic gun failure.

There have been "proof marks" stamped on firearms, and cannons since the Middle Ages. "Proof marks" on arms shows that the device has been fired with a double load of powder, to "prove" the soundness of the steel with an overpressure test. If the gun doesn't blow up or visibly crack, the small proof mark is stamped on the breech.

When I was young, I always wondered about invisible microscopic cracks, unaware of the concept of "what works, works." It worked, so they continued using the proof test. "Trial and error", but a better proof is a *second* test of the gun using a regular charge of powder to detect any micro-cracks.

"Proof" is also used to indicate the alcohol content of distilled liquor, like whiskey or gin. It is easy to tell in the modern laboratory by the liquid's density: water is exactly 1 g/cm3, and 100% pure (200 proof) ethyl alcohol is .8 g/cm3. The percentage of alcohol in water is based upon how close to the .8 value its density is.

The American Indians came up with a crude, but simple and effective method of their own that gave the name "fire-water" to an acceptable potency of alcohol. If the whiskey would ignite from a burning match (the alcohol burning with a non-luminous thin blue flame), it passed the test. If it didn't ignite, it was of inferior alcohol concentration.

Pretty ingenious of those "ignorant savages", huh?

> "If you want to learn something about the propaganda system, have a close look at the critics and their tacit assumptions. These typically constitute the doctrines of the State religion."
>
> -- "The Manufacture of Consent" (2002)
> Noam Chomsky

* * * *

Earthquakes, tornadoes, tsunamis, and landslides are unpredictable versions of nature undergoing catastrophic disassembly.

A simple and harmless example of what they call "catastrophic disassembly" is the plain ol' rubber balloon. You blow it up with air from your mouth, and it expands, stretching into a thinner and thinner layer of transparent rubber. But if you don't stop blowing air into it, at some uncertain point – "pop", it explodes with a sudden "bang". _Voila_. Catastrophic disassembly. The rubber – a long-chain synthetic polymer – thins under a tensile force stretches up to a point, where the weakest part, usually the thinnest area of the rubber membrane tears, and the whole stretched rubber bag of compressed air fails and falls apart in an instant, with an audible "pop" that children find amusing.

The classic example of catastrophic disassembly is the BOAC (British Overseas Aircraft Corp. -- now British Airways) "Comet", made by De Havilland, and introduced in 1952 as the world's first commercial jet airliner.

The plane's frame was covered by a thin, very light aluminum skin that made for a significant weight reduction, and thus considerable fuel savings.

At the Dachau Nazi death camp, they found through prisoner experiments that pilots couldn't live with the oxygen at 36,000 ft., but which would produce additional fuel savings. So, the plane had to be pressurized, as all airliners are today, for passenger "comfort" (survival) at the high altitude's jet's had to fly to save fuel in the upper atmosphere, and make long distance commercial aviation financially viable.

In 1953, a year after introduction, a Comet crashed, having unexpectedly exploded at 30,000' (13 km). A year later, another plane crashed under similar circumstances. All the planes were immediately grounded for safety reasons, while the crash was investigated, but the fleet was put back in service in March 1954. Bad idea...

Two weeks later another Comet exploded in the air and crashed. A formal Inquiry finally discovered the problem. It was metal fatigue of the thin aluminum skin, from repeated pressurizations and depressurizations. For its two years in service, that would be something like 700 inflations of the aluminum "balloon"-like skin on cabin pressurization, followed by deflation and aluminum skin contraction when the plane landed. Eventually after several hundred of these inflation/deflation cycles, the thin aluminum skin cracked and "failed" at 30,000 ft. The aluminum balloon "popped", as it were, and the plane self-destructed as a consequence, its wing covering stripped off, providing no aerodynamic lift. Then the electrical system wiring was broken by being exposed to the wind and speed of the airplane, and the fuel tanks might have had problems too.

It's the kind of story that engineers love, despite their sadness at its tragic consequences.

No one had thought of the consequences of metal fatigue on the thin fuselage covering. It was, except for the metal fatigue problem, a great light-weight design, that pioneered and made possible trans-Atlantic commercial flights, for a reasonable price, after all...

Rounded edge, rather than square passenger windows wouldn't have weakened the aluminum skin, and periodic circular internal structural reinforcement members would have limited skin tearing.

It took four years before the re-engineered Comet was back in service. In the meantime the Boeing 707 took over their market, having entered service in 1954, the year of the two Comet crashes.

The phenomenally good record for airline safety. It was a simple as high quality parts based on exacting specifications, and well-tested, of course. But all parts wear out, and systems always fail eventually.

The solution was simple as requiring a lot of paperwork documenting everything about manufacture, testing, and maintenance of the plane, **and** having a vigorously enforced schedule of early replacement of all parts well before their probability of failure was reached.

At least the two Comets crashed for the same reason, within weeks of each other, testimony to the reproducibility of the problem (pointing to a single mode of failure), and the tight specifications of the aluminum skin manufacture.

The ultimate safety solution is to set up a tightly controlled replacement program for all important parts, before they even come close to wearing out. Or experiencing catastrophic metal fatigue failure. If they had replaced the aluminum airplane skin every year, the accident would never have happened.

* * * *

"So tired that I couldn't sleep,
So many secrets I couldn't keep,
Promised myself I wouldn't weep,
One more promise I couldn't keep."

-- "Runaway Train" (1987)
Tom Petty/ Mike Campbell

Later in 1959, the first ICBM (Inter-Continental Ballistic Missile), the Atlas, fueled by cryogenic LOX (liquefied oxygen) and kerosene, had a rocket casing design of a pressurized, thin metal skin. But unlike the Comet, Atlas used stainless steel-skin, rather than aluminum.

The steel skin was the thickness of a dime coin, and was essentially a stainless steel balloon, which made the rocket very light, weight reduction being very important for it to achieve inter-continental range. But it would only be inflated once, upon fuel loading before firing at Russia, so there would no catastrophic disassembly of the delicate missile casing from metal fatigue failure.

Only from the nuclear explosion when its warhead exploded over Moscow.

* * * *

"What is Truth? Is Truth unchanging
Law? We both have Truths -- Are
mine the same as yours?"

-- Pontius Pilate
"Jesus Christ Superstar" (1970)
Tim Rice

The 1912 sinking of the Titanic ocean liner was more mundane a story. More of hubris than anything else. The steel hull made brittle by the sub-zero salt water. With 3300 passengers, the largest passenger ship then constructed, the "unsinkable" Titanic hit an iceberg on its maiden voyage, and hundreds of passengers drowned as a result.

First brittle steel, then gross negligence in the lack of enough life-boats – the old "double-fault scenario" that is the bane of engineers everywhere.

In the early 1990's, my mother, son, and I took an advertised tour of the DuPont Nylon factory near Cornwall, next to the St. Laurence River. I thought it was interesting, being an ex-chemist, so I asked her if she wanted to go.

While outside, surveying the network of tubes and tanks, I spotted a wind aerometer – the four-cupped spinning, measurement device that shows wind speed and direction.

I mentioned off-handedly to my mother, pointing at the device which was high up in the air, "That's to tell the wind speed and direction, in case there's a plant failure, and one of the chemical tanks bursts and releases a toxic cloud. So they can tell where it's drifting, and how fast, and who to warn to evacuate..."

"Really?" she asked in genuine surprise. She was shocked.

* * * *

"We are living the paradox of a society of workers
without work, where entertainment, consumption,
and leisure only underscore the lack from which
they are supposed to distract us [from]."

-- "The Coming Insurrection" (2007)
The Invisible Committee

The absolute weapon. The secret weapon, the surprise weapon

There was the U.S. Civil War; the Sharps repeating rifle, a breech loader, clobbered the muzzle loaders. But it was more mundane than that. The factories of the heavily industrialized North trumped the rural South. Railways and the telegraph allowed troop movements, and communications. The naval blockade of the South was quite successful.

WW1: The first use of Chemical Warfare and the "higher form of killing". Chlorine gas gave way to mustard gas

There was the WW2 discovery of napalm with its tremendous psychological effects. The Vietnam War: saw the widespread use of napalm. There was "Puff the Magic Dragon", the AC-130 Specter gunship. There were night rifle sights: Generation 3 microplate night sight, image intensifier Starlight scopes. And finally, laser-guided Precision Guided Munitions that were too late to win the war.

The atomic bomb: one plane, one bomb to destroy entirely a city

Jimmy Carter's 1979 Iranian hostage rescue mission failure: band-pass filter for helmet visor/night vision goggles had a problem with cracking. The band-pass filter had the light that allowed the pilot to read the lights of the control panels, without it causing "white out". They cracked on the mission. White out and crash...

High-tech weaponry vs. low tech guerrilla warfare. Urban vs. rural guerrilla warfare: the people as the ultimate weapon.

In a high-tech society, there is no defense. Infrastructure attacks are simple, easy, and very effective. And the "Cell system" to avoid police infiltration. Power grid, water treatment, phone system, arson attacks on fire dept. first. Commercial airplanes were the first to be attacked in 1970.

All I ask is that everyone stand up and be counted, _not_ stand up to be cut down.

Yogi, "that rough beast". Intellectual and ideologue. The full spectrum from super-polite to obscenely coarse. Polite and respectful or derisive, contemptuous and dismissive. Outrageously radical, yet hopelessly conventional. Intellectual and ideologue, agitator and activist, and seditious conspirator. Crudely racist yet virulently anti-discrimination. Cruelly anti-women misogynist, yet one of the original pro-feminists.

Yet irreverent jokester supreme. Cynic and misanthrope, masking ever hopefulness, and idealism, pessimist yet logically optimistic, fatalist yet dreamer.

Gentle and sensitive, yet swinging to violently extreme in ideas and solutions. And a man of honor as my ideal, always. A tendency to explosive emotion, yet with a religious reverence and commitment to logic in it purest form. A patient analytical mind that's inexorable and unstoppable when on the scent of absolute truth or a fascinating idea or puzzle.

A simple self-effacing man, with simple needs. Yet sometimes a raging egomaniac with pseudo-macho tendencies, when my passions have been aroused. A bloodied warrior, yet one who rejects his past and has tried to learn from it, preaching peace and non-violence at every opportunity (between rants).

In other words, a maze of entirely unpredictable contradictions, whose only commonality is delivery with a certain, dare I say, élan.

Quiet between rants. Quiet until I know the person. Garrulous, even loud, with friends and associates. I'm quiet because I'm basically shy. Or I'm quiet when I'm unsure, or fearful, or observing. Or bored. Usually the later. Or I just have nothing to say.

If only that was the motto for stupid people, there'd be a lot less noise in the atmosphere, disturbing my thoughts.

Wars end precipitously by the catastrophic disassembly of enemy forces. The success of a revolution occurs by the catastrophic disassembly of the country's leaders and internal security forces.

Dissent, screeds/polemics/verbal and written outbursts, protest, graffiti, vandalism, general arming of the radicals or the people, riot, attack, defensive/offensive preparation. Radicals appear from both sides. Commercial disruptions, union political activity, labor slowdowns, wildcat strikes, general strikes, labor disruptions -- deliberate or unintended (watching, distracted, concerned).

Individual symbolic acts, spontaneous, then planned, then sporadic incidents, scattered outbreaks, then organized, then mass events, as concern turns into a sweeping tide, as fence-sitters suddenly join in, choosing sides, and polarization clears the deck of moderates.

Sunsets. Decadence, corrosion, decay, corruption, and ultimately some combination of societal melt-down or catastrophic collapse. War, crop failure, economic dysfunction or collapse.

Earthquakes and the unpredictable release of energy from tectonic forces. The tsunami out of no where. Catastrophic disassembly.

The beauty of mathematics. Mathematical description of catastrophic disassembly as exponential decay & realization that it's a catastrophic disassembly and Serbs & the death of Yugoslavia. Catastrophic disassembly can be described mathematically. And so mathematics describes societal catastrophic disassembly.

40 years of nuclear MAD enforced peace: pressure cooker, failure of filtering out eugenic-effect of war: societal fault lines, social problems, decay and decadence.

>"My, my, hey, hey,
>Rock and roll is here to stay,
>It's better to burn out,
>Than to fade away

> My, my, hey, hey."
>
> -- "My, My, Hey, Hey" (1979)
> Neil Young

It was the reverse of T.S. Eliot. The end not with a whimper, but a bang.

I remember one time, I spent some free time considering what is known in engineering as "catastrophic disassembly". This occurs when a critical part or sub-system failure -- breakage, over-stress, over-heating, age wear out, simultaneous double failures, or similar -- introduces an irreversible, unrecoverable, and growing and spreading instability, which rapidly builds up to the point where the car, rocket, airplane in motion, or stationary bridge or building self-destructs rather quickly and spectacularly. Though a person killed by such failure would not view it as "spectacular".

The evil that men do -> demands of democracy that force war (Vietnam War and fragging) -> if only they knew; worse than civil war; neighbor on neighbor; night fighting; winter; scorched earth policy; soft targets and hard targets; 50+500 to destabilize; escalating tit-for-tat; rebellion, crackdown/over-reaction/government counter-attack, revolution, catastrophic disassembly & chaos, or see-saw, or the smashing of revolt, or overthrow of the government.

Hurley-burly: strife, commotion, tumult

The rationale for gun control in Canada is the hidden motive of Quebec; but actually an armed populace is a stabilizing influence.

Strategic warfare: bringing the war home. Terrorism: strategic urban guerrilla warfare against military, industrial, urban infrastructure or economic targets in the enemy's urban centers using a combination of tactics. The tactics are to politically destabilize the government, and attack the security and confidence of the general populace in the country's leadership and effectiveness, replacing these emotions with uncertainty, contempt and anger directed at the ineffectiveness of the government.

To affect inconvenience, and disrupt, weaken through direct attack, or distract the military and security apparatus into an ineffective mobilization, or sabotage and economically drain the financial base of the targeted industrialized country.

> "Ideas flow through [the] breasts [of the working classes] that will
> shake the basis of society: they say that everything above them
> is incapable and unworthy of governing; that the distribution of
> goods to the profit of some is unjust. ...
>
> When such ideas take root, they lead sooner or later...to the most
> terrible revolutions. We are sleeping on a volcano. ... Do you not
> see that the earth trembles anew? A wind of revolution blows, the
> storm is on the horizon."
>
> -- Alexis de Tocqueville (1848)
> speech to Chamber of Deputies

Prison Diary 59:   Pol Pot & Me
                                    ==========

> "Political society wants things simple. Political scientists know them to be complex... One could argue that, in part, the leftist impulse is so conspicuous among the educated and well-to-do precisely because they are exposed to more information, and are accordingly forced to choose between living with the strains of complexity, or lapsing into simplism."
>
> -- "The Schism in Black America"
> in "The Public Interest" (1972)
> Daniel Patrick Moynihan

> "Anger, impatience, a horror of cruelty, revengefulness, a romantic feeling for violence and a despairing (nihilist) feeling that only out of the ashes of destruction could the burnished phoenix of a new life arise..."
>
> -- "Lenin" (1971)
> Michael Morgan

I was there. And I saw -- or in this case, didn't see -- what I was expecting, and noted the fact.

And I shall bear witness.

The "New York Times" was known for its front page banner headlines of varying size of bold black letters, related to the importance of the event. I had collected in July 20, 1969 the front page "Men Walk on the Moon" headline.

I had saved with jubilation -- of final joyful comeuppance -- the larger, bold, black headline, "Nixon Resigns" in August 9, 1974, as he quit in disgrace, exposed as a common criminal, to avoid impeachment and removal for his massive criminality, and corruption of the Constitution he was sworn to uphold.

Then I waited for, and wanted to see the headlines "Phnom Penh Falls" on April 17, 1975, and then "Saigon Falls" two weeks later, on 30 April 1975, emblazoned across the top of the front page of the "New York Times" in big black type, to signal the final end of the Vietnam War, and the unification of North and South Vietnam under their rightful rulers, the Communists.

But it wasn't to be. Sixty thousand deaths of American soldiers for nothing but defeat at the hands of a third world nation -- North Vietnam. It was the third defeat of imperialism in history: Dien Bien Phu, the French defeat in North Vietnam in 1954, the Bay of Pigs, Cuba in 1961, the French defeat in Algeria in 1962, and again the U.S. defeat in Indochina in 1975,

But so profound was the humiliation, that the "New York Times" was suddenly mute on the issue. There was no banner headline. There was no headline at all.

* * * *

> "In 1848 Karl Marx published the 'Communist Manifesto in which [he wrote]...that human beings are not divided by nationality or tribal loyalty, but by their relationship to the means of production. Those who own wealth try to exploit those who do not. Government agencies, law courts, police and even the educational system are designed, in practice if not theory, to further the interests of the ruling classes."
>
> -- "Power" (1990)
> C.S. Nicholls, ed.

Hitler, Stalin, and Pol Pot are often damned as the three greatest mass murderers in human history.

But let's take a look at just one: Pol Pot.

Though Pol Pot's methods were extreme, he did have justification for his need to impose his New Social Order on Cambodia. And his damning by the Western World ignores the fact that Pol Pot merely used similar brutality that the West used in earlier times in their own history. The genocide of the American Indians, slavery and oppression of the blacks for over two centuries. Etc., etc.

Hypocrisy.

The Cambodian dictator who killed an estimated 20% of the population -- some 1.7 to 2 million people -- during his rule between 1975 and 1979, before being deposed by the armed intervention of the neighboring, now unified, and also communist, Vietnamese Army.

The Chinese, Vietnamese and Thai minorities were especially targeted by Pol Pot. Enforced starvation was the main cause of death, as well as being worked to death.

Though 200,000 were specifically executed, according to one estimate -- government executives and bureaucrats, military, intellectuals, and the educated. It was a purge of the old fascist order, the U.S. pawns and collaborators, and hangers-on.

The stage: A street in Phnom Penh, capital of Cambodia

The date: April 1975, and the victorious Khmer Rouge ("Red Cambodians") guerrillas have just entered the newly fallen city.

The picture: A guerrilla soldier is in the street. He is shouting at people unseen in the photograph, and his face is contorted with anger and hate. And in his hand, he brandishes a black semi-automatic pistol -- probably a Chinese copy of the Russian Tokarev, 7.65mm.

It was the emptying of the capital, Phnom Penh in April 1975 with the fall of Cambodia to Pol Pot's guerrilla army. All of Cambodia's cities were emptied by force; former urban dwellers made to relocate to the country side to work in collective farms and forced labor projects.

But it was the emptying of a city overflowing with rural refugees from the U.S. bombing of the Cambodian countryside. A bombing that killed 40,000 civilians.

But there was more to it than emptying the city of peasants who should now return to the countryside and their farms. Pol Pot's ideology dictated that cities were useless in a rural third world country. That cities were useless in an agrarian economy. Agrarian socialism, not merely the useless economic mass consumption only of urban concentrations.

So, by force, he ordered the cities emptied of their civilian population, towards a goal of "restarting civilization" at "Year Zero"

* * * *

> "October 1967, on special assignment, Con Tum Province, Kurtz staged Operation Archangel with combined local forces. Rated a major success.
>
> He received no official clearance. He just thought it up and did it. What balls! They were gonna nail his ass to the floor-boards for that, but after the press got a hold of it they promoted him to full colonel instead. Oh, man! The bullshit piled up so fast in Vietnam, you needed wings to stay above it!"
>
> -- Capt. Willard
> "Apocalypse Now" (1979)
> John Milius, Francis Ford Coppola

Khieu Samphan, one of the nucleus of the future Khmer Rouge leadership -- "Brother No. 5" -- in the future Pol Pot communist government, was awarded his Ph.D. in economics at the University of Paris in France in 1959, at which point he returned to Cambodia. His Ph.D. thesis was entitled "Cambodia's Economy and Industrial Development".

Pol Pot had already returned in 1954.

And when he rolled into Phnom Penh in 1975, he immediately began implementing a forced implementation of Samphan's Ph.D. thesis, which was an exploration of "Dependency Theory", and how it affected Cambodia.

Dependency theory is a social science theory developed by various intellectuals from both the First and Third Worlds that posits that wealthy nations have setup a system of control, perhaps subconsciously, that puts themselves in a superior position to less developed nations, which they then exploit in order to remain wealthy, while the impoverished countries remain poor and economically subservient.

The poorer nations provide cheap natural resources needed by the rich First World -- enforced by low prices and cheap local labor -- and in return are a destination for obsolete technology. As well rich countries can overwhelm and destroy local industry with cheaper imports, while siphoning off any actual money available, by exporting to them luxury goods for consumption by the small upper class elites of the poor nations.

The "Big Squeeze".

First World nations actively, but not necessarily consciously perpetuate a state of economic dependency and thus control through economic policy, superior financial resources, political initiatives, etc. Resistance by poor countries to break out of dependency are sabotaged by economic sanctions and/or military pressure or attack.

The solution to being at the bottom of the economic pyramid for the economic benefit of the West was to establish an agrarian state through land reform, nationalization and control of import companies. And there was little need for big cities.

The solution was thus through self-reliance, and an opting out of the world capitalist system from which it had been integrated in a subservient position. To achieve independent development to modernize for the needs of its own people first.

Pol Pot's absolute devotion to this theory led to extreme totalitarian cruelty, perhaps exacerbated by his years of deprivation in the cruel jungles of Cambodia fighting the Americans. Finally, his method's outraged his neighboring communist government, the Vietnamese, so much they finally put an end to his experiment in remaking Cambodian society, and they invaded, and drove him back to the jungle.

Yet, so bitter at being whipped was the U.S., that they condemned the Vietnamese and recognized Pol Pot's government, even after it was overthrown.

All hail the "killing fields"! What hypocrisy...

"[H]ere is the slogan of our camp: everything which can be destroyed must be so; only that which can withstand a blow is worthy of preservation; what falls into pieces is nothing but rubbish; in any case hit out left and right."

-- "Works" II (1897)
Dmitri J. Pisarev

"Few people can abide by the imperative of nihilism as Pisarev formulated it: to reject all traditional authority and convention and to examine each idea and institution in the cold light of reason and utility."

-- "In the Name of the People" (1977)
Adam B. Ulam

Prison Diary 60:   The **Real** Ronald Reagan
=====================

> "It's too late to apologize,
> It's too late,
> I said it's too late to apologize,
> It's too late."
>
> -- "Apologize" (2007)
> OneRepublic/Timbaland

> "Well, they passed a law in '64,
> To give those who ain't got a lot, a little more,
> But it only goes so far,
> Because the law don't change another's mind,
> When all it sees at the hiring time,
> Is the line on the color bar."
>
> -- "The Way It Is" (1986)
> Bruce Hornsby
> on the 1964 Civil Rights Act

When you have a two-party system, one dedicated to the betterment of the rich -- the Republican Party -- and one dedicated to the betterment of the people, the vast majority of the electorate -- the Democratic Party, the party of the rich must rule by lies and deceit, propaganda and gross manipulation.

Take Ronald Reagan for instance:

1) "He was as dumb as a stump". Quoted from the article, "The Stupidity of Ronald Reagan", "Slate" magazine, by Christopher Hitchens. Described by ex-diplomat Clark Clifford in 1981 as "an amicable dunce." Described by Canadian Prime Minister, Pierre Trudeau as an "imbecile". Described in 1988 by British P.M., Margaret Thatcher -- a supporter and ally -- this way: "Poor dear, there's nothing between his ears."

> "Reagan was surprised and shocked that the Soviets
> had taken his years of militant rhetoric and his
> massive arms buildup seriously."
>
> -- "Arsenals of Folly" (2007), p. 167
> Richard Rhodes on the early '80s

> "When Bill Clark became Reagan's second national
> security adviser at the beginning of 1982, he found
> that the president knew next to nothing about what
> was going on in many corners of the globe."

-- "President Reagan" (1991), p. 156-157
Lou Cannon

2) "If an individual wants to discriminate against Negroes or others, in selling or renting his house, it is his right to do so," -- Ronald Reagan (ca. 1964) "I would have voted against the Civil Rights Act of 1964", he stated, on one of the great black equal rights legislative accomplishments of the Lyndon Johnson Administration.

3) As Governor of California, on peaceful student demonstrators (the "People's Park" student occupation) he stated, "If it's to be a bloodbath, let it be now. Appeasement is not the answer." quoted in Los Angeles Times, 8 April 1970. Demonstrator blood was then spilt, and days of rioting and unrest resulted.

4) As Governor of California with regards to the successful Patty Hearst kidnapping demand by the SLA to the wealthy Hearst family to finance a free distribution of food to poor people, he stated, "It's just too bad we can't have an epidemic of botulism."

5) For six year after the first cases of AIDS were detected, as President, he refused to mobilize health resources as a social policy to get rid of undesirables such as gays and junkies. By the time Reagan first even mentioned the word "AIDS", the epidemic was in full swing, with 30,000 people already dead. Hemophiliacs were entirely wiped out in a single swipe, since a single injection of their required blood clotting agents was made from the donations of hundreds of people, making the probability of catching HIV/AIDS a certainty.

6) He never met a weapon system he wouldn't fund (See National Debt section): the B-1 bomber, the MX (Peacekeeper) ICBM, the W-79 neutron bomb, the Trident II SLBM (Submarine Launched Ballistic Missile), the Tomahawk GLCM (Ground-Launched Cruise Missile), the Pershing II IRBM, a 600 warship Navy, and SDI (the "Star Wars" defense system as one wag named it, and which became the accepted name).

7) He openly violated the 1972 ABM Treaty with the start, and then the laboratory testing of the "Star Wars" ABM system. Treaties are of the legal standing of the U.S. Constitution, as ordered by the U.S. Constitution, the bedrock foundation of U.S. law. An act of treason when committed by a top elected or appointed official (or should be).

8) His defense spending and tax policies caused a humungous budget deficit, *tripling* the national debt. The big lie of the Republican Party is to call the Democrats "tax and spend". Mark Weisbrot, co-Director of the Center for Economic and Policy Research, stated that Reagan's "economic policies were mostly a failure".

9) Invaded Lebanon in 1983, as part of a Multinational [peacekeeping] Force. Openly and publicly sided with the minority fascist Maronite Christian government and Israel, deadly enemies of the Druze/Muslim majority. Ignored grenade and rocket attacks on the U.S. Embassy in Beirut, and it was then blown up by a pickup truck loaded with 2000 lbs. of explosive. That was their warning. It was soon followed by the killing of 241 U.S. Marines sleeping in their barracks, by a dump truck filled with 12,000 lbs. of high explosives. Militarily stupid, and just plain stupid.

"Every self-imposed rule of the 'peace-keeping' force [had] been broken," wrote British journalist Robert Fisk. The Americans turned tail, and left Lebanon pronto. It was Reagan's most disastrous and stunning international defeat, and policy failure.

Oh, wait a minute. Let's not forget the training of Osama bin Laden...

10) He was the Butcher of Lebanon, the Butcher of Honduras, the Butcher of Guatemala, the Butcher of Grenada, the Butcher of Angola, the Butcher of Nicaragua, the Butcher of Namibia, the Butcher of Afghanistan, and the Butcher of the racist apartheid South Africa. He openly encouraged, supported, and supplied foreign dictatorships, and supported attacks against any government that didn't toe the U.S. line.

11) Supplied Iraq's Saddam Hussein with a mustard gas capability to use in its war with the in-the-right Iranians in chemical warfare, which Hussein then used against his own Iraqi Kurdish minority. "Theoretically" a gross violation of the six decade old (or so) Chemical Warfare Treaty.

12) The butcher of Afghanistan, he trained an army of Islamic extremists, including Osama Bin Laden, replacing the Communist enemy with the Islamist terrorist threat that directly resulted in the 9/11 World Trade Center attack, and the death of thousands of U.S. soldiers in G.W. Bush's wars in Afghanistan and Iraq (based on a total lie, and concerted propaganda campaign of lies).

Hitler, Stalin, and Pol Pot are damned as the largest mass murderers of modern times. But there's nary a mention of Reagan, who was responsible for 1,000,000 civilian deaths in Afghanistan alone, by funding and arming the rebellion against the Soviet-backed rulers.

13) A Vietnam War revisionist, before a student crowd at a university graduation, he bullshited out the line, "Young Americans must never again be sent to fight and die unless we are prepared to let them win."

I'll always remember the high-ranking American officer, who admitted long after the Vietnam War was over, that the NVA [North Vietnamese Army] was the finest in the world, in "combined arms" operations warfare.

> "Chekhov used to say that if a prisoner was not a philosopher who could get along equally well in all possible circumstances (or let us put it this way: who could retire into himself), then he could not but wish to escape and he _ought_ to wish to."
>
> -- "The Gulag Archipelago" (1985)
> Alexander Solzhenitsyn

Prison Diary 61   Damn Papists
                              ===========

> "[I]gnorance more frequently begets
> confidence than does knowledge..."
>
> -- "The Descent of Man" (1871)
> Charles Darwin

> "Objection, evasion, distrust, and irony are signs of
> health. Everything absolute belongs to pathology."
>
> -- "Beyond Good and Evil" (1886)
> Friedrich Nietzsche

I've been a supporter of the Catholic guerrilla organization, the Provisional IRA – the Irish Republican Army – since I was a young'un', in their struggle for independence against the British, and struggle to unify the 6 counties of Northern Ireland with the 26 counties of the Republic of Ireland.

"26 + 6 = 32", would be the slogan I would print a T-shirt with, if I lived in Ireland.

I guess that was the "Iron whore" Maggie Thatcher's -- the female ex-Prime Minister -- issue. It was just a **math** problem for her that caused her to struggle with simple arithmetic...

The Provo's -- Provisional IRA -- are officially a non-sectarian group. As committed leftist soldiers of the armed struggle, they are not interested in sectarian killings -- Catholics killing Protestants -- per se, but the UK occupying forces, the Irish government, and anti-Catholic discrimination.

That being said, let's talk about Catholicism.

* * * *

> "[T]he most magnificent combination of theoretical
> success and practical failure in...history."
>
> -- comment on the ALGOL68 (1968)
> (ALGOrithmic Language 1968)
> programming language

Jean-Paul II was the Polish Pope from 1978 to his death in 2005. He was the first non-Italian Pope in hundreds of years.

He was succeeded by Pope Ratfuck – I mean Pope Ratzinger -- a German fascist who renamed himself Pope Benedict XVI. One of his acts has been to beatify Jean-Paul II, causing controversy in some circles.

And he has repeatedly referred to JP2, as Jean-Paul II "the Great". But let's see "JP2's" official record, as pawn of capitalism and fascism, and ultimately Satan.

JP2 officially condemned "liberation theology" in Latin America, a priest-led movement with origins in the 1970s, for involvement in human rights issues in the face of government repression. He also refused the request of Oscar Romero, the Archbishop of El Salvador, to condemn the government of El Salvador for gross human rights violations and support of death squads, one of which martyred Archbishop Romero.

JP2 also pointedly criticized a Nicaraguan priest who was a member of the leftist Nicaraguan Sandinista government, which had replaced a fascist dictatorship.

JP2 shored up Vatican support for Ronald Reagan against the Polish communist regime, leading to its downfall brought about largely by a low class, ill-bred, farting, belching union leader Lech Walesa. He was honored by G.W. Bush for his bringing down communism in Poland.

In 1981 JP2 was shot four times with a 9 mm Browning Hi-Power semi-automatic pistol by 23-year-old, would-be right-wing, "terrorist" assassin "Grey Wolf" Mehmet Agca. Shot twice in the intestines by the supposed "expert Turkish gunman" he recovered completely. We should **all** be so lucky, with our medical care.

JP2 failed to respond quickly to the sex abuse scandal of pedophilic priests that first arose under his administration, and grew to epic international proportions. The scandal lasting and growing at least from 1980 to 2012. But he had time to beatify the WW2 Croatian Archbishop who was an active supporter of the Nazi Ustashe government that attempted the genocide of the Serbs, along with coerced conversions of some of them from Orthodox Christianity to Catholicism.

He thus continued the corrupt tradition of the Catholic Church since early Medieval Times, from the ludicrous "Papal Infallibility" doctrine, to their bribe-taking extortion, alignment with absolute monarchy, heretic burnings, the tortures of the Spanish Inquisition, centuries of murderous religious wars and crusades, to their scandalous anti-Semitism.

Bulwark and ally of the forces of primitive conservatism, as exemplified by the Church's alliance with the fascist General Franco's side during the 1936-1939 Spanish Civil War.

To their horrific anti-Semitic pogroms, and virulent Polish anti-Semitism, to the point where Jews who tried to return to their homes in Poland, after the Nazi defeat, were murdered by their new Polish squatter-inhabitants. (That's **three** Commandments that simultaneously crash and burned: Thou shalt not covet thy neighbor's... Thou shalt not steal. Thou shalt not kill)

Then there's suicide, divorce, contraception, and abortion. Grasping at straws, these issues are not mentioned in the Bible's New Testament Gospels, except divorce. Divorce, JC specifically stated was **allowed** for infidelity – which seems to be unknown to the Vatican.

* * * *

> "When [irony] is not employed as an honest device of classical rhetoric, the purpose of which no healthy mind can doubt for a moment, it becomes a source of depravity, a barrier to civilization, a squalid flirtation with inertia, nihilism, and vice."
>
> -- "The Magic Mountain" (1924)
> Thomas Mann

All in all, the Catholic Church brings forth to reality the Old Testament saying from Ecclesiastes: "What has been bent, cannot be made straight".

I have a solution. They should burn every Catholic Church to the ground, and shoot every priest in Canada **unless** they officially change the name of their church from "Catholic" to the "Church of the Holy Pedophile".

So there.

Fair solution, eh?

> "And thy doom I lament, thou grief-worn dog.
> One that same earth, which bore her, opening wide,
> Shall swallow utterly in yawning depths"
>
> -- "Alexandra" (ca.150 B.C.)
> Lychophon

Prison Diary 62:		Capitalism Deconstructed
			=====================

> "Words – words without a meaning – or with a meaning too flatly false to be maintained by anybody...Look to the letter, you find nonsense – look beyond the letter, you find nothing."
>
> -- "Origins of Totalitarianism" (1951)
> Hannah Arendt

> Professor Frankfurt defines a "bullshitter" as someone who doesn't care whether what he says is true or false.
>
> -- "On Bullshit" (2005)
> Harry G. Frankfurt

Capitalism likes to hitch its wagon to democracy. But nothing could be further than the truth. Capitalism has no master. It is the master.

In 1429, during Henry VI's English monarchy, an Act was declared that "great, outrageous, and excessive numbers of [men]...of small substance and no value" were voting at elections.

In the 1600's, revolution occurred in the Netherlands, England and later France, because the government had alienated or lost the confidence of the ruling class -- the nobles or aristocracy.

In England the "ruling class" was in conflict between the nobility and the upwardly mobile capitalist businessmen, who were thwarted in their ability to rise in class. Men of ability, means, and property. Tyranny of the incompetent and the corrupt, but powerful position-holders. Judges, administrators, clergy, monopolists, interference with court or administrative system from higher up. Financial crises, mismanagement, and corruption.

The first major reform since the time of Cromwell (1660 and the end of the English Civil War) was the Great Reform Act of 1832, which extended voting rights to the ten pound owner level, unsatisfactorily to "radicals". Lord Macaulay, on the other hand insisted than universal voting rights were "utterly incompatible with the very existence of civilization."

After a campaign that became more and more militant, disruptive, spectacular, and well-publicized -- with arrests, hunger strikes, and force feedings -- after over thirty years, women were finally given the vote in 1918. **If** they were over thirty.

The conditions of women at work at the turn of the century, as house-servants and in factories had been criminally poorly paid, abusive, controlling, and degrading. At the same time, men's trade unions were becoming more militant, and strikes were confrontations with police and the military.

*   *   *   *

"[E]lections were steadily coming to resemble

> totalitarian plebiscites..."
>
> -- "Politics and Vision" (2004)
> Sheldon S. Wolin

Capitalism: supply and demand: half the supply, double the price; the seller's income stays the same; double the demand, with constant supply, the seller's income is doubled and he gets richer, until the price starts falling.

The seller always comes out on top. The consumer always loses.

The "free market": competition. It causes innovation, promulgating consumerism and excessive consumption, or if short-circuited by what used to be called "trusts", then, monopolies, and now called cartels, and price-fixing, or covert agreements to effect the same, and keep prices high. And insider trading – more accurately called "informed trading".

Media control: by corporations, or from networks, independence of news divisions. Corruption of the democratic process by high ad prices and their overriding influence.

Taxation: lobbyists and campaign contributions to diminish taxes for the wealthy.

Welfare for white people: the CIA, the NSA, and other intelligence agencies, DoD officers, government policy contractors, weapons contractors and corporations, higher level government and police hierarchy and administration.

Boeing and General Dynamics sells 65% of its output to the government. Raytheon 70%, Lockheed 81% (in 1967; Galbraith).

In 1987-8, the 10 largest defense contractors: McDonnell – Douglas, General Dynamics. GE, Lockheed, GM, Raytheon. Martin Marietta, United Technology Corporation, Boeing, and Grumman.

Merely the class war of the wealthy upper classes versus the lower classes mutated in another form.

Ghettoization: inner cities as the new ghettos, suburbs for the middle class, and gated communities, and hidden areas for mansions for the wealthy.

* * * *

To be treated as an individual; like I was worth something as a person, as an individual. That everyone's worth something as an individual.

Capitalism **not**!

There was Brown v. Board of Education (1954), Rosa Parks (1955), the Civil Rights Act of 1964, the Voting Rights Act of 1965.

> "And he was rich -- yes, richer than a king,

And admirably schooled in every grace;
In fine we thought he was everything
To make us wish that we were in his place.

So on we worked, and waited for the light,
And went without the meat, and cursed the bread;
And Richard Corey, one calm summer night,
Went home and put a bullet through his head"

-- "Richard Corey" (1897)
Edwin Arlington Robinson

Prison Diary 63:   Lying Liars and Their Damned Lies
=============================

> "I still don't know what I was waiting for,
> And my time was running wild,
> A million dead-end streets
> Every time I thought I got it made,
> It seemed the taste was not so sweet."
>
> -- "Changes" (1971)
> David Bowie

> "Elvis was a hero to most,
> But he never meant shit to me, you see.
> Straight up racist that sucker was, simple and plain,
> Motherfuck him and John Wayne."
>
> -- "Fight the Power" (1989)
> Public Enemy

A pattern of misconduct. A pattern of misconduct. Repeat after me: a pattern of misconduct.

Clive Ponting, a British civil servant was prosecuted under the Official Secrets Act, when he released documents showing that the British government lied when it disclosed the direction the Argentine cruiser, the "General Belgrano", was sailing when it was sunk by U.K. torpedoes during the 1983 Falklands War, killing 368 sailors.

The jury refused to convict him, giving the finger to the fascist Prime Minister Margaret Thatcher. The sinking of the Belgrano was purely revenge for the sinking the British "Sheffield" warship earlier by a French Exocet missile by the Argentinians.

* * * *

> "[T]he Founding Fathers gave the free press the protection
> it must have to fulfill its essential role in our democracy.
> The press was to serve the governed, not the governors.
> The government's power to censor the press was abolished
> so that it could bare the secrets of government and inform
> the people."
>
> -- concurring opinion, "New York
> Times v. U.S." (1971)
> Supreme Court Justice Black

In what became known as the Stalker Affair, John Stalker, Deputy Chief Constable of Manchester, was tasked in 1984 to investigate 6 deaths in 1981 of unarmed IRA suspects (including a definitely innocent victim) shot by the Special Branch of the Royal Ulster Constabulary, and their killings covered up.

Stalker found that the U.K. was using death squads like some third rate South American dictatorship.

Stalker proved himself a dogged investigator, who would not whitewash the affair, so he was suspended on phony, trumped up charges of corruption, and removed from his Ulster investigation. He resigned in disgust soon after the charges against him were dismissed.

In 1988 three unarmed IRA members were shot dead in Gibraltar; the death squad practice continued...

In 1976, Thatcher abolished "special category" status for IRA detainees which correctly jailed them as violators of criminal statutes for political reasons. Henceforth, they were treated as common criminals.

After attempts to rescind the new treatment all failed, including things like the "blanket protest", IRA prisoners in H-block in Maze prison in 1981 began hunger strikes. Bobby Sand was the first of 10 hunger strikers to starve themselves to death. While on his hunger strike Sand was elected Member of Parliament, much to the embarrassment of the authorities.

Until they started dying, the English government thought they were faking starvation.

> "Our revenge will be the laughter of our children."
>
> -- Bobby Sand

* * * *

> "[In democracies] it is necessary to control not only what people
> do, but what they think... [T]hought can lead to action and
> therefore the threat to order must be excised at the source.
>
> It is necessary to establish a framework for possible thought that
> is constrained within the principles of the State religion. ... [It is
> best] that they be presupposed, as the unstated framework for
> thinkable thought. ...[T]he fundamental doctrine [is] that the
> State is benevolent, governed by the loftiest [of] intentions...
>
> Those who do not appreciate these self-evident truths, or who
> maintain the curious view that they should be supported by
> some evidence, simply demonstrate...that they are emotional
> and irresponsible ideologues, or perhaps outright Communists.
>
> Or more accurately, their odd views cannot be heard..."
>
> -- "The Manufacture of Consent" (2002)
> Noam Chomsky

Canada likes to cultivate its reputation as a neutral and peace-loving and encouraging nation.

A nation of Arctic hillbillies. Boobs in the boondocks. Some sort of Northern Banana republic. Canada is a frozen wasteland, but I didn't realize the phrase referred to Canadian morality, ideals, and intellect. A collection of intellectual boors.

uranium exports; Cruise Missile testing (honored by the Litton bombing); torpedo testing; BMEWS/DEW line; Canadian nuclear weapons; Canadian sonar work ASW: East and West Coast; Suffield Chemical Warfare nerve gas testing; "Scanlan's Monthly" seizure by RCMP;

Then there's Canada's legal faults.

Reporter's papers and news films can be searched. Denial of them as an independent source of truth. It's a real fucking disgrace, and makes cameramen reporter's the tools of the police, and thus attackable.

Police can ignore and keep asking questions after lawyer request. "Right to remain silent" BS as a result.

The police can lie to you, but you can't lie to them. A suspect presumed innocent has a right stripped away. And the seeds of corruption are planted in the police, through the acceptance of their routine lying.

In direct and damning violation of the Canadian Constitution, Islamist couldn't get papers to return to Canada. And nobody even mentioned that fact in the news media.

* * * *

In U.S. history, it was the most war-like and the most ferocious in their resistance, of the Indians that were well known and respected. The Apaches, and the Cheyenne, for instance. Crazy Horse and Sitting Bull, their names still remembered.

The Rehnquist-led Supreme Court (1986-2006). Their judicial logic relies on the "subjective" expertise of prison administrators and "deference" to their special knowledge. Same B.S. as national security exemption to sabotage any effective executive branch authority in violation of the Constitution.

To wash their hands of any oversight of the Executive Branch in direct violation of the U.S. Constitution.

In Bell v. Wolfish (1979), the Supreme Court in its opinion, ruled that pre-trial detainees were subject to the same oppressive practices as convicted offenders. Outrageous.

The police state had changed its name to the national security state.

G.W. Bush *deliberately* lied about Al-Qaeda being linked to Iraq in order to con the U.S. people into accepting a war against Iraq. As outlined in detail by former Vice-President Al Gore, (and winner of the popular Presidential vote in 2000, even if Bush's brother hadn't falsified election records). He provided no evidence at all, yet repeated over and over again that there was a link between Sadaam Hussein and Al-Qaeda not long after 9/11.

So an unelected, cowardly Vietnam draft evader took the U.S. to war, a war that the U.S. lost, based on a deliberate campaign of repeated, deliberate, unsubstantiated, bold-faced lies. And in which he became a vicious war criminal, and mass murderer of U.S. and Iraqis [Gore, "The Assault on Reason", p.108-114 (2007)].

The 1990 hit movie, "Pretty Woman", starring Richard Gere and Julia Roberts gives such a bad depiction of prostitution it's unbelievable and intolerable. The movie makes being a whore seem O.K. and pleasant. Wrong!

In 1989, off the Alaskan coast, when the Exxon Valdez oil tanker ran aground, creating the worst environmental disaster in U.S. history (until eclipsed by the B.P. oil rig disaster in 2010) they were quick to blame the captain, who was drunk.

But in reality, it was years of oil company pressure to increase profits, granted by the Reagan Administration, that led to an undermanned, over-tired, and under-qualified crew running into a well-marked reef on a cool, clear night.

The shoot-down of commercial Flight KAL007 and Reagan's lies and distortion to flay the Russian government in public. The U.S. and Israel have both shot down civilian airliners, and then shrugged their shoulders.

DNA testing became available in the 1980s. The first 250 convicts who were exonerated were convicted of rape or rape and murder (89%), and 9% of murder alone. About 10% were juveniles. Almost 10% were mentally disabled. And, most troubling, 70% were minorities, 2/3 of which were black.

76% of the first 250 exonerated were misidentified by an eyewitness, frequently pressured by the police, or using corrupt practices in the "lineup". 16% made false confessions, under police pressure, or torturous or violent, or extended police interrogations, or corrupt practices like providing them with details of the crime that only the perpetrator would know. More than half of the false confessions were from juveniles or the mentally disabled.

Rights and rules generally cover the procedural aspects of arrest and trial. Few rules, however, regulate accuracy rather than procedures. "Brady v. Maryland" was the Supreme Court decision that requires police and prosecutors to hand over to the defense exculpatory evidence. Evidence that is "both material and exculpatory" – points to innocence of the defendant.

Such matters that regulate accuracy are typically committed to the discretion of the trial judge." Such was the greatest crime of Ronald Reagan: packing the judiciary with fascist pigs like Judge James Ideman in Los Angeles, who couldn't follow even basic procedure, fairness, or justice. A disgrace to the Court, the bar, and himself. I mention him because I whipped him in open court for his bull shit.

Judges are supposed to be the impartial scales of justice. But all too often they are corrupt, incompetent, biased tools of the prosecutor, or simply lazy, disinterested, jaded or bored, or just don't care anymore.

> "What makes you a person then is the capacity to know the merit or demerit of your actions. Without accountability for past actions of good and evil, one can still be a 'rational creature'... But what makes a person...is responsibility: the capacity to *appropriate* these past actions 'to that present self by consciousness'."
>
> -- "The Law is a White Dog" (2011)
> Colin Dayan, commenting on
> John Locke (1700)

Prison Diary 64: "Unusually Harsh"

> "How many men had I already killed? There are those three I knew about -- close enough to blow their last breath in my face."
>
> I **took** the mission. What the Hell else was I supposed to do? But this time, it was an American...and an **officer**."
>
> -- Captain Willard
>   "Apocalypse Now" (1978)
>   John Milius

> "I've made mistakes that I can't erase,
> I've made mistakes.
> Though I'm sure to see the signs,
> That we are falling back in time,
> So far from where we started,
> So far from what we wanted."
>
> -- "Money Honey" (2007)
>   State of Shock

I wish I could put it on my resume. I wish it was a money-maker, instead of a friend loser. I admit it freely -- I have a sense of moral outrage that is surprising in its strength and deafeningly explosive venting. Moral outrage -- a visceral hatred of injustice. It's so ingrained in my consciousness, so uncontrollable a part of my personality, I believe, but can't prove, that I was born with it. Gifted with it, is my contention.

But I get the impression, that's it's not the majority opinion.

It is seeing the awful, unvarnished Truth of society's faults and flaws, and be willing to act on them, whatever the personal cost, for the collective good of society.

The non-believers -- the true infidels -- or the misunderstanderers, or the "don't carers", call it being "self-righteous". Or suicidal. Or a dead end. Or just misguided.

They are -- as usual -- wrong.

* * * *

> "But they had not come back the same men. Something had altered in them. They were subject to queer moods, queer tempers. ... Many of them were easily moved to passion when they lost control of themselves. Many were bitter in their speech, violent in opinion, frightening."

-- "Now It Can Be Told" (1920)
Philip Gibbs

I shall always remember her. She was kind, she was understanding, she exuded warmth and caring. And really -- isn't that what most people desperately want, need, and respond to?

But make no mistake, she was not gushy or gullible. I was not responding at all sexually to her -- make no mistake. She was just pure of heart. A decent heart. And that's probably the finest compliment I can give someone. She was my therapist and councilor, for four months, and she is the only person of the many mental health professionals I had seen, who I felt truly *cared* about me.

She said she shouldn't say it, but was going to anyway. At the conclusion of our formal therapy association -- the class graduation -- she said, under her breath, that I was her favorite "client" -- affable, magnetically curious, different, complex, yet lucid, all in all. Fascinating, as it were. Yet though these words were mine, they were obvious in her behavior and affect, and thus left unsaid.

For my part, I deeply appreciated and noticed immediately her interest in understanding, and thus helping me. Her patience, openness, casual friendliness, lack of judgment.

I've always said it. If you want to understand me, the first thing you have to do is _try_.

"Yogi," she said one time, with obvious diplomacy, "One thing is clear. I find you to be _unusually_ harsh in your judgments."

I looked at her directly in the eyes. I locked eyes with hers, and paused, not a trace of emotion on my face.

"No kidding..." I responded. I may have sounded bitter or derisive, but it was not directed at her. "And _why_ do you think I'm like that?" I almost added, contemptuously, but didn't. She was competent, and would well enough know the answer. She was just trying to get a message across to me. Spark an insight.

But it was an insight I was already aware of. Though I considered it "utter ruthlessness" rather than her therapist's non-threatening vocabulary of more diplomatic words -- "unusually harsh".

I was well aware of the extremes of my verbalizations, which concluded (with a bang) anything from a rant to vent emotion, to a calm, lucid expression of a complex political idea with an ultra-violent conclusion, as the answer to solve the problem.

A venting was my way to blow off steam and express my frustration, without anything but an obviously ridiculous blathering of mindless unrestrained hatred. My friends knew and understood this. Someone who had been listening to my political analysis, and capable of understanding it, and trying to understand me, could with only a slight stretch of their consciousness, grasp this, and move on.

The second, along with an over-simplified conclusion as to its solution, which I felt was fundamentally have a public spectacle component to signify society's revulsion at past human rights atrocities by those entrusted with only the power to improve society.

The third reason was to wake up the asleep with an outrageous "bang" of a violent solution. Just to be heard. Just to be listened to. By anyone.

* * * *

> "But there is anger...against all these people here that think so much
> of themselves and their smart talk; anger against this whole world, living
> here so damned cocksure with their knickknacks and jiggery-pokery, as
> though the monstrous years had never been when one thing and one
> thing only mattered, -- life or death, and beyond that nothing."
>
> -- "The Road Back" (1931)
> Erich Maria Remarque

Assassination. How does one reconcile one's religious -- Christian -- beliefs with the paradoxical belief in the necessity of selective killing of sufficient numbers of those in authority, in the name of the great improvement of society and its direction forward?

Violence begets violence. The corrupting influence of power. Evil to fight evil as bound to fail. The fruit of the poison tree is poison.

The _Cosa Nostra_ ("Our Thing") -- the Mafia -- and it origins as an innocuous home-grown Sicilian self-defense group turned into a monstrous evil provides the best example.

Yet, the killing of a few **can** be justified by the intransigent evil, for the collective good, and the collective benefit.

It caused me much hand-wringing and inner turmoil, made all the more because of my weapons expertise, and ability to carry out that which I advocated. I was an expert in weapons, guerrilla warfare. and clandestine operations. I was widely read in political extremism and the history of revolution. And I **believed**.

From hand-to-hand combat, to knife work, to small arms, to explosives and anti-personnel bombs, to nuclear weapons, I had become an expert in killing, even though I was a pacifist from the beginning, just sating my curiosity and fascination, and learning what "the enemy" -- the militarists -- were up to.

And I accepted Christ's message of love and peace. You can only fight darkness with light, not with more darkness. It's as simple as that.

Yet the world was clearly immersed in more and more darkness, with people only "happy" with the materialism that advancements in technology had brought society.

It was "Mirror, Mirror", the paradox in me, that I only recognized later in life. A peacenik who was a weapons expert, and armed to the teeth. A man who solemnly believed in anti-violence, and yet had a killer instinct ready to be unleashed at a moments notice. A man who was humble and polite, yet a political extremist. A man who was calm, laid-back, and easy-going, yet could be utterly ruthless with a killer instinct. A man who was normally unruffled, yet could go combat-ready into instant violent action.

The Angel of Death, within, to whom "unusually harsh" was *nothing*.

* * * *

> "When we contemplate the fall of empires and the extinction of the nations
> of the Ancient World, we see but little to excite our regret than the moldering
> ruins of pompous palaces, magnificent museums [and] lofty pyramids...
>
> [B]ut when the empire of America shall fall, the subject for contemplative sorrow

will be infinitely greater than crumbling brass and marble can inspire. ...here -- ah, painful thought! -- [t]he noblest work of human wisdom, the grandest scene of human glory, the fair cause of Freedom rose and fell."

> -- Tom Paine (1796)
> open letter to President Washington,
> published in a Philadelphia newspaper

It is the matter of the morality of political homicide.

"Another room in hell" is full.

A campaign of assassination, targeting those bearing the most guilt, the powerful, the effective, the deserving, with a carefully planned view to topple the government by attacking its top and middle-level officials, party leaders, multinational corporate executives, and the extremely wealthy. A strategy of destroying the government's ability to function and exercise power, create policy, effect decision-making, and plan and respond to the crisis, by excising the top levels on down of the leadership, their administration and ideologues, high-level bureaucrats, and key personnel.

To sow fear and paralyze the top levels of government, and show their vulnerability clearly to them, their colleagues, and especially the public at large. To dispense justice to those most guilty, most intelligent, most powerful, and most enriched by their evil system. A campaign of systematic, coordinated elimination as part of the other strategies to be implemented simultaneously to bring about the weakening, the destabilization, and finally the collapse of the government.

"Speaking truth to power" in the harshest and most direct terms possible.

* * * *

"Silent leger inter arma"
[The laws are silent in the midst of arms.]

Rome's proscription lists of those deemed Enemies of the present or new regime. The French Revolution's Reign of Terror, Stalin's purges, the Nazi Night of the Long Knives.

The Reign of Terror lasted a year in 1793 and 1794 and was a national purge of the aristocracy. It affected only a relatively small number of people -- the nobility, the rich, large landowners, and counter-revolutionary threats. Though it ended with a paranoiac purge of the National Assembly and the twelve member Committee of Public Safety, as it threatened to get out of control -- "the Revolution eating its own."

The Night of the Long Knives was the 1934 elimination of the leadership of the Nazi SA. The SA -- _SturmAbteilung_ were undisciplined, corrupt and unruly -- the Nazi Storm Troopers. They were to a large part street-level toughs and hoodlums -- dispensable to Hitler's needs once he had ascended to power, in favor of Heinrich Himmler's more loyal, professional, disciplined, and unquestioningly obedient SS -- truly useful. It was the treacherous, ruthless, decisive neutralization of a competing faction -- internecine warfare.

Only 60 or so leaders of the SA were purged. The SA storm-trooper organization was essentially decapitated to neutralize the SA as a possible breakaway faction that would threaten Hitler and the goals of the Nazi party through wasteful infighting, internal rivalry and divisiveness.

* * * *

> "There is a core of anger in the soul of almost every veteran,
> and we are justified in calling it bitterness, but the bitterness
> of one man is not the same thing as the bitterness of another.
> In one man it becomes a consuming flame that sears his
> soul and burns his body. ... It leads one man to outbursts of
> temper, [and] another to social radicalism..."
>
> -- "The Veteran Comes Back" (1944)
> Willard Waller

The government attitude to the Rule of Law and the expediency of political homicide is pretty clear. In the "dirty war" in Northern Ireland in the 1980's, British SAS commandoes were quite frank when questioned by a reporter about serious suspicions that had been raised that Her Majesty's "rules of engagement" had been secretly and illegally liberalized to permit government security forces use lethal force against known IRA guerrillas if they were armed, *suspected* of being armed, nearby to arms or bombs, or on an "active service operation". AND even if arrest was considered to be safe and risk-free.

"Big boys' games, big boys' rules," was the common attitude, freely admitted, amongst the commandos.

* * * *

> "[The] right to advocate lawlessness is, almost
> paradoxically, one of the ultimate safeguards of liberty."
>
> -- J. Michael Luttig (1998)
> U.S. Federal Appellate Judge

In January 1878, Vera Zasulich, a young Russian radical, on her own, shot and seriously wounded General Fedor Tepov, the St. Petersburg Prefect of police. Trepov was known to be particularly brutal and ruthless, even for his time, and his place -- the absolute rule of the Tsar, in the police state known as the Russian Empire.

Trepov's crime, she explained at her trial, was having a young student and political prisoner -- up to this time exempt from such punishment -- flogged one hundred times with the often fatal knout. He had failed to stand up in Trepov's presence, during an inspection by the Prefect. She -- like much of the Russian liberal public -- was outraged by this act of barbarity and offense against decency.

But with no legal recourse against Trepov, her only course of action was political violence against the offending party. The jury agreed, and acquitted her, though she found it advisable to flee the country. And Tsar Alexander II signed a decree abolishing jury trials for political offenses. He was assassinated in 1881, soon after.

The road to revolution had already started, and after years of violence, terrorism and instability, would end in 1917 with the Russian Revolution and the ascendancy to power of Lenin and the Bolsheviks.

"What should I teach you? Should I tell you that in twenty years you will be dried up and crippled, maimed in your freest impulses...?

Should I tell you that all learning, all culture, all science is nothing but hideous mockery...?

... Should I demonstrate how best to aim a rifle at such an incomprehensible miracle as a breathing breast, a living heart?"

-- "The Road Back" (1931)
Erich Maria Remarque

Prison Diary 65:   "Death's Twilight Kingdom"
=====================

> "Be careful how long you stare into the Abyss.
> For one day, the Abyss may stare back into **you**."
>
>   -- "Thus Spake Zarathustra" (1892)
>   Friedrich Nietzsche

> "We ate the food,
> We drank the wine,
> Everyone having a real good time.
> Except you.
> You were talking about the end of the world."
>
>   -- "Until the End of the World" (1991)
>   U2

Probably the German philosopher and prolific author's most famous quote, it was also uncomfortably prescient. For Nietzsche, like a disturbing number of historical genius-level intellectual figures, ended his career by going irretrievably stark, raving mad. I guess *he* himself stared into the Darkness infinite a little too long, the poor devil.

                       *   *   *   *

> "Don't get me wrong, Don Juan, I protested, "...[B]ut I
> also want to know everything I can. You, yourself
> have said that knowledge is power."
>
> "No!" he said emphatically. "Power rests on the kind of
> knowledge one holds. What is the sense of knowing
> things that are useless?"
>
>   -- "Don Juan: A Yaqui Way
>   of Knowledge" (1968)
>   Carlos Castaneda

The answer to a question I had not even thought of. Yet.

It was a quote I had seen and read many times in Internet messages posted over the years of the 1990's. It was a comment from well known, respected, and famed Internet icon John Gilmore (fifth employee of Sun Microsystems, creator of the alt.* hierarchy -- in the pre-web Usenet newsgroups -- one of the founders of the Electronic Frontier Foundation, etc., etc.), that he made in the early years (late 1980's), as the issue of censorship of Net groups or posts continually came up over and over again. He had merely said:

> "The Internet interprets censorship

> as damage, and routes around it, "

His remark was quoted in a 1993 "TIME" magazine article, and a "sound bite" was born. His statement was henceforth quoted endlessly, Gilmore's reputation adding to the technical validity of his comment -- recognized as iron-clad and unassailable.

I had heard the quote numerous times over the years, beginning in early 1992 when I first got on the 'Net, when the workstation network administrators at Bell-Northern Research (the precursor to Nortel Networks, the telecom hardware and software manufacturer that went from multi-billion dollar Canadian multinational to bankrupt in 2010), hooked us engineers up to this marvelous new world.

John Gilmore's clever line referring to the Net's architecture of computer hooked together to several other computers, that made for multiple, redundant links that provided the ability to resend message traffic along alternate paths, should a "node" -- computer -- in the fastest, direct network path "go down", break, or otherwise become unavailable.

But, more interestingly, Gilmore's remarks harkened to the origins of the Internet in the original computer network, the U.S. Department of Defense's ARPANet.

ARPA, the DoD's Advanced Research Projects Agency, had funded the first linking of different computers, in different cities, into a network, which eventually evolved into what became the Internet. Due the redundant links of the architecture, ARPANet was supposedly "nuclear survivable" – if during a nuclear attack a given computer was taken out by a nuclear missile, the other computers could still communicate, because the message traffic would automatically reroute itself to bypass the vaporized computer.

One day, this hit me like a flash. And there was more.

This chapter is the story of how the Internet -- the Net -- evolved from a small Department of Defense test network that was to be the "proof of concept" for a national network linking the 600 Minuteman III missile silos, and allow for the automatic dynamic re-targeting of the remaining -- unvaporized -- functioning missiles while under a Russian nuclear missile attack.

"Dynamic" re-targeting: it means "reconfigurable". It means "changeable in real-time in response to changing conditions". It means a current, up-to-date knowledge of remaining ICBM missiles, constantly updated in real-time by an "A-OK" signal from the computerized missiles, which were then kept constantly updated, if necessary, with a prioritized list of Russian targets, that would re-target the missiles in a nuclear war according to the balance of military priorities, with remaining, available missiles.

There would be only one purpose for a computer network that could survive damage from a nuclear attack: a computer network that linked the 600 MM3 Inter-Continental Ballistic Missile silos scattered mostly across the American Midwest, from Kansas to North Dakota. A computer network that kept track of each ICBM's targets and their priorities for nuclear destruction.

If a U.S. missile silo was hit by a Soviet nuclear missile warhead and destroyed, another missile silo could be reassigned the destroyed missile's targets. Basically the entire U.S. ICBM force would be networked for dynamic targeting updates during a limited or all-out nuclear attack.

The mainstay of the U.S. ICBM force since 1960 was the Minuteman, a marvel of cheap, reliable, efficient, mass produced mega-death, made possible by the development of solid fuels for long range missiles that could compete with the more powerful liquid fuels of the earliest Atlas and Titan ICBMs. By 1970, the Minuteman I had evolved to the increasingly more accurate and sophisticated Minuteman 2, and then the Minuteman 3.

With only 30 minutes flight time for an incoming Soviet ICBM, U.S. ICBMs would be prime candidates for any technology that would automate and speed launching to carry out U.S. targeting plans as effectively as possible while under attack.

* * * *

"Death's at the bottom of everything, Martins.
Leave death to the professionals."

-- Calloway
"The Third Man" (1949)
Graham Greene

Searching my nuclear document collection for confirmation revealed that a "Command Data Buffer" (CDB) computer device was installed on the first MM3 base in February 1975. Another document stated that the MM3 force was upgraded with the CDB starting in 1973 and completed in 1977, and the older, single warhead Minuteman 2 force from 1977 to 1979.

Further, the CDB allowed for re-targeting of the missile in "25 minutes", revealing perhaps, the maximum time for manual re-programming the last of the hundreds of MM3 ICBM's, or concealing a much lesser -- and classified -- time with installation of an Internet-like packet-switched -- automatic when under attack -- national computerized, all-out nuclear war, missile re-targeting system.

"Rapid re-targeting capability for Minuteman to improve [under attack] strike planning capability," confirmed more specifically another declassified document from the early 1970's.

* * * *

Nek: "You need only a button press --"
Kal: "And we will do the rest."

-- "Utopia Limited" (1893)
Gilbert and Sullivan
William S. Gilbert

It was Plato, of the city-state of Athens in ancient Greece, and the most famous philosopher of all time, who realized it over twenty-three centuries ago, when he wrote: "Only the dead have known the end of war."

But was he signifying war's inevitability? Was he merely indicating its intrinsic part of humanity's behavior? Or was it just his observation of the facts of his time, history of time's past, with an extrapolation of a predicted future? Which I suppose are three different ways of saying the same thing.

Was it a statement of finality? Or a message of warning -- an invitation of the difficult journey we would have, to identify and begin the inner change necessary to end our instinctive need for the regular, organized, and systematic collective violence and mass slaughter, commonly known as war?

And all the side effects of our acceptance of warfare, with random regularity, at the drop of a dime, that must poison man's collective and individual behavior even when not engaged in warfare.

But I'll tell you one thing. It is so misguided as to call into question the basic logic, thought processes, and maturity of thinking of an ordinary person who believes revoking the private right of a citizen to own a rifle, or such, is anything useful, positive or meaningful in a world where everyone else, but the citizens of some democratic countries -- and all dictatorships -- is armed to the teeth with all manner of firepower and fearsome weaponry.

It's a simpleton's simple solution to a complex problem. Or the solution of someone with the questionable motives of the hidden agenda.

* * * *

"Death is ordinary. Behold it, subtract its patterns and lessons from those of the death that weapons bring, and maybe the residue will show what violence is."

-- "Rising Up and Rising Down" (2004)
William T. Vollmann

Not most people, but some people, might know that I'm an inveterate quote collector. Or perhaps a somewhat mediocre computer engineer, unless interested by the work or inspired by teamwork.

Or a complete failure (or success) with women, depending on whether you talk to an ex or not.

What is less well known is my expertise in military technology, such as certain aspects of small arms, "specialized" ammunition, anti-personnel weapons, low and high explosive chemistry and weaponry, military strategic (ICBM.'s -- Inter-Continental Ballistic Missiles) rockets, incendiaries, nerve agents (chemical warfare), nuclear weapons technology, to name most areas.

Then there's organizational methodology, operational methods and security, clandestine operations, intelligence and counter-intelligence. There's warfare, military tactics, guerrilla warfare -- mostly urban -- sabotage, extremist ideology and history, and others.

It made interesting reading, my unofficial story -- my "black" resume. But it is misleading. I am an anti-war "peacenik", but had always have been, driven to "know" and understand the enemy. To study the ways and means -- the technology -- of the death-lovers, jingoistic, chauvinistic war-mongers. To understand war. For a good three decades.

Perhaps I now understand it too well.

* * * *

" 'God knows, I lie down to sleep so often with the wish, sometimes with the hope, that I shall not wake again.' "

-- "The Sorrows of Young Werther (1773)
Johann von Goethe

In 1936, British mathematician Alan Turing came up with the idea that was the theoretical basis that laid the ground-work for the programmable digital computer. Exposed as a homosexual in 1952, in less open times, he was criminally prosecuted and cashiered in disgrace.

Turing committed suicide in 1954 from cyanide poisoning -- a poisoned apple.

> (a reference to his favorite movie, the 1937 Walt Disney animated movie, "Sleeping Beauty and the Seven Dwarfs" -- based on the classic fairy tale published in 1687, made into the Tchaikovsky ballet first performed in 1890, and allegedly the inspiration of the PC computer giant Apple Computing's, corporate name, first computer name, and its corporate symbol).

But Turing's idea for digital computers lived on, and changed the world. So it goes.

The first digital computer, named ENIAC was built in the U.S. in 1945 for military purposes: automated artillery table calculations, and design of atomic bombs. The computer weighed 60,000 lbs. (27,000 kg.) and contained more than 18,000 vacuum tubes.

By the 1960's, with the introduction of individual transistors, computers were much smaller, but still massive in weight and volume, and were known as "mainframes", manufactured by industrial giants like IBM for corporate and government use.

The idea of small personal computers had not even occurred to mainframe manufacturers, even if it was technically feasible or financially cost-effective, both of which it was not. Home PC's would have to wait for the 1970's and evolution of transistors to the integrated circuit (IC), which put many transistors and associated components on a single -- and tiny -- chip. By the 1980's the price of the IC CPU had come down enough to begin the flourishing of the home PC. First came the IBM Personal Computer, or "PC", as it was quickly dubbed. The "accidental revolution" was about to begin. It was a technological revolution only, I'm sad to say, but so it goes.

* * * *

> "[I]n Italy for 30 years, under the Borgia's, they had warfare, terror, murder, and bloodshed, but they produced Michelangelo, Leonardo da Vinci, and the Renaissance. In Switzerland they had brotherly love -- they had 500 years of democracy and peace, and what did that produce? The cuckoo clock."
>
> -- Harry Lime
> "The Third Man" (1949)
> Graham Greene

But first there came a military need for small computers. Small -- in size, and particularly weight -- computers were needed for guidance control off the submarine-launched ballistic missiles (SLBM's) -- the Polaris Fleet Ballistic Missile -- under development starting in the mid-1950's for deployment in 1960.

Fast-forward to 1969.

Man's first landing -- and walking -- on the moon was Neil Armstrong and two other astronauts flying the Apollo 11 Saturn V three-stage, liquid-fueled booster (payload carrying) rocket. The Saturn V was the largest and most powerful rocket ever constructed -- to generate the "escape velocity" necessary to break free from earth's powerful gravitational pull. The rocket needed to be strong enough not just to get into earth orbit, but for the extra "kick" necessary to break out of earth orbit, and into outer space -- heading towards the moon.

The date was 20 July 1969, and my father allowed me and my brother to stay up past midnight --

and way past bedtime -- to watch on TV the fuzzy, murky image of man's first step on the moon by Armstrong.

But the technology base of the civilian Space Program was derived entirely from the then mostly secret -- or otherwise hidden from the public -- military *liquid-fueled* Inter-Continental Ballistic Missile (ICBM) program that started deployment in 1959 in the U.S. -- with the Atlas missile, almost immediately trumped in 1962 by the introduction of the reliable, storable, quick-launching Minuteman I. And within eight years the solid-fueled missile had reached its first pinnacle with Minuteman 3 land-based silo missile in 1970.

But for comparison purposes for the Guidance Control Computer, it is easy to take, for instance, the Apollo Guidance Computer (AGC). It was designed and developed at the MIT Instrumentation Laboratory, led by ICBM guidance expert Charles Draper, the details worked out by designer Eldon Hall. The contract for the development of the AGC was awarded to them in 1961 -- a 5 year development effort, since the first AGC flew in an Apollo 3 test flight in 1966.

AGC design drew heavily from the Polaris Submarine-Launched Ballistic Missile (SLBM) guidance system, also developed by (Eldon Hall of) the Draper group, and thus makes for a good comparison of the design and hardware of SLBM/ICBM guidance computer technology.

* * * *

> "Having wandered some distance among gloomy rocks, I came to the entrance of a great cavern, in front of which I stood some time, astonished and unaware of such a thing. ... [A]nd after having remained there for some time, two contrary emotions arose in me, fear and desire -- fear of the threatening dark cavern, desire to see whether there were any marvelous things within it..."
>
> -- "The Notebooks of Leonardo da Vinci" (ca. 1475)
> Leonardo da Vinci

The Apollo 11 moon-landing mission had one astronaut placed in orbit around the moon -- the Command Module --which released a smaller LEM -- Lunar Excursion Module -- to land on the moon with two astronauts, and when their moon walk and lunar sampling was complete, then blast-off to rendezvous -- rejoin -- the Command Module for return to Earth. The AGC computers -- one on the CM and one on the LEM -- were used for "critical mission events" -- such as the calculations of thrust for a soft landing on the moon and the even more precise calculations for rendezvous of the returning LEM with the Command Module, the first time such a feat was ever attempted in space.

The AGC was needed for when the crew of the Apollo could not rely on earth ground station state computers due to the time delay for transmission from earth to moon and from moon to the earth -- a double delay totaling almost 3 seconds for *each* question, data (information) transfer, for calculation requests and answers from earth to reach Apollo. Though seemingly a manageable delay, it was too long a delay for the necessary precision control for LEM space maneuvering where control measured in inches was necessary, as was required for LEM to CM rendezvous re-docking.

The Apollo system consisted of three major hardware pieces: the inertial guidance unit, an optical space sextant system, and the general-purpose digital computer -- the AGC -- programmed to work with the data input by the other two hardware pieces, and command queries punched in by the crew. Up till the development of these ICBM/SLBM computers, they had been massive pieces of hardware. These missile computers were tiny in comparison, heralding in the early 1960's what in 1980 would be the PC for civilian mass market use.

The inertial guidance unit -- using three gyroscopes to stabilize a platform connected to its accelerometer spacecraft acceleration measuring devices -- gave a continuous output of the Apollo's precise location in space, relative to earth, the spaceship's angle of movement (trajectory), and velocity.

The optical sextant made navigational angle measurements between a pair of stars, such as the sun and Polaris, the North star, and used to align periodically (error-correct) the inertial system platform based on star position, an innovation not used on the Polaris SLBM (though considered later for the Trident SLBM).

But by today's standards of the home or business PC – the year 2008 – the AGC of 1969 was a fossil.

The AGC weighed a hefty 70 lbs. (32 kg.), and had a volume of 1 cubic feet.

The AGC was a 16-bit computer and was based on one of the first military applications of the integrated circuit. Today's computers have 64 bit CPU's (the Intel Pentium 4).

The clock speed of the AGC's CPU was 2 MHz. Today, the Pentium 4 has a clock/ instruction speed of up to 3.8 GHz (a calculation speed of almost 2 million times as fast as the AGC) – at only 115 watts of power consumption, compared to 70 watts for the much less computing power of the AGC.

There was only 4 Kb of erasable (read or write) RAM memory. Today 2 Gigabytes of RAM is typical (500,000 times the amount of AGC RAM, as well as being smaller and lighter the older technology Apollo's 4 Kb).

Hard drives didn't exist at the time, and ROM (Read-Only Memory) was faster anyway. The computer's program was stored in about 74 Kb of ROM (Read-Only Memory), the memory of the magnetic core rope type. The AGC program allowed for only **34** calculations/instructions. Today, hard drives are used for permanent memory storage, and 500 Gb hard drives are common (almost 7 million times the amount of AGC ROM). Programs and applications (instructions) are many.

Input today is by alpha-numeric (alphabet-number) keyboard or mouse click, and output to the user is through a video screen. In the AGC, input was by the DSKY (Display and Keyboard), with the astronauts punching in one of only 34 calculations/ instruction, coded as 4 digits *numbers*. Output was also a number: three 5 digit number plus sign ("+" or "-").

Crew commands input were, for example, "display gimbal angles" (of the inertial guidance system) or "load star number". The input (crew request) and output (computer answer) was on a display and keyboard unit. Input and output was through coded numbers *only* -- a primitive user non-friendly interface that saved weight, complexity, and was consistent with the level of computer advancement of the time.

* * * *

"Have you ever seen any of your victims?"

"...Victims? Don't be melodramatic. Look down there. [at the crowds down below, tiny moving dots] Tell me, would you really feel any pity if one of those dots stopped moving forever?

If I offered you twenty thousand pounds for every dot that

> stopped, would you really, old man, tell me to keep my money,
> or would you calculate how many dots you could afford to spare?
>
> -- Martins to Harry Lime
> "The Third Man" (1949)
> Graham Greene

Pure research is about the mind, and its ultimate capabilities. It is on the edge of genius, or genius itself. It is about the mind being stretched to the extent of its true capabilities into the beauty of purely ideas and concepts. Gymnastics of the mental realm. Dreams of the mind awake.

Pure research is the stratosphere of scientific and technological development. Pure research generates new concepts as:

1) extensions or combinations of existing ideas or technology,

2) extensions of existing technology that require further technical -- hardware -- improvement, advancements, or development before they can be implemented, or

3) entirely new revolutionary ideas of developments unrelated to existing technology which may or may not require technological developments to catch up for implementation. Or that require human minds of those in positions of power to realize the importance of to accept, or find the need or the funding for.

After the lesson of WW2 of the importance of technology to the successful -- and easy -- prosecution to victory of warfare, a structure and administering bureaucracy was erected in the U.S. Pure military research became the realm of an assortment of well-funded groups, such as -- most famously -- the RAND Corporation, as well as Princeton's Institute for Advanced Studies, the Stanford Research Institute (SRI), the Institute for Defense Analysis, the JASON group of experts -- associated now with the MITRE Corporation -- and finally the government agency overseeing and financing such R&D, DARPA, the U.S. Dept. of Defense Advanced Research Projects.

ARPANET, the first wide-area computer network -- connecting distantly sited computers together so researchers could communicate, use each other's programs remotely, and exchange information -- was conceived, funded, and launched in 1969, but had fewer than a couple of dozen computers -- "nodes" -- connected together and communicating in early 1972 -- the first long distance computer network. Indeed, it was the first long distance functioning computer network *ever*.

The ARPANET -- built under a concept put forward by DARPA -- connected together a bunch of the large university mainframes in existence at the time, linking them together in a national network. And the simplest, quickest, cheapest, and easiest way to get this done was to piggy-back on an already existing national network -- the long distance telephone network.

So, they just rented a bunch of Bell Network long distance toll lines -- "trunk lines" -- connected them to new specially designed routing and message translation computers (called "IMPs" -- Interface Message Processor -- the first IMP was delivered to UCLA in mid-1969), and connected them to the different large mainframe computers at the various universities and institutions recruited for the "experiment" -- an ever-growing network, as more and more computer mainframes signed up after the concept was debugged of most of its initial -- and inevitable -- hardware and software problems, and stable and up and running.

The first IMP was delivered to UCLA in mid-1969, and it was used to hook UCLA to two other sites in what became a four "node" network: UCLA, Stanford Research Institute, UC Santa Barbara, and the University of Utah, Salt Lake City. By 1971, there were about nineteen nodes in the initial ARPANET, with thirty different university sites to be funded. The Internet was starting to take off. The original ARPANET was split off in mid-1975, and transferred to Defense Communications Agency.

* * * *

"Countries don't mistrust each other because of
armaments they build up, they build up armaments
because they mistrust each other."

-- Salvador de Madariaga (1973)

The network was the first implementation of a new type of communication type -- "protocol" -- called "packet switching" which was ideal for a computer network, as opposed to the only other type of network, telephone networks which used "channel switching". Channel switching requires the network to set aside a dedicated channel for the sole use of the two telephone users until they are finished and hang up.

The idea is this. Computers are very fast, and humans are very slow. Thus came the idea of a computer network based on "packets" sent down the pipe -- transmission line -- as opposed to having a dedicated -- and wasteful -- "channel", as in the telephone network.

The concepts worked thusly. A computer message is broken down into many individual, numbered smaller messages named "packets". They are then transmitted to the network, where individual packets take any route on the network, and a passed from network node to network node, until they reach their destination, where they are reassembled into the full message. The packet network is so efficient because unlike a telephone network, computers don't require a dedicated path -- the set-up of a continuous, dedicated, and expensive channel for communication, since they communicate in the occasional short, quick burst of messages.

* * * *

"Security can now only be achieved in common.
No longer against each other, but only with each
other [together], shall we be secure."

-- Egon Bar (1981)

In essence, the Internet was inevitable once computers became cheap and widespread. The concept of two users sitting at separate computers -- connected by an electrical cable across the room -- talking to each other is trivial. The work to make it happens is simply the details, the purpose, and the sophistication of what you want to do.

If you were an engineer working for a big corporation, professional chemist, physicist, or undergraduate science student who hung around the computer room around the late 1980's or early 1990's, and paid attention to the goings on, you were likely witness to a technological sunrise that was amazing to watch.

I was one of them -- a female colleague turned me on to "the Internet" in 1992, when I sat beside her in a long-forgotten company course we both took and showed me this amazing new thing on our Unix operating system computer "workstations" that changed the course of my life, by showing me a strange new world, that was being born, and made my life suddenly quite a bit more interesting.

That was one of the really cool days of my life, and I don't even remember her name.

What she had showed me on a UNIX workstation -- a $30,000 computer that I did my work on at Nortel, in Ottawa, working on telephone company telecommunications software for the gigantic computers that ran the public telephone network. What the fellow engineer had showed me was a Unix application called "Usenet", that allowed someone to "post", read and respond to messages on any of hundreds of topics ranging from computer software topics, hardware problems, to recreational groups -- guns, archery, weight-lifting, amateur photography -- to the frivolous -- jokes, cult movies, politics, both serious and outrageous.

Most are long gone now -- having moved on to bigger and better things --, but when I was new on the Net, you could see messages from the famous, and important -- the people who managed the net, were the gods of this expanding new universe, or revered names of "net gods" and "net personalities" who no longer posted, but who were still remembered by thousands, and honored silently by those thousands who knew their reputations, genius, and contribution to this, that, and the other thing -- all computer network-related.

And there were colorful characters, too. "Net kooks", eccentrics, the gregarious, and the entertaining.

And there were on-going controversies. Whether gay discussion groups should be allowed. "Yes" demanded the libertarian-types, as well as the "Live and let live", types. (And thus the response of John Gilmore quoted at the top of this chapter about "routing around censorship".)

"No", said the bigots, the rigid, the anti-gay, and the Christian fundamentalists, "Who cares?" said most, along with the ever-popular, "If you don't want to read a gay message group, don't read it. But what give you the right to prevent someone who wants to read a gay message group if they *want* to?" A system [local network] administrator -- sysadmin -- *that's* who. A sysadmin using his power to unethically impose his prejudices on the users on this computer network. But this is the real world, and it's a gutsy user who's going to complain to his boss that he wants to read a gay newsgroup, but the sysadmin is censoring it...

The controversy raged endlessly. Ain't freedom of speech -- the First Amendment of the U.S. Bill of Rights -- wonderful?

* * * *

"I wonder why I live alone here,
I wonder why we spend these nights together,
Is this the room I'll live my life forever?
I wonder why in L.A., to live and die in L.A.

I wonder why we waste our lives here,
When we could run away to paradise.
But I am held in some invisible vice,
And I can't get away, To live and die in L.A."

-- "To Live and Die in L.A." (1985)
Wang Chung, Jack Hues

She was 21 when I first met her -- and I was 29 and newly single. She was your classic "barbie" -

- exactly what I had always dreamed of -- young, tall, blonde, slim, stacked, and sexy, and she rocked my world in every way, and vice-verse. We seemed to be made for each other, and I was in emotional paradise for a time. But for heaven to exist, so must hell. In other words, I was going to learn the hard way, as one usually does.

She had a good mind -- she was intelligent, but not well read or informed -- not a crime or sin, by any means. But as a result, she lacked the accumulated facts that the foundation necessary for a good mind to properly achieve its proper performance and full potential.

But she was my woman, my love, and we lived together, and so I considered it my duty to slowly elevate her intellectually, as a pleasant part of my caring, sharing obligation to my mate.

So I began, in a low-key way, bit-by-bit, to explain and show her about the things I thought were important intellectually, as they came up -- such as on TV, in the newspaper, or in books. Over time I explained why I thought the things that I thought, and voiced frequently -- usually criticizing various views, and ideas.

I figured we should be on the same frequency, since I had some ideas and views on a lot of things, mostly novel, unconventional, elaborate, and well thought out, as far as I was concerned.

I wanted her to understand my belief system and political views, and why they tended to the extreme. My rabid criticisms of the world's institutions and elites. My rants typically were about blatant documented corruption, and all the faults and evil in the world. Why hypocrisy and lies filled the news. How the powerful kept their power through this bizarre systemic manipulation, violence and corruption.

Why I hated municipal police and called them routinely "pigs" in common conversation. Why I hated various groups and governments so virulently. Why I hated the U.S. as the fount of Evil in this world. Why I seemed enamored of the concept of armed revolution to overthrow an irredeemably corrupt regime. Why I accepted political change through carefully planned and targeted violence, including the assassination of government officials who were clearly abusing their power, so enthusiastically.

They had been entrusted with power and authority; they were abusing it and willfully causing death, great harm, and widespread injustice. Their acts were no secret, yet they broke their oath of office with impunity. If government oversight was non-existent, impotent, corrupted, or indifferent, then the collective failure of government power to police itself would be self-evident. Festering, with repeated incidents of abuse would show a possibly isolated incident to be systemic corruption and indicate a problem which threatened the integrity of the system itself.

I mean, if I didn't explain my admittedly radical views, she'd probably think I was crazy. (No big surprise here). Or just tune me out when I started ranting, like everyone else did. You see there was a simple, reasonable, logical pathway to all my beliefs, assuming one accepted the premise that it is reasonable to fight evil with violence. Governments everywhere had an even lower standard, I figured, so my beliefs were an improvement, or so I reasoned.

(Though I had pretty much renounced political violence in 1985 or so, *advocacy* of it when justified and unavoidable, as part of a mass movement, I still continued, on the grounds of free speech, emotional venting, and the fact that nobody listened to me, anyway..) And they didn't enforce the sedition statutes -- which are in violation of the Charter free speech clause, anyway.

So it became my little educational project to point things out to her as they came up. If I saw some error, or lie, or distortion on the TV news, or some blatant idiocy in the newspaper, I'd take the time and effort to show her what the issue was, prove my point quite quickly, and show her why the issue mattered.

Slowly but surely, point by point, I educated her in the political and ideological "World According to Yogi". I had an amazing memory, and an enormous database of accumulated facts and data in my head on a variety of subjects, all cross-referenced, and verified for correctness. I had read hundreds of non-fiction books, and had a good understanding of most of the technology that underpinned modern civilization, and an expert knowledge of an interesting and diverse subset of these.

I had realized that "they" -- people trying to keep power, influence, and status by hiding their corruption, manipulation, evil, and chicanery didn't bother to try to manipulate and distort technical data. And you could surprisingly often use technical data to uncover errors or deliberate distortions in news, historical, and other government or otherwise accepted "truth".

She always listened, sometimes asked questions, and seemed to understand. But I never really knew whether I was getting through or not, or whether she didn't really care, or was just humoring me, or what. After all, I freely admitted to myself that my views were radical in the extreme, and completely at odds with the world-view presented and believed by society at large.

Then one day, after six months or so of my low-key "indoctrination", I found out what she had learned. It was 1989.

"I have to talk to you, Yogi. It's important." she said one afternoon out of the blue. Taken by surprise, and curious, I sat down in the living room of my house, on the couch, as she stood in front of me.

"For six months I've listened to you as you pointed things out about the TV news, and in the paper. The errors, the lies, the distortion, the propaganda. Everything you say is absolutely true, and you've proved it all to me. You weren't exaggerating. Everything they say is a lie, deliberate bullshit, or just wrong because they don't understand what they're talking about. The government is completely corrupt, just as you say, and you've proved all of it to me, point-by-point.

"But the world you've shown to me is totally black and hopeless. But I've always been just a simple girl -- happy and carefree. But now I feel miserable and depressed knowing the Truth about the world.

She was right -- there was no light in my view of the world. It was bleak, barren, desolate, and hopeless, because that's they way I saw it. There was just darkness, and an atmosphere of a despairing present and future hopeless. It was an atmosphere that slowly poisoned one from the outside in, and completely absent any merciful side-effect of mental numbness or mental anesthesia.

It was the bleak reality of human history, with no improvement over the centuries other than more efficient killing methods, and a fuller and deeper understanding of the morality that we completely failed at exercising attaining it.

It was the desolate reality of human nature, and I could not blind myself to it. It was the ugliness of human nature that we had to see and recognize as a species, until it changed. Everyone could acknowledge the wondrous and fleeting beauty of a magnificent sunset, but would that stop war or end hatred?

I had always seen things about reality that others seemed to miss. So, I was inured to it. She wasn't.

I'd never heard her talk like this before, and I was captivated. Had I underestimated my Heather? That was a pleasant surprise.

Addressing me directly, and sincerely, she continued: "Yogi, I'd rather just be *happy* -- happy not knowing, the way I used to be -- so I have to get away, to be happy again. So I have to leave you. I'm sorry, but I have to leave."

And then she turned, and walked out, closing the front door behind her. I sat there stunned on the couch. Not moving, nor protesting her leaving. For once, I was speechless.

It was the finest compliment I have ever received...

"The fault, dear Brutus, lies not
in our stars, but in ourselves."

-- "Julius Caesar" (1599)
Act 1, Scene 3

Prison Diary 66:   "The Beauty of Our Weapons"
=======================

> "The true purpose [of Zen] is to see things
> as they are, to observe thing as they are,
> and to let everything go as it goes."
>
> -- "Zen Mind, Beginner Mind" (1970)
> Shunryu Suzuki

> "Search for the reality of each object -- that is, its real
> and only beauty...What is that vital thing -- in ugly as
> well as lovely things and places -- the thing that takes
> us out of ourselves, that draws and attracts us, that
> unnamable thing claiming kinship with us?"
>
> -- "Hundreds and Thousands:
>   The Journals of Emily Carr" (1931)
> Emily Carr

A technical expert may be defined as someone who doesn't know how much he knows on a given scientific subject. Another definition, I posit, is that a technical expert may be defined by the number of interesting or useful anecdotes he has in his mind.

Yet another definition, I also submit, is how easily he can simplify and expound complex ideas, so that they can be grasped by those of a somewhat lesser knowledge on the subject.

Here goes...

From an engineering perspective, the beauty of a firearm is that it is a mechanical device that automatically controls and harnesses the tremendous explosive forces -- heat and pressure -- produced by the firing of a bullet cartridge. And it does it reliably, and over and over again.

A modern rifle or pistol cartridge consists of a lead bullet, sitting on a thin brass cylindrical tubing -- the casing -- which holds the smokeless powder propellant, and with a primer seated in the brass casing's base, to ignite the powder, to burn and produced combustion gases that propel the bullet down the rifle or pistol barrel.

As the bullet is shot out the muzzle of the barrel at, say 300 m/sec. -- supersonic speed -- the soft brass casing at the other end of the barrel -- the breech end, or the chamber -- is stretched by the enormous pressure, and seals the hot high pressure gases in momentarily, until the bullet exits the barrel -- at which point the pressure inside the barrel immediately drops. And as it drops, the brass casing retracts, loosening in the chamber, waiting to be removed -- extracted -- and tossed out the ejection port.

Essentially, when the ammunition casing is fired, the thin, ductile, soft metal brass casing stretched and balloons, acting as a gasket to temporarily seal the breach chamber end of the gun, withstanding the extremely hot gas while maintaining a pressure of approximately 75,000 psi for a rifle, and for example, a maximum of 35,000 psi for a 9mm pistol chamber.

* * * *

> "One of the well-known peculiarities of
> modern civilized opinion is its refusal
> to acknowledge the value of violence..."
>
> -- "Movie Chronicle:
> The Westerner" (1954)
> Robert Warshow

It is truly one of the ironies of modern civilized society, that of its attitude to violence. It is a basic hypocrisy of society that it evolved from a violent, bloody, and barbaric past, that violence founded it, that violence consolidated its existence and sustains its equilibrium and power dynamics, and it is violence on which – alas – it seems to thrive.

In sum, the violence on which it depends in every way, and from which its bloody beginning was born, and to its fiery doom – or painfully slow decay, continues unabated. With a final "bang", and not T.S. Eliot's whimper. except, perhaps, from the desperate cries of its helpless victims.

* * * *

> "My tale was heard, and yet it was not told;
> My fruit is fallen, and yet my leaves are green;
> My youth is spent, and yet I am not old;
> I saw the world, and yet I was not seen;
> My thread is cut, and yet it is not spun;
> And now I live, and now my life is done[.]"
>
> -- "Tichborne's Elegy" (1586)
> Chidiock Tichborne

An amateur rocket powered by an intimate mixture of zinc dust and sulfur dust was propellant for a seven foot by one inch diameter, thin walled steel piece of straight electrical conduit. The rocket nozzle at the open end was a machined medium steel venturi shape, and the altitude reached of the vertically pointed device was 11,300 feet; the launch was in 1967, and the plan for the rocket was based on an article in "Scientific American".

Little has changed over human history in the basic concept of weaponry. We've been throwing things at other people for at least sixty centuries -- perhaps since time immemorial. It's only what they've been throwing that's changed and evolved. Soon enough this collective violence earned its own name. It became known as "war".

Enter the demand or independent invention of new weapons. Enter the weapon designer.

Archimedes was the first well-known weapons designer -- in his spare time between his usual occupation of revolutionizing science. Plutarch's describes the city of Syracuse being successfully defended from Roman attack by Archimedes' designed catapults throwing 1800 lb. stone projectiles.

Then in the Middle Ages came Leonardo da Vinci, another genius and part-time weapons concept designer. He came up years ahead of his time with the parachute (ca. 1485), giant cross-bow ballistas, the tank (ca. 1487-8), the first studies of flying machines (ca. 1478-1480), and the helicopter.

Galileo helped make a living for himself by tooling and selling spyglasses for the Florentine city-state.

* * * *

Richard Jordan Gatling first invented the precursor of the machine gun, a hand-cranked .58 caliber brass cartridge weapon that fired repeatedly, one-by-one through a circle of six rotating barrels (U.S. Patent Nos. 36,836 (1862) and 47, 631 (1865)).

But less than a decade later, though Gatling was famous, his gun was eclipsed by true fully automatic belt-fed machine guns, such as the Maxim and Vickers MGs.

Hiram Maxim, an American inventor, in 1884 produced the first practical heavy machine gun, the Maxim machine gun. It killed the most soldiers of all (even **more** than artillery, as well as rifle fire, and hand-to-hand combat) during the stand-still trench warfare it produced and made infamous in World War One.

John Moses Browning was a prolific genius when it came to gun design. During his stellar career he was issued 128 U.S. Patents for 80 different firearms. In one seventeen year period he designed 44 different firearms for Winchester, all of which they bought, but only ten of which were put into production.

Among the firearms he designed, that went into production, and success:

Winchester Model 1885 Single Shot Hi-Wall rifle, "the best single shot rifle [Winchester] had ever seen."

Winchester Model 1886 lever action shotgun, which sold for 71 years.

Winchester Model 1887 lever action shotgun, manufactured from 1887 to 1901, in 10 and 12 gauge. 5 shot; used by Arnold Schwarzenegger in "Terminator 2: Judgment Day"; became the Model 1901 in 10 gauge only, manufactured from 1901 to 1920, to handle the new smokeless powder (as opposed to black powder) cartridges.

Colt Model 1889, first U.S. gas-operated heavy machine gun.

Winchester Model 1894 .30-30 (30 grains of the new smokeless powder) lever action carbine (short-barreled rifle), the new load only requiring a new nickel-steel barrel.

Winchester Model 1897 pump-action 12-gauge shotgun, for 2.75" cartridges, short 20" barrel, 6 round tubular magazine, bayonet lug for sword bayonet, and air-cooled ventilated barrel shroud/hand-guard, along with take-down portability, became known as the "trench sweeper" for its use in WW1 trench warfare night raiding parties, and "riot" gun for civilian usage. Browning perfected the pump-action shotgun, first introduced in 1882 by Christopher Miner Spencer of Spencer rifle fame. Browning U.S. Patent 441,390 (1890). Popular for over a century, the millionth gun was made in the first sixty years.

After WW1 shotguns were banned from war by the Geneva Convention.

Browning Auto-loading Shotgun, the first semi-automatic shotgun designed; a five-shot long-recoil model produced by Belgian arms maker FN (Fabrique Nationale) in 1903, who sold two and a half million copies; widely copied.

Browning .32 ACP Model 1900, small and light and hugely popular Belgium blow-back semi-automatic "pocket pistol", over 1.3 million sold by 1914.

Colt .45 ACP, Model 1911A1, single action semi-automatic 7-shot handgun, still in use, though replaced as an officer's handgun in the early 1990s by the 15-shot Beretta Model 92F. It was adopted as the M9 as the standard sidearm by the U.S. military. The first gas-operated firearm, the principle invented in 1889 by Browning, and now standard in rifles and machine guns, such as the AK-47.

Browning M1917A1 water-cooled heavy machine gun and the Browning M1919 air-cooled heavy machine gun.

M1918A1 BAR (Browning Automatic Rifle) .30-'06 gas-operated 17 round, light machine gun. With a cut down barrel and stock, was a favorite escape tool, also used from his powerful V-8 car against pursuing police, by the Bonnie & Clyde outlaw gang.

M2 .50 cal. HB (heavy barrel), belt-fed, heavy machine gun, with 45" (114 cm) barrel, designed in the 1930's, an evolution of the M1917A1, in service from 1918 to 1968, an impressive half century. 2.5km range, the bullet will go through a house and everything inside.

Browning FN [Fabrique Nationale, Belgium] Hi-Power 9mm semi-auto pistol, 1935, in wide use, and still in service, with a 13-round magazine, when a maximum of 10 rounds was standard.

In 1931, Winchester introduced a new steel alloy that was hailed at the finest development known to gun manufacture, and won Winchester recognition as a leader in the treatment of alloy steels.

And finally there was Gerald Bull, a Canadian long-range, high altitude artillery designer and genius. Received his Ph.D. at the University of Toronto in 1951. Left Canada in disgust when his research started to look more and more military in nature, and they cut off his funding. Worked for a time for the Americans, who had no such qualms. But the U.S. Army got discouraged at the lack of military applications of his ground-breaking work and experiments carried out over the open sea from Barbados.

So he moved on, again. He designed the South African GC-45 gun, but got into trouble with the Americans because of the arms embargo against South Africa. Finally, running out of options to practice the love of his life – cutting edge artillery research, such as base bleed shells for extended range, and artillery-fired rockets-ass he finally go involved with the Iraqis under Saddam Hussein – then an American ally. But not for much longer. In Baghdad he got into his dream – the Iraqi "Super-Guns" -- "Little Babylon" and "Big Babylon" -- for launching satellites.

Baby Babylon was a 40m long prototype that was successfully fired. Big Babylon was 500 feet long, 1 meter in diameter, and weighed 2100 tons. It fired a three-stage rocket 3 feet wide, with a propellant change in the gun's 180 ton breech 20 feet long. With media publicity the giant parts were seized all over Europe as they were being smuggled in pieces to Iraq.

Bull was gunned down in 1990 – shot five times -- just outside his apartment in Paris by person or persons unknown, but there is a strong circumstantial case that he was assassinated by the Israelis.

* * * *

"Is this the promised end?"

"Or image of that horror?"

-- Edgar answers Kent
"King Lear" (1623)
Act 5, Scene 1

The first thing you notice when you pick up a center-fire firearm -- handgun, rifle or whatever -- for the first time, is how heavy it is compared to what you were expecting. The second thing you probably notice is how solid they feel, construction-wise. Hefty, robust, and not fragile or delicate in the slightest. (The M-16 -- designed in the mid-1960's to be ultra-lite -- is the one exception.)

In old firearms, sometimes I wonder how many owners it's had -- a properly taken care of firearm will easily outlast its owner -- and how many died in a war carrying the gun that survived them. And travel the world over, and sometimes through several armed conflicts, before it to ends up in the civilian market, and my hands. Thinking of these things is just a natural progression, in my opinion. It's not morbid or macabre -- it's just the reality of it.

One day, for no particular reason, I was looking over my deactivated Israeli Army (IDF), open-bolt, full auto, Uzi 9-mm submachine gun with folding stock. (You could partially cycle the bolt, but all the major parts had been either welded, or ground down to deactivate it completely.)

By chance, and once again, for no particular reason, I tried to stand it up on a table, with its metal stock unfolded.

I carefully stood it on the magazine -- the long straight rectangular cross-sectioned sheet metal tube -- the bullet clip/magazine -- that extends out of the bottom of the rear hand-grip.

To my amazement, instead of immediately toppling over, it felt perfectly stable, standing on the two square inch base of the 25-round magazine. When I carefully let go, the heavy steel Uzi did not fall over, but stood upright and perfectly steady on the table -- the 10 kg Uzi awkwardly balanced *perfectly* on the magazine sticking out several inches from the heavy SMG body.

It was no accident. That day, I discovered something special about the design of the Uzi. Its center of gravity was on the axis where you held it with your trigger hand, giving it intrinsically greater accuracy. This is of very significant importance in full auto weapons, SMGs being very hard to control during full auto fire, one of the reasons you have to fire in small bursts, rather than continuously.

The Devil in the details. The fascination I have in weaponry of all sorts, the awe at their elegant intricacy, the respect for their powerful beauty. It's hard to explain. It took me over three decades to understand myself, and I don't pretend to grasp it all. Perhaps it's a "guy thing". But that's too simplistic, as far as I'm concerned.

Beauty is obvious. It needs no understanding. It needs no help from me. It is perfect goodness standing proudly on its own. Simple or complex. A blink of time, or the universe of forever. A rare orchid, or the daily sunset. The smile of a young girl on a bright summer day.

It's the ugliness in life and in ourselves that needs our consideration and study. That needs our understanding. That we have to face as a species, in all its horror.

And after years of interest and dedicated study, years of research, learning, and finally experimentation, modeling, even in all its inherent spiritual ugliness, and ugliness of purpose, I could easily see a profound, sometimes awesome kind of beauty in even the most horrific of man's weaponry.

It's usually in the complexity and intricacy of their design, sometimes in their simplicity. Always in their power and strength. And if you don't believe me, listen to the words of Confucius, "Everything has beauty, but not everyone sees it."

* * * *

> "They had been corrupted by money, and he had been corrupted by sentiment. Sentiment was the more dangerous, because you couldn't name it's price."
>
> -- "The Heart of the Matter" (1948)
> Graham Greene

The first time I saw a firearm being fired -- a .38 revolver with a 2" barrel fired at a public-invited, university-staged lecture on forensics, when I was a kid -- the first thing I noticed was how *loud* a gunshot was, even from twenty meters or so away.

It was deafening!

And this was in a university auditorium, not a small, seedy apartment, like on TV -- where it would be even louder in the much smaller, enclosed space.

The first time you fire a handgun the recoil -- the kick -- is significant. Even with a moderate caliber, like .38 or 9 mm, it feels like getting hit with a hammer on your wrist. And with a heavy caliber, like .45 ACP, a novice firer cannot even control the aim, and the fired gun ends up pointing skywards. If you have thin arms, or like most people don't work out with weights, you'll always have difficulty controlling the .45 Colt M1911A1, the "standard" .45 ACP pistol -- and it's a heavy pistol, which lessens the effect of recoil and makes the gun more controllable on firing. Even when I worked out religiously, I couldn't fire 12-gauge 00 ("double ought") buck-shot -- the heaviest shotgun cartridge -- without flinching and ruining the aim and accuracy. I just couldn't master the recoil, no matter how hard I tried.

The shotgun is the deadliest firearm of all at close range. Deadlier than rifles, and much more deadly than handguns, with their medical wounding effect not to be "compared with other bullet wounds... [A]t close range they are as deadly as a cannon." ["Annals of Surgery" 177:174-5 (1973), "Gunshot Wounds of the Abdomen"]

Handgun gunshot wounds are 1.3 to 3 times more lethal than knife wounds. And a rifle or shotgun kills 5 - 11.4 times more often than a handgun wound. [Bull. N.Y. Acad. Med. 62:539-541 (1986)]

It doesn't take long before you realize that action movies and TV do not portray firearms very realistically. It's a short mental leap to realize that script writers are not weapons experts, and you're watching a movie not a documentary. It's entertainment, not a training film, I like to say.

There's a lengthy and classic shoot-out near the end of Michael Mann's crime movie, "Heat". It begins with bank robbery team member, Val Kilmer, opening up with a full auto assault rifle on a plain-clothes cop across a crowded, down-town Los Angeles street in broad daylight.

I immediately noticed, to my surprise, that the full auto fire sounded "different" from the usual movie sound-track. It sounded "real", not fake. Gunfire, particularly automatic weapons fire has a very distinctive sound. In the mid-1990's, I had the luck to get confirmation from someone who worked behind-the-scenes on the movie that indeed Michael Mann had spent the extra effort to make the audio of the gunfire on the soundtrack accurate sounding. One of the production guys for the movie -- this was the "old days" of the Net, when people with inside information and qualifications popped up from time to time -- made a Net posting confirming this.

* * * *

> "[T]he most dangerous moment for a bad regime is usually when it begins to reform. ...The abuses that are removed seem to reveal those that remain, and to make them more galling; the evil has lessened, it is true, but the people feel it more keenly."
>
> -- "_L'ancien Regime_ and
>   the Revolution" (1886)
>   Alexander de Tocqueville

Ammunition names are funny things. First of all there are thousands of different cartridges -- mostly rifle. They were developed over the centuries, and very few are completely obsolete -- in other words, unused and retired, because no working rifles exist to fire them.

At any one time, there are the few standardized military cartridges for machine guns, infantry rifle and handgun. One set for NATO -- the U.S. and its allies -- and one set for the old COMBLOCK -- the Communist Bloc -- the old Soviet Union and its Warsaw Pact allies.

And this is just for up to .50 caliber. Anything above -- greater in the bullet projectile caliber/diameter -- like 20 mm -- is considered, and called, a cannon, as is used for anti-aircraft fire, and jet fighter ground-strafing cannon fire.

Then there's commercial -- hunting and civilian target-shooting -- ammo, which is older and current military standard ammo, plus other common rounds still manufactured for sale to the public.

Then there's obsolete -- older -- and "wildcat" ammo -- non-standard, experimental loads developed by ammo and rifle designers, which either failed or developed a following and had just a few or many civilian rifles developed for it.

Last point. Now it gets strange. The bullet -- the generally lead-based projectile -- has no consistency as to naming and actually diameter. The classic example is the now militarily (but not commercially!) obsolete ".30 caliber ammunition" used up until partway through the Vietnam War in the mid-1960's, as the standard U.S. military cartridge, the 7.62mm x 51 NATO round.

The projectile for this "30 caliber" was the *same* diameter -- and the diameter was exactly .311" -- for .303 English, .30-.30, .30-06 Springfield, and .30 M1 Carbine, amongst many others. And also for the .308 Winchester/7.62 mm x 51 NATO (the commercial and military names, respectively, for the same bullet, with the same casing.

And the defense and deer hunting shotgun load, "00" buckshot, the heaviest 12 gauge load, fires six .30 caliber spherical lead balls, if I recall correctly. Which is why it's so effective at taking down a man, and has such a horrendous recoil kick.

How's that for non-standardized ammo nomenclature/naming convention?

* * * *

> "The king called up his jet fighters,
> He said, 'You'd better earn your pay!
> Drop your bombs between the minarets,
> Down the Casbah Way,'

As soon as the Sharif was chauffeured outta there,
The jet pilots tuned to the cockpit radio blare.
As soon as the Sharif was outta their hair,
The jet fighters wailed:

'Sharif don't like it, Rock the Casbah! Rock the Casbah!'

-- "Rock the Casbah" (1982)
The Clash
Jones, Mick, Mellor, John

It was back in the mid- 1980's, when you could keep loaded weapons available in the house. I had, shall we say, a mild, but controllable case of simple PTSD (Post-Traumatic Stress Disorder), and it gave me a sense of safety, security, and serenity to keep a short-barreled, extended mag Winchester Defender 12 gauge pump-action shotgun by the side of the head-board of the bed. To be ready for an unexpected nighttime attack.

Short-barreled, for easy aim in the close quarters of the apartment, and a wider shot dispersion for increased ability to hit the attacking human target. Extended magazine to give me an eminently respectable seven shot capability.

And the final touch: I loaded the magazine with alternating shotgun shells. The first, a slug, the next, "0" buckshot, and so on, repeating the alternating combination until the magazine was full.

The slug was a single piece of lead, weighing just over an ounce. It will punch through 1/4" thick mild steel. The buckshot was the second largest size of shot, something like eight, less than .30 caliber lead balls. I just couldn't take the heavy recoil of "00" buckshot without flinching just as I pulled the trigger -- which fucked up my aiming accuracy big-time. The kick was just too much.

As I crouched or leaned low beside the bed, the slug was to punch a clear hole through the drywall and wooden studs of the bedroom walls, to see approaching hidden attackers -- it would also take them down (and more than one, if they were lined up straight in a row) if they were visible through the bedroom doorway,. And the next round of buckshot was to take down the attackers now visible as they passed by the hole in the bedroom wall.

I guess I had too much time on my hands for thinking such elaborate -- but good -- strategies for events that would never happen.

About twenty years ago, I guess my PTSD finally healed spontaneously -- or just it was that enough time (a decade) had passed -- and I unloaded and put my guns away in a locked bedroom, converted to a storage/gun-room. The law -- and I -- had changed.

* * * *

"Place is a madhouse, feels like being cloned,
Beacon's been moved under sun and star,
Where am I to go now that I've gone too far,
So you'll come to know,
When the bullet hits the bone."

-- "Twilight Zone" (1982)
Golden Earring
George Kooymans

The .357 Magnum and .44 Magnum revolver cartridges and guns were made famous as "heavy loads" by the classic Clint Eastwood "Dirty Harry" movies of the early 1970's. As Clint would say -- in the second movie of the series -- in his role as detective, the .44 Magnum was the most powerful handgun load in the world. They were heavy bullets and faster than regular ammo, with a heavy gun to match the load, and make it more controllable/accurate in aiming.

But -- like the .45 Colt auto -- they were too heavy in the recoil department, and thus inaccurate for small armed or even just inexperienced shooters. They also catered to the hulking Neanderthal type shooter. You know -- the my gun and my penis are bigger than yours types -- that I thought were just loud-mouth blathering jack-asses.

The .357 (developed in 1935) was almost 50% more powerful than the .45 ACP round for the older ex-standard -- and eminently powerful man-killer -- military pistol, designed in 1911 for John Browning's .45 Colt auto -- the M1911A1.

The .44 Remington Magnum bullet -- to be precise -- was designed in 1954 and was twice the weight of the current U.S. 9mm military round, and faster, more than doubling its powerful hit over the 9mm. It is about 50% heavier and 50% faster than the .45 ACP round, making it almost three and a half times as powerful a round.

Not to be outdone, the .44 Magnum was no longer "the most powerful handgun in the world", exceeded by the .454 Casull Magnum circa 1973, which would fit the .45 Colt auto handgun -- and with almost four times the weight in powder propellant, blow up the gun if fired. It was about 50% more powerful than the .44 Magnum.

Finally, in 1983, there came the final commercially available, new "most powerful" handgun. It was the first gas-operated semi-auto handgun, the Israeli-designed (for the American commercial market) Desert Eagle, with its custom designed .5 AE (Action Express) ammo sporting a three-quarter ounce bullet, the gun has a ten inch (24.5 cm) barrel with muzzle break to reduce the recoil and increase the accuracy, and weighs a hefty five pounds (2.66 kg) (and giving it a more steady hold -- again increasing accurate fire), two-and a half times the weight of the two pound .45 Colt auto. The bullet had an incredibly high power that was half the power of the over-powered .30-'06 Springfield rifle bullet fired by the WW2 M1 Garand.

But as far as reasonable and eminently effective military use, there's the ultimate in one-man range and firepower: the Browning M2 HB .50 heavy machine gun. Fed by a linked metal belt, it pours its high-velocity, high caliber bullets down-range over 2.5 miles, and the bullets have high penetration, even more-so when they're AP (armor-piercing) rounds.

It is so powerful, it is inappropriate for use in civilian urban areas -- though this is widely ignored. It must be mounted, either by tripod, or -- most effectively -- bolted down to a jeep or pick-up truck's bed -- a mobile two-man (with the driver) death dispenser -- or other armored vehicle or military aircraft. The "HB" in the name means "heavy barrel", a way to resist barrel over-heating, along with new alloys, for the air-cooled barrel, so you can continue "hosing" the target with ammunition, without stopping before the barrel turns red hot and starts melting, bending downwards, muzzle first.

*   *   *   *

> "There is something that governments care for far more than human life, and that is the security of property, and so it is through property that we shall strike the enemy."
>
> -- speech at Royal Albert Hall, London

Emmeline Pankhurst (1912)

The problem of close order combat; house to house fighting, and room clearing; the FN P-90, HK pistol version, is the most modern, and the best.

The night raids of hand-to-hand combat in the trench warfare of WW1 evolved to the slow, grinding, and bloody urban house-to-house fighting in the dying days of Nazi Germany in WW2. Guns used evolved from the WW1 U.S. favorite for night-time trench raids, a sawed-off pump-action shotgun with extended tube magazine, heat shield (for overheated barrels) and a sword bayonet to the WW2 submachine gun.

The Soviet PPSh-41 SMG (Pp=machine pistol, Sh=Shpagin, the gun's designer, and 1941 was the year of introduction) came with a stick or drum magazine. The "burp gun" that equipped a million Chinese soldiers who poured across the Yalu River into Korea, to stop the American conquest of North Korea all the way to the border with China.

* * * *

"Civilization develops in man nothing but an added capacity to receive impressions -- that is all. And the growth of that capacity increases his tendency to seek pleasure in spilling blood. You may have noticed that the most enthusiastic blood-letters have always been the most civilized of men."

-- "Notes from Underground"
Fyodor Dostoevsky

For an even smaller automatic weapon than the SMG, enter the machine-pistol for the bloodiest of fighting: close-quarter combat, such as counter-ambush fire, house-to-house fighting, or, because it was eminently concealable, assassination.

Invented in 1896, and manufactured until -- an astounding four decades --1937, it was copied the world over, and preceded the SMG – also invented by the Germans – by twenty years (WW1). It was the German Mauser C96 "Broomhandle" pistol. It fired the 7.63 mm Mauser (equivalent in bullet weight and speed to the modern 9mm (9x19mm) standard U.S. pistol cartridge), had a fixed 10 round box magazine, and a 5 1/2" (14 cm) barrel. Full auto versions were available, to sweep a trench or room. It was quickly reloaded by a top-loading 10 round "stripper clip". You inserted the clip from the top of the empty gun, whose chamber conveniently held open, and with two fingers swept the 10 cartridges down into the gun in an instant, reloading it.

One later version of the Broomhandle was the 1920's Spanish Astra M901, selective fire with a **20** round magazine for much better firepower, and a slightly longer 6 1/4 inch barrel.

But the Broomhandle fired too fast, emptying its magazine with a trigger pull, instead of allowing for controlled bursts. As a result, the 10 round integral magazine was woefully inadequate. As well, its bullet had too much recoil, resulting in inaccurate fire, due to the rapid rising of the barrel on automatic fire. Unless your target was the ceiling, you wasted almost all of your shots.

These problems were ameliorated by a detachable stock that you locked in your armpit, and allowed steadier and more accurate fire. But this sort of defeated the idea of the machine-pistol: making a small, concealable pistol not so small, not concealable, less maneuverable, and slower and cumbersome to start shooting with.

The Czech VZ-61 (Model 61) Skorpion machine-pistol is the best machine-pistol design of the few that have been attempted. Of Czech manufacture, ca. 1960, it was a short 4.4" (11.2 cm) barrel, is compact, like small pistol, and allows one-handed firing.

It has a 20 round removable box magazine; .32 ACP (7.65 mm Browning short). smaller .32 better for machine pistol (though .32 ACP less powerful than some high-powered .22 long rifle rimfire cartridges), reducing the recoil that reduced the inaccuracy of automatic fire. Very concealable, close-quarter, quick draw weapon. Its design incorporated an ingenious counterweight mechanism in the butte to reduce the rate of fire, and a buffer to reduce recoil – essential for a machine-pistol.

The final contender was the famed Ingram MAC-10, 9mm or .45, designed in 1964 by Gordon B. Ingram, had a 32 round magazine, folding wire stock, was large pistol-sized, and designed for use with an optional screw-on Sionics silencer for assassination, covert, and commando operations.

The follow-up Ingram MAC-11 in .380 ACP (9mm Short), was smaller than the Model 10, the whole gun is about 10" long, has less recoil and thus more controllable/aimable in full auto since the .380 has a little over half the power of the 9mm, reducing the speed of the fired bullet, but with the same projectile weight. Made for the civilian market in semi-automatic as the Cobray M-11/9.

A dozen .45ACP rounds from a MAC-10 performed the 1984 assassination of Denver talk-show host Alan Berg, who had alienated a neo-Nazi extremist group, "The Order", broken up by the Feds in the 1980's. Bad politics, good gun...

The radio show of Alan Berg is the center of the movie "Talk Radio" (1988), the best movie Oliver Stone ever made. The last rant of the Jewish Berg, where he turns on his listeners, near the end of the movie, is absolutely priceless -- a brilliant social commentary soliloquy where Berg flays his audience for all their faults and flaws: their intellectual and moral bankruptcy, and general emptiness of their souls.

Lastly is the FN P90 from the 1990's smaller than an SMG, larger than a machine-pistol. It weighs 8.1 lbs. when loaded with a full mag of fifty 5.7mm rounds; barrel nine inches long, total length a little under 16" long and 7" high. Gas-operated with barrel compensator. Designed for ranges less than 150 m (490 ft.) but more stopping power and less recoil than AR-15/M-16.

*   *   *   *

"I, I can remember,
Standing by the wall,
And the guns shot above our heads.
And we kissed, as though nothing would fall,
And the shame was on the other side.
Oh we can beat them, forever and ever,
Then we can be heroes, just for one day."

-- "Heroes" (1977)
David Bowie

You always keep the inside of the barrel and internal mechanism of a firearm clean. And keep it clean in the jungle, by taping over the muzzle.

And **especially** you keep its insides well oiled. Gun oil, though any oil will do in a pinch. Engine oil is fine.

And for "storage" there's Cosmoline. It's a brand name for petroleum jelly -- Vaseline. You smear the weapon's metal parts -- internal and external -- with it, smearing it with a thick coating, including inside the barrel. Then you pop it into a plastic bag, tie it shut, and bury it or plaster it into a wall, or bury it in fresh concrete in the basement – preferably in reinforced concrete to be metal detector-proof -- for a rainy day. Don't forget the ammo.

For an improvised technically full auto rifle, you fit a shaped piece of paper-clip into the bolt face of a semi-automatic rifle, and the rifle will keep firing until empty. Don't do it too much or the barrel will melt!

For controlled full auto fire, those in the know, can take a shoe string, tie it to the trigger, loop it through the front sling catch, and tie it to the forward assist of the bolt. With the right length, and holding the string to the action, the semi-auto will fire full auto.

* * * *

"Society cannot exist without law and order, and cannot advance except through the initiative of vigorous innovators. Yet law and order are always hostile to innovations, and innovators are almost always to some extent anarchists. ...

Both temperaments are necessary, and wisdom lies in allowing each to operate freely where it is beneficent. But those who are on the side of law and order, since they are reinforced by custom and the instinct for upholding the _status quo_, have no need of a reasoned defense. It is the innovators who have difficulty in being allowed to exist and work."

-- "Individual Liberty and Public Control" (1917)
Bertrand Russell

The samurai sword: the ultimate technology and evolution of the sword, yet beauty as well as the perfection of a technology. The quest for the "absolute weapon". The quest for the surprise advance. New levels of barbarism and atrocity in the quest for victory.

Science, Religion and the Search for the Truth. Religion seems to come off the rails pretty early in the game, no matter how simple or obvious the message.

Infinity we cannot manage or understand, or comprehend. Out of chaos, comes order. Out of the black abyss, greyness, shadow, light, shades, and nuance. It is truly staring into the abyss.

But once again, reality cannot and should not be denied. It is, what it is. If you disagree with reality, rise up and steel yourself to begin to attempt to change it. But reality cannot be denied, and the messenger of reality is not the evil that reality may appear to be.

In the end, weapons are merely the extension of man's worst emotions, the expression of his vilest tendencies and the power to take them to the extremes of the technological capabilities of the day.

They are thus the inevitable exploitation of the tools available to the fullest advantage for what is justified as the "collective good", but which in fact are simply the base instincts -- call them primitive, savage, barbarian, or primeval -- that we don't seem to be able to acknowledge or even slow -- much less stop.

Firstly, with such an advanced democratic, prosperous society's with centuries of finely honed and polished religious, moral, intellectual, artistic, and philosophical stature, surely it is long past time to be honest with ourselves, and unite to practice these, rather than passing the buck to our caveman instincts.

To unite from the top, and effectively unite from the grass roots level to identify, publicize, isolate, and expose everyday evil. To openly establish a detailed plan to attack evil on a broad front, piece-by-piece, or individual deed-by-deed, as we open a second front to revise and revamp from the foundation up, advanced societies with just as advanced a preoccupation with morality and justice. And secondly, stop blaming the weapons. They are just the manifestations of evil. Nobody pulls the trigger, and nobody gets hurt. "The one-eyed man is king in the land of the blind," said H.G. Wells. The armed man is king in the land of the unarmed.

Do you fight obesity by banning spoons?

"Far along the world-wide whisper of the south-wind rushing warm,
With the standards of the peoples plunging thro' the thunder storm;
Till the war-drum throbb'd no longer, and the battle flags were furl'd
In the Parliament of man, the Federation of the world.
There the common sense of most shall hold a fretful realm in awe,
And the kindly earth shall slumber, lapt in universal law."

-- "Prophecy" (1833)
Alfred, Lord Tennyson

Prison Diary 67:   The Beauty of **Their** Weapons
=======================

> "Oh, curs'd device! Base implement of death!
> Fram'd in the black Tartream realms beneath!
> By Beelzebub's malicious art designed,
> To ruin all the race of human kind!"
>
> -- Orlando
>   "Orlando Furioso" (1532)
>   Ludovico Aristo

> "A monarchy divested of its nobility has no refuge
> under heaven but an army. Wherefore the dissolution
> of this government caused the [Civil] war, not the war
> the dissolution of this government."
>
> -- "The Prerogative of Popular
>   Government" (1658)
>   James Harrington

Sometimes things don't work as they're supposed to, especially in the field. And the more complex the device, the more likely is failure.

Take the ANVI (the Aviator's Night-Vision Imaging) system. It was developed in the late 1970's for helicopter pilots. Just in time for the 1979 Iranian Hostage Rescue gone-wrong. In the night, at a desert meeting point, amongst other catastrophes, a helicopter had crashed, and the bodies were displayed with glee by the Iranian authorities.

I heard the story about what really happened from within the "industry" -- I was working in high-tech in the mid-1980s. The ANVI system had a light filter over the pilots face that was a narrow-band pass filter construction: it blocked all external light, except for the frequency of the helicopter control panel lights, so that the pilot could look down and see the instrument panel lights – without washing out his vision with the 30,000 times light intensity -- and look up and see the night sky light intensified 30.000 times.

Well, it seemed that the filter was not completely perfectly designed and had a tendency to crack, blinding the pilot with "white out" from the 30,000x intensified instrument panel lights. Oops! Crash.

The limits of technology...

True story.

*   *   *   *

There's the Swiss Oerlikon. Belgium has Fabrique Nationale, Germany has Rheinnmetall. There's the Swedish Bofors. They're famous European arms manufacturers. Firearms, guns, cannons, and chain guns.

By conservative U.N. estimate there are 70 million AK-47's in circulation, making the Russian designed assault rifle the most popular firearm in the world, bar none. It is the weapon used by the COMBLOCK – the former Soviet Bloc and it East European Warsaw Pact allies, China, Middle Eastern governments, and African and Asian ex-colonial liberation governments, a total of the militaries of seventy-eight different countries. It is manufactured by at least nineteen countries on a large scale.

With its distinctive banana-curved ammo magazine, it is a battle-proven weapon, in use in various forms and variations for over half a century. Its loose specs make it -- surprisingly -- jam-resistant. It is cheap to make, and with its magazines double-taped together -- it has a good ammo capacity of sixty rounds. And its high caliber 7.62x39mm ammunition has range and power, with a bullet energy just greater than the quite powerful U.S. .30-'06 Springfield.

It's my favorite gun -- my weapon of choice if I had to make a choice of only one. (If I got to make two choices, the second would be the Beretta 92FS 9mm handgun, as my combat side-arm.)

The second -- distant second -- most popular military rifle in the world is the M-16, with an estimated 10 million in circulation, and the standard rifle of the U.S., its NATO allies, satellites, and puppet regimes.

It was the beautiful high-tech design in the mid-1960's of Eugene Stoner, chief engineer of Armalite Corporation (hence the designation AR-15 for the semi-automatic civilian version) (and hence the early Provisional IRA slogan, "God made man, Armalite made him equal."), but its one main problem is its short range, and that it jams easily if not handled with care, and kept clean of dirt and dust -- a fatal disadvantage on a muddy battlefield, or jungle swamp.

But its principle was great: short length for quick aiming from firing from the hip in an emergency, its light weight from a 7075 aluminum alloy receiver (the "body" of the gun, to which the steel barrel is attached) and plastic stock and barrel shroud was less tiring for the soldier to carry. A very high velocity bullet and terminal ballistics (it tumbled or fragmented when it hit its target soldier) made it especially lethal, and especially – especially -- its light weight 5.56x45mm NATO ammo allowed the soldier to carry much more of it than was previously possible with its 7.62 mm NATO predecessor.

* * * *

"At the very core of all this evil that has burst at last in world disaster lies Kruppism, this sordid, enormous trade in the instruments of death."

-- H.G. Wells (1914)
on WW1

Then there was the Soviet RPG-7 -- Rocket Propelled Grenade, Model 7 -- which didn't actually fire a grenade, but an armor-piercing anti-tank HE shaped charge, that could also be used against enemy troops. It served the purpose of three separate U.S. heavy arms: the WW2-era bazooka, the early Vietnam-era (1960) M-79 -- that resembled a single shot sawed-off shotgun with a 12" barrel, but fired a single 40mm HE anti-personnel fragmentation round, and the late Vietnam War-era M-72 LAW, a disposable general-purpose "Light Anti-tank Weapon".

The Soviets were first with:

1939: the truck-mounted, multiple-tube, rocket launcher. Dubbed "Stalin's Organ" the highly mobile unit fired a sequential barrage of small, double base, solid-fueled, short-range Katyusha rockets, with a terrifying sound that was loud and seemingly unending.

The Katyusha was 71" (180 cm) long, 5.2" (13 cm) in diameter, and weighed 92 lbs. (42 kg) with a 48 lb. (22 kg) warhead. Range of 3.4 miles (5.4km) in barrages. Several launchers could deliver 4 tons of HE across a 10 acre (4 hectare) area. [Ford (2011)] A shotgun effect of "Cluster rockets", like the concept of the later cluster bomb, but much more inexpensive and easy to produce.

1943: PPSh-41 for all infantry, as well as the even cheaper PPS-43 SMG: they had portability, but only with cheap mass-produced PPSh-41 for everyone -- all infantry soldiers. Over 5 million produced by 1945; it was the gun that won WWII. 71 round drum magazine; also stick mags. 7.62 mm Soviet ammo.

Not "Strength through superior firepower", but "Strength through massive firepower"!
    It: 1) became truly effective (SMG not belt fed, so massed SMGs required!)

    and   2) won WW2, rolling back inexorably the Eastern Front, by overcoming
    Nazi small arms of much lesser number and ammo capacity.

1950: the first advanced design, miniaturized a-bomb; double the yield and half the weight of the U.S. "Fat Man".

1953: the first deliverable H-bomb.

1956: the first submarine ballistic missile capability (the Zulu Class, Zulu V).

1957: Sputnik, the first orbiting satellite.

1957: the first ICBM, much superior than the American B-52 bomber for strategic nuclear forces.

1959: the first nuclear ballistic missile submarine (Hotel Class sub).

1960: the first nuclear SLCM (Submarine-Launched Cruise Missile), in the Echo I class nuclear submarine.

1969: the first titanium-hulled nuclear attack submarine, the Alfa Class, which enabled deepest diving (2,000-2,500 ft./600-750m) capability, light and fast, and anti-magnetic properties.

1969: the first working liquid metal-cooled (lead-bismuth alloy) submarine reactor, for the Alfa Class sub, with the fastest speed of all subs of 43 knots submerged.

1970s: The COMBLOCK Czech Skorpion, the first truly effective machine pistol.

1977: the first real attack helicopter gunship.

Ca. 1980: the ZSU 23 ("by far the best antiaircraft gun in the world." [Hadley. p. 267 (1987)])

* * * *

> "Constitutional government is chiefly concerned with civil liberties, revolutionary government with public liberty. Under constitutional rule it is almost enough to protect individuals against the abuse of public power. Under revolutionary rule, the public power is obliged to defend itself against all the factions that attack it."

>   -- speech before National Convention (1793)
>   Maximilien Robespierre ("The Incorruptible")

Then the was the WW2 Nazi medium machine gun, the MG-42: the Grand Guignol of firearms; stamped, mass production, and cheapness that goes with it, and a revolutionary roller-locked bolt design that made it the best machine gun design of its day and for many years after.

1200 rpm of 7.92 mm ammo. Renamed, but still used by Modern Germany, M60 of U.S. with different caliber (7.62 mm NATO), bipod-equipped, released in 1963. When the U.S. during WW2 originally tried to copy a captured MG42, the new weapon was a failure. (So much for U.S. WW2 ordnance engineering competence...)

The Nazi MG-42 evolved to the 7.62 x 51 mm (.308 Winchester) CETME assault rifle of 1958; 30 round slightly curved box mag. Spanish CETME (Center for Technical Studies of Special Materials) assisted by German engineers like Ludwig Vorgrimmler who had escaped to Spain.

Work began in 1949. Heckler and Koch founded in 1948, and bought the CETME patent in mid-1950's. G3 work started in 1959, using 7.62 x 51 mm cartridge (HK91, the civilian, semi-automatic version). The roller bearing design is also used in the Heckler & Koch MP-5 SMG, which also has a nifty, modular break-down, design construction.

Krupp went to Bofors in Sweden for the "88". MG-42 designers went to fascist Spain and the arms firm CETME. Gerry Bull went from Canada to South Africa to Iraq

Krupp & Bofors for P.175, 1.6" Bofors L/60 AA gun; P.174 .8" AA gun; most widely deployed AA gun in WW2.

Alfried Krupp, of the famous German Ruhr Valley arms manufacturing dynasty, designed the Flak (Flieger Abwehr Kanone) anti-aircraft cannon, the famous "88" -- it was 88 mm caliber (3.5") and used first as an AA gun (height range of 32,000 feet), then as an anti-tank weapon, and finally as a tank main armament gun itself.

There were the Krupp Flak 18 & 36 & 41 AA guns. Then, famously -- in its anti-tank role -- known as the "88", with a high velocity (3,335 fps) 88 mm (3.5") 20.7 lb. HE ammo,

A derivative was made into anti-tank gun in 1943, the PAK43 (Panzer Abwehr Kanone -- anti-armor canon), 88mm caliber, the ammo had a mild steel skirt at base that by the muzzle squeezed -- the "squeeze bore" -- the round down to 88 mm, for a super-high velocity of 3,700 fps, with a second innovation, a sixteen lb. tungsten-cored ammo which could penetrate 2.2" steel armor at half a km, and feared by Allied tank crews as a result.

In 1942, it took on its third role when the "88" was made into the main turret gun -- the 88mm KwK 36 -- for the deadly Tiger tank.

* * * *

> "Those who wonder what motivates American gun
> owners should understand that perhaps only one
> word in the English language so boils their blood
> as 'registration', and that word is 'confiscation.' "
>
>   -- NRA fact sheet (2010)

The U.S. did have some firsts.

The U.S. was first with a WW1 light machine gun, the BAR – Browning Automatic Rifle. It was first with the Colt 1911A1 pistol for WW1: a high-caliber, 7-shot semi-automatic. In WW2, the U.S. was first with the innovative introduction of semi-automatic rifle (the 8-shot, .30-06 M1 Garand) for general use of the infantry; vs. the much slower to reload bolt action rifles in general use.

The Gatling gun and Hiram Maxim's Maxim MG, revolutionized the wars of the 20th Century by the first invention of the heavy machine gun, though the Gatling gun was a relative failure in adoption and use, until resurrected in the late 1960's as the AC-130 Specter gunship, where massive firepower was helped by the barrel cooling action of multiple barrels.

Thompson M1 SMG had a smaller box mag in the military model (it had a high capacity, but less reliable, drum magazine too for civilian model),

The 1918 BAR was before its time and too heavy; .30-'06 heavy; the 1911 Colt auto pistol was too heavy and of too high a caliber, though it was robust, and the M1 Garand was too heavy and noisy (the enemy could hear the "ping" sound as the empty cartridge holder ejected, and know that your rifle needed reloading), and the .30-'06 had too much recoil. And the M3 "Grease Gun submachine gun was too heavy, and again the .45 was too big a caliber.

M16 problems, originally had no forward assist for when it jammed caused many deaths; they finally got it right with the M-16: light & light high-velocity ammo, but then they fucked it up, too! It jammed too easily when dirt or sand got in the mechanism.

Ingram MAC10 .45 with Sionics silencer, helical alum. gas disperser and slower down and too late; way too powerful, like .45.: recoil issue. and ammo heavy. A later model was reduced to the lighter recoil .380 ACP caliber, for more effective fully automatic fire.

There was the beehive artillery round shooting a rain of hundreds of steel "flechette" mini-arrows to cut down storming massed enemy forces trying to overrun the U.S. base camps in Vietnam.

The U.S. was first with the "daisy cutter" FAE bomb, napalm for WW2, Claymore anti-personnel mine, and the cluster bomb.

* * * *

"A complex weapon makes the strong stronger,
while a simple weapon...give claws to the weak."

-- "You and the Atomic Bomb" (1945)
George Orwell

People lie. But technology never does.

The European's took Maxim's Heavy Machine gun and built a plethora of firearms, machine gun, artillery, and anti-aircraft guns for their countries and both World Wars.

There were many famous European firearms manufacturers at one time: Krupp and Rheinnmetall (Germany), Bofors (Sweden), Skoda and Brno (Czechoslovakia), Fabrique Nationale (Belgium), and Oerlikon (Swiss). Now eclipsed by the Cold War and the victor, the U.S. as the biggest arms exporter in the world, with Russia a distant second.

I liked my gun collection. But I **loved** my submachine gun and light machine gun collection -- all legally deactivated. It was Yogi's "museum of SMG and LMG technology". One of each type that

I could get my hands on, and could afford -- they cost $400 to $800 dollars a pop for just a hunk of steel. But, unfortunately, the belt-fed, tripod-mounted heavy machine guns and Nazi guns were over $1000 - $2000 a piece -- well over my budget.

And then one day I was "playing" with my SMG's, and I realized something profound.

I laid on my bed next to each other, the English 9mm STEN SMG, the U.S. .45 M3 SMG -- commonly called a "Grease-gun" for its comparison to an auto grease-gun. And finally, I laid down the Russian 7.62 x 25 mm Tokarev, drum-fed PPSh-41.

It was a comparison of submachine guns that were used by the Allies during WW2.

The STEN gun jams easily, is cheap, and cheaply made, a thrown together crude design for the war by an economic cripple the U.K. was, It cost something like $13 to make at the time (about $250 in 2012 dollars).

The U.S. "Grease Gun"; was overpowered at .45, and a very solid heavy steel gun -- when lightness is critical in combat -- but solidly built.

The PPSh-41 was light weight, small caliber, high capacity (71 round) drum magazine, made with inexpensive sheet metal pressings. Cheap enough to manufacture and equip all their soldiers, unlike the Allies and Axis forces.

The Russian answer was to make the gun cheap enough, and simple enough, to mass produce and equip all the soldiers with it, instead of just officers like the enemy -- or specialized crews with heavy "light MG's", like the rest of the Allies. It was pure fire-power alone, much less with a unit of soldiers all equipped with them.

> "Liberty came to the freedmen not in mercy, but in wrath, not by choice, but by military necessity, not by the generous action of the people among[st] whom they were to live, and whose goodwill was essential to the success of the measure, but by strangers..."
>
> -- "Autobiography: Life and Times of Frederick Douglass" (1881)
> Frederick Douglass

Prison Diary 68:   "Things Fall Apart"
=============

> "[T]he [Iraqi] roadside bombers had **their** own 'International business model', with weapons tested in Sri Lanka, engineered in the Balkans, retested in Indonesia, and finally aimed at [NATO] troops in Afghanistan.'"
>
> -- "Arms and Innovation" (2008)
> James Hasik

> "Brave new hates, brave new bombs, brave new wars. The beautifully purposeless process of society's suicide."
>
> -- Charly Gordon, on the world's future
> "Charly" (1968)
> Stirling Silliphant

The name IED, Improvised Explosive Device, belies the brilliance and planning that enabled its use for effective guerrilla warfare against (mostly) U.S. forces (2003-2012 and continuing, as of this date) in Iraq, and then Afghanistan in the wars started by George W. Bush, and which we're not sure if he won.

No, we're sure. America lost.

IED's – the Road Side Bomb – consist simply of two or three 6" or 8" diameter artillery shells (or less effective, a mortar shell), and optionally a squat propane cylinder. The shells are wired for remote detonation (by radio -- a modified cell phone -- or by a long wire pair with a switch and a person observing, on the end). Placed above-ground by the side of the road to maximize their effectiveness against a passing vehicle, they are extremely effective against the side of any armored vehicle, even tanks. The HE (high explosive) blast does most of the "work". A shower of steel artillery shrapnel, and a fireball covers all the bases.

The standard U.S. 8" (203 mm diameter) artillery shell for howitzers weighs 200 lbs. (90.7 kg) and contains 25% by weight of HE (TNT) -- two or three shells producing a tremendous and lethal explosion and a shower of sharp fragmented high speed shards of steel.

By way of rough comparison, a much smaller mortar shell (81 mm dia. -- 3.2") killed 79 civilians when one was fired and hit dead-on a crowded Sarajevo open market-place, during the civil war that tore apart Yugoslavia during the early 1990's.

The 81 mm HE fragmentation mortar round has a lethal radius of up to 40 m (135') and a casualty (wounding) radius of up to 190 m (630'). It is much less powerful than an artillery round, and suitable for anti-personnel use only, as opposed to the anti-armor capability of the artillery shell.

It was well known that prior to the invasion of Iraq by G.W. Bush in 2002 that Hussein had littered the cities with small store-houses of arms and ammunition, including artillery shells. The obvious reason was that he realized he had no chance of opposing the U.S. attack in a conventional war, so he prepared for popular guerrilla warfare against the Americans.

And including the artillery shells with the small arms and ammunition in the small, covert storehouses? The IED was the reason why.

The troops that survived IED explosions suffered concussions which produce memory loss, headaches, and confused thinking -- all permanent. The blast wave caused traumatic brain injuries, blindness, deafness, or mental impairment. Some soldiers were unable to stand or even think.

* * * *

"Dynamite! Of all the good stuff, this is the stuff. ....
In giving dynamite to the downtrodden millions
of the world, science has done its best work."

-- "Alarm" (1885)
T. Lizius

The conquest of nature began when war began taking more lives than disease plagues. New weapons permitted greater separation between friend and enemy -- dropping bombs from high above, firing rockets from greater and greater distances. It was the end of chivalry, individual bravery, and honor.

The Marines were just cannon-fodder on D-day, a sunny day in June of 1944, overwhelming German defenses with waves of landing craft full of men, until the Germans crumbled from the continuous onslaught.

So it was the end of dog-fighting and "top gun" passed into mythology as fighters in the 1950's became mere supersonic air-to-air missile platforms, first with their missiles equipped with proximity fuzes, then under radar control, then IR targeting. Each more effective than the last.

* * * *

"I once asked an American scientist in 'defense biology' what
they were looking for. He said, 'A cure for metabolism.' "

-- unknown (~ 2005)

Most people – especially the police – don't know that my original "love" in weapons was not guns. They were just a legal hobby – a side-line for minor amusement (and what is a revolutionary without a gun – an armchair revolutionary...).

My original love was explosives. It all started in Grade 6, when I was ten year old and the Canadian government banned firecrackers, because of a few injuries of unsupervised children.

Well, I was having none of it. How dare they? My favorite holiday weekend – firecracker day – gone. So I figured out how to make gunpowder, by looking it up in "How It Works", and tried to (unsuccessfully) make home-made firecrackers.

I tried tightly rolled up paper. I tried walnut shells glued together. Nada – just a flare. Amusing, but not a firecracker. Then I started using the smaller metal tomato juice cans for my flares. Then I moved up in the gunpowder game, and began using the "incendiary mixture" of potassium chlorate and sugar, which had a more powerful, faster burning, white hot flame.

And, much to my surprise and fear, the "flare" exploded in front of us (me and my school-mates). We later found the empty casing. Its top had blown off, and the bottom lid was stretched out into a concave shape. I had accidentally made my first improvised bomb.

But my experimental days only lasted till Grade 8, and then I realized I was getting into dangerous territory, and got my vicarious pleasure on the topic of explosives only by reading and imagining.

* * * *

"Permanent revolt by word of mouth, in writing,
by the dagger, the rifle, and dynamite."

-- "Le Revolte" (1880)
Peter Kropotkin

I was playing around with a U.S. patent database, one day, when I uncovered a patent that that had a difference of an unusual and amazing thirty years between its filing date and its issue date -- a sure sign of a patent that had been held under a secrecy order that had been rescinded. It had finally been declassified.

It was U.S. Patent 4,673,430 issued in 1987 to Inco -- the International Nickel Company -- a miner in Sudbury, Ontario, and refiner of the metal nickel. The patent concerned using compressed air on molten nickel to create a fine powder of tiny, perfectly spherical balls of pure nickel metal of different sizes – 3-7 μm in diameter.

Bingo!

Though it never mentioned anything about its purpose, I immediately recognized it, since I remembered that weapons grade uranium production used porous nickel barriers for the older uranium hexafluoride enrichment process. It was the raw material for manufacture by sintering of the nickel dust into the nickel barrier for production of weapons grade (93.5% U235) highly enriched uranium (HEU) for nuclear weapons. (A process since replaced by ultracentrifuge enrichment, which was more efficient, and used less power, and thus much cheaper.)

* * * *

Please to remember, The fifth of November,
Gunpowder treason and plot.
We know no reason, Why gunpowder treason,
Should ever be forgot."

-- Children's Nursery Rhyme (ca. 1606)
anonymous

The detonation of a chemical high explosive is characterized by the release of a relatively large amount of energy in a small volume of space, in a very short period of time.

In other words, the definition of a high explosive is a meta-stable chemical compound that is capable of releasing its stored chemical energy in an extremely short time, when its stability is upset by a mechanical -- a blow -- or electrical shock. Or, more typically by the explosion of a small amount of a more sensitive explosive in a blasting cap.

In simple -- and understandable -- terms, high explosives do not "explode", _per se_. High explosives just burn very, very fast, and this leads to all sorts of weapons-usable properties. By definition, a detonating high explosive burns with a supersonic speed. And the supersonic shock wave is just another form of the sonic boom heard by people on the ground after a plane like the Concorde civilian jet, or a military jet passes by.

To give you an idea how fast is "fast" burning, take a typical military (i.e., powerful) HE, such as RDX -- the HE used in the plastic explosive C4, consisting mostly of the HE RDX (also known as cyclonite or octogen). If you were to lay a small string or strip of C4 9 km long -- about 4 miles -- long on the ground and detonate it at one end, it would take about a second to burn to the other end -- the entire 9 km length.

This parameter is known as the detonation velocity, and is simply the (very high) rate of reaction of the explosive. There is also the heat of the reaction, and the amount of gaseous products produced -- the higher the better, and the more the better, respectively. With the heat released, the gas released expands to double its volume for the first 275 °C increase (500 °F) in temperature, triple its volume for 550 °C (1,000 °F), etc.

These are the three parameters that define the effectiveness of the high explosive.

HEs do their work by creating a pressure wave of several hundred pounds per square inch (psi). One atmosphere is about 50 psi. Exposure to this over-pressure will kill half the people exposed to it by pulping their lung tissue, which is obviously a very delicate membrane. Concrete walls need even less of an overpressure to be fractured – 3 psi.

The low pressure for collapse is due to the horizontal force of the blast on a wall, which is designed only to be strong against the vertical forces pressing down on it.

But the rate of decrease of pressure over distance is governed by what's called an "inverse square law", which means, for instance, that if you double the distance from the explosion, the pressure falls by two squared – which equals four. Triple the distance, and you have one-ninth the pressure.

This is why HE is used to throw off shrapnel – since grape-shot cannon shells in the 1800's -- which have a much longer lethal radius than the explosive pressure wave of pure HE.

* * * *

"Is it conceivable that a man's weakness can be so
strong, that such evil can overpower...& exhaust me
to the point that I become evil too?"

-- Journal entry (1948)
Margerie Lowry, wife of
Malcolm Lowry

For centuries the only explosive was gunpowder, a finely powdered mixture of potassium nitrate, charcoal, and sulfur. It made a better propellant for cannons and small rockets, but it could also be used for blasting in mining or bombs if confined in a metal casing.

Guy Fawkes was the first political extremist to use a bomb. In 1605 in London, he tried to blow up the King, and both houses of parliament simultaneously with barrels of gunpowder secreted in the basement of the parliament building.

But his conspiracy, the "Gunpowder Plot", was exposed, and his plans – a religious dispute – came to naught. He was executed, and the "Guy Fawkes Day" is still celebrated with fireworks, and firecrackers in England.

The evolution of high explosives followed a simple path, all (but FAE's) involving nitrogen gas compounds -- nitrates and nitro group chemicals produced from nitric acid by various methods:

gunpowder (black powder; blasting powder) -> gun-cotton (1846) -> nitroglycerin (1847) -> dynamite (stabilized nitroglycerin) and the blasting caps to detonate it (1867) -> first picric acid (1873), then TNT (1902), introduced the era of a plethora of nitrated organic compound HE's -> RDX (1938) -> plastic explosives (1939) -> cluster bombs (1943) -> the HE shaped charge (1944) -> HMX (1960) -> and finally culminating in FAE's (Fuel-Air Explosives – humorously referred by Vietnam War soldiers as the "daisy cutter") (1965).

* * * *

> "There seems to be a curious American tendency to search...for a single external center of evil to which all our troubles can be attributed, rather than to recognize that there might be multiple sources of resistance to our purposes and undertakings, and that these sources might be relatively independent of each other."
>
> -- George F. Kennan (1985)

The relevant **numbers** (just three) concerning some popular modern military explosives are:

	Detonation Velocity (meters/second)	Heat of explosion (kcal/gram)	Volume of Gas Produced (liters of gas/kg explosive)
TNT:	6,950	1.1	1000 --> 4.0×10^6 --> 2.8×10^{10}
RDX:	8,950	1.4	800 --> 4.1×10^6 --> 3.7×10^{10}
PETN:	8,300	1.4	780 --> 3.5×10^6 --> 2.9×10^{10}
HMX:	9,100	1.2	900 --> 4.0×10^6 --> 3.6×10^{10}

1) The detonation velocity determines the strength and pressure of the explosive shock wave of the explosive and the explosive rate of gas production (1 mole of any gas equals 22.4 liters, almost 6 gallons of volume, if you remember your high school chemistry. Or if you remember high school...), which is about 1 liter (1/4 gallon) of gas per ounce of HE.

2) The heat of explosion determines the rapid expansion of the very hot gaseous products of the explosion: every roughly 300 °C of heated gas doubles the volume of gas. If heated to 600 C, the volume is tripled.

3) Combined with the number of moles of gaseous products (22.4 liter/mole gas) produced and the velocity of detonation, this determines the rate of explosive expansion of the gas cloud.

4) Then there's the significant added heat of the "after-burn" of the mostly gaseous reaction products (like carbon and carbon monoxide), but let's keep things simple.

Suffice it to say that these properties result in high explosives being able to create a hell of a lot of mischief. Officially or unofficially...

* * * *

"There was a young chemist in Ealing,
Who with trinitrophenol was dealing.
But, he added red lead,
And, the truth must be said,
They found him a splash on the ceiling."

-- Frank Hawke (ca. 1980)

The discovery of the pale yellow, oily liquid, nitroglycerin, in 1847 by the Italian chemist, Ascanio Sobrero, added a revolutionary new tool to the art of war. It was the first -- soon of many -- nitrated organic chemical high explosives. They produced a powerful explosion even when unconfined -- unlike low explosives such as gunpowder and smokeless powder (aka guncotton, nitrocellulose, or chemically speaking, cellulose trinitrate) which had to be confined in steel pipe or other such casings -- and high explosives were *much* more powerful to boot.

Hold a thin metal bottle cap with a pair of needle nose pliers. Place a small drop of NG in the bottle cap, and detonate it by touching -- keep your face away -- the liquid drop with a hot screwdriver tip that has been flame-heated. The one drop of NG will blow a hole through the bottom of the metal bottle cap. The "decomposition reaction" will impress, no doubt.

But "nitro", as it came to be known, was dangerously unstable: heat, acid, percussion (dropping it, or hitting it) caused it to explode spontaneously and unpredictably. This made its use -- and even manufacture -- too dangerous and unreliable. It took another twenty years for Alfred Nobel -- of Nobel Prize fame -- to find a solution.

In 1867, Nobel made nitroglycerin safe, stable, and thus usable, by mixing it with one-third its weight of the white powder and liquid absorbent, kieselguhr -- "diatomaceous earth" or "Fuller's earth" -- which consisted of the silicon dioxide (sand and quartz are also silicon dioxide), hollow spherical skeletons of tiny, ancient diatoms, deposits of which are found all over the world.

Dynamite was thus, 75% NG, 24.5% kieselguhr, and .5% sodium carbonate, the latter added to neutralize the dangerous destabilizing effects of the acid formed by nitro's easy decomposition by even low heat.

Dynamite was one example of what is called the science of hydrodynamics. The perfect symmetry of the tiny hollow spheres of kieselguhr disperses any applied force evenly with its tiny, even matrix of the hard and hollow, spherical, inert sand, and renders the nitroglycerin stable against being hit or dropped accidentally, and thus safe for commercial and military use.

In 1875 Nobel mixed NG with collodion (nitrated cotton – gun-cotton – dissolved in alcohol and ether) to form the water-proof explosive he named "blasting gelatin" – and also the first new ammunition propellant since gunpowder, it came to be called a "double-base powder", because it combined **two** explosives, one high, and one low.

Blasting gelatin had a detonation velocity of 8000m/s and consisted of 93% NG + 7% collodion cotton (a 9% alcohol/ether solution). It was a stiff jelly – the first plastic explosive. It could be kneaded and shaped and molded depending on its intended use. This made for a more efficient and versatile explosive, for example in cutting a tube neatly – another example explainable by the then unknown science of hydrodynamics, the science of material properties under extremely high pressures.

Blasting gelatin was easily detonated by a low powered No. 1 blasting cap. A military explosive, Nobel made what he named "gelignite" for commercial civilian uses. It consisted of 60% NG, 4%

collodion, 8% ground wood powder, and 28% potassium nitrate and was less powerful than blasting gelatin.

A third invention, in 1888, was "double-base powder", not an explosive, but with a slower detonation velocity, so that it only deflagrated (burnt quickly) instead of exploding. With increasingly large proportions of NC in the NG gel, a tough, horny solid colloids is formed that became the ammunition and artillery propellants known as ballistite or cordite.

The 1880s to 1890's saw the use of dynamite "explode" amongst the Fenians (the precursor to the modern IRA), French anarchists, and Russian nihilists, as radicals realized that science had given them a new and effective tool that could advance and make effective revolutionary war which had stalled at the French and American Revolutions, with the rapid urbanization, factory work and other societal, social, and economic ills brought about by the capitalist Industrial Revolution.

"Long live Dynamite!" became the cry of the newly invigorated radicals, and their tool became the crude bomb, either thrown like a crude grenade, or made with a longer fuse for a short time-delay, for the bomber to attempt to escape before the explosion.

Joseph Conrad's "The Secret Agent" (1907) is based on a failed revolutionary bombing attempt.

First made in 1863, trinitrotoluene (TNT), was not used as an explosive until 1902. With its high stability, ability to be safely melted and cast (at less a temperature that the boiling point of water), its power, and low expense, it became the most used military explosive of all time, standard with armies the world over.

It was perfect technological timing: the simultaneous availability of high explosives to burst a brittle steel shell casing into deadly high velocity fragments, a new higher energy propellant, smokeless powder, and high performance longer range, accurate, steel artillery. Messed together perfectly with the mass production of the factories of the Industrial Revolution.

War had made a giant leap forward, with especially artillery, but also land and floating sea mines, submarine-fired torpedoes, and aerially-dropped bombs.

During WW1 and WW2, the list of nitrated organic explosives added and used after TNT, included picric acid, tetryl, RDX, PETN, and the most powerful of all the common explosives, HMX – or hexogen. HMX is too expensive for use in anything other than implosion atomic bombs or H-bomb primaries.

* * * *

> "Something went very badly wrong, either in somebody's head or in some piece of machinery or in the execution of commands that never should have been given...This technology really depends on people and machinery. I don't know if people really understand that or not. Both can make trouble."
>
> -- Stephen Hanauer (2003)
> AEC official on 1961 SL-1
> Idaho Fall reactor accident

The beauty of militarily useful high explosives, like TNT, is that not only are they very powerful explosives, but that they are ordinarily quite safe and stable. Thus, they are called secondary explosives, and require a more sensitive primary explosive in a blasting cap to set them off.

Examples of primary explosives are: lead azide, lead styphnate, mercury fulminate, lead picrate, and fulminating gold and silver: You may notice that primary explosives are compounds of heavy metals, and it is no accident. The heavy metal atoms when subject to percussion (a blow or hit) or heat, "bang" into the rest of the explosive molecule and set it off with their strong momentum.

* * * *

> "It's a nitroglycerin base. It's a bit more stable. I learned to make it when I was a kid. Make sure there's none on the threads. Like this. [...]"
>
> "You must have had a fun childhood..."
>
> -- Reese to Sarah
> "The Terminator" (1984)
> James Cameron, William Wisher

TNT was great because it was very easy to produce in industrial quantities, very stable, and could be safely melted at low temperatures, and cast to shape, such as a shell. But for demolitions use, it was typically available only as solid, square block.

So finally there was developed "plastique" or plastic explosives. Soft, moldable explosive mixtures for military use, when it was found that pressing and forming explosives greatly increased their efficiency of effect, as well as the utility of them for uses such as cutting down trees, or neatly severing heavy steel plates.

You plasticize TNT by adding a small quantity of beeswax or paraffin candle wax to it, and kneading the mixture at an elevated temperature until the two components are thoroughly mixed.

WW2 added RDX – aka cyclonite -- to the inventory, as the best military explosive at its price. C4 is the plastic explosive version of RDX. Home-brew C4 consists of 8% silicone grease lubricant and 92% by weight RDX, mixed thoroughly.

A more professional method of making home-brew C4 is as follows. Carefully mix 500 ml. of diesel fuel with half a kg. of melted paraffin to get a clear liquid with a temperature just above the melting point of the wax. Then slowly stir in 6 kg of finely powdered (ground) RDX, to form a sticky, dough-like mixture suitable as plastic explosive. Wrap in newsprint for storage or use immediately.

The ComBlock version of C4 was named Semtex, and had a detonation velocity of 7600 m/s. It consisted of the cheaper explosive PETN with styrene-butadiene copolymer as the plasticizer. PETN is slightly less powerful than RDX/C4.

It was formerly of Czech manufacture. Semtex became known to the general public, when Libya's leftist dictator, Colonel Gaddafi donated **three tons** of Semtex to the Provisional IRA in 1986, which they used extensively in a bombing campaign in their armed struggle against England.

* * * *

> "You daren't handle high explosives; but you're all ready to handle honesty and truth and justice and the whole duty of man, and kill one another

> at that game. What a country! What a world!"
>
> -- "Major Barbara" (1907)
> G.B. Shaw

As tanks got more and thicker armor to defend against anti-tank weapons, anti-tank weapons evolved upwards too. The first anti-tank weapons were AP (armor-piercing) rounds fired from rifle.

Germany adopted in 1938 the Swiss-made, Solothurn S18-1000, quick-change barrel, recoil-operated anti-tank rifle. Solothurn was owned by the German arms firm Rheinmetall to get around the restrictions of the Versailles Treaty that ended WWI. The S18/1100 variant had a barrel almost **five** feet long (1.5m), and muzzle velocity of 850 m/s (2800 ft./s – almost Mach 3) for its 20mm (.8") cannon round (full "name": 20x138B mm). It was obsolete before it hit the field, with a penetration of only an inch of armor plate at 500 yards.

The Boys Anti-tank rifle, designed in the U.K. in 1936, had a five-round box magazine, a 36" long barrel, and fired a .55" diameter bullet -- a cannon round -- made of tungsten-steel.

The 1939 Finnish Lahti L-39, semi-automatic, gas-operated anti-tank rifle, with a 1.3m (51.2") barrel. its 20mm (German 20x138Bmm, or Swiss 20x138mm Solothurn Long) ammo a large rimless (B for belted) cartridge with an almost 6 oz. armor-piercing hardened steel bullet 800m/s (2500 ft./s).

Successful during the Finnish-Soviet "Winter War", it was rendered obsolete for anti-tank use by the later Soviet upgrade to the more heavily armored T-34 tank in the run-up to WW2.

1941 Russian PTRD Anti-tank rifle, with a 48 1/4" (122.7 cm) long barrel for higher velocity, with a projectile diameter of 14.5mm (.6").

2.8 cm sPzB 41. This German artillery piece introduced in 1941 was novel in several ways. It introduced the Gerlich "squeeze bore" design, firing a 2.8 cm round consisting of a tungsten core surrounded by a malleable metal shell, which could be squeezed down to 2 cm diameter in the muzzle to give it a higher muzzle velocity of 1402 m/s (4600 ft./s), without increasing the weight of the barrel, by using higher quality alloy steel.

It introduced dense tungsten alloy rounds for increased penetration by concentrating the energy of the projectile's weight and velocity on a smaller spot on the target -- the kinetic energy projectile. Tungsten is much harder than steel, and is 2.4 times as dense as steel, so you can pack a much heavier bullet without having to make the barrel caliber larger. And the hardness and higher density gives the projectile a much higher force on the area of armor it hits, allowing it to punch through more easily.

A variant of the gun with a light tubular frame-work of half its total weight was made for use by blitzkrieg paratroopers.

A later variant of the squeeze bore tungsten shell artillery piece was the 75mm Pak 41, which could penetrate 17cm (6.73") of armor. But machining tools had a higher priority for tungsten, effectively ending its career when the ammo ran out.

The British used a different technique for their tungsten or tungsten carbide artillery designs. Rather than the squeeze bore, they achieved higher velocity by using a discarding sabot system with a tungsten core.

The German "88" -- 88 mm -- wheeled gun was much feared by Allied tank crews. It fired a tungsten squeeze-bore high velocity round that could punch through their armor easily: it could penetrate more than 3" (7.5 cm) of armor plate at 1000 yds. (1000 m).

But the best of all was the shaped or hollow HE charge, which exploded to produce a forward projecting jet of high velocity metal, which bored through armor, or caused its failure by generating a large shock wave that shattered the metal.

The U.S. M1A1 "Bazooka", introduced in 1942, was recoil-less, 54" (1.37 m) long, firing a shaped charge of 3.5 lb. (1.54 kg) HE, 60 mm (2.4") in diameter, integral solid fuel propellant fired projectile.

The 8.8 cm Raketenpanzerbuchse 43 ("RP43") was copied from a 2.34" U.S. bazooka captured in 1943, but the shell enlarged to 3.5". Its shaped charge rocket warhead weight was .65 kg (1 lb. 7 oz.), and it could penetrate 8.25" of armor.

* * * *

> "Everyone who is aware of the possibility of mankind's
> self-destruction must resist that possibility to the utmost
> .... The resistance against the self-destructive consequences
> of man's technical control of nature must come through
> acts which unite religious, moral, and political concerns,
> and which are performed in imaginative wisdom and
> courage."
>
> -- "Pulpit Digest" (1954)
> Paul Tillich

The shaped charge is based on the "Monroe effect" which showed that the explosive force of HE could be directed instead of just going outwards in equally in all directions with equal force. This discovery eventually led to the "shaped charge" warhead by WW2. In 1944 as the Nazis crumbling under Allied invasion, developed for civilian use a hand held, one-shot, recoil-less _Panzerfaust_ ("armor fist") weapon equipped with a shaped charge warhead for anti-tank use, being able to punch a hole through a maximum of – incredibly – eight to eleven inches (20-28 cm) of steel tank armor – far superior in penetration-power than the U.S. bazooka and other Allied anti-tank weapons. Its warhead was propelled by three ounces of gunpowder.

The shaped -- hollow -- charge had a 6" diameter, a 60 degree cone-shaped copper cavity liner, surrounded by an HE charge of 3 lb. 7 oz. of Composition B (a mixture of RDX and TNT).

* * * *

> "If the road to hell is paved with good intentions,
> it is partly because that is the road they generally
> start out on."
>
> -- "Altruism and altruistic love" (2002)
> Stephen Garrand Post

In the 1960's the British developed "Chobham composite armor" for tanks, the exact composition of which is secret. But the modernized variant appears to be a Kevlar fiber micro-mesh supporting, and separating a mesh of very hard silicon carbide ceramic tiles that are small in area, but one inch thick, surrounded by a titanium front and back plate. The silicon carbide is probably a matrix of air and hollow circular balls of silicon carbide, using the impedance difference to absorb the concentrated energy of the kinetic energy of the projectile. It is effective against both KE projectile and shaped charge anti-tank rounds.

From 1962 to present the Russians introduced the cheap, mass-produced, and now ubiquitous, shoulder-fired 40 mm caliber projectile, shaped charge HE anti-tank warhead, the RPG-7 (Rocket Propelled Grenade-7), having only a 95 cm (37 1/4 ") barrel length -- making it small and portable, and a range of up to half a kilometer, but only really accurate up to 100 yards, except with practice.

An RPG-7 took down a million dollar Blackhawk attack helicopter in Somalia in the 1980's by repeated firings of dozens upon dozens of rounds, until one hit it at the right angle so that it did not to bounce off (an incident made famous by the book and Ridley Scott movie, "Blackhawk Down").

* * * *

> "It is a scandal in contemporary international law, don't forget, that while 'wanton destruction of town, cities, and villages' is a war crime of long standing, the bombing of cities from airplanes goes not only unpunished but virtually unaccused. Air bombardment is state terrorism, the terrorism of the rich. ... Something has benumbed our consciousness against this reality. ... '[W]anton destruction' is *just* the term for it.
>
> -- in "The Nation" (1994)
> C. Douglas Lummis

But best of all was the buried anti-tank mine. Its ultimate variant was patented in 1954, and classified for almost thirty years, it was merely a flat slab of HE, covered by a .5 or 1" thick, circular steel "flyer" plate, which was hurled up by the explosion at high velocity of 1-2 km/s through the thin armor on the bottom of the tank, punching a big hole in it upon impact.

Finally there was FAE's -- Fuel-Air Explosives -- which were developed in 1969. It was known for decades that grain silos, flour mills, and strangely enough, even raw cement powder storage silos could explode spontaneously with hugely destructive violence.

It was eventually found that extremely fine dust, stirred up into a cloud, and triggered by a random spark of static electricity would actually ignite and explode with tremendous force if the volume was large enough.

It made a cool scientific demonstration to wow public auditorium crowds using a sheet metal paint or aluminum milk can, with the sealing cover loosely placed on top.

As a fuel, either a teaspoon of flour -- or even better -- lycopodium powder (the spores of a mushroom-related fungus) was used, blown by the demonstrator from a funnel attached to a rubber tube running through a hole punched in the side of the can.

FAX's are merely the weaponized, highly optimized version of the same phenomenon, when someone thought it would make a neat weapon of huge destructive proportions, that was dirt cheap to boot.

A peacenik book on weapons had disclosed to me that the current U.S. military FAE was typically a bomblet of 70 lbs. of ethylene oxide. But why ethylene oxide?

The answer I found in the "CRC Handbook", while bored one day. The "CRC Handbook" -- one of the two bibles of the chemist, the other being the "Merck Index" -- had a neat little table of numbers -- one of many -- that gave the explosive limits of a list of chemicals.

The answer soon became obvious. An obscure chemical -- conveniently liquid when kept under pressure or below 11 °C. -- called ethylene oxide had the second widest range of all -- 3 - 80% (by volume) -- so that when dispersed it immediately formed a vapor cloud in air which was then ignited (with a 125 ms to 4 sec. ignition delay, depending on the conditions), to produce an explosion with just about any concentration of ethylene oxide vapor.

The wide range made it easily explosive under such varying concentrations of ethylene oxide vapor, when dispersed by a small explosive charge set off within the bomblet, that broke apart the bomblet outer casing, and scattered the liquid, which immediately vaporized and scattered into a vapor cloud that mixed with the surrounding air.

And the ethylene oxide -- cheap and widely available industrially -- in vapor-air mixture was easy to ignite. As a double bonus it burned into methane and carbon monoxide, both of which themselves burned into carbon dioxide and water, and carbon dioxide respectively, producing five times the volume of the original ethylene oxide vapor in the time it takes to burn. Including the heat of explosion (7.1 kcal/g) the total expansion of the gas is over 5,000 times the original volume, producing a massive shock wave and explosion from the rapid expansion of the gas, and then a "sucking" vacuum effect when it contracted upon immediately cooling, following the explosion.

Used in the Mayaguez crew rescue in the 1975, a U.S. Marine said of the Cambodian soldiers killed, "It sucked their lungs right through their noses. It was ugly."

One version the CBU-82 (Cluster Bomb Unit), containing 100 lbs. of liquid fuel, produces a cloud 56 feet in diameter and nine feet high, which is exploded, causing damage many times greater than a 100 pound charge of dynamite would produce.

* * * *

"It isn't that they can't see the solution.
It is that they can't see the problem."

-- "The Scandal of Father Brown" (1935)
G.K. Chesterton

(Acetylene gas also has a wide limit of inflammability -- 2.5 - 80% -- and explosive limits of 3.0 - 65%, but as in welding tanks, since the gas explodes when compressed, to store it safely for use it must be dissolved in acetone. Under 12 atmospheres pressure -- about 175 psi --1 volume of acetone dissolves 300 volumes of acetylene.)

(The FAE would be even more powerful if mixed with liquid oxygen, to "super-charge" the explosive effect over just the 20% oxygen content of air. In fact, a mixture of compressed hydrogen gas and pure oxygen would have explosive limits of between 4.7% and 94%, as well as having the highest energy release of all)

The delay timer ignition source (or sources!) ignites the inflammable cloud of fuel vapor\air, the flame spreading rapidly and coalescing into a giant exploding fireball and explosion. And anything within or near the fireball is incinerated.

But a blast wave -- an overpressure -- extends beyond the fireball, and kills humans via the overpressure of the wave -- about 25 atm. (360 psi or 2.5 MPa) over a large area unlike typical solid HE's -- more than enough to pulp and destroy the delicate oxygen-transfer membrane that line a person's lungs, allowing him to breath. The person asphyxiates, foaming at the mouth a mash of pieces of lung tissue. That is the **real** reason explosives kill, other than shrapnel, like the infamous improvised "nail bomb".

FAE's were full circle to gunpowder -- a simple burning of a mixture of fuel and an oxidizer. But instead of a mixture of finely ground powders, being a mixture of two gases, capable of mixing at the molecular level for almost perfect efficiency and tremendous speed of burning.

A crude FAE was used by an anti-abortionist to blow up a doctor's house. A hole was drilled in the wooden frame of a window, and propane gas was pumped into the house through a tube connected to a propane BBQ cylinder. (Though propane is a technically a poor FAE, being an explosive only with 2-10% propane to 90-98% air, the propane being heavier than air, it filled the basement on up, and with air circulation on the first floor, worked.)

After the house was partially filled with the resulting air-propane mixture, a random spark in the house – the fridge compressor motor probably -- detonated the FAE, demolishing the house.

One thing you can say about home-brew, right-wing American extremists is that their ingenuity in such matters is stellar, and much superior to those on the left. However, this is offset by the superiority in simplicity of leftist government ordnance: the AK-47, Semtex, the RPG-7, the helicopter gunship, for instance.

* * * *

> "American legal processes are hardly discretionary. Under them there's no provision for moral anguish, moral conviction, moral resistance. ... Law has become its own end. So why should I submit to those processes? [It] seems to me that 'consequences' are suffered today by any man who tries simply to become a man."
>
> -- "America is Hard to Find" (1972)
> Daniel Berrigan

The evolution of civilian extremist home-brew explosives followed a predictable evolution. I discovered it myself, going through Urbanski, one of the explosive bibles. I was looking for HE's that were made from simple ingredients or from ordinary commercial (density = 1.4 g/cm3, 70% HNO3) nitric acid, rather than the much harder to buy (d=1.5 g/cm3 100% HNO3) white or red fuming nitric acid.

There was methyl nitrate – a liquid -- urea nitrate, nitrourea, guanidine nitrate, nitroguanidine, and mercury fulminate. And ones (very dangerous to use) that didn't even require nitric acid: nitrogen trichloride, nitrogen tri-iodide, fulminating silver.

I only found out later about two others: HMTD and acetone tri-peroxide.

Of course, I knew about ANFO, long before it was used. I also realized that instead of ammonium nitrate, you could even substitute potassium nitrate – saltpeter.

And it was actually fairly easy to make 100% fuming nitric acid with potassium or ammonium nitrate and sulfuric acid that are required for the really cool HE's – PETN, TNT, or RDX (cyclonite).

[Later, methyl nitrate, urea nitrate were used, in order, for hijackings, and "terrorist" bombings, which I could have predicted, requiring only the ordinary, commonly, available, commercial nitric acid. It was an escalating series. I predicted – to myself – that the only thing left was crashing the commercial airliners, but the 9/11 World Trade Center attack was beyond my predictive powers, alas.

It was brilliant and unique and full of meaning. Not a bombing, or crash. But the use of Western Technology against economic targets in a simple non-technical way: simply paying to train a few suicide pilots at a commercial aviation school!]

* * * *

"I will never forget how it felt to hold a loaded gun for the first time and lift it and fire it, the scare of its animate kick up the bone of your arm, you are empowered there is no question about it, it is an investiture, like knighthood..."

-- Billy Bathgate
"Billy Bathgate"
E.L. Doctorow

The first crude IRA attempt at IED manufacture became known as the "sugar stick": a potassium chlorate/ icing sugar mixture, confined in an iron pipe – the old-fashioned pipe bomb. An anti-personnel device of limited destructive power.

This evolved to "Coop Mix", another binary explosive, this time a high explosive made from a mixture of potassium chlorate and nitrobenzene, and invented in 1889. Then there was gelignite, a commercial dynamite stolen from quarries and construction site magazines. Then ANFO for truck bombs in the 1990's. The Irish Revolutionaries were a little slow, which is perhaps why it's the longest revolutionary fight in history: five centuries.

In the 1970s the Provisional IRA in Northern Ireland relied on sniper attacks on security forces. They changed to bombs by 1980 due to the development by the British army of anti-sniper proficiency.

A VCR timer was used for the Brighton bombing of the Conservative Party Conference in Grand Hotel, Brighton in 1984. unsuccessfully targeting Margaret Thatcher and her entire cabinet.

Finally there was Semtex, a Czech-made PETN-based plastic explosive, three tons of which were provided courtesy of Libyan dictator Gaddafi in 1986, who was sympathetic to the IRA liberation cause, like many decent people.

There was the 1987 "horse" bombing incident at Enniskillen: the dead horses were of more outrage to the public than the eleven people – mostly mounted soldiers – killed...

Then there were the home-made mortars for a new IRA campaign, begun in 1990, with an attack on 10 Downing Street, the Prime Minister's residence.

And finally the ANFO fertilizer truck bombs in England in the London stock exchange. and financial district (1990), at the Baltic Exchange in London (1992 – 800 million pounds damage), at the London Canary Wharf Hotel/Complex (1996 – 85 million pounds damage), a 1996 van bombing in the center of Manchester (causing 100 million pounds damage), which finally ended in 1997 the 500 year guerrilla war.

The lesson (courtesy of Wiley Sutton, American bank robber): go where the money is...

* * * *

> "Why? McVeigh told us at eloquent length. But our
> rulers and their media preferred him as a sadistic,
> crazed monster...who had done it for the kicks."
>
> -- "Perpetual War with
>   Perpetual Peace" (2002)
>   Gore Vidal

One of my axioms is that "terrorism" is defined as a private (civilian), as opposed to governmental, act of minor, and usually justified – or at least explicable – political violence.

One of the axioms of realpolitik is that, "The enemy of my enemy is my friend."

Gulf War veteran, and American white supremacist, disaffected right wing extremist Timothy McVeigh in 1995 used a unique binary – two-part -- HE mixture for expressing his deep-seated alienation by exploding a truck bomb at the Federal Building in Oklahoma City, Oklahoma.

I don't know what all the fuss was about. It was a legitimate military target, it was an impressive success, and technically-speaking, without the moral implications, it was a brilliant designed improvised device, especially considering it was done by only one person, working essentially alone.

And frankly, I've never heard **anyone** complain that there were too few Oklahomans in the world...

His bomb was a supped-up variant of the well-known, but relatively low velocity explosive ANFO – ammonium nitrate plus fuel oil. While ANFO is cheap, and simple and easy to make, and ammonium nitrate is sold in 50 lb. bags as fertilizer, and easily obtainable, his variant was just a little more expensive, and required a little more knowledge on explosives, showing that McVeigh had done his homework, and was serious about his "project".

It was classic American ingenuity. Instead of ANFO, as universally reported at the time – and still is – he actually used "ANNM" – ammonium nitrate with nitromethane substituted for the normal diesel oil. Nitromethane is a model airplane and drag racing fuel. It's the stuff called "Nitro" in drag racing competitions to increase the acceleration of their specially modified racing engines.

He used 5,000 lbs. (2,300 kg) of ANNM, stored in closely packed, and stacked 55 gallon drums, each filled with the high-test, high velocity, powerful modified HE, which blew away half the large office building, and killed 168 people in the largest act of domestic "terrorism" in history, until eclipsed by bin Laden and the 9/11 World Trade Center attacks in 2001.

Of the 168 people killed, he whacked eight federal law enforcement agents, which allowed the government to seek and get the death penalty at McVeigh's trial – he was caught shortly after the blast, trying rather amateurishly to escape the scene.

He was tried, convicted, and sentenced to death. He was executed in 2001 by lethal injection, brave and unrepentant to the end, like the True Believer he apparently was.

McVeigh's stated justification was that the bombing was revenge for the federal government's botched, probably unjustified, and illegal assault on a religious cult's compound in Waco, Texas.

The cult was run by a charismatic, apocalyptic "Christian" religious kook named David Koresh, but the BATF (U.S. Federal Bureau of Alcohol, Tobacco, and Firearms) raid was outrageous overkill, and executed with gross incompetence. After bungling the initial police assault, taking a number of casualties, a second assault with a tank started a fire that burned the compound's house to the ground and killed everyone inside, including Koresh, and his followers, which included many innocent women and children.

Yet the government propaganda machine was sure to mention the few children killed in a daycare center for government employee children in the McVeigh revenge attack on the Oklahoma City Federal Building – a legitimate military target by any standard...

McVeigh stated shortly before his execution that the U.S. Government was the ultimate bully.

He made no final statement at his execution. He just said something gruffly like, "Get on with it."

* * * *

"Comrades, I am Father Purge,
Pharmacist to humanity.
My pills are sure to give the urge,
Prescribed by Doctor Equality.

My shelves are stocked with fulminate,
Saltpeter too you'll surely find,
Gunpowder, nitro, and picrate,
Enough to clean the poisoned mind."

-- popular French anarchist
drinking song (early 1900's)

ANNM consist of 60% AN and 40% NM by mass, producing a wet slurry. Optionally, to prevent separation of the ingredients of the slurry, it can contain a gelling agent of cellulose acetate (acetate rayon or cellophane), ethyl cellulose, or other chemicals. (The original patent for ANNM is U.S. Patent #3,419,444, by the way.)

One of the most powerful variants of ANNM (according to U.S. Patent 4,093,478, by Dr. Gerald Hurst – who I know/was a Net friend, who I "met" in the newsgroups sci.chem, rec.pyrotechnics, and alt.eng.explosives) consists of 1 kg of AN, 185 g of NM, an 84 g of methyl alcohol.

Optionally, ANNM can contain for increased explosiveness 3-10% metal (aluminum, magnesium, zinc, or silicon) powder/dust.

Or as a complete substitute, ANNM can be replaced by a mixture of sodium perchlorate and aluminum powder. Available commercially as "Binex", the sodium perchlorate is in solution, probably dissolved in methanol.

A book, "The Turner Diaries", was widely blamed by the government and media as McVeigh's inspiration and motivation for his blowing up the government building, even though it was an obvious and logical target, and an obvious and legitimate military target, consistent with McVeigh's extremist anti-government philosophy/hatred.

In "The Turner Diaries", a 1978 fiction book of a right-wing extremist guerrilla soldier fighting in the future against the U.S. Government, was written by Andrew McDonald, pseudonym for extremist, right-wing white supremacist William Luther Pierce.

In the book, amongst other things, Turner blows up the FBI building in Washington, D.C. with a truck bomb filled with ANFO. Yet McVeigh traveled widely, driving all over the Mid-West, at least. His beef was with the BATF, not the FBI (so much for inspiration). Truck bombs with ANFO were a well-known, cheap and easy guerrilla technique, and guaranteed wide publicity and would be a grand coup, and historic symbol. And "Diaries" does not give sufficient details to actually make an ANFO truck bomb – which is trivially easy to construct.

And most relevantly, McVeigh, using nitromethane instead of fuel oil, used the much more sophisticated and tremendously more powerful ANNM, not ANFO like in "Turner Diaries", and sold books and videos at traveling commercial gun shows on home-made explosives that described the mixture – such as "Homemade C4 – A Closer Look" (1991), by Ragnar Benson, put out by right-wing publisher Paladin Press (as well as hawking "The Turner Diaries"). McVeigh was no explosives expert and his use of ANNM was definitely as a result of his viewing and reading of the semi-underground explosives book and/or video.

And his alleged use of some sort of shaped charge for the barrels of ANNM to cause the building's massive destruction is untrue and impossible. High office buildings are built for vertical strength to hold the building up. They are weakly constructed for horizontal, sideways forces, which the builders don't expect. The ground-level sideways force of the explosive blast was ideal for taking out window, the floors, and even the reinforced concrete pillars, causing the collapse of the front of the building.

The thing that might have caused the promotion of the "Turner Diaries" as the evil inspiration – and thus "the cause" -- for McVeigh's act was its anti-Jewish writing, and the fact that at the end of the book, Turner gains control and launches U.S. nuclear missiles at Israel, exterminating them and making for a second Holocaust.

In effect – and to Pierce's probable delight – Israel essentially becomes a giant crematorium. An oven for incinerating the Jewish race, akin to Hitler's death camps. A horribly nice touch...

With many Jews in position of power and wealth in the U.S. and U.S. Government, they no doubt have noticed -- and been enraged by -- this too, and thus promoted the rather obscure "Turner Diaries" as the McVeigh bugaboo to reap political points with, and create a rallying point against the "virulence" of anti-Semitism, in general. Violent propaganda strikes, by any other name is...

This reason, rather than the more mundane explanation: exotic explosives books that would be too esoteric, and not be as understandable as anything to get excited about, or that would not create a stir, or controversy in the press or public.

"The historian may condemn, condone, admire:
[but]it is essential that he should understand."

-- "The Roman Art of War
Under the Republic" (1939)
F.E. Adcock

Prison Diary 69:   The Zen of Napalm
================

> "You have the glittering beauty of gold and silver,
> and the still higher luster of jewels, like the ruby
> and diamond; but none of these rival the brilliance
> and beauty of flame."
>
> -- "The Chemical History
>   of the Candle (1860)
>   Michael Faraday

> "You smell that? Do you smell that? Napalm, son.
> Nothing else in the world smells like that.
>
> I love the smell of napalm in the morning.
>
> You know, one time we had a hill bombed...You know
> **that** smell -- that gasoline smell. The whole hill! It
> smelled like...**victory**...
>
> Some day this war's gonna end..."
>
> -- Col. Kilgore
>   "Apocalypse Now" (1979)
>   John Milius

The _lamed vovnik_: "lamed vov" is Yiddish for 36. The _lamed vovnik_ comprise an ancient Hasidic legend. A fable of Jewish mysticism.

The thirty-six are the only just, righteous men whose existence in the world is the only thing that stays the hand of God from sweeping away and destroying humanity totally for its unending wickedness.

No one knows the identity of the lamed vovnik, including themselves, but upon their compassion, empathy, and feeling for the human suffering they see around them, rests the fate of the world. They are the hidden, secret pillars of humanity, and for them the spectacle of the world is often that of an unspeakable hell, as they absorb the unending hurt, and suffering of those in pain.

* * * *

> "A wall of flame, a mile high, five miles wide, traveling
> ninety to a hundred miles an hour, hotter than a
> crematorium, turning sand into glass."
>
> -- Peshtigo, Wisconsin (1871)
>   description of natural firestorm

The human heart. The human mind. They each are vital to the human condition.

But weapons technology produces a dichotomy -- a separation between emotion and logic. It is to be found in the dark, hidden, arcane world of weapons design, the successful application of scientific principles to available technology for the purpose of war, and ultimately the organized, efficient, mass killing of human beings.

And that required time and deep concentration -- devoid of any emotion -- of the mind to the problem at hand, the analysis, the mix-and-match of old ideas with the new, and the conceptualization process, especially.

To allow emotion to enter the mind is to poison the logical, dedicated process of focused thought. To bar any emotive thought is to not allow the imagined screams of the dead to impinge or interfere with the imagined brand new shiny weaponry.

Victory *and* death were the goals, but that was for later. The design process came first.

And *my* frequent goal, for curiosity's sake as much as anything, at first, and as I gained experience and knowledge through sheer fascination -- was to delve deeper and deeper into understanding these thought processes of weapons technologists -- mostly dead or retired -- and in effect, reverse engineer the weapon. To achieve not just a rote recitation of the written word, but a true, deep understanding. To be a knowledgeable, but solitary "peacenik". A peace activist.

And all by my lonesome, I did well. Because books and reports were really just superficial and inadequate, limited by the author's inexperience, indirect (secondary source) experience, or lack of time, of understanding, or teaching skills.

The real answer, of course, was that I had to understand the Zen of Napalm.

> "Truth lies in no place, but in the course of the pursuit of Truth itself."
>
> -- "Hidden by the Leaves" (1716)
> Yamamoto Tsunetomo

* * * *

Chemical warfare – chlorine gas – came first, in WWI, and was affirmatively described by the inventing scientists as "a higher form of killing". It provoked hysterical opposition from both military and, of course, civilians.

However napalm, the first burning fire chemical weapon – introduced in WW2 by the Americans, produced no such distress until the Vietnam War twenty years later. Apparently "premature cremation" was more acceptable, though not to those burnt alive. And they were genocidal fascist, war-mongering Nazis, anyway. They got what they deserved... Though Third World peasants fighting a war of liberation against a superbly armed U.S. foe were another matter...

Burning is the heat-producing process of the chemical reaction between a fuel and oxygen, or a compound of oxygen. But there is more to it than that. There is the "rate of burning" that makes all the difference.

The rusting of steel is the very slow burning of iron. So slow that it is not even noticeable in our time frame.

Then there is "thermite" -- a mixture of a metal and a metal oxide – typically aluminum powder and ferric oxide powder – rust. Aluminum thermite burns so slowly that it doesn't explode, yet it burns relatively fast – though slower than gunpowder – with a temperature on the order of 3,000 °C. Enough to nicely melt steel.

Then there are high explosives, which burn at a rate of up to 9 km/s, producing effects of great interest to the military in artillery shells, or bombs dropped by airplanes.

And once the military got around to it, the slower burning, high energy compounds like thermite resulted in a new class of military weapon, a mix of regular HE bombs, mixed with thousands of small air-droppable "incendiary" bomblets. And then there was a new inflection of modern war. In WW2 the "firestorm" was perfected. It incinerated entire cities and their civilian populace.

Incendiaries had come to the toolkit of Total War. Fire was an old weapon – you'll remember the Ancient's had "Greek fire" -- made new again.

> Fire, fire, burning bright,
> Twinkle, twinkle all alight.
> Cities burning in the night
> Watching -- horror -- at the sight.

Earth to earth, ashes to ashes, dust to...

* * * *

> "Let me ask you one question, is your money that good?
> Will it buy you forgiveness? Do you think that it could?
> Oh, I think you will find, when your death takes its toll,
> All the money you've made, will never buy back your soul."
>
> -- "Masters of War" (1963)
> Bob Dylan

In World War 2, massive bombing raids on German cities were begun in 1942, as the war began to turn in the Allies favor. The thousands of high explosive bombs dropped on the German city cores left the city's buildings and houses smashed, with bricks and wood scattered about the streets. Winston Churchill's desire to illegally use – in violation of Treaty – chemical weapons had been thwarted, but his desire to attack German civilians was implemented with vigor.

The night-time "thousand bomber raid" had been invented to try and overwhelm rapid fire anti-aircraft Flak guns. But the Allies were still not breaking civilian morale or industrial war production with the invention of carpet-bombing and massive urban civilian attacks.

So a new tack was tried. Three kg incendiary bomblets -- magnesium-cased thermite powder – dropped by the thousands along with the 600 lbs. HE bombs ignited the wooden rubble scattered by the HE. Thousands of little fires sprung up, grew, and spread, finally combining into one, and making the city a giant bonfire. Fire stations and fire hydrants -- what little they could do – were put out of action by the massive bombings.

Thus the firestorm -- as it was called -- was discovered. Like a fireplace sucking air from the living room of a house, and the wood's flames sending the smoke and burnt gases upwards into the chimney, the fire enveloped the city, and sucked the air at ground level from all around the flaming city, creating ground-level winds with speeds of up to 150 mph (240 kmph), sucking civilians like fallen leaves into the towering flames of the man-made maelstrom of hell.

As well, the tremendous concentrated heat of the firestorm -- around 800 C. (1,500 F) -- melted the asphalt streets into a sticky tar that trapped street-crossers, and baked to death the occupants of underground concrete air-raid shelters, who had previously been protected except from a direct bomb hit.

* * * *

"A science is said to be useful if its development tends to accentuate the existing inequalities in the distribution of wealth, or more directly promotes the destruction of human life."

-- "A Mathematician's Apology" (1940)
Prof. Godfrey H. Hardy

The "U.S. Strategic Bombing Survey" was part of the start of the detailed application of scientific analysis to warfare in a timely fashion.

German AA Flak was a storm of steel, too effective and requiring higher and higher bomber altitudes. The latest of the famed "88", the 3.5" (8.8 cm) Flak 41 had a 21lb. HE shell, and with a 258" barrel length, had an effective vertical range of over 35,000', and a ceiling of 48,000'.

There were 10,000 Flak units in service by the summer of '44 which had fired more than 3.4 million AA rounds, taking an incredible toll on Allied bombers. And the "88" introduced the U.S. Army to the metric system, which they've used ever since, I think.

Bombs sights left a lot to be desired – despite propaganda about the accuracy of the Norden bomb-sight. Night bombing was essential for protection, but made bombing more inaccurate. Fearful aviators – alarmed by their high rate of attrition and their suicidal dispensability -- dumped their bombs short of their target. And black-outs of city lighting made night bombing difficult and impotent.

The analyses was a revision of bombing methods: the Allies could bomb at high altitude if they used carpet bombing or even firebombing to create firestorms, when this "natural" phenomenon was first discovered from the bombing of Hamburg.

The firestorm may have been first noticed in the bombing of Hamburg, but it was perfected on Dresden. And when the war in Europe ended in May of 1945, it worked even better in Japan, which had mostly wooden homes and buildings. There was Yokohama and Tokyo, and then there was even a firestorm after the atomic bombing of Hiroshima in August 1945.

The firestorm of Tokyo killed more people than the nuking of Hiroshima, making the decision to use atomic bombs easier to understand within the cold calculus of war. One bomb and one bomber, as opposed to tens of thousands of bombs, and a thousand bombers, and ten thousand crew members.

* * * *

"Holy fires blazed, fed by sinful fires,
sacred and profane flames were merged."

-- "Chronicles of the Roman Calendar" (8 A.D.)
Ovid

The aerial bombing of civilians in cities as a technique of war was first introduced crudely by the Germans in WW I -- randomly dropping huge (210 lb./ 94 kg) long-range (81 miles/ 130 km) artillery shells on Paris (fired by the "Paris Gun", of course), and bombs dropped on England by Zeppelin air balloons.

But the bombing of cities reached perfection in WW2, this time by U.K. and U.S. airman, with the development of the strategy to produce lesser losses of planes, of huge propeller-driven Lancaster "heavy" bombers massed in large squadrons. Thus was born the "thousand bomber raid".

Navigation by bombers to the cities of the enemy, was one use for the newly invented radar. For massive destruction, and reduced plane losses from "Flak" – German anti-aircraft rounds – the thousand bomber raids overwhelmed enemy anti-aircraft fire – losses reduced further by high altitude "carpet bombing" of German enemy cities, beginning with Cologne in May 1942.

The "strategic" bombing of enemy home country cities, leaving utter destruction -- the city was razed to the ground -- was the ultimate goal to win the war. A direct and **deliberate** attack on civilians to break morale vigorously promoted by the Winston Churchill, who cut his teeth in killing in his overseeing the continuation – and active promotion – of brutal and racist British imperialism. And, surprisingly military factories were hardly damaged, it was only the civilian population that fried.

> [Churchill also actively promoted chemical warfare – banned by international treaty – but was fortunately, and with difficulty, over-ruled. The Nazis had invented nerve gas unknown to the Allies, and had tons of it. It had a massively increased lethality over Allied CW technology, and couldn't be smelt, but Hitler's hand was stayed because he thought the Allies might have invented it too. All Churchill was, was a great and pompous speech-writer. ...Big deal.]

In July 1943, the first "firestorm" was caused by three consecutive nights of bombing of Hamburg. The new phenomenon killed as many as 50,000 civilians.

In 1944, the firestorm was tested and perfected in Dresden, Germany by the U.S. with British and American planes. After two waves of bomber attacks dropping high explosive bombs to destroy and scatter the remnants of buildings, and destroy and hamper fire-fighting ability, the third bomber wave blanketed the city with small incendiary bomblets.

The firestorm completely consumed and destroyed the entire city, and its hurricane-like winds killed thousands by preventing their escape from the inferno. After the defeat of Germany soon after Dresden's destruction, there was only Japan left. And with the costly, bloody, and painful defeat and dismemberment of its Pacific empire by U.S. forces, there were only the two main Japanese islands left under Japanese control -- the homeland. Strategic bombing again was the plan, with attacks launched from nearby captured islands.

The Japanese were the perfect target. With homes in their cities built entirely of wood, the U.S. Air Force began its second campaign of strategic bombing, devastating them with incendiary-started firestorms. Hundreds of thousands were killed, and the cities attacked -- including Tokyo -- destroyed.

And the last firestorm -- with the new touch of post-mushroom cloud, post-attack firestorm, and radioactive "black rain" -- was Hiroshima, started with just one bomb, instead of thousands. With one bomber dropping one bomb, 140,000 died immediately or soon after, the death toll eventually claiming an additional 150,000 victims.

The atomic bomb was unleashed to herald the beginning of what was called the Age of the Atom, but was really, simply a new age of war -- and a new age of inhumanity.

* * * *

> "Fiery the angels rose,
> and as they rose deep thunder roll'd.
> Around their shores: indignant
> burning with the fires of Orc."
>
> -- "America, A Prophesy" (1793)
> William Blake

And finally, there is the "Molotov cocktail". It was introduced by leftist Loyalist Forces -- armed by Stalin -- in the lost cause against fascism of the Spanish Civil War, 1936 - 1939. It was then adopted in WW2 by anti-Nazi partisan guerrilla forces, and resistance fighters. And 1956 anti-Soviet Hungarians (former Nazi Allies) as the only effective, improvised, cheap, and easily available weapon against Soviet tanks out to crush their revolution. Successfully... (By the way, to be perfectly successful weapon, you have to drop it or throw it from an inner-city multi-floor building – an apartment building is perfect -- while the Soviet tank is lured into a narrow street. (And the building should have a back exit...)

V. Molotov was Prime Minister under Stalin and Foreign Secretary by WW2. (Ultimately, he was cashiered and expelled from the Communist Party in 1956 by Khrushchev, successor to Stalin.)

In 1943, the tide of WW 2 began to turn as the Soviet retreat and "scorched earth" policy became an ever stronger behind enemy lines resistance campaign, and finally reversed into Napoleon's winter rollback, as the new Eastern Front formed from a resurgent Red Army, and Nazi forces finally began crumbling under the onslaught.

Russian resistance to Nazi-occupied Russia was credited with killing over half a million _Wehrmacht_ and _Waffen_ SS soldiers of the Nazi war machine. They burned 11,483 supply & troop trains, destroyed 897 storehouses, blew up 6,763 bridges, destroyed 1,742 tanks, 20,490 trucks, and 122 airplanes. [Ref: "Violence in Our Time", Sandy Lesberg]

And a favored weapon for partisan guerrillas, and resistance forces was the simple, easy to make, improvised Molotov cocktail. It consisted a regular glass bottle (a wine or liquor bottle of about a liter or a quart volume was perfect for throwing and smashing) filled with two-thirds gasoline and one-third motor oil lubricant mixed in as a thickening agent to make it stick was used – or, in a pinch diesel fuel could substitute for the lubricant.

Latex was also used, which was even better a thickener. And 12% tar/asphalt was even better. One half a liter of the mixture would burn for 20 minutes when the Molotov cocktail was thrown and smashed on the hard target -- through a window of a building, on an enemy vehicle, etc.

The last part of cocktail manufacture was the ignition system, consisting a cotton or woolen rag stuck securely into the narrow neck of the gasoline-filled bottle, and wired and duct-taped, (or carefully sealing firmly in the bottle neck a partial plug of melted candle wax). Tipping the bottle soaked the rag, which was ignited with a match, and then thrown at the intended target.

(Other thickening agents worked better than motor oil, gelling the gasoline into improvised napalm in the modern Molotov cocktail. Examples are asphalt, latex, synthetic rubber, and rubber cement (with optionally hexane -- lacquer thinner added)).

The cocktail was also used most effectively in WW2 and after, as a civilian-used anti-tank weapon in built-up urban areas where tanks had limited mobility in streets and alleys. Such use was highly effective, but only if the engine located in main tank body behind the turret was hit. This destroyed the engine, disabling the tank (along with mobility and turret rotation, by loss of electrical power with battery destruction). Modern tanks now have counter-measures that make cocktails inadequate to take them out.

Che Guevara (in his widely available 1960's manual "Guerrilla Warfare"), the famous Cuban revolutionary, described a method to modify a shotgun for lobbing a Molotov cocktail over a distance using a broom stick seated in 12 gauge shotgun shell emptied of its shot at one end, and chambered in the shotgun, with the Cocktail mounted on the other end of the broom stick projecting out the end of the muzzle. It was used as an improvised mortar for use for attack against government anti-guerrilla forces, and other targets.

The Molotov cocktail -- aka "petrol bomb" [UK], aka gasoline bomb, aka firebomb -- is classified as an "IID" -- improvised incendiary device. Under U.S Federal Law it is classified as a "destructive device" -- like hand grenades -- and legally administered federally by the draconian National Firearms Act.

The NFA Act is only draconian if you violate it. But weapons laws are the purview of the States, so the NFA Act is only a tax law. The law itself allows for the possession of sawed-off shotguns, Molotov cocktails, and other serious weaponry on payment of a $200 license fee -- with other requirements, such as finger-printing, and State approval -- and is administered with vigor by the BATF -- the Federal Bureau of Alcohol, Tobacco, and Firearms -- who imprisons violators with penitentiary time.

* * * *

> "What would the sun be itself, if it were a mere blank orb
> of fire that did not multiply its splendors through millions
> of rays refracted and reflected...?"
>
> -- "On Wordsworth's Poetry" (1845)
> Thomas De Quincey

Before the invention of the car, and before widespread electric lighting, gasoline was thrown out, as a useless, highly volatile, and dangerously flammable and explosive by-product of kerosene manufacture (for lamps) from primitive crude oil refining.

By 1920, when car ownership (and Edison's light-bulb) took off with the U.S. public with the introduction of affordable cars like the Model-T, and mass production techniques, and the waning days of kerosene lamps with the widespread introduction of electricity, gasoline was suddenly the main, and highly desirable product of petroleum refining.

Incendiaries, as I discussed above, started coming into their own in WW2. There was magnesium-cased thermite for small, mostly six lb. (2.7 kg) aerial bomblets, there was the gasoline, compressed air-powered flame-thrower (used to great effect by U.S. forces in the Pacific War of World War Two, 1944-1945, against dug-in Japanese forces), and historically most important, the invention of napalm.

Napalm -- thickened, actually jellied, gasoline -- was invented in 1942, by a team at Harvard University (WW2 was the beginning of the induction of American Universities in military research contracts, a practice that continued up till now) led by Dr. Louis Fieser, too late to have much effect in the European War.

Napalm's development was inspired by a wartime shortage of latex -- natural raw rubber. The

Harvard group was part of a larger U.S. research effort looking into better alternatives than latex for jelling gasoline.

Napalm was much cheaper than thermite – using derivatives of coconut oil, and was more effective -- even though of lower heat generation power -- because as a thickened, gelled liquid when dropped from low-flying planes onto enemy targets in special bombs, and ignited by delayed ground hit detonators, it spread to cover and incinerate a wide area.

* * * *

"To defend yourself in that savage world was impossible.
To go on strike was suicide. To go on hunger strikes
was useless.

And as for dying, there would always be time."

-- "The Gulag Archipeligo" (1985)
Alexander Solzhenitsyn

But what exactly was it that was so special about napalm over regular gasoline as a fire-bomb? The explanations I read didn't really help explain in an understandable form why napalm was superior. I had heard about "thickened", and "gelled", and "sticky" -- but that didn't really give me a logical physical explanation of its effects as a weapon.

I took a first pass at **really** understanding its mechanism of action by looking at it closely, comparing the two agents -- napalm and gasoline -- and with a deeper level of thought with a view to actual physical properties, and the mechanism of burning. And persistent thought about the issue. Persistence, the forgotten key to most problems.

And here is what I came up with.

Napalm would wet and stick to the target in a thicker layer, instead of dripping off like regular gasoline. The thickened gelled liquid would also splash off less -- once again, a waste of gasoline's burning properties as far as the weapon scientists were concerned.

Most importantly, with the slower combustion of napalm it would burn longer on the target, with less vaporization and "flash off". Flash off was the impressive-looking but heat-dispersing, and thus wasteful, "fireball effect" -- the totally fake, artificially-induced fireball of gasoline vapor, that looks spectacular in movies but that heats the surrounding air, and not the target which you want to burn up.

Gasoline and the "safe" kerosene have about the same heats of combustion. You'll be surprised to read that the only differences are gasoline's much lower boiling point than kerosene, resulting in a higher rate of burning, and its vapor having the ease of forming, and exploding if it encounters a spark or a flame.

* * * *

" 'Do you know anything about Jungian psychology?...One of his
ideas was that everybody had a dark side to his personality,
which he called the "shadow". The shadow contained all the
unacknowledged personality aspects – the hateful parts, the
sadistic parts, all that. Jung thought people had the obligation
to become acquainted with their shadow side. ... As Jung saw it,
if you didn't acknowledge your shadow side, it would rule you.' "

-- "Sphere" (1987)
Michael Crichton

Michael Faraday, the Englishman, was one of the greatest scientists of all time. Starting in 1827 he began giving a yearly -- and very successful -- children's Christmas lecture at the Royal Institution of London, of which he was the director for more than forty years. In 1860, at the age of sixty-nine, Faraday gave the last of his Christmas Lectures, "The Chemical History of the Candle".

Candle wax doesn't burn. You hold ordinary paraffin candle wax to a flame, and it merely melts to a colorless liquid, and drips off. You need a wick to make a candle. You put a flame to the wick -- a braided piece of string -- it takes alight, burns down to the wax, and melts it into a pool at the base of the remaining blackened wick. Then capillary action draws the liquefied wax up the string, where the hot wick vaporizes the liquid. Only then does the wax burn, because only vaporized wax is actually flammable.

And so it is with gasoline. It is not the liquid gasoline that burns, as it is the gasoline vapor that takes aflame, heating the liquid gasoline below the flame, to produce more thick gasoline vapor, which rises, mixes with the oxygen in the air, and begins burning. The flame is thus above the object on which the gasoline is pooled, and the tremendous heat of the burning gasoline is mostly dispersed in the air above.

I had solved the first piece of the puzzle. The thickened, jellied napalm burned close to the object it coated and actually set it on fire and incinerated it much, much more effectively than with pure gasoline.

The WW2 napalm patent (U.S.P. #2,606,107 by Louis Fieser of Harvard) was declassified in 1952. The patent disclosed that napalm was made by preparing an **aluminum** soap, using aluminum hydroxide (instead of the sodium hydroxide used in normal household soap) and the fatty acids, naphthenic and palmitic acids.

(These fatty acids gave napalm its name and were obtained easily from petroleum and coconut oil, respectively. It was later found that lauric acid, also from coconut oil, could also be used).

The powdered aluminum soap was added with stirring to gel the gasoline -- gelling occurred in 3-20 minutes -- into a thick and sticky, firm gel with decreased volatility, but still highly inflammable when ignited.

With its increased viscosity, lower volatility (evaporation rate), and slower burning time, napalm substituted for regular gasoline increased a flamethrower's range from 30 yards to 150 yards, and was thus much more effective in burning out dug-in and fanatical Japanese forces in the bloody island-by-island U.S. battle victories of the Pacific War.

Napalm in aerial bombs was first used in July 1944 in Europe by U.S. forces, and then later in the Korean War, and then most controversially in the Vietnam War.

An improved napalm was made by gelling gasoline with aluminum octoate.

* * * *

"Knowledge is a fine thing, lads. These trees and this
meadow find a voice to teach the finest knowledge of
all – how to forget what one knows."

-- "Barberine and Other Comedies" (1892)
Alfred de Musset

But after WW2, research continued on incendiaries, and a new, cheaper, and vastly superior formula for napalm was developed. It used a plastic named polystyrene -- now widely available to the public as "Styrofoam", a solid air foamed form of polystyrene used in disposable cups and as household wall and attic insulation -- to produce a more viscous, jelled incendiary.

> [The only patents I could find on this subject were U.S. P. 3,957,550 (1966; declassified in 1976) by Tannenbaum of Thiokol Corp. on "Explosive Napalm", that used polystyrene; and Meyerholz, U.S.P. 2,966,401 (1960, unclassified), the first to use polymers as gasoline thickeners. But it used ABS polymer as the thickener, not polystyrene.]

The new formulation, called "napalm B" was also dubbed "super napalm", and consisted of 50% polystyrene, 25% gasoline, and 25% benzene, another hydrocarbon, like gasoline, to aid in dissolving the polystyrene – a Styrofoam cup is 100% polystyrene – used as the new thickener/gelling agent. (It is useful to know that 50% by weight, Styrofoam will produce a white gel with 50% gasoline alone, since benzene is not easily available, especially since it was identified as a carcinogen and removed from most uses.)

Polystyrene-thickened napalm was much more effective -- it stuck better and burned hotter and longer than the original formula napalm, and of course pure gasoline. Regular napalm combusted for only 15-30 seconds. Napalm B burned for up to ten minutes. The same heat released over longer time, seared and then burned into the target even better.

It generated temperatures of up to 1,200 °C (2,200 °F). As well, polystyrene consists of long carbon chains which makes the burning napalm B hydrophobic (hard to wet with water), and thus makes it hard to put out the fire with water.

And the usual desperate measure of rolling on the ground quickly to put out the burning napalm was ineffective, because of the thickness of the coating, and the presence of flammable gasoline vapor (there's still a little).

Immolation – severe and massive burning of the body flesh -- and third degree skin burns over much of the body surface were the major killing mechanisms of napalm. The vocal cords were also injured by all the screaming the victim would do while being cremated alive...

Lethal effects also included asphyxiation due to rapid loss of oxygen (consumed by the conflagration of burning gasoline), heat stroke, smoke inhalation, and carbon monoxide poisoning, a product of incomplete combustion of a burning mass of gasoline

The napalm was deployed in the 500 lb. BLU-118 bomb, a cigar-shaped, thin-walled aluminum canister dropped on suspected enemy positions from a low-flying fighter plane working close air support for on-the-ground infantry units engaging "Charlie" -- Viet Cong guerrilla units ["Victor Charlie", the radio operator mnemonic], and later (1968-1975), the NVA (North Vietnamese Army).

The 165 gallon (630 liter) canister, dropped from a low-flying aircraft produced fire damage to a jungle area of 2,500 square yards -- an area the size of less than half a football field, but longer and narrower in shape.

The contract to produce the new napalm for the massive quantities needed for the Vietnam War by U.S. aerial forces was won in 1965 by Dow Chemical, the inventor of Styrofoam and the only manufacturer of polystyrene. For thousands of pounds of polystyrene, a cheap plastic, but a lucrative government contract.

However Dow Chemical Company (motto: "Better Living Through Chemistry") executives were horrified by the bad national publicity generated when their manufacture of napalm was publicized by student anti-war groups. Burning people to death is not a popular way to wage war to civilian side-line observers. (Shooting is acceptable, though.)

And Dow's case was not ameliorated with things like a quote picked up by journalists from a Dow Chemical PR clown's protestations over Dow protesters' placards showing burn victim photos:

> "There are no skin burns from napalm. It burns right through [into the flesh] or kills by concussion and suffocation."

When the napalm manufacturing contract came up for renewal and rebidding in 1969, Dow allegedly spiked their bid, to deliberately let another company win it, for PR purposes, though they still supplied the new contractor with polystyrene with which to make the napalm, having the only plant large enough to supply the needs of the U.S. and South Vietnamese war criminals. And they wondered why the North Vietnamese treated shot down P.O.W. pilots poorly in the "Hanoi Hilton"...[also a violation of international treaty, by the way]

* * * *

> "Photographers snip, snap,
> Take your time she's only burning
> This kind of experience
> Is necessary for her learning.
>
> If you'll be my flotsam,
> I could be half the man I used to be
> They said you were hot stuff,
> And that's what baby's been reduced to."
>
> -- "Baby's on Fire" (1973)
> Brian Eno

In 1972, a Pulitzer Prize winning picture of a nine year old South Vietnamese girl, Kim Phuc, running in terror and pain, naked -- her burning clothes burnt or stripped completely off -- from a napalm-attacked zone, was splashed across U.S. and international publications, creating mass horror amongst the public world-wide, and increased opposition to the already unpopular U.S.-sponsored war.

* * * *

> "The light that burns twice as bright, burns half as long.
> And you have burned so very, very brightly, Roy! ...
>
> "I have done questionable thing..."
>
> "Also extraordinary things! Revel in your time!"
>
> -- Dr. Tyrell to Roy Batty

"Blade Runner" (1982)
Hampton Francher, David Peoples

And then one day, I was thinking about napalm B, and how and why it was such an improvement over the older version of napalm. I emptied my mind of all emotion, and considered the basic science.

I was lucky. I had achieved the requisite, magical state of mind.

I knew, I understood, the Zen of Napalm.

There was a second -- not secondary -- effect of the polystyrene in burning napalm, other than as a better, cheaper thickener of the gasoline, I realized. A very important one.

Polystyrene is a solid plastic -- a polymer (Styrofoam is foamed polystyrene and is 99% air). It is a hydrocarbon mainly composed of very long, repeating carbon chains and rings -- molecules of styrene strung together in a linear string. The burning napalm would heat, melt, and then carbonize the plastic -- decomposing it to pure carbon -- and would thus fill the napalm gel with a layer, and particles of carbon. As the napalm burned down, the target would eventually become coated with a layer of hot carbon.

Now carbon has a very high melting point -- 3,550 °C. (6,442 °F) -- well above the burning temperature of the napalm -- up to 1,200 °C., as I said earlier. So as the rest of the gasoline burned down, it would heat the carbon coating the target to incandescence, efficiently scorching the target. Human beings or military equipment wouldn't stand a chance.

Because when the gasoline was all burnt off, there remained a thick -- and glowing white then red hot -- carbon coating layer sticking to whatever was hit by the burning napalm -- houses, machinery like tanks, the jungle vegetation, or just humans, military or civilian. The destructive heat therefore would last long after the last of the gasoline had burnt off.

But the solvent benzene is hard to get, since it was declared a carcinogen. But other aromatic hydrocarbon solvents such as toluene, xylene (as well as acetone, turpentine, and MEK (methyl ethyl ketone)) are some substitutes. Toluene -- commonly available in hardware and paint stores, incidentally -- in particular would the best substitute, being a cheap, aromatic, hydrocarbon solvent, very closely related chemically to benzene.

You dissolve the polystyrene in the aromatic solvent, and then dilute the solution formed with an aliphatic hydrocarbon -- the gasoline. The polystyrene precipitates partially, and forms a gel-like mass of the whole.

"One or two napalm attacks can change the fighting spirit of a whole [Viet Cong] company."

-- Lt. Commander Fitch
"Air War: Vietnam" (1967)
Frank Harvey

* * * *

"The _bodhisattva_ shined his light about him so that everyone could see as he could see, giving them the opportunity to see the deathless nature of the ultimate."

-- "Opening the Heart of the Cosmos" (2003)
Thick Nhat Hanh

The above quote is a comment on the "Lotus Sutra" (ca. 100 A.D.), 23rd chapter, where the Bodhisattva Medicine King shows the selfless nature of the body by setting himself aflame, extending the "light of the _Dharma_ [Buddhist enlightenment] for 1200 years."

It inspired the following incident -- the only show of protest possible by a true member of a religion committed to non-violence, as Buddhism is -- a protest of self-immolation.

And so a variant of the cocktail was used in June 1963, when Thich Quang Duc, a senior Buddhist monk -- the first of a number of such incidents, and which grabbed international headlines -- burned himself to death in public in Saigon to protest government discrimination and repression against the Buddhist religion -- subscribed to by the majority of the civilian population.

It was repression of religious freedom done for no color-able reason or justification by the U.S.-installed puppet dictators of South Vietnam, who just happened to be opportunistic converts to Catholicism, the religion of their former French colonial masters.

So the monk choose to express his protest of the situation in the only non-violent way a man of peace can, with an individual, dramatic, symbolic, self-sacrificing, one-man way, with a very public, and very spectacular suicide.

("Did you know your messy death would be a record-breaker?" went a lyric from the title cut of the first, and wildly successful rock opera, and later film and Broadway Play – in 1971 – "Jesus Christ Superstar")

So the monk soaked himself in a mixture of gasoline and motor oil -- the thickened fuel of the classic Molotov cocktail -- and lit a match to set himself -- fatally -- on fire, as he sat legs folded together in the classic Buddhist pose on a busy Saigon intersection, in his monk's loose, white cotton cloak.

His shaved monk's head saved him the pain of his hair catching on fire and burning off, and the third degree burns over his body burnt off the pain-causing nerve endings, making his ordeal much less painful. And, anyway, he was burned deeply over so much of his body's skin surface that he died almost immediately.

> [How a senior Buddhist monk knew that you had to thicken the gasoline with motor oil was not explained. I was only 11 when I learned that trick, and the 1960's had just ended...Perhaps he met the Japanese fighting monks who invented Karate, and defended town walls with weapons, and were armed with swords until they were banned.]

Numerous international TV news and newspapers the next day carried the story and showed the photograph snapped -- by AP's horrified Saigon bureau chief, Malcolm Browne -- of the burning monk sitting calmly in a lotus position – a picture that won him a Pulitzer Prize, to assuage his trauma.

Such bad publicity no doubt contributed to American government alarm. This was communicated to more suave factions of the South Vietnamese military. In November 1963 -- less than five months later -- a _coup d'état_ was successful, there was a new puppet dictator, and the former dictator, Ngo Diem and his brother, were conveniently shot dead during the military takeover.

The same month, coincidentally, U.S. President Kennedy was assassinated.

Whatever goes around, comes around...

> "As I was walking among the fires of Hell, delighted with
> the enjoyments of Genius; which to Angels look like
> torment and insanity. I collected some of their Proverbs."
>
> -- "The Marriage of Heaven and Hell" (1793)
> William Blake

Prison Diary 70:   Confessions of a Weapons Expert
===========================

> "I will do such things --
> What they are I know not, but they shall be
> The Terrors of the Earth."
>
> -- "King Lear"
> Act II, Scene 4

> "Still falls the rain --
> Dark as the world of man, black as our loss --
> Blind as the nineteen hundred and forty nails
> Upon the Cross."
>
> -- "Still Falls the Rain" (1940)
> Edith Sitwell

Fiat nex. [Let there be slaughter.]

Being wrong seems to be the only human activity requiring effort that neither tires people out, nor that they get tired of. But there has always been a solution at the ready -- the proper and consistent application of logic.

Logic is what gave rise to "Game Theory", developed by John von Neumann in 1928, and applied to the development of the first atomic bomb in the years from 1943 - 1945.

The perversion of logic is what gave rise to "realpolitik" from the Germans in the late 1800s and exemplified by the statesmanship of Bismarck. A rigid and uncompromising government political plans, and strategy based on practical realities and realistic goals, ruthlessly implemented opportunism, and selfish material needs, absent any consideration whatsoever of morality and ideals -- the death of "the right thing to do".

Realpolitik was a foundational principle of Hitler's actions leading up to and including WWII.

Logic isn't always pretty. Not by a long shot. It's sometimes terminally depressing. But don't shoot the messenger.

*   *   *   *

> "Function: Retrained (9 Feb 2019)
> Pol. Homicide."
>
> -- Zhora the android's
> official file card
> "Blade Runner" (1982)
> Hampton Francher, David Peoples

"This is Zhora. She's trained for an off-world 'kick'-murder squad. Talk about beauty and the beast – she's both," adds police honcho Bryant ominously, as he briefs human police assassin Rick Deckard on one of the four files of the targeted android "replicants" Deckard is assigned to hunt down and "air out".

Then there's the lovely Pris, with a function defined as "Military/Leisure". "Yer standard pleasure model" leers Bryant, to accentuate her official job as a prostitute for earth's inter-galactic soldiers. Her creation date is listed as 14 February – Valentine's Day – one may note in passing.

More ominous, is Leon's occupation: "Weapons Loader (Fission)". Earth in the year 2019, rules a galactic empire (colloquially known as "off-world") with an iron fist, that makes conquers, or makes war on rebellious outer space colonists with atomic weapons.

But the dark and dreary dystopian world of "Blade Runner" is more than just an Evil Empire. The world on earth of Rick Deckard is a demonic police state. When he finally tracks down Zhora the official android assassin, he calmly shoots the fleeing girl in the back on a crowded downtown street, then merely flashes his police squad ID to police officers attracted to the commotion, before he's on his way again to the next "hit".

* * * *

"The function of government is to direct the moral and physical force of the nation towards the purpose for which the government is instituted. The aim of constitutional government is to preserve the Republic.

The aim of revolutionary government is to found it. The Revolution is the war of liberty against its enemies. The constitution is the rule of liberty when victorious and peaceable."

-- "Twelve Who Ruled" (1941)
R.R. Palmer

I was trying to send a message to an ex- who was causing me a lot of trouble about 25 years ago over the break-up of our host-parasite relationship. I waited until the temperature outside was -30 or -40 C. (-40 F) and said, "Let's go for a drive!" We dressed lightly -- considering the temperature, jumped into the car, which I had warmed up, and roared off into the setting sun off into the country.

Out of the city, out into the country, miles into the fields and forests. When we were miles from any house or habitation, I slowed to a stop. "Let's get out for a second," and we both did. We wandered away from the car, and then I suddenly sprinted back to the car, jumped in, locked the doors, and roared off.

At the next high point in the frozen gravel road, I hit the brakes. She was just standing where I had left her, and saw the brake lights go on. I waited a few seconds, then turned around, and slowly cruised back.

When I reached her, I rolled down my window a bit and said, "If I had kept going, you would have frozen to death in about 20 minutes or less. And no one would have been able to prove a thing against me...Now get in."

We drove home, and I engaged in no further conversation with her.

* * * *

> "I still don't know what I was waiting for,
> And my time was running wild,
> A million dead-end streets.
> Every time I thought I got it made,
> It seemed the taste was not so sweet
>
> -- "Changes" (1971)
> David Bowie

I told you that I was – and am – a "peace-nick". Anti-violence. But I was seduced by the technology of death, and so over the years, though still believing in non-violence I became an amateur weapons expert. First explosives, from experimentation (by accident), then reading. Then guns. It had been by reading, until I was old enough for a firearms license. And then more reading.

And when I read everything about guns and explosives, I was hooked on weapons, and started branching out. Incendiaries, chemical warfare nerve gas, knives, nuclear weapons, and hand weapons, and finally lethal hand-to-hand combat. There were the sciences: hydrodynamics, nuclear physics, mechanics, organic chemistry, anatomy. And let's not forget toxicology...

It was, as usual, one of the paradoxes of my personality, the non-violent weapons expert. And one of my driving forces was the absolute contradictions of "reality" with the Truth. The lies of books and T.V., and movies drove me onwards to the real truth, and how horrible it really was, and what a lie "TV reality" -- as U2 sang – was. Not just because they wanted to not disclose harmful information, but because they were ignorant and stupid, and their sources wanted the truth hidden.

I used to love the CBS investigative reporting TV show "60 Minutes", when it had Mike Wallace on it. It was my source of truth, shown with vigor, succinct, and with the humor of Mike chasing his guilty target almost running away, all on camera.

But then I noticed the quality of the show slipping. It was past the time that Mike became too well known, so he couldn't chase anyone into a door slammed into his face on camera.

They were doing a piece on former Iraqi dictator Saddam Hussein. It was a firing squad with multiple people having been shot. Mike stated as the camera panned over the executed victims, "The commander couldn't resist shooting the executed victims in the back of the head with his pistol."

I couldn't believe it. It was a "former U.S. ally turned enemy" propaganda piece. Hadn't Mike Wallace or his researchers heard of the _coup de grace_ administered following a firing squad by the commander of the firing squad?

The 12 man firing squad aims at the victim's heart. The coup de grace is a shot to the back of the head to make sure the victim is dead, and end his suffering, if he isn't. It's standard operating procedure the world over, amongst professional armies.

And – little known – a shot to the cerebellum (the brain stem) at the back of the unconscious victim's head, avoids seeing the involuntary, unconscious twitching movements of the body – the death throws – of the nervous system's reaction to dying, and that can go on for a few minutes during the physical process of death, and are extremely disturbing and traumatic to observe.

Thank God, I've never seen it in person, myself.

I stopped watching "60 Minutes" after seeing that bullshit piece.

* * * *

"We must realize that today's Establishment is the new George III. Whether it will adhere to its tactics, we do not know. If it does, the redress, honored in tradition, is revolution."

-- "Points of Rebellion" (1969)
Justice William O. Douglas

The street-fighting man comes equipped with steel-toed boot or shoes. You kick a guy in the shin, or preferably directly to the front of one of his knees, or knock him down with a hammer-blow to the side of the head, and push him hard to the ground. Then – fast -- you kick his head with your boots till he stops moving. Case closed.

Then there's marching. Ever thought about military marching?

It's done in a way to produce self-hypnosis. Less tiring for long marches.

Lobster arm. Always hold a bag in my left hand. It feels more comfortable. But then I realized I was keeping my fighting arm – my right hand – free. Subconsciously. **And** I was exercising my left arm, to avoid "lobster arm", whereby your right arm is visibly more muscular than your left.

Fighting with someone who is left-handed is more difficult, too. It confuses the right-handed fighter, who is not skilled. Or someone who's ambidextrous, who switches his lead fighting hand, is even more disconcerting, and leaves the unwary fighter off-balance, and defeatable.

* * * *

"Crier havot!"

[Cry, havoc!]

-- 14th Century Teutonic
command to begin pillaging

Bullets are noisy and traceable and knives are bloody – a real mess. And it's hard to keep the blood off you.

An ice pick hard in the ear from behind, or deep into the eye, or up the nose requires a steady, accurate hit, and is easily noticeable, but is effective.

A hard fall down a long staircase headfirst, being thrown over a balcony, or a bridge from a height of at least 75 feet, are all much superior methods of murder. Knocking out the victim, and throwing them at night from a height, or a car accident, hit and run with a stolen car, and car dumped into river or lake. Or even a reported accident by someone with no connection to the victim.

More direct means are severing the spinal cord of the victim in the cervical region of the backbone (neck area) with a light blow of an ax or hatchet across the upper neck length-wise, or a hard strike with the point of a knife, which requires more accuracy and expertise.

* * * *

"A man who's been trained to ignore pain, ignore weather,
to live off the land, to eat things that would make a billy goat
puke. In Vietnam his job was to dispose of enemy personnel.
To kill, period! Win by attrition. Well, Rambo was the best!"

-- Col. Trautman on John Rambo
"Rambo: First Blood" (1982)

The hit man is not a mercenary, like TV or the movies would have you believe. He is the product of organized crime. The Mafia, or Organized Motorcycle Gangs, which is the only place he can safely carry out his profession with limited fear of betrayal, and solid organizational backup, and payment.

The mafia's old weapon of choice was a silenced semi-auto .22, with multiple shots quickly through the eyes. Multiple shots – emptying the magazine -- with hollow-point .22 long rifles into the upper torso – lungs and heart – will do in a pinch. Reload with a spare mag, if practical.

Body disposal is actually often a bigger problem than "termination with extreme prejudice". Cutting up the body is messy and time-consuming and leaves evidence. The bones are problematic. Acid to dissolve the body is good, but the nitric and sulfuric acids are hard to get in volume and traceable.

Once again we turn to organized crime. The old drum in the ocean, or a deep lake second best. The drum is punctured to let out the gases of the decomposing body, and weighted down with steel chains or concrete. They learned to puncture the steel (now plastic) drum the hard way, when the gases from the decomposing body caused a drum to float up from the bottom of the sea floor, and it was found. How embarrassing...

In a pinch, a sleeping bag stuffed with the body and weighted with a hefty length of chain works for the "deep six" body disposal. That was a Canadian Hells Angel's trick. Didn't work, for other reasons (informant).

There's the car-crushing machine with a body in it. Crushed car goes to a recycling blast furnace...

Then there's the "double burial", if the mafia had access to a cemetery. Bury the body, before the funeral. The casket for the actual funeral is lowered over the buried victim's body unnoticed.

In 2011 Joseph "The Ear" Massino, N.Y. mafia boss turned informant, explained that by co-operating, he was violating a sacred oath he took during a 1977 induction ceremony to protect the secret society. It was the protocol about a "hit". It was understood, he said, that "once a bullet leaves that gun, you never talk about it."

* * * *

"My spirit is too weak -- mortality
Weighs heavily on me like unwilling sleep,
And each imagined pinnacle and steep
Of godlike hardship, tell me I must die
Like a sick Eagle looking at the sky.
Yet 'tis a gentle luxury to weep
That I have not the cloudy winds to keep[.]"

~~~
                            -- "On Seeing the Elgin Marbles" (1817)
                               John Keats
~~~

But most weapons were developed initially for fighting, and most recently for the mass destruction of the foreign "enemy" in war. For the internal enemy a different theory and method of "execution" was necessary.

Eventually in my study of weapons, unfortunately I developed an intense fascination with the minutiae of death -- an obsession with death and the horrors of war. And with PTSD it made for a bad combination: unhealthy intrusive thoughts about hanging, killing, and destruction. Thinking about death.

When I spotted a 1992-05-26 "New York Times" P. B1 piece, "'Spotting a Hidden Handgun" by variations in a pedestrian's gait, I realized that there was something really, really wrong going on in society.

* * * *

"I have detailed files on human anatomy."

"I'll bet you do. Makes you a more efficient killer, right?"

"Correct."

~~~
                            -- The Terminator robot to Sarah Connor
                               "Terminator 2: Judgment Day (1991)
                               James Cameron, William Wisher
~~~

In the dying words of Socrates, in 399 BC, he said that he had never engaged in or advocated, or supported violence, and yet he was sentenced to uncalled for, extreme state violence, and sentenced to execution. In fact, forced to the indignity of executing **himself** by drinking a slow-acting poison.

Certainly, poison hemlock -- containing the toxin coniine -- is an essentially painless poison, so that's not the issue. But the fact that it's slow-acting, and the head remains clear until the last few minutes, is a mental torture no doubt beyond belief if you think about it, as the poison slowly and progressively numbs and paralyzes the toes, the feet, the knees, the thighs. Then the numbness and paralysis rises up the torso, reaches the shoulders, breathing becomes difficult and labored. Then the heart fails.

Blackness, unconsciousness, then death...

It was Plato, Socrates' teacher who recorded the symptoms in detail, as he displayed his loyalty by standing by physically and accompanying his teacher, mentor, and friend's last day on earth.

It's a story only a pharmacologist could love, and one of my Bachelor of Science degrees is in pharmacology.

Then there's cyanide. Hydrogen cyanide is the gaseous form, while there are many solid "salts" in use for electroplating and other industrial uses, such as potassium or sodium cyanide. You mix the salt with an acid – say sulfuric acid – and you get the gas hydrogen cyanide, which if inhaled is a quick and painless death. The blood absorbs the HCN and goes directly to the brain, which shuts down. Unconsciousness is immediate and death soon follows. With swallowing of the salt, however, death is slower and painful, as the salt reacts with the stomach acids and is generally absorbed in the body.

Dr. Wallace Carothers was a chemist – the chemist who discovered Nylon in 1935 – **not** a biochemist – and a depressive. So when he committed suicide, not knowing cyanide's mechanism of death, he swallowed a salt of cyanide (rather than relatively easily making, and then inhaling, the quickly lethal hydrogen cyanide gas). He died alone, and in agony, dying in a hotel room he had rented for the purpose of suicide, and was discovered by the noise complaint from the neighboring room, from his dying convulsions as he banged into things in his room in his death-throws.

* * * *

> "Two ideas gave purpose to the struggles in France during the Revolution, the rights of the individual and the sovereignty of the nation. Revolutionary philosophers saw no conflict between them.
>
> Through the sovereignty of the nation the individual made his rights effective, freeing himself from the old restraints of customary law, monarchy, class, church, guild, and corporation, as well as from domination by foreign powers. Individual liberty depended on national sovereignty.
>
> But the balance [between the two] is not easy to achieve."
>
> -- "Twelve Who Ruled" (1941)
> R.R. Palmer

And then there's government-sanctioned execution. Capital punishment by firing squad. It's not barbaric, or state terrorism. It's fast, cheap, efficient, humane -- no muss, no fuss -- unlike the elaborate, lengthy, torturous methods of the "free world". The barbarism of the neck-snapping hanging, the electric chair, the gas chamber.

In communist countries they just shoot you. And the reasons are far different, and much more reasonable than here: political crimes, then economic crimes -- bribery, corruption, theft of corporate funds, economic fraud. Crimes against the State and thus Crimes against the People.

"You have been weighed in the balance and found wanting."

The Chinese communists are a little more elaborate. They do it publicly, in a filled stadium. The condemned is forced to kneel, his hands bound behind his back to receive a single pistol shot to the back of the head. And a bill for the ammo cartridge is sent to his family, to make a point of it.

A little history of the firing squad. It was the WW1 punishment to be shot at dawn for cowardice in the face of the enemy. **Making** you face the bullets. Or being late – even just a little bit – from "leave". Execution by firing squad for "desertion". The iron discipline of the time.

The firing squad would be 12 men picked at random – like the twelve apostles. And one man would be given a blank, symbolically, so that no one knew who killed the man – though the lack of recoil was a dead giveaway. The target was a piece of paper pinned over the man's heart.

After the eleven shots through the heart, the officer would deliver the _coup de grace_, a pistol shot to the back of the head. A shot to the medulla – the brain-stem – to finish off the victim, and avoid the convulsions and twitching of the man's "death throw's".

"It is time that communities in this country understand that such breaches of public confidence [sheltering Confederate guerrillas] are to be followed by such terrible consequences as to deter the people from their repetition. We must end the war as we go, either by parole or devastation; and where paroles are rendered useless the alternative is a terrible military necessity."

-- U.S. Major General Samuel Curtis (1864)
in a letter to a Missouri congressman

Prison Diary 71:   Illegal!
=====

> "[Our program calls for] terrorist activity to remove the most important personalities belonging to the [Government] Administration...This will have as a general aim the weakening and demoralization of the Administration....the strengthening of popular belief in the party's ultimate success, and finally the inculcation of a fighting spirit."
>
> -- "The Will of the People" (1879)
> first urban guerrilla organization
> Russia

> Freud thought that the lost love object of the depressed person could be either another person, or a homeland, or even an intangible idea.
>
> -- "Mourning and Melancholia" (1917)
> Sigmund Freud

Assassins! Assassins with sniper rifles to pick off their target from miles off! Assassins with submachine guns to riddle their targets! Hit Men with Silencers! Sawed off shotguns! Urban street youth -- sawed-off gangsters -- gang members with **attitude**!

Bullshit!

*   *   *   *

> "At home there are seventeen year old boys,
> And their idea of fun,
> Is being in a gang called the Disciples,
> High on crack, toting' a machine gun."
>
> -- "Sign O' the Times" (1987)
> Prince

Machine Guns -- actually submachine guns -- are useless for any civilian or even criminal use. You can't even brandish them or show them off to friends, the chance of being informed on is too great. Much more than usage, given the harsh sentence dealt out for even possession, much less usage, the civilian usage of SMG's is nonsensical. It is pure idiocy and asking for terrible trouble unless you're leaving the earth with the final bullet.

You can -- at least -- cement them into the basement walls or foundation for use with the end of civilization, civil war, or roaming mobs in a race war.

Machine guns are for war. Period. To end mass charges from a distance, of soldiers against your position. What civilian or criminal use can you use that for? And the criminal penalties just seal the inadvisability. It's a bust just waiting to happen. You can't, you must not -- especially --

tell or show them to your woman.

Submachine guns are for war. Almost period. Nothing justifies the heavy prison term you'll get if you're caught with one, when there are legal substitutes that are even better.

High capacity magazine, semi-automatic pistols or carbines (like the semi-auto mini-Uzi), or short-barreled (but legal) shotguns with an extended tubular magazine are much better, and controllable fire, and thus aimability.

As opposed to the spray room-clearing of the SMG which just requires arm strength as you sweep the room. The usual SMG firing technique, however, requires a lot of expert training, and up-to-date practice to get it right, as well as very good arm strength to try and keep the bucking bronco of a firing SMG on target. Trigger control is essential most of the time, three-round bursts or something of a quick trigger pull and release. Otherwise your 20 - 32 rounds -- or so -- doesn't go very far, and is fatal to the SMG firer with people who can shoot back after the surprise of sudden close-range attack has worn off.

* * * *

"Everyone gets everything he wants. I wanted a mission -- and for my sins, they gave me one. Brought it up to me like room service. It was a real choice mission -- and when it was over, I'd never want another."

-- Capt. Willard
"Apocalypse Now" (1978)
John Milius

It's the same with silencers -- or even worse. In court, they'll call it part of an "assassination kit". Possession is another instant and lengthy prison term.

The physics of silencers is at least interesting. And essential, since you'll probably have to design **and** make it yourself. One idiot made a silencer that *increased* the gun's report -- the fired bullet's sound -- at least he got off, when he was charged, though he could never look his gun friends in the eye ever again...

You can only silence semi-automatic pistols which use sub-sonic ammunition, such as .22 Long Rifle, .45 ACP, .380 ACP, and certain 9 mm cartridges, such as the 147 grain bullet 9 mm round. You can't effectively silence revolvers. They're too leaky of the bullet gases. You can't silence rifles, because rifle ammunition is all supersonic, and has the supersonic crack. You **can** reduce the muzzle blast, but the best you can do is mess with the direction people hear the rifle shot from.

So it's only semi-auto type pistols. Screw-on silencers are the best, but require custom threading of the end of the pistol's barrel.

The purpose of any silencer is to capture the hot, very high pressure gas, produced by the burnt powder that pushes the bullet out of the barrel. This high temperature, high pressure gas is released into the atmosphere as the bullet just leaves the barrel, and produces the gunshot sound, or report.

The silencer works by allowing the hot gas to slowly expand to atmospheric pressure, as well as cool as much as possible before it leaks out into the surrounding air.

Maximize volume, and maximum length are the key parameters of the silencer, and copper or silver construction for maximum heat conduction

Delrin or rubber wipers: flat plastic doughnuts with hole slightly smaller than bullet diameter.

Baffles to create chambers to capture, disperse, and reduce energy of the bullet gas.

Mitchell L. Werbell's III's "Sionics" silencer for Gordon Ingram's MAC-10, designed in 1968, used two aluminum helical spirals of opposite spiral, to swirl gas around the helix to slow it down. Attached to the threaded end of the MAC-10 barrel.

Construct filling of aluminum or copper or bronze fine mesh, wet with water for maximal cooling capacity. Or a filling of fine steel wool. Or mesh washers

They used to sell an adapter for 2 liter plastic soft drink container. Packed with fine fiber-glass wool. It actually worked, was simple, fast to make, cheap, and disposable. The government having no sense of humor banned the adapter, under the existing anti-silencer law.

It also resulted in the U.S. Federal Government banning silencer parts without a license, and $200 per wiper change, and in the process wipers were swept up in the ban. All parts had to be permanent, and not wear out like wipers.

Silencers are legal in France, at least. But they've kept tightening the law, so that when I last checked in 1990, only a single-shot, bolt action .22 pistol was legal with a silencer.

* * * *

"Did he live his life again in every detail of desire, temptation, and surrender during the supreme moment of complete knowledge?

-- "Heart of Darkness" (1902)
Joseph Conrad

The sawed-off shotgun. Take one shotgun, one hacksaw, and some grit-paper, and you're all set.

The double-barreled is classic for looks. But the multi-shot -- single barrel only 18" long, slide action, has a 5 round pump capacity, plus one in the chamber.

It was invented for U.S. WW1 night-raiding parties for the bloody insanity of trench warfare. Equipped with a sheet metal perforated head shield to cover the over-heating barrel, and complete with a long bayonet in case you ran out ammo before you ran out of enemies, it was a mean and effective weapon. The bayonet lug is still part of some modern Winchester models, a historical artifact of its "trench broom" days.

It is illegal because it a pure weapon, with no other legitimate use but against another man or men. Specifically -- like the SMG -- it is ideal for close-quarters room clearing. Wider dispersion of shot with the shorter barrel makes aiming easier -- shooting from the hip -- and from a further distance than room-clearing, the taking down more than one guy is possible in the spray of lead shot.

The shortened barrel also allows for trench-coat conceal-ability, which is a big plus. And the shot is non-traceable, since the barrel is smooth-bore, which bugs the forensics guys no end. Use it with buckshot or BB shot for man-killing, though the recoil is a bitch, and takes someone with muscle not to flinch and screw up the aim.

Then there was the "Enforcer", popular during the 1960s. It was based on the vintage, surplus U.S. military .30 M-1 carbine. Semi-automatic, 30-shot magazine, with the barrel cut down below the 18" legal limit, and the stock cut down, and replaced by a pistol grip, violating the law that a rifle's overall length had to be at least 27". (Older models used to be convertible with an easily available parts kit to the M-2 fully automatic rifle, but the BATF put an end to that, requiring modifications to the M1 to make this too difficult – the spoil sports.)

The Enforcer's aim, of course, was increased firepower over the sawed-off shotgun while maintaining conceal-ability – with a long coat, and the rifle hidden underneath, supported by a leather or improvised rope sling. Its purpose was either urban armed robbery or a "hit".

So there you have it: easy to make, easy to aim and fire, room-clearing action, deadly, multi-shot, concealable, and untraceable. What more could you want?

* * * *

> "[T]he primary reason each of these patients had originally sold out to the demonic was from loneliness. They were not only lonely people, but were accustomed to being lonely, and when they came to exorcism each was still basically a loner. Their courage in doing so may be all the more apparent when it is realized that neither was basically a trusting person. A major reason that the team was crucial in each exorcism was that the team gave them very first experience of a true community."
>
> -- "People of the Lie" (1983)
> Dr. Scott Peck

A switch-blade is merely a folding knife with a spring-loaded knife that pops open at the press of a button -- basically a one-handed opening pocket knife, with a two-edged or dagger-like longer blade. Thus the negative traits from a legal perspective were smallness -- conceal-ability -- and quick single-handed opening to justify their banning, in order to attack through weapons laws a marginalized group.

They were popular with 1950's slicked back hair, urban punk street gangs, and this connotation only led to their prohibition.

But a long bladed knife is even faster, requires no manual dexterity of trying to position the handle in your hand, and push the small opening button, and just as concealable with appropriate care. And in switch-blades the blade snap-opening spring can easily break, so they're not completely 100% reliable, and reliability is a prime consideration in human combat weaponry. They *are* impressive and scary looking to have one brandished at you, but that's it.

The modern butterfly knife (a Philippine Balli-song), is also classified in Canada as a switch-blade, since it is opened by "centrifugal force". As well, it requires superior skill to master its use, making it even more worthless for usage.

The modern Spyderco (and its many imitators) is not so intimidating, but functionally the same as a switch-blade. It has a large hole as part of a hump visible on the closed/folded blade that you use your thumb to open, single-handed, just like a switch-blade.

Then there are brass knuckles – also illegal to possess. The legal replacements: heavy brass, square jewelry rings worn on neighboring fingers, and the older – and much more effective -- lead powder sewn into the knuckles of a heavy leather, fingered glove – called a "sap glove". There are aluminum versions of "brass knuckles", and brass knuckles that have smaller holes with thin, somewhat sharp edges. All you need is a file to grind off the excess brass, so that they fit the fingers, and dull the inner edges. Legal to possess, illegal to modify.

* * * *

'The horror! The horror!' "

-- "Heart of Darkness" (1902)
Joseph Conrad

Then they're the hand weapons that require too much skill, and training -- continual -- to master. Even so, many are illegal in Canada, with its anti-weapons bureaucracy ever vigilant for new weapons to ban before they become popular. Or something like that...

Butterfly knives (which are not commonly known to be illegal "prohibited weapons" in Canada). They are the nunchuks of knives, and though impressive to see in use by a trained person (in a movie), are dangerous to the user (like nunchuks) if he fucks up in the twirling action. They happen to fall within the definition of "gravity knife", which falls under the definition of the banned "switch-blade" under the Canadian Criminal Code.

"Nunchuks"/nunchakus/ were popularized by the Bruce Lee martial arts movies and then banned in Canada. They are two eight inch rods made of hardwood or metal, joined at one end each by a chain or cord. Very effective, but require a lot of training to use.

Throwing Knives; though not illegal, of dubious weapon quality, usually poorly made, require mega-training, and have a high miss and failure rate from the target being of varying distance from thrower.

Another Japanese medieval weapon like the nunchuk is the shuriken (a circular saw blade falls under the legal definition as one, the law was so poorly written, and is thus technically illegal), or throwing star, requires little training. It's a hand-sized piece of sheet steel with sharpened flat knife-like projections on its circular circumference. It is a much better, effective and easy to use version of a throwing knife. It was banned in Canada at the same time as nunchuks, even though it was little known or used.

Small wooden baseball bat with a number of holes drilled (*not* nailed -- the wood will crack) all the way through the end of the bat. Push a threaded screw with nut on end, and bolt head on the top end. Repeat with a bunch of screws until you have a simple home-made medieval to WWI mace.

With a cord, hang it under your arm under your jacket for conceal-ability. Illegal, but cheap, throw-away, and highly effective at fracturing someone's skull with one easy to place, reasonably hard swing. Use, then burn or throw away -- preferably in a far-off dumpster, if it's near collection day.

* * * *

"Civilization develops in man nothing but an added capacity to receive
impressions -- that is all. And the growth of that capacity increases
his tendency to seek pleasure in spilling blood. You may have noticed
that the most enthusiastic blood-letters have always been the most
civilized of men."

-- "Notes from Underground"

Fyodor Dostoevsky

Then there are the zip guns, which originated in the 1950's, and the crude teen-age street gangs, no doubt with pompadour hair styles.

A telescoping radio aerial, snapped off a car, for a barrel, taped to a piece of wood, with a door latch as a firing pin.

Then there's a single-shot, plastic-barreled pistol flare gun. With a cartridge just smaller than a 12-gauge shotgun shell, the flare cartridge is good, and can be modified.

Then there's the semi-automatic starter pistol. There are several kinds, and several ways to modify them to shoot real bullets. Bore a small hole in the barrel, cover the gas port with chewing gum – the appropriately named "Bazooka" gum is good. Load the blank cartridge with a steel phonograph needle, or sharpened sliver of steel, or a short piece of sharpened bicycle spoke.

There's also an aluminum, single-shot, tubular flare with a spring-loaded trigger mechanism that can be drilled to fit the bullet.

Air rifles can be modified, too. To fire the sharp pellets, after sharpening the pellet. You can modify the air gun's barrel, and with more difficulty double up the compressed air cartridges with a little work, or machining, too.

You can make a pistol according to the "Improvised Munitions Handbook", an Army Technical Manual pretty easily. Or a WW2 gun.

"Today we were unlucky, but remember, we have only
to be lucky once. You will have to be lucky always.
Give Ireland peace and there will be no more war"

-- Provision IRA (1984)
communiqué after Brighton bombing
that failed to assassinate Margaret
Thatcher

Prison Diary 72:   Nothing but a Dreamer
==================

> "Many noble dreams are dreamt by small and voiceless men,
> Many noble deeds are done, the righteous to defend,
> We're here today, John Brown, to say, We'll triumph in the end,
> That's why we keep marching on."
>
> -- "Move on Over" (1965)
> Len Chandler

> "You see things, and you say, ' Why?' But I
> dream things that never were, and I say 'Why
> not?' "
>
> -- Robert F. Kennedy, quoting
> "Back to Methuselah" (1921)
> by George Bernard Shaw

In the classic 1965 David Lean film, "Dr. Zhivago", the stunningly beautiful young woman, Lara (played by Julie Christie) is discussing her fiancé, Pavel Antipov, a young, idealistic Russian revolutionary, with the rich, older, and successful, but totally corrupt, Komarovski, who is her lover. It is 1917, or so, just before the Revolution.

Komarovski contemptuously describes Lara's intended as one of the two types of men in the world, and at the same time discloses the philosophy that has turned his corruption into wealth and power:

> "There are two types of men and only two. And that young man
> is one kind. He is high-minded. He is pure. He's the kind of man
> the world pretends to look up to, but, in fact, **despises**."

Ah, dreamers. Idealists. The morally pure and uncompromising. The ideological perfectionists. The refreshingly innocent and undisillusioned. The believers. The visionary intellectuals. The creative theoreticians, who expect that the details and bumps can be worked out later.

God bless them and their purity. They are the keepers of the flame. They keep hope alive, through the darkness of the night, or the eclipse of God.

*   *   *   *

> " 'These [little children] believe in me. It would be best for the
> person who causes one of them to lose faith to be drowned
> in the sea with a large stone hung around his neck.' "
>
> -- Matthew 18:8

I read an old report, a little while ago. It was over a decade old. A close intimate had been interviewed and was asked to describe my childhood behavior. She described me as "a dreamer, a boy who lived in his own world, and was constantly disappointed by people." It was a fair, and well-meaning characterization, then -- and now.

Some people -- with good reason -- fear the "True Believers" above all. I fear the disillusioned idealist as even more dangerous.

On a broadcast in 1969, following the first moon landing by human astronauts, Kurt Vonnegut with Walter Cronkite – CBS news anchor – Gloria Steinem, and others were asked to comment. Vonnegut reiterated his view that the $33 billion spent could have been put to better use "cleaning up our filthy colonies here on Earth."

The network got letters...

* * * *

"I quite understand children liking beautiful
fairy-tales. But I ask, is it proper for a
serious revolutionary to believe in fairy-tales?"

-- "Collected Works" (1918)
V.I. Lenin

In the 1975 movie "Rollerball" [the original Norman Jewison version, starring James Caan], Jonathan, is the laid-back world sports uber-champion, turned accidental nascent revolutionary by a one government world that has killed love. He is summoned to the hospital to sign-off on pulling the plug of his brain-dead team-mate and closest friend, the hulking giant of an athlete, "Moon-pie".

"Does he...um...does he dream?" Jonathan asks the impatient Japanese doctor.

"No, there is no brain wave at all. No sort of consciousness. Just a deep coma...a vegetable. No dreams...nothing," says the doctor with finality.

But in his first, simple act of rebellion, Jonathan refuses to do what he's told, and walks out without signing the authorization form, leaving the doctor sputtering, "You *must* sign!".

* * * *

"All men dream: but not equally... [T]he dreamers
of the day are dangerous men, for they may act
their dream with open eyes, to make it possible."

-- "Seven Pillars of Wisdom" (1918)
T.E. Lawrence

In the 1985 Terry Gilliam film, "Brazil", Sam Lowry the quintessential spineless and chicken-hearted minor bureaucrat claims in exasperation at dinner to his mother to have no ambition, hopes and, "Not even dreams!" before he disappears back into his fantasy dream-world.

Martin Luther King's 1963 "I have a dream" speech was truly inspired, and considered one of the best speeches of all time.

"To sleep, perchance to dream", says Shakespeare's Hamlet, speculating that death might possibly even be a pleasant reverie. It's in Shakespeare's famous "To be or not to be" soliloquy, where Hamlet, Prince of Denmark, is contemplating suicide.

> "Let us learn to dream, gentlemen, and
> then perhaps, we shall learn the truth",

said chemist August Kekule in a speech in 1865. Kekule, who advanced organic chemistry immeasurably when he solved the mystery of the structure of the solvent benzene. He had fallen asleep and had a dream of six snakes that bit and held onto each others tails, forming a circle of the six snakes. He woke up and realized that benzene was a *ring* of six carbon atoms.

Kekule had solved a problem that had stumped the best efforts of the chemists of the time. It was produced by heating coal and cooling the gas that distilled off. It had six carbon atoms, but so did hexane, a completely different liquid. Benzene was a simple ring of six carbon atoms, while hexane was a straight chain of its six carbon atoms. (Though I find it hard to classify a snake-filled dream as anything but a nightmare!)

OK, OK, just *who* is this "Yogi" fellow. Can we pin him down to the four-dimensional universe, or will he forever be the moving ghost target that is the dreamer?

I have something to declare.

In paradise, we will eat only asparagus tips, artichoke hearts, and pre-shelled shrimp.

I declare that the three most significant inventions in modern history were low-cost, high quality steel (late 1800s), the cheap, mass-produced machine gun, the German WW2 MG-42 (1942), and the electronically amplified electric guitar (late 1950's) that permitted stadiums filled with people to hear the rock music played by the band.

And as an engineer, I declare that, "Claude Shannon Rules!" So, there!

They always tell you, "Watch out for the quiet ones." Now I know why. Because they're dreaming of a better world.

And as even the cynic knows, dreams die hard.

> "It is time to mark clearly the aims of the Revolution. ... We wish an order of things where...commerce [is] the source of public wealth, not simply of monstrous riches for a few families.
>
> We wish to substitute in our country morality for egotism, probity for a mere sense of honor, principle for habit, duty for etiquette, the empire of reason for the tyranny of custom, contempt for vice for contempt for misfortune, pride for insolence, large-mindedness for vanity, the love of glory for the love of money, good men for good company, merit for intrigue, talent for conceit, the grandeur of man for the triviality of grand society..., all the virtues and miracles of the Republic, for all the vices and puerilities of the monarchy."
>
> -- speech before National Convention (1794)
> Maximilien Robespierre

Prison Diary 73: "Remember My Chains"
                  ==================

> "[T]he bewilderment of the eyes are of two
> kinds...either from coming out of the light
> or from going into the light..."
>
> -- "The Republic" (360 B.C.)
> Plato

> "Whatsoever I feared has come to life,
> Whatever I've fought off, became my life.
> Just when every day seemed to greet me with a smile,
> Sunspots have faded, And now I'm doing time.
> 'Cause I fell on Black Days."
>
> -- "Black Days" (1994)
> Soundgarden

[By the way, the chapter title is Colossians 4:18]

And then the nightmare fog lifted, the clouds drifted away and started to fade. I was going to be FREE!

"Could I borrow your cell phone? I need to call Osama Bin Laden and get my new orders," I joked excitedly to my female guard, who told me I wasn't very funny -- though she wasn't mean about it.

It was the end of February 2008. It had taken six months to get my second bail hearing -- in front of an actual judge. He cut through the bullshit, ignored most of the testimony and lawyer/crown submissions, and cut to the chase.

He said he looked through my stack of gun permits -- the old style for restricted and prohibited firearms -- and read off a list of Chiefs of Police of the Ottawa Police over the years, their signatures stamped on my permits, as the authorizing official. In my 25-odd years as a legal gun collector, there had been several men in the top slot. He was making the point that I had been squawking about to my ass-hole lawyers all along.

Next, he stated that the RCMP National Security Unit had exonerated me of any security/ terrorism concerns, eliminating any public safety concern over granting me bail.

Finally, he stated that all that they had on me was two unregistered rifles that I was entitled to possess. Furthermore, he stated that having been found after a warrantless search of my house, the Crown thus had no case against me (because without a search warrant, the two unregistered rifles were illegally obtained evidence, and so couldn't be used at trial).

He was hinting to the Crown Prosecutor that this case shouldn't get to trial, as he granted me bail. I was free and on the street within less than 30 minutes.

Two days later, I was diagnosed with complex PTSD (Post-Traumatic Stress Disorder) -- Rambo's problem -- as the mask of togetherness crumbled, and I could completely fall apart without endangering my safety from the predatory nature of the inmate culture of savagery, where a state of vulnerability, or otherwise perceived weakness was grounds for violent exploitation.

So it goes...

"If you live in a neighborhood where there are no police, and everybody has guns and lives in constant fear of being attacked, then there is going to be a lot of shooting.

This is the sort of 'neighborhood' that all the countries of the world live in. We call it 'war'."

-- "War: The Road to Total War" (1983)
Gwynne Dyer

Prison Diary 74:   Catching a Break, Breaking a Catch
================================

> "Procul, O procul este profani"
>
> -- Sibyl to Aeneas
>   "The Aeneid" (VI, 258) (ca. 25 BC)
>   Virgil
>
> ["Keep your distance, you, the uninitiated."]

> "Let us not 'assassinate' this lad further, Senator. You have done enough. Have you no sense of *decency*, sir, at *long last*? Have you left, *no* sense of decency?"
>
> -- Joseph Welsh (1954)
>   Army-McCarthy Hearings

Injustice is a sin against God and Man. And men have a lot less patience than God.

I thought that "Western civilization" was supposed to be an actuality, and not just some sort of vague intention. Or allegation. Or empty myth.

A few days after my release on bail, I returned to OCDC to retrieve my "property" -- clothes and writings. And one more thing. To try and pass -- successfully -- a message amongst the guards, thanking profusely the guards who were kind and merciful to me. I'm sure it would be a first for an inmate to do *that*.

In the parking lot the chaplain saw me, and intercepted me, and we briefly talked. He knew me from our friendly association, from my involvement in the weekly Christian visitor meetings, and my Zen Buddhism visitor. "Don't come back," was the last thing he said to me, kindly.

I then continued on to the reception area and told the on-duty guard of my request for my property. He said to wait in the entrance corridor, while they got my stuff. Finally one of the nice young guards came out lugging my bag of stuff. "Great!" I thought, he would be perfect for my message.

After the property transfer was completed, and signed off, I asked the guard if he could pass a message to all the guards. I asked him sincerely, "Could you pass a message of gratitude from 'Yogi' to all the guards who treated me with decency and mercy? And that I really appreciated it, and wanted to say 'Thanks'."

"We're not all **assholes**, you know." he responded, which I thought was a cute and rather amusingly frank disclosure and acceptance of the actual reality of the guard-inmate dynamics.

* * * *

> "She loved me for the dangers I had passed;
> And I loved her that she did pity them."
>
> -- "Othello" (1604)
> Act I, Scene 3

I lasted a month or so, on the street, before I was thrown back in jail. I was "breached" -- I was found in violation of one or more of the many conditions attached to my release on bail, and so my bail was summarily revoked, and I was back in jail. It is illegal by court precedent to attach conditions unrelated to the offense, but so it goes.

The words of the court precedent are:

> "The courts should require that the Crown show a compelling reason why [the] basic rights of an individual to do what is lawful should be curtailed."

and

> "The rights of an accused cannot be restricted on a speculative concern of danger."
>
> -- R. v. Collins (1980)
> C.R. (3d) 283 (Ont.Co.Ct.)

I should have seen it coming, though. I was "making the rounds" at a local housing project. "Slumming", literally. I visited a female friend, then dropped by a male friend's apartment. A female that I knew showed up. She invited me to her place cheerily.

I knew Jenn was a hooker. I didn't care, but I was unhappy, miserable, and lonely. I agreed, though I had no interest in her at the moment.

But I had been set up. A pig spent 15 minutes threatening me with serious violence for *no* reason that I could figure out. I had been set up by an informant for no discernible reason. She had invited me over to her place, when we happened to run into one another. I was depressed and lonely, and it was the best offer I had received all evening. So I agreed to drop in.

When the pig's physical threats had no effect on me, it appeared he was actually getting ready to start assaulting me. I calmly looked him dead in the eye, and said matter-a-factly in a clear, but low voice, flat, and without emotion:

> "I've been kicked around most of my life.
> This is just another 'bad day'."

My eyes remained locked to his for a brief moment. And then I closed my eyes, clenched my jaws a bit, and waited for the physical pain to start. It would hurt less if you didn't see it coming. After 15 seconds, and nothing happening, I opened my eyes. Maybe nothing was coming?

Then it started. I was sitting on the edge of the bed, and he began scraping the front edge of his black leather police boot soles up and down against the front of my legs. He spent two minutes on me and left both of my shins scrapped into a bloody mess from my knees down to my ankles. Up and down, from to top of my ankle of each leg, to the bottom of my knee. Over and over, stripping the skin off, eventually stopping, after the front of my legs were bloody, and scrapped raw of skin. I just sat there, unmoving. Physical pain meant nothing to me. I began to get bored waiting for him to finish, as he scraped and bloodied my shins thoroughly. After a while he stopped.

After a time, they let me go. They couldn't arrest me after such visible, bloody, physical abuse. I slunk home, first to a pay-phone, beaten down, and embarrassed to be calling my mother to come and pick me up.

When we got home, I stripped off my clothes, and flopped onto the bed, the blood covering my legs having clotted. I suspected the cops would wait a week to try and "breach" me. But I wasn't sure.

* * * *

> "Do dreams have lessons? Do nightmares have themes, do we awaken and analyze them and live our lives and advise others as a result? Can the foot soldier teach anything important about war, merely for having been there? I think not. He can tell war stories."
>
> -- "If I Die in a Combat Zone" (1973)
> Tim O'Brien

The cops showed up at my parent's house a week later at 11:30 pm on Friday. I was supposed to be in by 11:00 pm every night, and so I was in violation of my bail conditions. I actually did show up the next morning, having spent the night at my regular home. I spent a nervous weekend wondering when the axe would fall, and they would show up again to arrest me.

Monday morning I appeared with my lawyer in Quebec court in Gatineau, to get the Quebec warrant canceled. Since I was in jail in Ontario, I missed my original Quebec court date, and a warrant was automatically issued for "Failure to Appear". The cancellation was routine, with an appearance.

They showed up again early Monday evening to arrest me. It took at least six cops to affect my arrest, being the dangerous character I was. I was handcuffed and placed in the back of a cop car, and we drove off.

"We're DART," one of the two cops said to me, as we drove to wherever they were taking me. It didn't mean shit to me, so I said nothing. "Do you know what DART is?" he asked.

"No," I responded.

"Direct Action Response Team," he informed me. I said nothing. "Big fucking deal. Who cares?" I thought to myself, though I kept my piece. No gain in antagonizing these pricks.

For some reason, the same DART clown mentioned Alex, a crack-head girl I knew. I was supposed to be impressed that they knew one of my associates, I guess.

I held my piece once again, but I felt like saying, "The only thing that I have to say about Alex, is that she's a better fuck than your wife." Instead I perked up and said, "Oh, you know Alex? How's she doing?" in a jovial affect.

He also mentioned the Quebec warrant that I had had canceled with my lawyer that morning. Kafka, again. I was going to jail either way. If I had been home the previous Friday, I was going to jail for the BS warrant, and if I wasn't home, I was going to jail for not being home. And I was going to spend at least the weekend in jail, since they came on a Friday.

And then the prison "doctor", who had done **nothing** for me, took me off my prescription Clonazepam, an addictive tranquillizer of his own accord, "Because I don't want you to take addictive drugs." So for a week, I lived the hell of cold-turkey Clonazepam withdrawal, and the horrible mood swings it produced.

Kafka, again? On bail, one of my conditions was that I take all drugs my psychiatrist prescribed to me, but in jail an incompetent doctor thought he knew better than my shrink and took me off one of them, unilaterally.

> "If we do not act, we shall surely be dragged down the long, dark, and shameful corridors of time reserved for those who possess power without compassion, might without morality, and strength without sight."
>
> -- speech at Riverside Church (1967)
> Martin Luther King

Prison Diary 75:   Who Mourns for Odysseus?
=====================

> "A hundred bloodthirsty badgers, armed with rifles,
> are going to attack Toad Hall this very night, by way
> of the paddock.
>
> Six boat-loads of rats, with pistols and cutlasses, will
> come up the river and effect a landing in the garden;
> while a picked body of toads, known as the 'Die-Hards',
> or the 'Death-or-Glory' Toads, will storm the orchard
> and carry everything before them, yelling for vengeance."
>
> -- "The Wind in the Willows" (1908)
> Kenneth Grahame

> "If this which he [avows] does appear,
> There is nor flying hence nor tarrying here.
> I 'gin to be aweary of the sun,
> And wish th'estate o'th' world were now undone.
> Ring the alarum bell. Blow wind, come wrack,
> At least we'll die with harness on our back."
>
> -- "Macbeth" (1606)
> Act V, Scene V

Victorious or routed, test of arms, and test of will.
The war is over, the field of battle still.
Yet war lives on in heart and mind,
With war so cruel, shouldn't peace be kind?

There are two types of soldiers, I told my son once, good soldiers and dead soldiers.

I was always looking for a good fight. Something to take a stand for. Then I mellowed out, and ceased looking. But if something to take a stand on presented itself to me, I wouldn't hesitate. And finally I lost faith, and then lost my nerve.

All that was left was a feeling of profound emptiness, the numbness that I used to require, but which now just leaves me sometimes feeling like an empty shell. All that was left are the instincts that don't fade, and undependable reflexes, and the decay of with its loss of steely will. And the deep scars and wounds that never heal -- one of which most people call "regret".

But, if anything, I had always wanted to be Rimbaud, not Rambo...

* * * *

> "Torture might last a short time, but
> the person might never be the same."

-- Amnesty International report (1970)

In World War I, they called it "shell shock". English officer and poet, Siegfried Sassoon, probably its most famous afflictee. In World War 2, they called it "battle fatigue". Or "the thousand yard stare". After Vietnam, they finally called it Post-Traumatic Stress Disorder (PTSD).

"Complex" PTSD, in my case. A constant state of being combat ready, a complete destruction of social trust (the "complex" part of Complex PTSD -- the most severe form of PTSD), an exaggerated startle response; hyper-vigilance; recurrent, deeply disturbing nightmares, and flashbacks, to name a few.

The most famous ancient work of literature is the Greek epic "The Odyssey" by Homer. The story of the war hero attempting to return home. His name: Odysseus in Ancient Greek, Ulysses in Latin.

> [The most famous work in English literature -- as best that I can
> ascertain -- is "Ulysses" by the Irish literary giant James Joyce]

The "Odyssey", I found out from a book on PTSD by a psychiatrist, is the essential story of a veteran's return from war. It takes him ten long years of trying to succeed, and the adventures he goes through mirror the symptoms of PTSD.

*   *   *   *

> "Everybody says I'm someone else,
> That I'm sick and there's no cure.
> Dammed if I know who I am,
> There was only one place I was sure,
>
> When I was still in Saigon,
> Still in Saigon.
> I am still in Saigon,
> In my mind."
>
> -- "Still in Saigon" (1982)
> Charlie Daniels Band
> Dan Daley

Colonel Kurtz of "Apocalypse Now" (1979) represented the absolute Truth driving men insane, like the Siren Song of the Homer's "Odyssey". Driving them insane with PTSD.

PTSD first really entered public understanding and popular culture with the 1982 hit movie "First Blood", based on David Morrell's 1972 novel of same name, and starring Sylvester Stallone as John Rambo. At the end of the book, after destroying the town where he had been abused by the local sheriff, Rambo shoots himself in the head. Fade to black...

But in the **movie** – the ending changed because of the insistence of test audiences – his former commanding officer, Colonel Trautman is called in to helicopter him away, a sort of modern _deus ex machina_. ["First Blood" was so successful, by the way, that it spawned three sequels, all of them bad.]

"Still in Saigon" (1982), by the Charlie Daniels Band -- a country & western, rock cross-over hit -- whose title originated from part of Capt. Willard's opening narration in "Apocalypse Now" ("Saigon. Shit! I'm still only in Saigon.").

In "The Falcon and the Snowman" (1984), about the mid-1970's Chris Boyce spy satellite espionage case, a drunken NSA (National Security Agency) "black vault" cipher clerk goes off, gone surly and mean, after having gotten drunk at a bar drinking with his work-mates after work.

"Spare us, Gene -- war's over," pleads a female co-worker. But Gene's on a roll.

"I killed thirteen of them myself," he says, taking another gulp of beer. Then his anger flares, as he spits out his story: "Communists! Special Ops! Laos! Close-range! Right between the eyes..." Glaring at his teenage or so, co-worker and friend, actor Timothy Hutton as Chris Boyce, he says in a mixture of cynicism, bitterness, and drunkenness:

"Don't believe me, do you? Believe it, college boy," Gene sneers drunkenly.

In the movie, "Highlander" (1986), a crazy ex-Marine, Vietnam Vet named Matunas, who drives around in an old Camaro all night, with a trunk full of guns and ammo, sees something going on as he cruises the New York City streets. It's playing "Hammer to Fall" by Queen in the car, when he sees two guys in a NYC back-alley sword-fighting. He stops to investigate, getting out of his car armed with an open bolt (full-auto) Uzi.

The Kurgan -- an evil Immortal -- is sword-fighting the black Immortal Kastagir, and eventually decapitates him. Whereupon Matunas riddles the Kurgan with bullets -- to which he's immune -- from his Uzi, and is thereupon impaled by the Kurgan on his sword, but ends up not being killed, only hospitalized.

Finally there's the rock hit, "Copperhead Road" (1988) by Steve Earle:

> "Now the DEA's got a chopper in the air,
> I wake up screaming, like I'm back over there.
>
> I learned a thing or two from ol' Charlie,
> don't you know.
>
> You better stay away from Copperhead Road.
> Copperhead Road."

* * * *

> "Apocalyptic predictions require, to be taken seriously,
> higher standards of evidence than do assertions on
> other matters where the stakes are not as great."
>
> — "Foreign Affairs" (Winter 1983-84)
> Carl Sagan

PTSD "combat-ready"; makes it difficult to relax, makes it easy to isolate, makes you only truly comfortable in your "oasis" -- my oasis was my bedroom, upstairs in my parent's three-bedroom town-house. Originally, I alternated between rage and tears in the privacy of my room.

And you have sometimes severe difficulties with short-term memory loss.

The reason for short-term memory impairment, I realized, was that in "combat", there is no past. There is no future. There's only here-and-now. And that's it. The here-and-now that is your only focus, because it's the best chance of winning what you view as a game of "death or survival", with nothing in between.

There is no need for a past, so there is no need for wasting your effort on memory creation. There is no future, because with only death or survival, you can "delete" the future as an inefficient waste of brain processing needed and better used in worrying about here-and-now.

Other complex PTSD symptoms:

- " 'I trust no one,' is the voice of complex PTSD." wrote the shrink's book ("Odysseus in America") on PTSD **twice**.

- Towards strangers (men) a hair-trigger excessive fight-or-flight reaction to a "perceived" (to the afflicted) threat

- Escalating sequence of anger to rage to violence with specific triggers by intimates (close family members who you feel should understand, but don't)

- rapid unpredictable mood swings; rage to despair. Bouts of crying in private

- a free-range mild to severe depression, a sense of impending doom, hopelessness

- feelings of no meaningful future or past, only a blighted present

- frequent, persistent suicidal ideation, and detailed planning, and thoughts

- profound sense of alienation, leading to social withdrawal and self-isolation

- impulsive, reckless behavior without regard for danger or judgment

- greatly reduced attention span, impaired ability to focus and concentrate

- easily distracted, tendency to ramble verbally or go off on tangents, or rant

- loss of sense of time; being late or unaware of scheduled appointments

- severe impairment of ability to motivate oneself (awaiting only combat?)

- impairment of normal patterns of sleep, hygiene, presentability, consideration of other's opinions or normal conventions of behavior (combat ready)

- recurrent violent nightmare or nightmares so that you feel like you can't escape from your awful sense of consciousness even with sleep

- hyper-vigilance; walking in the middle of the road at night to avoid being jumped/ambush-attacked from behind bushes, or tall hedges if you walk along the sidewalk

- hyper-reactive; to certain loud, sudden and unexpected "banging" noises, like a knocked over glass, dropped ash-tray, or suddenly turned on nearby police car siren

- a nervous tick -- a leg movement when seated -- either spontaneously, around groups of people, or when under stress and last but not least:

- extreme susceptibility to substance or alcohol abuse to kill or dull the mental pain.

Welcome to my world.

All in all, my life could be broken down into "good days" and "bad days" -- and even the good days weren't so good, usually involving "flight", or "escape" from my sad reality.

I have been heavily medicated since my diagnosis – minor tranquillizers, major tranquillizers, dream-suppressants, anti-depressants.

* * * *

"It is royal to do good and be abused."

-- "Meditations" VII, 36 (ca. 175 A.D.)
Roman Emperor Marcus Aurelius
quoting Antisthenes

"Long periods of clarity," were the words I heard from a psychiatrist once, years ago. I suppose they were meant to be reassuring. I closed my eyes for a second, and then as I felt a sinking feeling come over me, and an "Oh, no..." escaped from my lips.

"Madness in great ones must not unwatched, go," warned an observer in "Hamlet".

R.D. Laing, a Swiss psychiatrist wrote that there are three things that men fear: death, other men, and their own minds.

He also wrote extensively in the 1960's of his belief and theory that insanity was the reaction/ over-reaction of perfectly reasonable and normal people to an insane world. A world that was a farce, and an utter abomination in form, substance, foundation, and practice. The implication was that conventional understandings of normality should be defined as a collectively insane human race, but that produced a society "functional" in some bizarre sort of way, that could probably described as "lurching along in a generally random way".

R.D. Laing has been criticized for his philosophy, as denying the real suffering that the mentally ill go through.

* * * *

"Democracy is the only system in this world that puts a premium on human dignity. Even though people differ in all kinds of ways, for example, in ethnicity, proclivity, and even ability, the democratic system considers them equal in dignity."

-- "Categorically Incorrect" (2006)
Alan Borovoy

Some relevant comments on how others have looked at, or interpreted, insanity:

"The sleep of reason begets monsters."

"Madness is one of the means man has
of losing his freedom."

-- resignation letter (1956)
Dr. Frantz Fanon

Or, as the alcoholic private-eye protagonist in the 1987 semi-animated comedy, "Who Framed Roger Rabbit?" concludes,

"He's one seriously disturbed [Car]'Toon!"

And finally, Mr. Spock states with typical insight, and finality, in the 1967 "Star Trek" TV episode entitled "The Alternative Factor":

"Madness has no reason, nor purpose -- but it may have a goal."

*   *   *   *

Violence, deceit, and lawlessness are
natural functions of the State, any State."

-- "The Manufacture of Consent" (2002)
Noam Chomsky

But finally, all in all, to paraphrase "Antek" the _nom de guerre_ of a Jewish fighter (and one of the few survivors) in the 1944 Polish Warsaw Ghetto uprising against the Nazis (as quoted in his interview in the 1985 movie "_Shoah_" ("The Destruction", the Hebrew word for "The Holocaust"):

"If you could lick my heart,
you would taste poison."

Prison Diary 76:   "Out Where the Buses Don't Run"
==========================

> "They called me mad, and I called them mad,
> and -- damn them -- they outvoted me."
>
> -- Nathaniel Lee (1684)
> Bedlam asylum inmate

> "When you go to have a tooth pulled, you're
> frightened -- it'll hurt -- you prepare yourself.
> But this isn't a tooth, it's the whole of you --
> your whole life being pulled out. And what
> does it mean? Nobody knows.
>
> And I am sick at heart and terrified."
>
> -- Anna Ivanovna
> "Doctor Zhivago" (1957)
> Boris Pasternak

Enemy of the State. Prisoner of the Law. Prisoner of Politics. Prisoner of the Heart. And now, I was horrified to find myself a Prisoner of the Mind.

Madness indeed, takes its toll.

Two days after I was released on bail, I received an emergency appointment with my psychiatrist -- who I liked. I came into his office sat down, and launched into a rant of screaming insults at Canada, white people, the police, everyone.

It didn't take him long to diagnose me with PTSD -- Post-Traumatic Stress Disorder -- from my exposure to violence in OCDC, the Ottawa jail.

They had finally broken me, after six months, and he immediately put me on anti-depressants, three different tranquillizers, nightmare suppressants – mood alterants. Being "heavily medicated" worked after a time. The recurrent, repetitious nightmares ceased immediately. And after three months under medication, I was no longer continuously suicidal, amongst other afflictions.

An escape I'm not sure that I value...

* * * *

> "Madness, insanity. Living profanity. Then
> some punk sayin' they un'erstandin' me?"
>
> -- "Colors" (1987)
> Ice-T and Afrika Islam

After I was released on bail, in the spring [no pun intended] of 2008, I attended a meeting scheduled by my new disability lawyer to discuss a pending civil suit against my employer. At some point, as she sat behind the desk in her large and finely-appointed window office, she pronounced what I considered a singularly unhelpful platitude: "Mental illness is not well understood."

"Alienation, social rejection, isolation, misery, and despair. What's so hard to understand?" I felt like responding.

(Actually, I ended up really liking and respecting her -- and possibly vice-versa -- so it's all good.)

Fortunately most people aren't insane -- they're just stupid. Lucky them. I had now discovered -- to my surprise and consternation -- that there's a worse state of consciousness than stupidity. I guess it was about time. And anyway, stupidity and ignorance are not _per se_ sins, as annoying as they are. It's immoral behavior that is.

With Post-Traumatic Stress Disorder (PTSD), like many mental illnesses, you don't really understand what has happened to you and why. There's really no one to talk to, because no one really understands or can comprehend what has happened to you, what is going on inside your head, and the inner torment you're going through. They can't really understand why you're behaving the way you are, even if they cared to look beyond the insularity of their tiny little lives of boring, finely planned certainty.

And I couldn't blame them. There was no way they could see, or understand, and that there was really no way they possibly *could* understand. But that insight didn't really help my serenity and peace of mind much at all, not surprisingly. Alienation, is one of many torments of post-traumatic stress disorder, and it's a known factor that amplifies the pain, distrust, and suspicion that are some of the many components of the condition. As a result, drug abuse and suicide become major risk factors as the wounded sufferer seeks easy solutions that aren't.

And you find out soon enough, that most expressions of compassion extend only until people realize not just that your erratic behavior is an *inconvenience* to them, but that you make them and everyone around ill at ease -- even spooked -- as your struggle to stay afloat plays out in a disconcerting spectacle of a floundering human being, one that they once knew, but know no more.

And have been in varying states of closeness or intimacy with you, they'd rather not share the pain of watching you drown close up and right before their eyes, with not even an understanding of why.

Who can blame them? Even I could understand their reaction, despite being its target.

Of course, like I used to joke in saner times, "To understand a spinningly complex, paradoxical, and unusual person such as Yogi, there's a simple first step. First you have to try."

And in all who observe you -- if they're strangers -- they briefly stare, then just pretend to ignore you, perhaps shaking their head as they hurry away to the safety and anonymity of distance.

And of those who observe you, and worst of all, have seen you in better times and know you as an acquaintance or closer, you're a shock -- as the subconscious, ever-present doubt of everyone's own tenuous hold on sanity, and an uncomfortable reminder of their suspicion of its fragility.

It's the randomly erratic behavior of the mentally "afflicted" that unfortunately is most disconcerting to others that is the problem. It's the unpredictability of your disquieting behavior that people just can't seem to deal with, in my estimation. And it is fundamentally unsettling to the sense of order and stability of the "normal" person.

And a (usually) irrational fear, soon leads to an instinctive withdrawal of social contact, ultimately leading to a shunning of the mentally-suspect individual. It's the ultimate paradox and -- in a sense -- the ultimate insult that the predictability of the Fool is tolerated, while the uncertainty of the mentally troubled is not.

* * * *

> "May the state fence in the harmless mentally ill solely to save its citizens from exposure to those whose ways are different? One might as well ask if the state, to avoid public unease, could incarcerate all who are physically unattractive or socially eccentric."
>
> -- Justice Potter Stewart (1975)
> U.S. Supreme Court opinion
> in O'Connor v. Donaldson

"Mr. Jones" (1993), starring Richard Gere, is a charming little movie on the travails of a man trying to deal with bi-polar mental illness – manic-depression. There's a great scene towards the end where Gere, playing an intelligent, charming, but manic-depressive ex-concert pianist reduced to day-laborer carpenter, is angry at his psychiatrist.

Opening up for once, he vents at her, ranting about the awful social cost of his affliction -- his great love finally rejecting and leaving him -- and the armor he hammered together to protect himself from the emotional pain he knew was going to be a permanent part of what he assumed would be a blighted future:

> "Too much trouble...I was always too much trouble...
>
> You know I got this little trick, this really good, little trick. You see, you're not human anymore -- **none** of you! You're not human! You're like -- you're like goldfish! **All** of you!
>
> One dies, I get another..."

Marvel at the range of Gere's acting ability as, in the full scene of his emotional outburst, Gere seamlessly shifts through a range of distinct emotions: explosive anger, bitterness, rage, intense rationality, indignation, vindictiveness, despair, and finally self-pity, all within about 40 seconds or so.

When I finally noticed the stream of emotional change, I understood finally what is meant by "great acting".

* * * *

> "Run to the bedroom, in the suitcase on the left,
> You'll find my favorite ax.
> Don't look so frightened, this is just a passing phase.
> Just one of my bad days."

-- "One of my Times" (1979)
Pink Floyd / Roger Waters

"Out Where the Buses Don't Run" (1985), is the title of my favorite episode of the iconic mid-1980's TV cop show, "Miami Vice", directed by the genius of Michael Mann. It was a series which re-invented weekly action shows with a quality upgrade and redesign that included creative and intelligent plots, hard-hitting scripts, superb acting, and colorful and sophisticated sets, impressive cars and uber-fashionable clothing.

The fast-paced hour long series typically focused on human failings and the minutiae of current criminal issues, using an edgy and stylish cop show as a captivating backdrop.

The two vice squad partners the show was revolved around wore stylish and expensive pastel designer suits and drove expensive sports cars. And, for the very first time, a weekly TV show paid the mega-bucks needed to use recent original artist hit rock songs appropriate to each show's plot. The sound track of "Miami Vice" was thus another mind-blowing innovation to a show the was the very definition of innovation.

The plot of "Out Where..." involves the sudden appearance of a legendary ex-Miami vice cop cashiered a decade before under mysterious circumstances. His amusing and insightful, yet completely insane shenanigans, culminate in unveiling a man whose burden of guilt over murdering a drug kingpin years before, finally becomes too much for him to bear.

As he loses touch entirely with reality, his riddles finally resolve as he exposes the mummified remains of his victim, stuffed behind the dry-wall -- "sheet rock" in the U.S. -- of a now abandoned, and soon to be demolished, residential Miami house.

The final epitaph comes from his conscience-impaired ex-partner – now a Federal Agent -- who comes clean with an admission of having helped his partner dispose of the body of the slain cocaine drug baron, who had just been acquitted at trial.

Abducted by the two cops, determined to see vigilante justice finish the job that rules and process couldn't, the show concludes with the Dire Strait's classic, "Brothers in Arms", playing, as the ex-partner pleads for understanding from the other stunned cops showing in full color reality, the tight bond between men facing a common foe, the problems facing men in a society that pretends it isn't at war with itself, and survival in an environment with everything at stake, corruption, gun play, and no fixed rules, and the moral quicksand waiting and ever ready.

> "I helped him build the wall... He was my **partner**.
> You understand? He was my **partner**."

* * * *

While Colonel Kurtz -- played by Marlon Brando -- in "Apocalypse Now" (1979) was one of the first popular culture portrayal of PTSD, it's not the movie's focus. So as a subtext of Kurtz's consciousness, it eludes the viewer's attention, distracted by his brutal, cryptic, but insightful intellectual ramblings:

> "I watched a snail crawl along the edge of a straight razor.
> Crawling...**slithering** along the edge of a **straight** razor...
> and surviving. That's my dream -- that's my nightmare."

In "First Blood" (1982), with Sylvester Stallone playing John Rambo, a deeply troubled Vietnam vet, his PTSD is front and center in the plot, triggered into its violent stereotype, when Rambo is gratuitously fucked over by a country bumpkin sheriff -- a refugee from Central Casting, if there ever was one -- who doesn't realize he is messing with the wrong guy, **and** a Congressional Medal of Honor honoree.

Rambo is targeted because in the self-righteous sheriff's hypocritical world -- he's actually just a bully with a badge -- Rambo is an "undesirable", a "drifter", a worthless vagabond. Rambo is not viewed as what he really is -- existentially lost, condemned to the biblical punishment of permanent wanderer.

Rambo is not just an aimless wanderer – he is the ultimate loner trying to be alone, accomplishing his goal of solitude by keeping on the move, in the country in which he once felt was home. But his strategy fails in the end, an example of fate. Ever-fickle fate. He cannot escape the world and its random unfairness. And the uber-warrior within unleashed, he single-handedly razes the sheriff's town -- overkill masquerading as a kind of righteous nemesis and catharsis that audiences revel in, as vicarious revenge and resolution for every wrong ever done to them, real or imagined.

Rambo wins the battle, but not the war, once again, and the movie originally ends with Rambo shooting himself in the head. His only victory can be one of symbolism – as he controls even the time and manner of his own death, and suffer bravely the same thing he dished out to his enemies. A final unmistakable display of his rejection of the world, a final act of rational choice, and that *he* controlled his reality and destiny, and would not accept anything less.

Resistance and individualism, and a refusal to yield or surrender, and to deny his persecutors anything but a bad taste in their mouth, and prove that if a final victory was to elude Rambo, so would it elude them, in being unable to capture and prosecute him. Not that he cared that they understood this, or even believed they were capable of doing so. That **he**, himself understood was all that was important.

But American test audience reactions were so insistent and strong that this ending was unacceptable, and be changed to a bogus Hollywood happy ending. So it goes. Realistic endings can be kind of depressing -- and it **is** supposed to be entertainment...

* * * *

> "A long time passed. Weeks, perhaps months. The account of this period is less full than the author would have liked.
>
> Facts are scarce, and even the notebook, which has provided much information, is suspect. We cannot say for certain what happened to Quinn during this period.
>
> For it is at this point in the story that he began to lose his grip."
>
> -- "City of Glass" (1994)
> Paul Auster

There's a well known association between genius and insanity. Back in the days before I was crazy, I was more miffed at the link between genius and the economic exploitation of genius, as well as the link to social alienation.

But then I realized that chasing skirts required no social skills -- it actually seemed to be a prerequisite. And suddenly things started looking up for my general happiness. I also came to the conclusion that keeping your woman happy and content required no effort. Though she did seem to complain a lot...

And that only working when the boss was watching was a viable strategy for success. Successful work avoidance... In fact, it also became clear that merely looking concerned was a successful solution to most of life's problems and the general vicissitudes of human existence.

* * * *

"You should never argue with a crazy mind,
You oughta know by now."

-- "Movin' Out/Anthony's Song" (1977)
Billy Joel

Paris, France -- I've been there many times -- is like an open air museum. At random, I took a picture of a street sign, presumably of a Frenchman of note. It was Rue Auguste Comte, in the 6th Arrondissement, of Paris' Left Bank (think "Eiffel Tower"), South from the Latin Quarter's St-Germain-des-Prés, near the large park called the Luxembourg Garden.

I later found out that Auguste Comte, French philosopher and mathematician (1798 - 1857), theorized that there were some things Man would never understand -- such as how the sun and the stars worked -- and then went nuts.

And, by the way, we've known since the 1930's how the stars work. Hans Bethe won the 1934 Nobel Prize in physics for working out the thermonuclear fusion reactions involved, and unintentionally and indirectly lit the fuse leading to H-bombs, the Cold War, and a world a launch code away from Total Nuclear War and human extinction from the Nuclear Winter that would follow.

The Marquis de Sade spent most of his adult life (during the late 1700's) confined in insane asylums, including the famous Bastille Prison. One July 14th, he spotted a peasant mob outside through the bars of his cell window and began yelling that the guards were massacring the inmates, triggering the start of the French Revolution. There's a Paris subway stop named after the Bastille now. The Bastille itself, is no more, razed to the ground.

The Marquis, however, is known more for his rather bizarre sexual proclivities, rather that his mendacious initiation of a defining moment in French history, world history, the start of state reform by violent revolt of the masses.

On the other hand, there was Dr. Wallace Carothers, an alcoholic chemist who was afflicted with bipolar mental illness, who would "disappear" from his job at chemical giant DuPont Corporation for weeks on end, carried a vial of potassium cyanide in his pocket, and committed suicide at the age of 41-- with a cocktail of lemon juice and potassium cyanide -- leaving behind his pregnant new wife behind.

Yet Carothers, a gifted polymer chemist who invented not just the synthetic rubber dubbed Neoprene, but capped off his brief -- but stellar -- career and troubled life by inventing Nylon in 1938, just in time to provided a plethora of wartime (parachutes, and the first generation "bulletproof" jackets – the ballistic nylon "flak jacket") and consumer products, and earned DuPont, revenues of $6 billion a year for at least a couple of decades.

His tragic, but little known story was finally chronicled in the 1996 ACS (American Chemical Society) book, "Enough for One Lifetime". Good title, and a story that needed to be told -- that needed to be documented. But I found the book uninspiring. A chemist was the author...what can I say?

And as the final postscript, it's interesting to note that Carothers' erratic behavior wouldn't be tolerated by modern corporations. There's a lesson unlearned in there.

* * * *

> "Sickness will surely take the mind,
> Where minds can't usually go.
> Come on the amazing journey,
> And learn all you should know."
>
> -- "Amazing Journey" (1969)
> in Rock Opera "Tommy"
> The Who / Pete Townsend

Jacobo Timerman was a Jewish newspaper editor in democratic Argentina. In the mid-1970's, with U.S.-backing, a coup d'état by officers of the Argentinian Armed Forces overthrew the democratically-elected government, and established a military junta -- dictatorship by a committee of generals. The end result of the brutal fascist rule they established became known as the "Dirty War" -- a systematic campaign of round-up and extermination of an estimated 30,000 citizens with leftist or otherwise suspect beliefs, including suspected militants or radicals, political opposition workers, organizers or other activists, sympathizers, union supporters or leaders, students, writers, reporters.

Conform or be noticed. Questioned and be suspect. Defy and disappear. Believe or die.

Timerman was one of the many arrested and merely jailed. (Pandemic anti-Semitism in Argentina, and the blatant anti-Semitism of his tormentors, did not help.) It was the Day of the "Disappeared" -- and it was a very long day. It was part of the body count of the U.S. Cold War. And to be a citizen of a country within the U.S. sphere of influence was to be a citizen of a country where democracy and freedom were sacrificed to the stability of dictatorship. Where compliant dictatorship provided U.S. economic stability, and unimpeded access to resources for the undeclared economic war of attrition against the Main Enemy, the Soviet Union (USSR).

After his eventual release, Timerman wrote "Prisoner without a Name, Cell without a Number" (1980), in which he described the pain and hopelessness of his ordeal -- the unfairly imprisoned, indefinitely imprisoned, and the harshly imprisoned. And he described how it works its evil sorcery, especially in a fascist police state, which operates through pure force of arms, absent the constraints of the rule of law or any such basic human rights and freedoms.

As his detention and suffering continued with no end in sight, first you pray for death, he wrote -- but death does not come. Then you pray for insanity. And soon enough, insanity hears your prayer. But there is no escape in madness and desperation the relief of your suffering has not served you well, in welcoming it into your soul.

> "_Quos vult perdere Jupiter dementat_"
> ["Whom the Gods destroy, they first make mad."]

warned Euripides, Ancient Greek play-write, in about 450 BC.

* * * *

> "I saw the best minds of my generation destroyed by madness,
> starving hysterical naked,
> Dragging themselves through the negro streets at dawn looking
> for an angry fix,
> Angel-headed hipsters burning for ancient heavenly connection
> to the
> Starry dynamo in the machinery of night."
>
> -- "Howl" (1955)
> Allen Ginsburg

While the validity of the evil entity known as the "demon" has passed into mythology, according to the Christian New Testament, the demon is real -- a fugitive from Hell, who must find a living human host to inhabit, in order to escape being returned to the torments of the Pit.

And Jesus made almost a side-line out of freeing the possessed from their demons, since he could see and talk to the demons, and could expel them from their victims with little effort -- being the son of God, and all that.

> "My name is Legion. Because
> there are many among us."
>
> -- Mark 5:9 & Luke 8:30

So says the apparent "spokes-demon" representing the 6,000 or so demons -- a Roman Legion numbered 6,000 soldiers --- talking to Jesus, who has encountered a local "madman" possessed by this legion of demons. The man is uncontrollable, gifted by the demons with super-human strength -- but a social outcast, his home a cemetery away from his town, alone to endure the insufferable torment that he does not know the cause of.

Jesus exorcises the man of the demons, returning the grateful man to normalcy. The demons come to a bad end, the founding of the U.S. Republican Party still being a couple of millennia in the future.

This story is recounted in two Gospels (Mark 5:1-20 & Luke 8:26-39), and I had always found it memorable. Now -- give or take a few demons -- I felt like I was living it... There's only so much suffering a man can take. And the weight of my burden just seemed to keep growing.

* * * *

> Defining the difference between a neurotic and a
> psychotic, he described the first as "the man who
> builds a castle in the air," the second as "the man
> who lives in it." And, he added, the psychiatrist is
> "the man who collects the rent."
>
> -- Lord Webb-Johnson
> physician and wit

The "score card" -- the snowball -- as it stood:

1) A completely false criminal allegation was made to the local municipal police by an individual

with diagnosed mental problems, about a man with whom she was involved in some sort of relationship. There was no independent evidence, injuries, or witnesses, history of previous discord, or any police questioning of the accused, yet an arrest warrant was issued, and suddenly the enormous power and resources of the state were brought to bear on an individual with no criminal record, and a reputation of good character and stability.

A word from a woman routinely results in the machinery of the state going into action, arresting a man without any reasonable grounds, violation of several clauses of the Charter -- the Constitution -- and showcases Canada's judicial system -- from cop to Supreme Court -- as unable to recognize, much less stop, a scandalous record of hundreds of cases of bumpkin-style kangaroo court Gilbert and Sullivan judicial antics by a bunch of overpaid imbecile boobs.

Like a Gilbert and Sullivan play without music, lyrics, script, acting, or entertainment value, leaving only a bunch of buffoonish, highly paid officials drunk on the own self-importance, stupidity, and ability to abuse their powers at the own discretion, with no fear of accountability, and for no useful public purpose.

The government was not involved in a personal, private, domestic dispute, on a man of sound and long-establish reputation with just a phone call from a woman who wasn't even a citizen, and had a lengthy record of disputes with neighbors, family, boyfriends, employers, who had finally been diagnosed with a major personality disorder, and freely accepted the diagnosis as valid.

2) A totally illegal warrantless search of my house in Ottawa was conducted by Gatineau Police in flagrant violation of basic Charter rights. No tip, information, probable cause, or other reason existed. And without a search warrant, they were engaged in grossly illegal misconduct. Their reasoning apparently was that they just felt like it -- and that this justified their search.

3) A blatantly false arrest by Ottawa Police, called in by the Gatineau officers after they noticed my legal collection of deactivated machine guns -- even though I showed and explained the deactivation in detail -- numerous welds and destruction of essential components, that could only have been on racially-based, witch-hunt grounds. The false arrest occurred after I had shown the deactivation details of the machine guns, and Ottawa constable Laplante assaulted me, and further compounded his illegal malfeasance by failing to answer my request for the details of the charge he claimed he was arresting me for...

4) A police press conference for the media is held to display a big pile of my deactivated machine guns -- which they freely admit to reporters are perfectly legal. In fact the only items bearing any taint of criminality were two unregistered rifles -- with everything else displayed to reporters being legal.

What the reason was, and what gave the police the right to display a citizen's legal property to the media was not a question asked by anyone.

The only reason I could see was a deliberate smear campaign to cover-up the misconduct of the Ottawa Police against me: my false arrest, and an illegal warrantless search. By manipulating the media into being the vehicle for the police smear job, their bungling would be lost in the smokescreen of suspicion, innuendo, and an atmosphere of public fear.

And the motive seemed to not just involve defaming my reputation, trashing my privacy rights, and putting on a show for an already paranoid public. They were setting the stage for bail denial using an atmosphere of racially-motivated witch-hunt politics to camouflage with further malfeasance the previous police misconduct.

5) Illegal leaks to the press by the Ottawa Police Guns & Gangs Unit using confidential information obtained from the illegal search, once again to smear my reputation.

6) A rubber-stamp denial of bail release by an unqualified justice of the peace spooked by the public attention focused on him, extensive police perjury, and a shrill prosecutor engaging in falsehood and mischaracterization, using the entirely baseless "terrorist" bugaboo.

7) Exoneration of all security/terrorist concerns by RCMP National Security Unit, yet continued objections by Crown prosecutor to release me on bail.

8) The delay of my trial to presumably deliberate delay my release and continue my coercive Incarceration in an environment of regular but unpredictable violence. I had been physically assaulted twice, and choked unconscious ("by accident") in a third incident.

9) Continued incarceration for six months under violent, torturous conditions. Daily abuse and degradation and nonfeasance by guards, and malfeasance by the jail administration.

10) Disingenuous -- read "Bullshit" -- cancellation of February trial date, on the date of the trial, by prosecution, in order to waste my money, and to keep me in jail longer, coupled with refusal to consent to bail.

11) Visitor calls from my mother to me were listened to by Crown. As a pretrial detainee the only reasonable and justifiable reason for call monitoring was security of the institution. It is illegal and unconstitutional in the United States.

12) Monday 3 March 2008; Release on bail after six months in jail, and being transferred all around the province in retaliation for filing numerous administrative requests on the provided jail form -- request that dental floss be provided, request for a Canadian Criminal Code, request for a prison law library, etc.

13) Wednesday 5 March 2008, diagnosed with "complex" PTSD -- PTSD with complete destruction of the afflictees social connection with everyone -- by my psychiatrist at an emergency appointment. I started screaming at my shrink as soon as I sat down for the appointment... "Fuck the pigs! Fuck the legal system of this country! Fuck Canada! Fuck Canadians! Fuck white people! I mean, fuck white men. You're all corrupt, evil murderers and hypocrites. Damn you all to hell!"

14) Good Friday, 21 March 2008. Being set up by an informant for no reason and then kicked bloody for absolutely no reason by an Ottawa Pig in gross violation of the law, human rights, and human decency. My shin was still scarred two years later, and I was bloodied from my ankle to the bottom of the knee on both legs.

14) Friday March 28, 2008 One week later, "Guns and Gangs Unit" shows up to arrest me for violation of illegal bail conditions illegally imposed by Crown for this purpose.

15) Monday March 31, 2008: Arrested for breach by DART

16) Wed. 8 April 2008: Arbitrarily denied my prescribed tranquilizers by the prison "Dr." Tate, resulting in a week of terrible mental suffering due to the resulting cold turkey withdrawal.

" ' If people do this to a green tree,
what will happen to a dry one? ' "

-- Luke 23:31

* * * *

My second lawyer called me, "eccentric" for using expressions like "tissue of lies", writing books about nuclear weapons, reading books about the Irish War of Independence (and he was Irish!), and spouting 1960's-era leftist slogans, in response to the remarks of his that I felt were stupid and a betrayal of beliefs he once shared with me decades before.

But "eccentric"? That sounded to me just like a diplomatic way of calling a paying client "nuts" to his face.

Not fooled, and disappointed and amazed by, among other things, his attitude, belief system, lack of professionalism, and betrayal of the leftist ideology that governed his teenage years when I knew him in high school, I fired him before my next Court appearance.

I may be crazy, but stupidity, I try to avoid. And I don't pay good money to people too dumb to avoid irritating me, especially when I'm in a tight corner, and dependent on their professional skills. But his fate was sealed when, after failing miserably to win my release at the bail hearing, he tried to get me to "plead out" -- plead guilty and get out with a sentence of probation, immediately, to avoid spending six months in jail awaiting trial to hold out for acquittal I would win, and deserved.

I rejected his suggestion, without reservation. Expediency is no substitute for principles. I didn't need to pay him $2,000 to plead guilty after he lost the bail hearing. A free legal aid attorney could have done no worse.

I told him, "I would stay in jail as long as it took, rather than plead guilty. It was a matter of principle." And for me, that wasn't just a cliché or platitude. He was going home soon, while I was going back to my cell. I would be taking a stand on principle. I would be the one paying the price to be able to hold my head up high. And I would be the one showing that my actions matched my words whatever the cost. I would be the anvil living out the truth and ringing with the forge hammer blows.

Freedom is not free, they say. And this was a mere parking ticket of suffering compared to the price others have been willing to pay.

And self-sacrifice is the greatest, most noble thing a man can do to show what it means to be a man of principle. You can take a stand, or stand for nothing. You "can die on your feet, or live on your knees."

"They **did** teach you about principles in law school, **didn't** they?" I asked my lawyer. And then I added contemptuously, "Or did you miss that class?" The sour look on his face, in response to my remark, was priceless. Who did he think I was? And what kind of lawyer was he?

* * * *

"You know, if your body is hurting, people send cards and flowers. But if your mind is hurting, people throw rocks,"

to paraphrase Richard Berendzen.

A troubled mind is a troubled life, suffering that sometimes seems to never give you hope of ever permanent relief. It hobbles or cripples uncounted aspects of your life, consciousness, existence, and future, adding the indignity and inner torment of knowing how public your struggles are, and that they are not viewed as struggles, but cracks of a broken mind.

Too bad, so sad...

But there is also a certain freedom and release I noticed. Your reactions and behavior and abilities may be at best unreliable, and at worst irredeemably flawed and laughable. But your view of what's important, your focus on assorted aspects of reality and the insights and revelations that result, show that you *can* think and understand, but with a terribly frustrating, but not fatal, inability to express it understandably, or sometimes even coherently. I miss that treasured coherency terribly.

People have always frequently disagreed with me, but I viewed that with contempt over the logic and counter-arguments.

And understanding of what matters in life, and what's important and new reflections on our values, goals and priorities and how poorly our judgment sometimes serves us. It eliminates the emotions of pride, arrogance, ego-mania, and shows you the way to compassion, humility, and the serenity their practice brings. And finally, by allowing your mind to be free of many distractions, and other judgmental, repressions, social inhibitions and other dynamics, sometimes the raw, unvarnished truths that evade many seem to be suddenly so clear and obvious.

History, and power, and where we stand as a race at this point in our "evolution" doesn't seem to show that the rule of the sane has produced sane results.

But of this I'm sure: the edge of insanity has the best view.

> "When it's too hard to catch a crook,
> Just slip a good guy on the hook.
> The boss just counts your numbers,
> And the cop ranks are full of bunglers.
>
> 'Cuz the pay is pretty good,
> Screwing good guys, screwing hoods
> 'Cuz Gino's been working for the Gun Squad,
> Gino's been bustin' for the Gun Squad.
>
> Then there came the night of the biggest ever raid,
> They arrested every gun that had ever been made,
> And they didn't stop the bust,
> 'Till the jails were all stuffed.
> 'Cuz Gino's been working for the Gun Squad,
> Gino's been bustin' for the Gun Squad."
>
> -- "Gino's in the Gun Squad" (2009)
> Yogi Shan, with apologies to
> "Julie's in the Drug Squad" by
> The Clash (1978)

Prison Diary 77: "Rock on, dude!" -- Liberation Day

> "Free at last! Free at last! Thank God almighty, I'm free at last!"
>
> -- "I Have a Dream" speech (1963)
> Martin Luther King, Jr.

> "Revolution and reformation have in common a compulsion to reject the existing order, with all its perceived abuses and follies."
>
> -- "Deadly Gambits" (1984)
> Strobe Talbott

Canadian rock group icons, "The Tragically Hip" are known for their arcane and cryptic lyrics. In this respect, they're like "Steely Dan" and "REM", except without the widespread fame and fan-base, except in the relatively minor music market in the country of their birth -- and mine, too -- Canada.

In the classic 1992 Hip song, "Wheat Kings", about the freeing of David Milgaard, arrested by Saskatchewan police at the age of 16 as a patsy for the rape-murder of a Saskatoon nurse. It was a crime that the harmless Milgaard, obviously had nothing to do with, yet he was railroaded to conviction, and wrongly imprisoned for over *twenty* years, with astounding ease. The pigs who framed him were promoted, and a complete failure of the court process that is supposed prevent such an injustice set in concrete, as a useless welfare system for white boobs to feather their nests while periodically attempting to satisfy public fears by committing evil.

But the big question is what was the root cause of the Milgaard miscarriage of justice?

Was it caused by the heinous nature of the crime -- in which case the courts regress to little more than a lynch mob, and exposes a complete failure of the justice system to carry out its basic function, completely abdicating its responsibilities, and abandoning any pretext of being a civilized organization engaging in civilized behavior?

Was it the intense publicity of a high profile case, causing unreasonable pressure for police to solve the case quickly and thus demonstrate police effectiveness and competence, as well as calm public fears not used to such a random violent and horrible crimes to the point that they would top the evil of the original act?

It was a complete failure of the guarantee of "innocent until proven guilty". It was the exposure of judicial incompetence and laziness which renders the system utterly impotent as nothing more than a rubber stamp for the police. It exposed the complete compromise or abuse and corruption by the Police of the evidential and investigatory process for mere expedience and a corrupting and cynical atmosphere of career preservation or enhancement, which corrupts the legal process from the start -- apparently irreversibly -- and which corrupts completely the adversarial process resulting in a collapse of fundamental justice guarantees; and a failure of judicial independence (in this case "spine") a major pillar of fundamental justice.

But the millions of -- tax payer -- dollars he was given in compensation, was essentially meaningless, a gesture that couldn't compensate for his tremendous suffering and a young life stolen and destroyed.

A scape-goat. When a horrible, high-profile crime has occurred, someone apparently has to pay, to "maintain public confidence and calm their fears". Which is a nice way of saying, "to cover up the complete ineffectiveness, incompetence, and moral bankruptcy of the pigs". Apparently they are too busy committing crimes and assaulting citizens to do any real work.

And the second reason is to prevent the exposure of the complete failure of the justice system to do the very thing that is its essential job. The job of the courts is to expose injustice and free the innocent. It's mistaken to think that its primary purpose is to establish guilt. That only really requires the police, if they did *their* job with any sense of duty, responsibility, or professionalism.

But then again, they're essentially a state-sanctioned criminal organization engaged in gang warfare with rival crime groups. A way for society to reduce crime, by keeping busy the sub-class consisting the large number of white people whose sense of entitlement to power and high pay has been derailed by their limited intellect, poor education, and are likely to be trouble for society unless they are bribed with high pay and their violent and thuggish tendencies redirected away from "proper society" and towards criminal and other sub-classes that are considered of low value, marginalized, ignored, or otherwise expendable.

Or that can't help but be kept "in line", silent, and usefully off-balance and in their place by the atmosphere of fear generated by random police oppression.

A man utterly destroyed for no reason other than the expediency of closing a high profile case by simply railroading a random victim with a minor criminal record – harmless but useless, by society's standards – who wouldn't be missed, but would *finally* perform a useful function for society, by keeping up the appearance of society functioning smoothly, efficiently, and effectively.

It all becomes suddenly crystal clear why the police resist accountability, and oversight, and transparency. And why the government tolerates the status quo, when they could implement these effective solutions immediately and face no reaction that could not easily be dealt with.

The reason is that they do not wish to expose society's failings and their responsibility for not dealing with these failings.

A Canadian made-for-TV movie's sad and poignant ending shows the freed Milgaard, twirling 'round and around purposelessly in the street, drenched in the Saskatoon rain falling around him. He's stoned completely out of his mind, the ten million in provincial government compensation way too late to help his fatally wounded spirit and crushed soul.

The symbolic scene showing that his maladaptive but understandable solution to his destroyed life and unending pain is obliterating his consciousness with soft drugs.

* * * *

> "[The watch] was Grandfather's and when Father gave it to me he said, Quentin, I give you the mausoleum of all hope and desire; it's rather excruciatingly apt that you will use it to gain the reductio absurdum of all human experience which can fit your individual needs no better than it fitted his or his father's.
>
> I give it to you not that you may remember time, but that you might forget it now and then for a moment and not spend all your breath

trying to conquer it. Because no battle is ever won he said.
They are not even fought.
The field only reveals to man his own folly and despair, and victory is an illusion of philosophers and fools."

-- "The Sound and the Fury" (1929)
William Faulkner

What is truly astounding is the casual attitude to committing what surely could at least be termed "careless Evil". But is not careless, but actually quite deliberate, routine, and pervasive. And the reckless damage and suffering it causes is enormous, despite our ignoring of it. Jesus was quite charitable when in the midst of terrible pain and suffering he called out,

"Father forgive them, for they know not what they do."

-- Luke 23:24

Steven said much the same thing later in the New Testament (Acts 7:60), as he was being stoned to death. A double lesson. (Or just a single object lesson that you don't have to be the Son of God to not only forsake hatred, but even practice forgiveness and in the face of horrible suffering, and show that forgiveness transcends everything.)

Three things are necessary for reform to take root in the fight against evil. To see the Truth, to refuse to deny or accept the denial of the Truth, and realize that only through the spread of the Truth is there a chance that light will spread and darkness retreat.

Otherwise there is no chance at all, and Evil continues. And you are complicit in its continuation. Failure is not a sin or anything to be ashamed of. It is the silence from not caring that has these traits. And I now understand that where it is possible, it is better to reform society with thousands of persistent voices, than with political violence and the feeling of revenge and righteousness and the satisfaction and empowerment they produce.

When I was brought to Court on Tuesday morning, 22 April 2008, and all the criminal charges were "stayed" [withdrawn by the prosecution], the courtroom public gallery was empty of anyone interested in my case. There was no crowded gallery, no newspaper headlines or stories, TV news or radio station news reports.

And no apology at the final court hearing from the Prosecution or Judge. No one even looked at me, as I sat in the prisoner's box. The Crown would not glance at me. He ignored me completely. I stared right at him daring him to look at me and read the scorn in my eyes, but he never did. And the Judge ignored me, too.

My eight months of agony had ended with a whimper, not a bang. Because, of course, I guess it was just business as usual.

Unwilling to leave it at that, yet shy and uncertain as ever, I settled for a symbolic gesture. An innocuous footnote. Just before I disappeared from the Court Room, my right arm shot straight upwards as I gave the clenched fist salute, noticeable only if someone in the Court Room was looking directly at me as I exited the Prisoner's Box through the door in the wall that led back to the bowels of the basement prisoner areas.

Twenty minutes later, the holding door cell was unlocked, I was led to a basement elevator, and left by my special constable courtroom guard to go to the Main floor and go home a free man.

The nightmare was over, and I was free! Finally free!

"The world that man makes for himself is a world of error. The history of human institutions shows us that man is morally depraved and intellectually the prey of illusion. There is something of the Yahoo in every man, and in the realm of thought and imagination he dwells on appearances, not realities."

-- "Jonathan Swift" in
"18th Century Poetry & Prose" (ca. 1750)

Prison Diary 78: Fighting the Good Fight
==================

> "It was a good fight, anyhow..."
>
> -- Michael O'Rahilly (1916)
>   IRA fighter, in note to his wife as
>   he lay dying on a Dublin street
>   following the failed Easter Rebellion

> "Look! This was never about money for us. This was about **us** against the System. **That** System that kills the Human Spirit. We **mean** something to those dead souls inching along the freeways in their metal coffins. We **show** them that the Human Spirit is still alive!"
>
> -- Bodhi
>   "Point Break" (1991)

Yes, I **am** my brother's keeper.

I have turned the other cheek. I have ignored provocation without regret, or concern of how I looked. I have turned, walked away, or run, to avoid useless and unnecessary violence. And I have stood up, if I had to fight.

But a "fight", does not just mean a physical confrontation, the coarse incivility of fisticuffs, the meaningless brawl, the crudity of the Hurley-burly. It can also be a metaphorical fight -- a serious, irreconcilable conflict over direction, or action, or open resistance on a matter of principle, for instance.

Dante Alighieri, of "Divine Comedy" fame, said that the hottest place in Hell was the reward reserved for those moral cowards who tolerated evil. Ben Franklin, scientist and Founding Father of the American Revolution, said, "Those who would sacrifice freedom for security, deserve neither." And **get** neither, I would add.

A German Protestant Minister named Martin Niemoller, in his 1968 oft-quoted remarks, was a frank admission regretting his failure to take a stand on principle against the Nazis.

> "When Hitler attacked the Jews I was not a Jew, therefore I was not concerned. And when Hitler attacked the Catholics, I was not a Catholic, and therefore, I was not concerned. And when Hitler attacked the unions and the industrialists, I was not a member of the unions and I was not concerned. Then Hitler attacked me and the Protestant church -- and there was nobody left to be concerned."

George Bernard Shaw wrote in "The Devil's Disciple" (1901), he expressed the same sentiment in more moral terms:

> "The worst sin towards our fellow creatures is...to be indifferent to them; that's the essence of inhumanity."

And:

> "The only thing necessary for the triumph of evil is for good men to do nothing."
>
> -- modern paraphrasing of
> "Thoughts on the Cause of
> Present Discontents" (1770)
> Edmund Burke

The bottom line is that the toleration of evil is itself evil.

The problem with white people is that they think that because they subcontract out their killing to the government, there's no blood on their hands. Jesus Christ had a different view of the matter.

> " ' A lot will be expected from everyone who has been given a lot. More will be demanded from everyone who has been entrusted with a lot.' "
>
> -- Luke 12:48

* * * *

It was while watching Richard Attenborough's fabulous biography, "Gandhi" (1982) that I had an epiphany, realizing one of the essential messages of Gandhi's strategy of non-violence. Which was that injustice must be fought each injustice by each injustice, not letting injustice be tolerated and suffered unopposed, accumulating over time the resentment that leads to an eventual explosion of rage and violence from the oppressed.

And I also assert that the toleration of evil, by those aware of its unjust effects, who benefit from it, or even are aware, but unaffected by it, is **itself** an act of evil. The toleration of evil by those who are aware of it, acquiesce to it, become part of the essential dynamic that evil counts on to win, and spread its darkness.

As fictional serial killer Jonathan Doe -- "Jo[h]n Doe", played by Kevin Spacey -- in the classic "Seven" (1995), starts explaining his rationale for a seemingly senseless killing spree, as it approaches its crescendo:

> "Only in a world this shitty could you even try and say these were innocent people, and keep a straight face. But that's the point...
>
> We see a deadly sin on every street corner, in every house, and we tolerate it. We tolerate it because it's **common** -- it's trivial We tolerate it morning, noon, and night. Well, not any more."

People don't seem to realize that it is a moral failure to be lulled by the comforts of a materialistic society, into the toleration of injustice, inequality, or evil in any of its many manifestations. The fifty years of the Cold War between the U.S. and the Soviet Union was a nuclear stand-off imitating "peace" to the average citizen, and which ended the financially ruinous cycle of Total War the world had previously been locked into. Peace became merely the rearmament interlude of the latest evolution of warfare, a war of attrition dubbed "Total War".

The Cold War's false peace **did** produce the economic prosperity previously denied even to Total War's "winners", as modern warfare between rival powers became so expensive and destructive that it was economically hobbling even to the victors.

With the end of the Cold War twenty years ago, though the U.S.-led bloc was now without rival, and the West's ascendancy unopposed. As the U.S. economic empire consolidated its power and domination, its true self soon emerged, revealing democracy as mere window-dressing hypocrisy. Prosperity and comfort has evolved to include laziness, indifference, and moral decay, and the usual realpolitik cold war indifference to the immorality of others.

Don't tell me that nothing can be done, or that it's pointless -- a waste of time and effort. Those are excuses, rationalizations, "cop-outs", deflections of responsibility, validations of inaction, laziness, sometimes the avoiding of risk of controversy or retaliation, avoidance of standing up for what's right, and what you **know** is right.

It's easy, and there is no excuse. Because the minimum action that morality requires -- demands -- is for one to speak up, and continue speaking up **at least**. There is no risk in that, and little effort. To take a stand by voicing your vote for morality, decency, goodness. That is what's meant by the clarion call about solving entrenched systemic problems by realizing that everyone is either "part of the solution, or part of the problem."

The effort it takes to take a stand, is simply standing up...

To act against it in any way possible, takes more effort, and invites the self-sacrifice of consequences -- noble in theory, less attractive in practice. And to continue seeking ways to act against it. To plan. To organize. To lead.

Isn't that what life is meant to be for? What it is to be a "man"? The stuff that heroes are made of? And that principles were not just noble words and empty ideals, but a tangible individual manifestation of the collective good, requiring standing on them, as appropriate, promoting, and ultimately in knowing when they need to be defended.

Interestingly, Norse mythology is instructive and relevant, since it is unique. Northmen, Norsemen, Norman (as in Battle of Hasting and the Norman Conquest), Norwegian -- all speak of the same Viking lineage. And Norse mythology -- the Norse religion, when it was current -- spoke of Asgard, the Norse heaven and home of the gods. A heaven containing a hall of honor known as Valhalla. But the Norsemen believed and accepted that **all** -- Gods and Man -- would face defeat and ruin, and inevitable, unavoidable doom, as Evil would finally overcome them all in one final battle.

The concept being that true heroism comes not from the triumph of defeating Evil, but more so from the honor, bravery, and iron will to fight on in the face of inevitable and certain defeat. You fight not for victory, but for principle. And that is the mistake of humanity's history. Victory is just the icing on the cake, if you're fighting the good fight.

Perhaps, though, it's even simpler than that. We **all** face eventual death. We are all mortal. We all lose in the end. Which establishes that fighting the good fight is a reasonable and meaningful option, even if there is a strong possibility of failure. It's the choice of a living an honorable life, even if it means an early death. An honorable early death, rather than a cowardly eventual death. But this is an extreme example.

* * * *

> "Not because of any reward does a free spirit take his stand for a great truth, nor has such a one ever been deterred because of fear of punishment."
>
> -- "The Failure of Christianity" (1913)
> in "Mother Earth" Magazine
> Emma Goldman

But when you put it in perspective, it all becomes clear.

You're _supposed_ to take the hard road, if you realize God has given you the tools and strength to do it without going straight into the ditch.

I've had more choices than many, and you have to play the hand you're dealt -- or have dealt yourself -- as best you can. Or, in the vernacular: "you pay your money, and you take your chances." And I may come to the proverbial "bad end". But no one lives forever, so -- really -- we all come to a bad end sooner, or later.

"To thine own self be true," is a better philosophy than the spray-painted "Our motto: Apocalypse Now".

It was an epiphany. It was a clarion call to action.

A bugle! A bugle! To arms! To arms! It was suddenly all clear. Unavoidable. Immutable, inflexible, incorruptible. An imperative.

But more importantly, God had clearly endowed me with various gifts, especially my mind. The mind to identify and fight the good fight on behalf of the collective good. God had selected **me**. God had given **me** an assignment. I, and probably I alone, had the spirit to take on such a task. I had the justification and the means to accomplish such a formidable, challenging goal. I had the tools, the resources, the time to accomplish it.

But most of all, I had the fight and the feist, the emotional strength, the experience for the long haul persistence to see the fight through. The good fight! It used to be my specialty, almost my motto.

There is nothing in life, for a man, more honorable, noble, exhilarating -- liberating -- and rewarding than "fighting the good fight".

[Sorry, it's mostly a "guy thing". But we welcome recruits, baby!]

* * * *

> "Error is the force that welds men together. Truth is communicated to men only by deeds of truth."
>
> -- "My Religion" (1885)
> Leo Tolstoy

It all boiled down to **me**. What else could it be? What else? If not me, then whom? If not now, then when?

A moral question had been posed in such a way as to plainly make it obvious to the one person knowledgeable, motivated, and feisty enough to deal with the answer. But more importantly, it was the duty -- the moral obligation -- of someone of the intellectual capacity to see the problem, and see the route to a solution. And the intellectual capability to have a decent shot at success.

Data, logic, and analysis, of course do most of the work. Emotions mislead treacherously. But yet, the final decision -- the answer -- was so simple, yet hard, and indeed could only be found by reaching deep into the human heart -- "to thine own self be true".

I had my marching orders. I accepted them. Now would come the hard part. Because it was uncharted territory, and a tall order. It sounds like an exaggeration, but it was an unexpected surprise -- a shock -- for me to realize the situation as well.

It was about time that Canada, my country of birth, drop the false front, take off the mask, quit pretending, quit faking it, and actually **govern** by the widely known and accepted ideals that make a free, democratic society based on the Rule of Law. The ideals that it professed to honor and respect, but that I found out didn't. We disgrace ourselves as a nation, hoping no one will notice.

It was time for Canada to **actually** join the ranks of dynamic and vigorous democratic civilization.

Rubber makes a very poor moral fabric.

"You are not expected to finish the job,
but neither are you free to lay it down."

-- The Talmud

" ' I have fought the good fight. I have completed the race. I have kept the faith.' "

-- 2 Timothy 4:7

Prison Diary 79:   Heroes of the First World -- Disciples of Satan
====================================

> "When Johnny comes marching home again,
> Nobody understands it can happen again.
> The sun is shining, an' the kids are shouting loud,
> But you gotta know it's shining through a crack in the cloud.
> And the shadows keep falling, when Johnny comes marching home."
>
> -- "English Civil War" (1978)
> The Clash

> "To come right down to it, if I take the kind of things in which I believe, then add to that the kind of temperament that I have, plus the one hundred percent dedication I have to whatever I believe in... These ingredients would make it just about impossible for me to die of old age."
>
> -- "The Autobiography of Malcolm X" (1965)
> Malcolm X

I stopped wearing a poppy on Remembrance Day, 11 November, that honors the fallen of WW1, WW2, and the Korean War. And here's why.

The human cost was horrible, and sad, but why should I give a damn about a bunch of dead white men fighting wars against mostly other white men for conquest, and power, and imperialist empires based on the exploitation of non-whites?

While at home were fascist or authoritarian governments, and virulent racism ruled as the order of the day. And had I been alive at the time, I would have been by official government policy a second-class citizen -- a non-man. The British – Canada's ex-colonial rulers – were a racist, imperialist, murderous society at the time. British is a fascist society top to bottom, even today.

I'm sorry but the gutter cannot be considered the moral high-ground. And a fight between racists, capitalists, and fascists is not much of my concern.

*   *   *   *

> But I know I'm alone in this fight,
> I load my rifle and jump into the night.
> There's a cold black hand in the center of the world today,
> Why are we so uptight?
> Let's push the button, aren't we always right?
> You gotta be cruel if you want to run the world today."
>
> -- "Nuclear Boy" (1981)
> 20/20
> Ron Flynt/Steve Allen

Now let's take, for instance, Winston Churchill, politician, World War Two Prime Minister of the U.K., and its Empire, and deconstruct him:

1. All he did was flap his gums. Just a windbag with talent as a wit and orator, but what he said was meaningless hot hair. He would have been executed if the Nazis had conquered England, the difference to the toiling masses of conquered England would have been little.

2. An alcoholic. An arrogant, deeply flawed man, he protected his own status as a leader and rich person in the social strata of England

3. Leo Szilard & Neils Bohr; Bohr the god of physics and Copenhagen spirit and last hope of man to avoid the arms race and the cold war; when Churchill found out about Bohr's lobbying against secrecy, he tried to have Bohr arrested and detained. Roosevelt ignored Churchill's advice and merely had Bohr put under surveillance.

4. Confirmed white racist and supremacist, leading imperialist of the British Empire that enslaved millions, advocate of chemical warfare and carpet bombing of German civilian urban areas -- total "total war", during WW2 against the Nazis....

5. Called Gandhi -- a living saint -- a "half-naked fakir". During the terrible Bengal famine of 1943 - 1944, he sent a cable demanding to know if food was so scarce, why wasn't Gandhi dead yet. Oh, boy, what a wit!

* * * *

"Power never takes a back step -- only
in the face of more power."

-- "Malcolm X Speaks" (1965)
Malcolm X

There are lots of men will like -- leaders of -- and glorified and honored still in spite of all the evil they caused. I cover Nixon, Reagan, and George W. Bush elsewhere. But other U.S Presidents who were evil, amongst them were Theodore Roosevelt, and Andrew Jackson.

"Power concedes nothing without a
demand; it never has and never will."

-- Frederick Douglass (1857)
speech

Prison Diary 80:　　　　　　　"A Terrible Beauty"
　　　　　　　　　　　　　　=============

> "Lots of guns. They've got all the
> fucking guns in the world..."
>
> 　　　-- Angie
> 　　　　"Falling Down" (1992)

> "I don't know why I'm wasting my time or breath. But what the hell? [...] I don't believe in the hypocritical, moralistic dogma of this so-called 'civilized' society. I need not look beyond this room to see all the liars, haters, the killers, the crooks, the paranoid cowards -- truly trematodes of the Earth, each one in his own legal profession. You maggots make me sick -- hypocrites one and all.
>
> And no one knows that better than those who kill for policy -- clandestinely or openly -- as do the governments of the world which kill in the name of God and country, or for whatever reason they deem appropriate. I don't need to hear all of society's rationalizations. I've heard them all before -- and the fact remains that, what is, is.
>
> You don't understand me. You are not expected to. You are not **capable** of it. I am beyond your experience. I am beyond good and evil. Legions of the Night – night breed – repeat not the errors of the 'Night Prowler' and show no mercy. I will be avenged!
>
> Lucifer dwells within us all."
>
> 　　　-- Richard Ramirez (1989)
> 　　　　29 year old serial killer --
> 　　　　dubbed the "Night Stalker" --
> 　　　　at his sentencing hearing

Facing the face of evil. Yet, Richard Ramirez's words have an astonishingly hypnotic eloquence. The beauty of his words, with a chilling and contemptuous message of pure ugliness. And haunting truth -- truths and echoes -- that are undeniable. Even with the sound of metaphors mixing...

A tragic soliloquy, perhaps more for us, than its speaker, who really didn't care about anything -- including himself.

Richard Ramirez called himself the "Night Prowler" -- taken from the AC/DC rock band song. And, as usual, the band took a lot of heat, as society looked for someone to blame, and as always, nailed with irrelevancy the undeserving.

The mother of Richard Ramirez worked in a factory that exposed her to leather processing chemical fumes, which caused her to faint in the fifth month of her pregnancy. He received a brain injury when he was accidentally dropped as an infant, began having grand mal seizures, and at the age of six was diagnosed with temporal lobe epilepsy. But his blighted life continued to take even more hits.

He was exposed as a preteen to contact with a cousin -- a seriously disturbed ex-Green Beret Vietnam War veteran, whom he witnessed abuse and finally kill his wife. He began sniffing glue in his teens, a practice that is well established to cause progressive central nervous system damage. His progressive and chronic drug abuse continued, and evolved, until he ended up a regular cocaine user.

At the age of 24, he began his six month killing spree, leaving at least 20 murder victims -- mostly women, who he also raped -- in the Los Angles area. He was caught, convicted of 13 murders, and sentenced to death. "See you in Disneyland," he commented to the courtroom gallery -- laughing off his death sentence -- as he was being taken from the courtroom.

He remained on death row, until his death in prison in 2013, at the age of 53 of liver failure.

* * * *

> "Oh Sikhs, be ready for self-destruction!... Our lands are about to be overrun, our women dishonored. Arise and once more destroy the Mogul invader. Our Motherland is calling for blood! We shall slake her thirst with our blood and the blood of our enemies!"
>
> -- Tara Singh (1947)
> speech

Suddenly, the Devil appeared, on horse-back. I had been sauntering aimlessly along the sidewalk, reveling in the solitude and anonymity of an evening twilight walk. I was no longer surprised by such bizarre occurrences and undesirable, inevitable, but inopportune meetings.

"Let me guess?" I addressed the black rider. "My choices are -- Good or evil... Good or Evil, right?" There was silence, except the rustle of a tumbling, crumpled ball of waste paper in the gutter, as the wind brushed it along.

I pulled a quarter out of my jeans, and thumb-flipped the coin over-and-over into the air. I stuck my hand out flat, to intercept the trajectory of the falling piece of metal. What can you buy with a fucking quarter these days, anyway?

"Oh -- it's heads. Sorry, man -- I man 'beast' -- 'Good' won. Now, hit the road, sucker, before I sick Jesus on you!"

"Begone, wastrel!" I reiterated at the slowness of any response or reaction, as I started to get bored. The dark horse gave a loud, rough snort, tossing his head up and down. And then "Poof!", the horse and rider disappeared, and I was alone again with just my thoughts for company.

"Next time send a *blonde*, asshole," I muttered to myself. But only the wind could hear me.

* * * *

> "There is a great deal of [government propaganda] about terrorists. They are frequently described as mindless, irrational killers. But terrorism for the most part is not mindless violence....Terrorism is violence for effect. Terrorists choreograph violence to achieve

maximum publicity. Terrorism is theater."

-- "Will Terrorists Go Nuclear?" (1975)
Brian M Jenkins
Rand Corp. Report P-5541

Extreme cruelty, like ruthlessness may not be something I practice anymore, or even consider. But not forgetting is harmless. And forgetting might make people more comfortable, and make me seem like I still espouse practices when it's only bad memories and nightmares I express with just words. And at a terrible cost in self-torment, to forget is to deny a truth that still exists in the world.

At the very least, my duty is to bear witness. At a cost, apparently of being misjudged terribly.

I don't really like studying human history any more. It just gets so depressing after a while,
and it starts to get to you not just the tiresome, horrid repetitiveness, and a towering, overwhelming sense that there shear magnitude of the problem allows not even a solution, but only silly bandages that delude you into thinking there's at least relief in knowing you can do at least do something and feel .

And then you come to big emotional dead end they call "despair". Better about our overall impotency. If you've been able to last this long, staring right in the face of horror.

Looking into the face of Evil itself, and seeing their own death in my eyes. I don't know what that look is -- that very special look emanating out from my eyes, but I can feel the emotions and the darkness welling up out of my soul, and the muscle contraction and tone that the look has as it freezes into a cold and penetrating look that's a stare, but not a stare. And it's not even eye-to-eye contact, because I'm not really looking at them directly. But I'm visualizing a death with no corresponding emotional release of any sort. Just more coldness and a feeling like there's an unstoppable sequence coiled tightly that's about to spring alive.

"She is older than the rocks among which she sits,
Like the vampire, she has been dead many times,
And learned the secrets of the grave;
She is a diver of the depths of the abyss,
And she lets the twilight surround her."

-- In "Fortnightly Review" (1869)
Walter Pater

Prison Diary 81:   "It's Not Easy Being Green"
====================

> "What you got against white folks [?]"
>
> "Ain't got nothing against them. I goes
> my way and lets white folks go theirs."
>
> -- "The Sound and the Fury" (1929)
> William Faulkner

> "The courts have acknowledged that racial prejudice
> against visible minorities is...notorious and indisputable...
> [It is] a social fact not capable of reasonable dispute."
>
> -- R. v. Spence (2005)
> published Canadian judicial ruling

I don't know who came up with the phrase "racial profiling" when a more accurate description of official police discrimination would be "**racist** profiling".

"All right, we are two nations" wrote John dos Passos famously ("The Big Money" (1938)) about the judicial railroading and eventual execution of Italian immigrants Sacco and Vanzetti, two radicals swept up during a U.S. anti-immigrant witch-hunt during the Depression.

*   *   *   *

> "[P]eople who coo over [guns], they stroke and they cuddle and cradle
> them. I've seen them do it, and it's as embarrassing a sight as any
> you've ever seen. But I think there is something else going on here.
> There's a concentrated, immediately accessible amount of power in
> those weapons, and that's where the attraction lies."
>
> -- Gwynne Dyer (2005)
> interview

Regimes come, and then they go.
Power ebbs, and power flows.

But with the fall of the old Order, the *real* work begins: the consolidation of Power.

[And, incidentally, Adolph Hitler was the all-time master of that game, and it is this, that was his greatest crime against humanity -- the consolidation of power in Nazi Germany. The _Shoah_ -- the "Destruction" -- was just a financial side-show, and sub-set of his consolidation.]

A thorough "cleansing" becomes the main focus of political change with the defeat or collapse of the _ancien regime_ and the triumph of the new political Order, typically as they over-run the former center of power. Centralization of power, of course *is* the consolidation of power.

But the business of power never rests. The mad scramble to fill what's called a "power vacuum", assuming that the fallen political order had multiple militarily viable opponents. Splits can occur from within the "new management", as factions with varying differences vie for dominance and control of the New Order's direction, priority and leadership posts.

Alliances -- coalitions -- between these "factions" can fall apart or hold united, at least until the vacuum is filled. Or the alliances can fall apart, as each faction decides it can attain monopoly control of the state. Their political interest have, in other words, "diverged", and a power struggle results between roughly equal military forces who believe superior tactics, strategy, or position will decide the contest in **their** favor.

That work is the consolidation of power, whose purpose is to firmly entrench the winning, forces, expand and deepen their power to its maximum controllable limits. The losing forces must be militarily neutralized completely, rolled up, broken up.

The coup d'état is an internal revolution, lightning fast in execution, as these things go, typically right-wing, and typically to terminate and roll-back political administration and policies back to the conservative/right-wing direction.

Electoral victory is a relatively narrow choice, with radical change only theoretical. The consolidation of power in democratic states seems to have been restricted to the unacknowledged and/or unspoken acceptance that continuity of a narrow spectrum of choice is the unwritten rule of allowing people a democratic choice.

The consolidation of Power becomes the main focus of any political change from initial electoral victory to collapse of the _ancien regime_ and the triumph of revolutionary forces, as they overrun the center of power of the _ancien regime_. Victory is irreversible, and all that remains is cleanup -- the crushing of remaining enemy resistance.

* * * *

> "Your problems will never be fully solved until and unless ours are solved. You will never be fully respected until and unless we are also respected. You will never be recognized as free human beings until and unless we are also recognized and treated as human beings.
>
> Our problem is your problem. ... This is a world problem, problem for humanity. It is not a problem of [Negro] civil rights but a problem of human rights."
>
> -- O.A.U. conference memo (1964)
> Malcolm X

In the year 1898, the sunset of the remaining tribes of the North American Indian was almost completed. The consolidation of the American White Man's unchallenged monopoly of power over essentially the entire North American Continent had been completed in just over one hundred years, with annexation of British-controlled Canada their only misstep – their only failure.

From a pre-white man population of about twelve to fifteen million, the Indian had been genocidally reduced to 200,000.

The more successful genocide of the primitive Indian tribes was not a planned strategy, but just the side-effect of a moping up operation. Tribe by tribe would be penned up in concentration camps they called "reservations", or if they resisted this destruction of their way of life, and seizing of their lands, crushed. The schedule would be decided by the white man's needs, exploitation of resources, settler safety concerns, and the white man's capabilities and need for economic expansion -- inexorable, and unstoppable.

With no other impediments to effective continental hegemony, success was guaranteed, and just a matter of time.

* * * *

"They've always said I'm anti-white. I'm for anybody who's for freedom. I'm for anybody who's for justice. I'm for anybody who's for equality.

I'm not for anybody who tells me to sit around and wait for mine."

-- Malcolm X (1964?)

In 1803, with the Louisiana Purchase, Napoleon sold the French territory consisting of the western part of the Mississippi River valley. All 900,000 square miles of it was handed over to the thirteen Original States, ending the possibility of French competition and eventual conflict, and allowing for massive westward expansion of the growing U.S. into the Midwest.

In 1819, Florida was purchased from Spain

A brief military misstep -- the War of 1812 -- failed to annex the British colony of Canada (really just Ontario & Quebec at the time), comprising the largely uninhabited frozen north of the Continent, mostly above the 49th Parallel in latitude.

The huge landmass of what would become the state of Alaska would be bought from the Russians, giving the U.S. an Arctic presence in North America.

Mexican independence from Spain was achieved after the successful Mexican revolution in 1821 brought independence, but was followed by thirty years of political instability as a result of a see-saw of inconclusive class warfare.

Taking advantage of this weakness, the U.S. continued its territorial consolidation of the continent when Texas achieved independence by swallowing over half of Mexico, six weeks after the fall of the Alamo in 1836, a convenient justification without any merit. Soon Texas was part of the U.S., along with the U.S. Southwest to the Pacific -- Arizona, New Mexico, and California.

The Confederate War of Independence was a costly fratricidal war of economics and attrition, that was -- finally -- brutally crushed in 1865. What is now called the U.S. Civil War was a distracting, divisive, and wasteful internecine conflict instigated by an alliance comprising the resentful -- but short-sighted -- agrarian Southern States, who felt marginalized by the U.S political power-base of the strong, industrialized Northern States.

But with the end of hostilities, a united and stable America could now free her energies and resources for westward expansion of the White Man's dominion to its ultimate goal of a united country stretching from the Atlantic to the Pacific.

By the 1890's, white hunters had exterminated the once 60 million strong wild bison herd, making way for cattle ranchers and their fenced-in grassland ranges. The bonus was the simultaneous destruction of the nomadic way of life of the Western Indian Tribes, who depended on the buffalo for food, clothing, and housing.

But also starvation, alcoholism, disease, and savage reprisals for sporadic and ultimately ineffective incidents of Indian war – the Apache, for instance – rebellion, and resistance, had almost completed the unofficial genocide -- a word unknown at the time -- of the "savage" aboriginals.

By 1890, no land had Indian title anywhere in the U.S. Ultimately, the fraction of Indians that had survived disease, starvation, and military defeat, were herded passively onto government-administered "reservations". Starving, confused, and dispirited, their once thriving culture and lifestyle in disarray, the reservations were a miniscule fraction of the expanse of their old ranges, and consisted of land worthless to prospectors and the white settlers pouring in to colonize for farming, mining, and cattle ranching of the old Indian lands.

In the year 1898, as the sun set on the race of North American Indians, there appeared a vision to an Indian mystic, Wovoka, who introduced a new purification ritual named the "ghost dance". It spread rapidly to the Indian survivors, a broken people, reduced to groping desperately for spiritual relief.

Reality had failed them in a new world ascendant to the fall of theirs, systematically destroyed by organized, complex, and vastly superior forces the Indian "primitives" could not begin to understand, much less repel.

Wovoka's vision had promised that one day their faith would be rewarded if they danced the proper ritual dance without stop. All day, and all night, sometimes for four days on end, the dancers eventually felled by exhaustion.

The promise of this new "ghost dance" was that one day they would wake up, their fallen warrior brethren and ancestors resurrected, the buffalo herd returned to their beloved Plains, and the ruined and stolen land of their ancestors theirs again, reborn to bounty and fertility.

And lastly, if they were faithful to the Ghost Dance until complete exhaustion, they would wake up with the White Man gone, never to return.

"I' yehe! My children -- Uhi' yeye' heye'!
I' yehe! We have rendered them desolate --
Eye' ae' yuhe'yu! The Whites are crazy! --
Ahe' yuhe' yu!

-- Arapaho ghost-dance song (1898)

Prison Diary 82: White – I mean Might – makes Right
==============================

> "The press has too long basked in a white world looking out of it, if at all, with white men's eyes and white perspective."
>
> -- "Kerner Commission" Report (1967)

> "This is...about extraneous persons. Subordinated and expelled from society, they take on new shapes: humans, things, dogs, and spirits... Their transformations prompt us to think about what it means to be considered in terms of [the] law. ,,,
>
> [M]etamorphosis...[is] some of the manifold ways that law dwells on, messes with, and consumes persons. It is through law that persons, variously figured, gain or lose definition, become victims of prejudice or inheritors of privilege. And once outside the valuable discrimination of personhood, their claims become inconsequential."
>
> -- "The Law is a White Dog" (2011)
> Colin Dayan

Race makes a difference. Even though it is frequently subconscious, it's there. And it's a powerful hidden force in the First World, with the death – or paralysis, more appropriately – of overt racism.

U.S. Canada and England are united by a common ethnic – racial – background. Almost always united in political actions, though their policies may differ. With Australia and New Zealand the same ethnically, but a little more divergent by the separation of such a great distance from both the U.S. and mother England.

French, Italians, and Spanish are the next closest. Germans the next, but made closer by their Protestant ethic.

Assimilated Jews fit in with the Americans

Next are the East Europeans: Polish, Hungarians, Romanians, Bulgaria, and the Balkans

Russians are the break, Yugoslavia: Slavs

Then Mexico, Central America, and Latin American's – Spanish contaminated by Indian blood ex-slaves of the Spanish Empire, never having recovered from their inferior status.

Then the Middle East, Asia, and the Indian sub-continent.

And finally, Africa at the bottom: poor, frequently starving, constantly at war, unstable, tribal, ex-slaves, savages. The most different racially of all from whites. And the first advanced civilization – so old history only has a dim, forgotten perception of it. But it was there. I know it.

* * * *

> "[The AK] is more than a weapon; it has become a symbol. The Maxim [heavy machine gun] represented the power of the imperial armies, while the AK has become an icon for many of the anti-establishment insurgent, freedom fighter, and terrorist organizations that exist today."
>
> -- Frederick Ezell, in foreword to
> "The Social History of the
> Machine Gun" (1975), by Ellis

What **is** this shit about white supremacy? It is time to "deconstruct" white supremacy, as they would say in academia.

Sure, we are in a period of world history where white civilization dominates completely. Technologically, scientifically, socially, culturally, and, of course, militarily. But let's take at look at the history of the world's great civilizations.

Chaldean civilization:	6300 BC – 539 BC	(5800 years)
Ancient Egypt:	6000 BC – 667 BC	(5300 years)
Mexican civilization:	5000 BC – 1000 BC	(4000 years)
Ancient East Indian civilization:	3000 BC – 1800 AD	(1200 years)
Assyrian Empire:	2300 BC – 608 BC	(1700 years)
Sophisticated African civilizations:	2000 BC – 1700 AD	(3700 years)
Chinese civilization and Empire:	1500 BC – 1800 AD	(2300 years)
Persian Empire:	800 BC – 450 BC	
Carthage:	800 BC – 146 BC	
Mayans of Mexico and Central America:	300 AD – 1000 AD	(700 years)
Attila the Hun and the Hunnic Empire:	370 AD – 454 AD	
Japanese civilization:	500 AD – present	(1500 years)
Muslim Spanish Moors:	711 AD -- 1492 AD	(800 years)
Seljuk Turks (of the Crusades era):	1071 AD – 1325 AD	
Genghis Khan and the Mongol Empire (the largest contiguous empire in the world):	1206 AD – 1368 AD	
Ottoman Empire:	1302 AD – 1911 AD	
Inca civilization and Empire:	1200 AD – 1533 AD	

Ancient Greeks (only city-states), including Alexander the Great:	800 BC – 300 BC	
Roman Republic and Empire:	500 BC – 400 AD	(900 years)
Eastern Roman/Byzantine Empire:	395 AD – 1450 AD	(1000 years)
European Empires:	1500 AD – 1943 AD	(450 years)
Russian communism:	1917 AD – 1991 AD	(70 years)
U.S. economic "Superpower" empire:	1945 AD – present	(75 years)

* * * *

> "Every society needs order if it is to survive. To have order, a society must have values. Those values, expressed most vividly in the changes within education and family life, have become fragile ...
>
> There has been a breakdown in moral transmission from one generation to the next. ...[T]hey display a failure to recognize the need for clear moral judgments, discipline and punishment..."
>
> -- "The Female Eunuch" (1971)
> Germaine Greer

So, let me get this straight.

The white race was one the last of the world's peoples to achieve "civilization". In its latest – and most evil – incarnation, it has lasted only 60 years. And it is now in a period of decline: militaristic and armed to the teeth, wars, terrorism, major financial scandals and crises, a corrupt and paralyzed government, crushing national debts, oceans fished out, oil almost gone, environmental disaster, world's resources mined out, global warming.

After a second kick at the can – after the Romans. It appears the Western civilization is once again in sunset, perhaps taking mankind into a new Dark Age. Or extinction.

And you're proud of yourselves to the point of contempt, derision, and racism for the other non-white people's of the world whose civilizations rose and fell thousands of years before you stopped sodomizing each other in caves?

OK, that was a cheap shot...

And like, say the oldest civilizations that rose and fell – the Africans, Egyptians, and Mexicans – who fell completely into disarray, and are held in contempt by the modern world, **so shall you fall** and become the new "niggers", as another different civilization rises. Or human's become extinct from the white man's depredations of the earth and its environment.

Enjoy yourselves for the ride downhill. We **are** living in interesting times...

> "And after having had the revolution of freedom and the counter-revolution of glory, you will have the revolution of public consciousness and the revolution of contempt."
>
> -- Alphonse de Lamartine (1847)
> speech at banquet

Prison Diary 83: "You hypocrites!": The **Real** Canada
===============================

> "Cooperation in any shape or form with
> this satanic government is sinful."
>
> -- Mohandas Gandhi (1919)
> speech

> "Is **that** what this is about? You're angry because
> you got **lied** to? Is that why my chicken dinner is
> drying out in the oven?
>
> Hey! They lie to **everyone!** They lie to the fish!"
>
> -- LAPD Sgt. Prendergast
> "Falling Down" (1992)

Though I have focused mainly on the U.S., with Canada only a willing participant and acquiescent of the American military and economic Empire, perhaps it is time to shift my microscope to the true nature of Canada.

We are a nation of alliterate Arctic hillbillies ("alliterate" -- can read, but don't.). Some sort of frozen banana republic, as the following selection of examples will show.

* * * *

> "Alienation is when your country is at
> war and you want the other side to win."
>
> -- cover caption (April 1969)
> "Ramparts" magazine

Canada the "peaceable" nation. From 1957 to 1967 the U.S. Strategic Air Command with Canadian Government approval deployed atomic and TN weapons at Goose AFB (Goose Bay, Labrador (Canada)) until technological improvements and two widely publicized B-52 accidents (in 1966 in Palomares, Spain, and in 1968 at Thule, Greenland) made foreign bases superfluous.

T. Cousins, DREO (Defense Research Establishment Ottawa) wrote a classified report on neutron bombs and tactical nuclear warfare in 1989 (DTIC # ADA212 748), showing continuing research on U.S. defense/offensive nuclear research.

* * * *

> "It is easier to remain a pacifist if you've never felt the force of the State."
>
> -- "Active Transformation"
> (Feb/March 2000 issue)

> interview with "Black Block"
> anarchist

Richard Therrien was in 1971 -- when he was a 19 years old law student -- convicted of aiding the FLQ -- but was pardoned in 1987. In 1996 he was appointed to be a Quebec judge.

As he had been pardoned, he did not disclose his expunged conviction to the Selection Committee. When they found out, he was stripped of his judgeship on a complaint by the Quebec Minister of Justice. Therrien, to get his judgeship back, sued all the way up to the Supreme Court and lost the case in 2001.

Apparently a pardon is not a pardon, twenty-five year old convictions are still important, and the FLQ is still feared and unrecognized for its – albeit in an unintended way -- role in liberating Quebec from four centuries of damnable English oppression.

* * * *

> "Go thou and fill another room in hell"
>
> -- Richard II, on stabbing someone
> "Richard II" (1597)
> Act 5, Scene 5

Canada is a Country of Pedophiles: Catholic priests, Boy Scout masters, and then there was the case of the Cornwall (Ontario) pedophile ring scandal. According to Health Canada data 30% of boys under 18 have been sexually abused, a figure that may be low.

In the Cornwall case, a public inquiry was convened which lasted four years and cost over $50 million, issued a 1600 page report, and which only succeeded in jailing for contempt of court for six months an ex-Cornwall policeman of integrity who refused to testify before the inquiry, calling it a fraud and a joke that would do nothing. The inquiry did nothing, and was a fraud.

* * * *

> "[We are] billion year old carbon."
>
> -- "Woodstock" (1969)
> Joni Mitchell

Mulroney was the most popularly despised Prime Minister in history. In the 1980's after he left office he was pursued by the new Liberal government of Jean Chretien, who considered him criminally corrupt. Mulroney sued for libel for $50 million in 1995, after the Chretien called him a crook in papers sent to obtain evidence from the Swiss. In 1997, the Chretien government had to fold, because they had no solid evidence.

Almost thirty years later it turned out that by his own testimony in 2011, Mulroney was paid thousands in cash by a German, Karlheinz Schreiber, in the so-called "Air Bus affair", and declared it only 6 years later, just shy of Canadian Revenue [taxation] Agency rules requiring the declaration of such money to avoid income tax evasion only a few months later.

* * * *

> "The makers of our Constitution...recognized the significance of man's

> spiritual nature, his feeling, and his intellect. They knew that only a
> part of the pain, pleasure, and satisfactions of life are to be found in
> material things. ... They conferred, as against the government, the
> right to be let alone -- the most comprehensive of rights and the
> right most valued by civilized men."
>
> -- "The Right to Privacy" (1890)
> Louis Brandeis and Samuel Warren

David Milgaard, Guy-Paul Morin, and others found to be completely innocent, and released after conviction and jailing for years under proper judicial procedure. No changes were made to the judicial system, and no judicial officials punished or even investigated.

Mahar Arar, Abdullah Almalki, and others were spirited to countries where Canada knew they would be tortured, even though there was no evidence against them and they were completely innocent of Al Qaeda connections or terrorism.

* * * *

> "[T]he right of privacy reaches beyond any of its specifics. It is,
> simply stated, the right to be left alone; to live one's life as one
> chooses, free from assault, intrusion or invasion except as
> they can be justified by the clear needs of community living
> under a government of law."
>
> -- unpublished opinion (1966)
> Judge Abe Fortas

Louise Arbour, resigned from the Supreme Court to join the International Criminal Court in Europe, where she disgraced herself by ignoring the blatantly illegal and outrageous attack on Yugoslavia (over Kosovo) by NATO in violation of UN rules, and thus international law.

Yugoslavia was the last communist European nation (independent from Russian dictates during the period of Russian communist supremacy); she thus became an ignorant, unknowing tool of U.S. realpolitik in attending to its destruction.

All I can say about Louise is: "What a whore!"

* * * *

> "Among the many misdeeds of British rule in
> India, history will look upon the act of depriving
> a whole nation of arms as the blackest."
>
> --"Gandhi: An Autobiography" (1927)
> Mohandas Gandhi

The Canadian government has over the past twenty year or so, attempted to disarm the Canadian citizenry by progressively harsher and harsher laws, more and more intrusive laws, and greater and greater difficulty getting a gun permit. They instigated at least a six month waiting period to prevent citizens from arming themselves in the case of civil unrest.

It took a 2011 Supreme Court case to exonerate a citizen who defended his house and home from an intruder with deadly force, rather than exiting the house like a coward, just because it was an option, and possible -- the prosecution's argument.

Meanwhile, Canadian officials and bureaucrats are protected by a legion of heavily armed police, ready to attack with deadly force at the slightest threat or possibility of a threat.

* * * *

> "If there is any fixed star in our constitutional constellation, it is that no official, high or petty, can prescribe what shall be orthodox in politics, nationalism, religion, or other matters of opinion or force citizens to confess by word or act their faith therein."
>
> -- West Virginia Board of Education v. Barnette (1943)
> Justice Robert Jackson

The Cult of the Empty Gesture. Keeps finding more and more causes to liberate with laws against discrimination, without any effort to actually end the discrimination of previous groups. Considers the right of a man to sodomize another man a matter of supreme importance.

A "right" that is condemned as a grievous sin by every major religion in the world. (Personally, I prefer the reduction in competition for women. As well, what a man does consensually with anyone else may be disgusting, but as far as I'm concerned that's none of my concern. And homophobia doesn't means "fear" of homosexuals. It's not fear, it's disgust.).

* * * *

> "T-Bird an' George let their gimmicks go rotten,
> So they died of hepatitis in Upper Manhattan.
> Sly in Vietnam, bullet in the head,
> Bobby O.D.'d on Drain-O, on the night that he was wed.
> They were two more friends of mine.
> Two more friends that died."
>
> -- "People Who Died" (1980)
> Jim Carroll

The AIDS scandal: it was three years after the U.S. -- in 1983 -- recognized and publicized widely the danger to the blood transfusion supply that Canadian doctors who ran the Canadian equivalent noticed and rectified the situation. For this incompetence, detailed in a Public Inquiry -- the Krever Commission -- many Canadians were infected needlessly with HIV/AIDS and Hepatitis C, and died.

And, by the way, the U.S. response was slow and lackadaisical, as detailed in the bestseller "And the Band Played On" by Randy Shilts, a San Francisco Chronicle reporter. Reagan didn't care because he viewed it as a disease of gays and drug addicts.

In response to a suit filed by the victims of the contaminated blood supply, the Supreme Court ruled that the Canadian officials guilty of negligence were not to be punished, criminally or by civil suit, astoundingly, because "incompetence" is not actionable against public officials in Canada.

Or Supreme Court justices...

* * * *

> "A righteous person may fall seven times,
> but he gets up again seven times."
>
> -- Proverbs 24:16

The Canadian Criminal Code, encompassing all Canadian criminal was created in 1892 (Confederation was in 1867, you'll recall), and subject to a thorough review and revision in 1948. It is now 2011.

This is a joke of a criminal justice system, and an international disgrace for a nation that considers itself the epitome of an advanced, peaceable example of civilization for the world.

* * * *

> "The Kalashnikov isn't just a gun; it's a
> legend, a currency, a symbol of liberation."
>
> -- "The Legacy of War" (1998)
> Jacklyn Cock in
> "War and Peace in South
> Africa"

Canadian law allows the extradition of Canadians based on provably fraudulent evidence to France (and any other country with an extradition treaty). France, in return, doesn't allow French citizens to be extradited for any reason and with any evidence. (This came out in 2011 concerning a 30 year old Palestinian bombing case in Paris).

What kind of moronic idiots rule our country, and what kind of _cretins_ negotiate important legal instruments like treaties?

* * * *

> "So let it rock, let it roll,
> Let the bible belt come and save my soul,
> Hold on to 16 as long as you can,
> Change is coming 'round real soon,
> Make us women and men.
> Oh yeah, life goes on,
> Long after the thrill of living is gone."
>
> -- "Jack & Diane" (1982)
> John Mellancamp

In 1992 the Northern cod fishery near Newfoundland collapsed, falling to 1% of its previous yield. Overfishing was the cause, synchronized with the government's inability to show leadership and stop Newfoundlanders from destroying their major industry and employer, in spite of loud, repeated, and desperate warnings from scientists in the Federal Fisheries Ministry.

The abundance of cod was noted by English explorer John Cabot in 1497, who stated that the schools of fish were so crowded in the water that they could be scooped out from the surface by men in his ship with baskets, and that their numbers were "so thick [that] they slow the ship".

* * * *

> "I often wonder whether we do not rest our hopes too much upon
> constitutions, upon laws, and upon courts. These are false hopes,
> believe me... Liberty lies in the hearts of men and women; when it
> dies there, no law, no court can even do much to help it. While
> it lies there it needs no constitution, no law, no court to save it."
>
> -- "The Spirit of Liberty" speech (1944)
> Judge Learned Hand

Canada is guilty of war crimes in Afghanistan for (for all that is known) allowing the torture of Afghan detainees, many completely innocent. Canada is guilty of tolerating the even worse war crimes of the U.S. in Afghanistan and Iraq -- which is also a war crime. Victor's justice...

And Canada is guilty of aiding the U.S. war criminals who refused to accede to International Criminal Court jurisdiction -- also a moral crime, and a war crime. (The U.S. attitude to war crimes is that they are free to commit them with impunity. It is summed up by the 1968 My Lai massacre during the Vietnam War. 500 dead innocent civilian old men, women, and babies. It was covered up for 18 months, until exposed by reporter Seymour Hersh, who received a Pulitzer Prize. Serving his life sentence, Calley was spotted by a reporter wandering around his base, instead of confined to the Leavenworth brig. Calley was pardoned by ace war criminal Richard Nixon after three and a half years of "house arrest".)

George W. Bush visited Canada after he left office, making Canada guilty of harboring a fugitive war criminal. They failed to arrest him and hand him over to the International Criminal Court.

* * * *

> "[B]ut he has a good right to be anti-American
> because during the 1960s we became a highly
> unlovely country and anybody in fact who is pro-
> American at this time has an addled wit, to say
> nothing of an unserviceable amoral sense."
>
> -- "Firing Line" T.V. program (1967)
> William F. Buckley, show host

In the late 1980's it was announced that the Public Service Commission was taking over the government student summer employment program because they had found that 97% of the student jobs were being funneled to relatives of government civil servants – blatant and system-wide nepotism and "soft" corruption.

This scandalous situation was disclosed in a small column article near the back of the front section of the "Ottawa Citizen"...

* * * *

> "_Solventur risu tabulae, tu missus abibis_"
>
> "The case will be dismissed with a laugh.
> You will get off scot-free."
>
> -- "Satires" II.i.86 (ca. 20 B.C.)

"Wee Willie" Wilbert Keon, an Ottawa heart surgeon who achieved fame and accolades by performing Canada's first heart transplant – **25** years after in South African, Dr. Christian Barnard, did the first one in 1967.

Oh well, better late than never...

He gladly accepted an appointment in 1990 as Senator by Prime Minister Brian Mulroney, helping him pack the Senate in order to pass the hated GST (Goods and Service Tax) which Liberal Senators had stalled.

But Wee Willie was also a whore-monger, who was caught in 1999 soliciting a street prostitute who was actually an undercover female cop, and arrested. In lieu of facing criminal charges, he was lucky enough to be diverted to "john school" – a course for men facing their first solicitation charge -- but his name was leaked to local newspapers.

With his moron wife standing beside him, he held a press conference where he denied soliciting the fake whore.

Yet what should have been noted was that you have to admit you're guilty of the charge of solicitation to be eligible for the "john school" the diversion course. But, of course, you're guaranteed anonymity by the police for the john school attendees, and that promise was compromised. So the police couldn't say anything.

And the incompetent journalists who were happy to betray him to the public – sordid scandal sells newspapers -- couldn't figure out the facts of the story, so his lies went unpublicized.

* * * *

> "[T]he Quakers have done their share to make the country what it is...[and] many citizens agree with the applicant's [pacifist] belief, and that I had not supposed...that we regretted our inability expel them because they believe more than some of us in the teachings of the Sermon on the Mount."
>
> -- dissent in U.S v. Schwimmer (1929)
> Justice Oliver Wendell Holmes

The Sermon on the Mount (Matthew 5:3-12), to remind you, says in part:

> "Blessed are those that hunger and thirst for God's approval. They will be satisfied. ...
> Blessed are those whose thoughts are pure. They will see God. ...
> Blessed are those who are persecuted for doing what God approves of. The kingdom of heaven belongs to them."

Lt-General Romeo Dallaire – was a "toy general" who had never fought in any major war or conflict. In 1993 he was appointed the U.N. Force Commander in Rwanda, and in 1994 witnessed the attempt at genocide of its Tutsi minority (800,000 out of 1.2 million killed) by the Hutu majority and did nothing, but make phone calls to U.N. HQ asking to be able to do something, and being refused. He lacked the guts and fortitude to defy the denial of his requests to HQ, and stop the genocide and also, tellingly, the murder of the fellow Belgian peace-keepers under his flawed, weak-willed, and cowardly command.

During the debacle, ten Belgian paratroopers were brutally executed by Hutu extremists, after Dallaire told the paratroopers over their radio to not resist and surrender their guns to the extremists.

After he returned home, Dallaire was found drunk on his knees in Jacques Cartier Park -- a known gay hangout -- across the river from Ottawa in Hull/Gatineau.

When it was publicized, his story came out, and in a dramatic reversal of fortune, he was eventually lionized by the media and public as a "victim" of the whole affair, and afflicted by post-traumatic stress syndrome. The mighty general was even appointed to the Senate.

However -- not publicized in the Canadian media, who love the sordid and love heroes -- the Belgians had a different take on Dallaire's conduct and leadership.

A Belgian Senate Commission investigation highly criticized his conduct that led to the deaths of the Belgian paratroopers.

Dallaire avoided traveling to Belgium because the Belgians said that they would seek a criminal indictment of him. The Belgian Minister of Foreign Affairs even called him a coward. And former Canadian General Lewis Mackenzie also criticized his conduct in Rwanda.

* * * *

"What's the go o' that?"

-- James Clerk Maxwell (1834)

In the U.S., more than 10,000 public officials have been jailed in the ten years since the year 2000 for corruption: bribe-taking, solicitation, tax evasion, tax fraud, influence peddling, election fraud, and other crimes of power.

In Canada, the number of convicted politicians is closer to zero, because of incompetent, lazy, and look-the-other way prosecutors, and the fact that what is considered grievous criminal misconduct in the U.S. is known as patronage in Canada, and "the way business is done" here.

* * * *

"Well, there's 97 crosses planted in the courthouse yard,
And 97 families who lost 97 farms.
I think about my grandpa, my neighbors, and my name,
And some nights I feel like dyin' like that scarecrow in the rain.
Rain on the scarecrow, blood on the plow."

-- "Scarecrow in the Rain" (1985)
John Mellancamp

Canada is not an ally of the Superpower, the U.S. It is just a satellite. A close and convenient source of raw materials for America's war machine and economy.

Canada is a country rich in mostly finite natural resources: oil, natural gas, hydroelectric power, aluminum production, wood/pulp/paper production, uranium, nickel, wheat/corn/tobacco, asbestos, and a bountiful cod fishery until it was destroyed by a dithering government that allowed over-fishing.

When our environment is raped and stripped bare of its natural resources, all that will be left is too many overpaid nitwit officials and bureaucrats, an expensive and draining welfare state, and Canada's moral bankruptcy, presiding over a frozen wasteland.

A nation emptied of natural resources, full of empty babble on human rights, and no future, except bankruptcy and poverty.

Canada is just a lot of talk and blather, and no action, totally disconnected from its own reality and accurate self-perception.

A nation of jackasses rules by jackals.

"In the course of all of it we are learning the fundamental principle that ethics is everything. Human social existence, unlike animal socially, is based on the genetic propensity to form long-term contracts that evolve by culture into moral precepts and law."

-- "Consilience: The Unity of Knowledge" (1998)
E.O. Wilson

[For a fascinating and insightful book on the pathetic failings and misdeeds of Canada, the Canadian Government, and its moronic people, see Alan Borovoy's book, "At the Barricades", listed in the Bibliography section at the back of this book.]

Prison Diary 84:   Conspiracy of the Individual
                             ====================

> "What is the process by which you judge what is
> what, so you're left utterly off the fucking hook?"
>
>   -- Mickey
>   "Hurley-Burly" (1998)
>   by David Rabe, based
>   on his 1984 play

> "As every cell in Chile will tell,
> The cries of the tortured men.
> Remember Allende, and the days before,
> Before the army came.
> Please remember Victor Jara,
> In the Santiago Stadium,
> _Es Verdas_ -- those Washington bullets, again."
>
>   -- "Washington Bullets" (1980)
>   The Clash

It's the **real** global conspiracy.

Do you remember the science fiction special effects extravaganza, "The Matrix" (1999) (starring Keanu Reeves, and Laurence Fishburne as "Morpheus", and written, and directed by the Wachowski Brothers)?

The main idea of the movie is several decades old, and they merely borrowed it. But, more to the point, what I'm about to detail is the actual "Matrix" -- the real world we live in and act our part in, rather than the fantasy dream-world we think exists, but is just a dream- construct scripted by others, just to fool and control us, but allow the successful and orderly continuity of society -- like in the "Matrix" movie.

Our actual fantasy dream-world is a planetary acting stage, where we are moral, and good, and evil is our boss, or the government, or other wicked countries -- their evil leaders and citizens, that it is okay for us to bomb, kill, and destroy.

It is *not* an evil cabal conspiracy to deceive, manipulate, and control every aspect of our lives -- and our collective future.

The answer is as simple as it is obvious -- a drum-roll please:

                             It's **us**.

It's us. All of us, collectively.

It's every individual, acting separately and unknowingly to each other, but somehow unknowingly acting in concert by their similar behavioral patterns -- driven by their similar "morality" -- and resulting in a sort of collective unity of corrupt self-interest.

When you cheat on your taxes, or fail to declare your proper income, reducing helpful government programs. When you buy a hot (stolen) item of merchandise, you encourage shop-lifting, or theft. If you file a false or inflated insurance claim, you increase the honest person's monthly insurance bill.

When you file a false allegation with the police or children's aid society solely -- and cowardly -- for the purpose of vengeance on an innocent person. When you lie in court. When you lie to the traffic officer and deny that it was your mistake that caused the car accident.

When you steal credit for someone else's work at the office. When you accept a promotion, when you know a colleague is more qualified, or would do a better job. When you accept a job, through connections or from a friend, you deny the principle of merit, and further the continuance of racism in employment opportunities.

More importantly, we are a part of the conspiracy when we pursue a lifestyle of materialism, and the resulting income inequalities and apolitical attitude that results. When we feed our pet(s), while people in the world starve. When we keep up with fashions, of clothing etc., at great cost. When we buy luxury homes and luxury cars, and take expensive over-seas vacations.

When you vote for conservative candidates that reduce welfare programs, tax the rich and upper middle class less, and support our country's wars, when you work for a giant multinational corporation, do weapons research or manufacturing

Or when we eat meat (9 kg of corn feed makes 1 kg of beef) or drive our car, or turn up the house heat instead of putting on a sweater -- or otherwise contribute to the continued environmental destruction that will eventually lead to our extinction – merely out of convenience, comfort, laziness, or self-interest.

It goes on and on, these individual acts of unknowing individual acts that are arguably at least soft "evil".

> "This rotten corruption ... is built on telling them a series of lies. The educational system conspires to make them believe that they have achieved when they have not. It sponsors individual self-delusion and national myopia. ...
>
> self-esteem does not emerge from lies and self-delusion. ... anything that causes effort or pain is forbidden ... failure is a banished concept altogether. But children are not stupid. They know when they are being sold a pup. ... What is more, it has produced a spiral of madness."
>
> -- Melanie Phillips

Prison Diary 85: "It's Not Over 'Till the Fat Lady Sings."
=============================

> "I tried to play 'husband'. I tried to taste the life of a
> 'simple man' -- it didn't work out... [So,] because I
> envy your normal life, it seems that envy is **my** sin."
>
> -- Jonathan Doe
> "Seven" (1995)
> Andrew Kevin Walker

> "Wherefore the Law hath no right to accuse
> me, or to hold me any longer..."
>
> -- "A Commentary on St. Paul's
> Epistle to the Galatians" (1531)
> Martin Luther

A man charged with a criminal offense, but who's completely innocent of the allegation(s), should not need a defense lawyer to establish this to a judge, assuming it is a simple, uncomplicated, straightforward case. It was as simple as that, I decided.

That's it, that's all.

And it would be the perfect "experiment" to prove this simple, and logical idea by defending myself against the last two outstanding charges against me -- the original Gatineau, Quebec domestic dispute charges that started the whole chain of events that snow-balled into my nightmare of the last two years.

My life in ruins, I had nothing left to lose, anyway, I told myself.

Of course, there was more to it, in the complex universe known as "Yogi's thought process". For instance, I'd had it with criminal defense lawyers anyway. I was fed up with criminal lawyers after my experiences with them in Ontario. Liars and deceivers, over-priced, lacking competence, error-prone, inefficient, ethically-impaired, and flawed in ways too numerous to mention, was my general feeling, and I just wasn't going to put up with it anymore.

And the process itself was flawed garbage. **I** was present during the alleged incident, and was therefore most competent to question and refute testimony or statements that were lies, exaggerations, untrue, or nonsense. The lawyer knew fuck-all because he wasn't there, and read a bunch of police reports, and that was basically it. **That** was really ridiculous, and allowed the prosecution to introduce all manner of crap that the defense lawyer would not be able to refute or address properly.

As well, the process poisoned fairness if the accused exercised his right to not testify.

So, I humbly asked God to help me in my time of need, and hoped that he would hear my plea. I was worried -- he hadn't been very nice to me for about a year and a half -- but it was an act of faith, I guess. My first, actually.

* * * *

>"And if you love him, oh, be proud of him,
>'Cuz after all, he's just a man.
>
>Stand by your man,
>Give him two arms to cling to,
>And something warm to come to,
>When nights are cold and lonely."
>
>>-- "Stand by Your Man" (1968)
>>Tammy Wynette, Billy Sherrill

I had forgiven her, and it actually felt good, even after all she had done to me, for no reason. She was mentally ill -- how could I blame a mentally ill person, especially as close to me as she was. Being my "Baby mother 2" and all that.

It felt surprisingly good having forgiven her. And I felt more serene, as a result.

I had studied her specific affliction as much as I could, in preparation for the trial, and for other reasons. Just to understand what had happened and why, at least. I had heard of, but knew nothing about "borderline personality disorder", which she had told me over a year ago she had been diagnosed with by a psychologist. And the stuff to read about it didn't help much in synthesizing an understanding of. Of matching the written word, with her bizarre actions, culminating in her destroying my life.

False allegations were linked to the disorder. Borderline -- they were on the borderline between neurotic and psychotic. But it still didn't "gel" in my mind into a coherent understanding.

Then I finally hit the bull's-eye -- at least one of them. When under "stress", as the borderline defined it, they actually become temporarily psychotic -- and thus delusional. She had slipped over into psychosis, and actually believed that I had hit her, and threatened her with death.

It took me a bloody six months of reading, and observation of other BPD females -- it's a female affliction by a huge majority -- to learn this. Thanks a lot, BPD book and paper writers!

I learned more about BPD from hanging with a girl I learned was a BPD case, who was a whore and a "friend" (and attractive and a reasonable bang, to boot!), for several hours until I fucked with her head (and body) enough to spur a highly negative psychological reaction, and split before she blew (no pun intended)!

* * * *

>"The soldier loves his rifle,
>The scholar loves his books,
>The farmer loves his horses,
>The film star loves her looks."
>
>>-- "Foxtrot from a Play" (19**)
>>W.H. Auden

And there was an important decision that I had to make. I had exercised my right to remain silent for the Ontario weapons charges -- the bail hearing. It had cost me, but I did not want to testify and possibly be torn apart on the stand on cross – cross-examination by the prosecutor.

But for the Gatineau charges, I was going to have to take the stand and testify. There was absolutely no evidence against me other than the words of my daughter's mother. So, I made the strategic decision that to ensure acquittal, it was best that I refute her charges by testifying.

But there were two ways I could take, and I wasn't sure if honesty with warts showing would be respected, given my experiences in Ontario. The argument went on in my head, back and forth, until at the last moment, I finally stopped dithering.

I'd be honest and straight-forward. I was relieved at my decision, in a way, because that was what Christ said you should do -- and what I have always tried to do, even before my recent religious "conversion".

I decided that I wouldn't conceal any warts and personal flaws that were bound to be brought up. It was the only way. It's openness and sincerity that works, rather than lying and covering-up. Appearing evasive and deceptive, which is what usually happens when people lie under competent cross-examination, are what the prosecutor and judge are practiced at detecting.

Act of faith, or just another high-stakes gamble? Or finally I just had had enough, and said "Fuck it..." – I'm taking control of my fate, my destiny, and let the chips fall where they may. But I had never done such a thing before, and didn't know the Canadian legal system very well, so the risk was there, and it was real.

* * * *

" 'None of us can help the things life has done to us. They're done before you realize it, and once they're done they make you do other things until at last everything comes between you and what you'd like to be, and you've lost your true self forever.' "

-- Mary Tyrone
"Long Day's Journey Into Night" (1956)
Eugene O'Neill

On Monday May 25, 2009 at just before 9:30am, I walked into the Court Room No. 12, on the third floor of the Gatineau, Quebec _Palais de Justice_ -- Court House -- for my trial before Judge Real R. Lapointe.

The charges were, that on 17 August 2007 in her suburban home in Gatineau, I assaulted and made verbal threats of death against the mother of my daughter, the complainant. The "assault" charge can mean threatening to hit, causing fear of being hit, touching without consent, or applying force to the person without causing actual physical harm, or a mark.

The Crown prosecutor -- _procurer_ – was Marc Philippe, a pleasant but serious-looking, short, young French-Canadian. Throughout the case, I had represented myself, and now at trial, I was going to defend myself. After a couple of meetings pre-trial, I thanked the _procurer_ for his treatment of me with respect and dignity, and offered him my hand. He took it, and we shook hands.

Marc had estimated a trial length of two hours. I had said I thought it would take two days. If they were going to waste my time, I was going to waste theirs, I joked to myself. Besides, I had a lot to say. And I was going to say it.

Marc and I entered the Court Room at the same time, and sat down across from each other, in front of the court reporter, and Judge. The public gallery was empty.

Based on our comments to each other, and general demeanor, the Judge commented that he had never seen Prosecutor Philippe and the defense lawyer get along so well. I smiled. I **did** get along with him.

Despite her saying that she would not, CM showed up as a prosecution witness and testified against me. (My daughter AC didn't; CM claimed she refused to allow her to be called to testify.)

I became flustered and disorganized, and totally bungled cross-examination of Christine's testimony. I had stopped taking my PTSD medication, to clear my head for the trial, but now I was grossly impaired by the PTSD. I was inexperienced and stressed, fading out and back in, as far as being organized and in control.

My mother testified on my behalf. She was **too** good. I commented out loud, "Even I don't believe your positive testimony about me! But thanks."

Next, I was to testify. But there was a little problem. I had to be sworn in -- on the Bible. I informed the Court that Jesus Christ in Chapter 5 of the "Gospel According to Matthew", specifically says you are not to swear oaths. That they were the work of the Devil. So it would be an act of sacrilege to make me swear an oath before testifying, and a violation of my Christian beliefs and obligations.

The prosecutor and Judge were confused. The trial ground to a halt in disarray.

I had brought my Bible. "I'll look it up, and read it for you." I quickly found the passage. "Here it is: Matthew 5:33-37. I'll read it to you to save time."

> " ' You have heard that it was said to your ancestors, 'Never break your oath, but give to the Lord what you swore in an oath to give him.'
>
> But I tell you don't swear an oath at all. Don't swear an oath by heaven, which is God's throne, or by the earth, which is his footstool or by Jerusalem, which is the city of the great King. And don't swear an oath by your head. After all, you cannot make one hair black or white.
>
> Simply say yes or no. Anything more than that comes from the evil one.' "

"So there it is. Swearing oaths is soundly condemned by Jesus Christ," I told the Judge. I paused. "Doesn't anybody read the damn thing sitting in every court room in the land?"

I "harrumphed" in mock disgust.

The prosecutor and Judge were still confused. I had successfully derailed my trial for ten minutes. We were at an impasse. But I had made my point. So I was let myself be sworn in. JC will understand, I'm sure.

First, I was questioned by the prosecutor. I denied that anything at all had happened as CM had stated.

I commented that the first thing she said in her written statement was that she hit me, so why was she not facing charges?

I commented that she said in her written statement that I threatened to stab her, yet she testified that I threatened to shoot her. She said I carried a knife all the time, yet I didn't use it, or brandish it at her, but only **threatened** to stab her. There was a contradiction in there.

I had the "911" call tape played in Court, in which there is **no** background noise of screaming, shouting, or an altercation. The only sound is my daughter asking my mother what she should say, after she got through to the 911 operator.

After the prosecutor was finished, the Judge began questioning me. Boy, he was good. He got right to the crux of the matter in no time. My daughter, AC.

"I made a 'command decision', with the facts that I had at the time, when she told me she was pregnant. And I knew whatever I did, someone was going to get hurt. Sometimes you're fucked -- excuse the language -- whatever you do. But a decision had to be made. So I made it." I was on the verge of choking up, but I got through it.

"And when I found out I was wrong twelve years later, I tried to make things right."

"You're emotionally fragile, aren't you?" asked the Judge.

I was shocked. He hit the nail on the head. "Yes. Yes, I am... How did you know?"

* * * *

"Brandy bottled in 1783. 1783 was a very good year.
Mozart wrote his great mass. The Montgolfier brothers
went up in their first balloon. And England recognized
the independence of the United States."

-- Connor Macleod
"Highlander" (1986)

I was able to get my shit together, and made what I though was an excellent closing argument -- a list of all her contradictory and illogical written statements and testimony, and the lack of any corroborating evidence.

* * * *

" Do, or do not. There is no ' try'."

-- Yoda
"Return of the Jedi" (1980)

When the Judge returned shortly, he acquitted me on both counts. The judge enumerated all of my arguments concerning the unreliability and contradictions of the testimony and written statements of the accuser against me, the lack of any corroborating evidence supporting her claims, and the absence of any audible dispute in the background of the 911 call audio recording -- and thus why, therefore, the mother had asked my daughter to make the 911 call.

It was a total victory for me. "Not guilty," on both counts. All charges were tossed out, and I was finally free of any accusations. My legal entanglements were finally at an end. I was so relieved, it was as if the proverbial 1000 kg weight had been lifted off my shoulders. I was smiling ear-to-ear!

I thanked the Judge in open court for the Quebec Legal System, for their respectful treatment of me, and their treatment of me with dignity. I commented on the Ontario Legal System which, by comparison, had abused me terribly.

But I had one final comment to make to the Court. I offered the statement, "Isn't it about time the government got out of the marriage counseling business? The justice system is a sledgehammer, not a relationship mediator. You're just not helping, and great mischief is being done.

'First, do no harm' should be the justice system's commitment to the public, just like the Hippocratic Oath a new doctor swears."

> "All the dreams we held so close,
> Seemed to all go up in smoke,
> Let me whisper in your ear,
> Angie, Angie,
> Where will it lead us from here?
>
> Angie, don't your weep,
> All your kisses still taste sweet.
> I hate that sadness in your eyes,
> But Angie, Angie,
> Ain't it time we said good-bye?"
>
> -- "Angie" (1973)
> Mick Jagger/Keith Richards

Prison Diary 86:					Home Is Where the Hurt Is
					====================

> "Long were the days of pain I have spent within its walls, and long were the nights of aloneness; and who can depart from his pain and his aloneness without regret?"
>
> -- "The Prophet" (1923)
> Kahlil Gibran

> "Happy families are all alike; every unhappy family is unhappy in its own way."
>
> -- "Anna Karenina" (1877)
> Leo Tolstoy

I was born in Ottawa, Canada in 1959 and grew up in Ottawa's East End -- my early childhood was spent growing up in New Edinburgh, to be precise. By 1971, my father, a National Research Council theoretical physicist, had moved up in the world, and we now lived in a three-bedroom two-story brick house in nearby -- adjoining to the New Edinburgh neighborhood – to solidly middle class Lindenlea on Acacia Avenue.

It was important, because Acacia Avenue was on the border -- which side we were on was not discernible to the casual observer. But, we were on the side of Acacia that was the middle class Lindenlea, and on the other side of Acacia Avenue was the rich enclave of the mansions of Rockliffe...

* * * *

> "Is it surprising that prisons resemble factories, schools, barracks, hospitals, which all resemble prisons?"
>
> -- "Discipline and Punish" (1975)
> Michel Foucault

Alienation. Rejection. Repudiation. Estrangement. Realms of darkness.

"Sunshine" (1999) is a wonderful movie, Canadian-produced, with the depressing conclusion that neither preserving your ethnic culture, nor assimilation works in a foreign country. Separation is the only solution. Accommodation, assimilation, co-existence is a hopeful, but invariably flawed, profoundly inadequate solution.

Everything you stand for. Everything you represent. Everything you believe, means nothing to virulent racial discrimination.

Starring Ralph Fiennes, "Sunshine" follows the ups-and-downs of three generations of a well-to-do family of Hungarian Jews during the times of anti-Semitism of the Austro-Hungarian Empire, through WW1, through the inter-war years, WW2 and the concentration camps (Hungary was a Nazi ally, bears the ignominious honor of having had exterminated the most Jews of all the European countries under Nazi rule), followed finally by Communist authoritarian liberation and rule.

In 1956, the Hungarians revolted against communist rule, and were "dealt with". Appropriately, I believe for a bunch of racist, former Nazi pig, mass-murderers. The Cold War was in full swing, so they were called "freedom fighters" in the West.

How soon, conveniently, they forgot...

* * * *

"Today we worship Christ, but the Christ in flesh we crucified.
Stoning prophets and erecting churches to their memory
afterward has been the way of the world through the ages."

-- "Collected Works" (1947)
Mohandas Gandhi

I was proud to be a Canadian. Sometimes I was really proud to be a Canadian. I have seen acts of pure goodness, and original brilliance. I have seen acts of charity and compassion that show the intent of sincere worthiness and compassion and decency.

My home and native land. Born here. It's obvious I am a native born. If you don't have an accent you are either Canadian born or you immigrated at an early age, that should be obvious, but is not widely known. I was born, raised, and educated here.

I have rather strange beliefs and opinions that are considered rather "different" to say the least. People usually talk to me about politics **once**. I had views of religion that used to be jokes classifiable as heresy or sacrilege. Now that I've read the New Testament, and a little of the Old, I claim to be an amateur theologian with a list of clearly erroneous – other people's -- heresy that needs to be stopped by logic or a campaign of ridicule to expose idiocy masquerading as religious dogma or belief.

But culturally speaking, I'm thoroughly Canadian in all or most ways, other than the way I'm viewed – and less often treated – by some white male Canadians. Things are a **lot** better than they used to be – and for that I'm oh-so-grateful. The harsh, and overt racism, and the entry-level professional employment.

And I made the conscious, considered decision when I was young to chose cultural assimilation, because I considered what made a "Canadian" Canadian to be the obvious choice, and an easy choice over the alternative of maintaining an ethnic culture and identify that I knew nothing about, and considered inferior anyway, over things like its rigidity, caste system, and their abuse of women, who were allotted second class status.

I knew little about these ways anyway, but I did not consider maintaining my ethnic identity viable, practical, acceptable, or even close, because I rejected some really obnoxious and detestable things about my father's culture that made Canadian flaws seem trivial.

But. either way, though, I was an outsider whichever choice I made, who didn't fit in anywhere. It was, and is, a hard road to take.

* * * *

> "Manliness consists not in bluff, bravado, or lordliness.
> It consists of daring to do the right and facing
> consequences, whether it is in matters social, political,
> or other: it consists in deeds, not in words."
>
> -- "Teachings of Mahatma Gandhi" (1945)
> Mohandas Gandhi

Disillusionment and betrayal.

I used to be proud to be a Canadian: the welfare state to help the poor, and so nobody starved. Free medical care, so the sick would be taken care of, even if they were poor.

I was proud of our government's liberal, peacenik, and leftist world view: the acceptance with open arms as immigrants of American Vietnam War draft dodgers, evaders, resisters, and deserters in the late 1960's, early '70s.

*   *   *   *

> "A popular government, without popular information
> or the means of acquiring it, is but a prologue to a
> farce or a tragedy; or perhaps, both. Knowledge
> will forever govern ignorance. And a people who
> mean to be their own governors, must arms
> themselves with the power knowledge gives."
>
> -- James Madison (1822)
> Letter to W.T. Barry

Casus Belli. The Cause of a War.

It'll be a short war that we'll win in short order, no problem. We're strong, therefore we're right. We're aggressive and dominant and that's good, so let's fight. We'll win because we're right -- and we're always right. We have to fight, because our honor is at stake, and that's important enough to go to war for. It's someone else's kids who'll die, for our enrichment, so let's go to war.

It's a democracy and if you elected officials can't get what we want, we'll find someone who will, to elect. So war becomes a first, mandatory option of elected officials, to give the people what they want by making war for economic reasons. It's a big flaw of democracy.

Or just: we'll win. Conquer the bastard inferiors.

Casus Belli works pretty much the same as minority group discrimination. And racial or ethnic discrimination is either a hot or cold war against the minority.

PET. Pierre Elliott Trudeau. The greatest of a pissy lot of Canadian Prime Ministers, ranging from the mediocre to the idiots, to the fascist. Trudeau: who publicly rejected the U.S. Vietnam War, and repealed the old and backward sex laws against conduct between consenting adults in private, saying "the law has no place in the bedroom".

Lastly, Trudeau brought Canada's Constitution home in 1982, and established the Charter of Rights and Responsibilities -- essentially equivalent to the 1778 U.S. Bill of Rights.

Well, better late than never...

But what was most important at the time to me were Trudeau's new policies of fantastic liberalism and reformation. What I call the "Trudeau Spring", the end of the English domination and entrenched discrimination against the French -- and other minorities, including me. (When I grew up in the 1960's and 1970's, Canada was a profoundly, overtly racist country -- I was called names, insulted, mistreated, and beaten up with regularity over my non-whiteness by the other kids until Grade 6.)

The Trudeau Spring began following the end of the October Crisis at the end of 1970, when Trudeau showed a strong hand -- in theory -- against terrorism by arresting people willy-nilly, and calling out the Army. But, then when the Crisis was over, he ended the systemic and systematic discrimination against the French, and hired them into the government by the thousands. And other minority groups rose along with the rising tide, and discrimination seemed to withdraw and disappear.

It was a new dawn for me, and the way I was treated. Sunrise, finally.

It was the only racial/ethnic conflict problem that I know of that was solved. And it was solved without civil war, or separation, and no government or communal violence -- majority ethnic civilian slaughtering of, or violence against the minority ethnic group.

It was one of the main times when I was proud to be a Canadian. When my heart was warmed in that delightful way. And all was right with the world through my eyes.

I had left Canada for its systemic racism in disgust for the U.S. in 1980. However, when I came back in 1984, I couldn't believe it! Everything had changed for the better. I later called it the "Trudeau Spring". Discrimination against French-Canadians had disappeared. And with it, discrimination against non-whites like me had gone too.

I could finally get a job, and it was a good job. And the government helped me find it, and helped fund it initially. I did well. I was happy and content.

* * * *

"The most dangerous enemies within our country are...
statesmen who turn people into nihilists and cause
indignation and revulsion against the government."

-- "Diary" (ca. 1870)
A. Nikitenko

Then there was the time -- in the 1990's, I believe -- that the U.S. was seeking the extradition of a couple of Canadian crooks who had fled back to Canada. They were wanted for defrauding hundreds of people -- including many gullible seniors -- in a disgusting telemarketing scam.

The extradition would have been routine. *Then* the mouthy state District Attorney crowed on camera on the 6 o'clock news how he reveled at the idea of getting his hands on them through extradition, convicting them, and then their being *raped* in state prison!

Canadian officials took issue with a U.S. prosecutor bragging about prison rape, and refused the extradition request.

I thought the crooks were scum, but I was so proud to be a Canadian **that** day.

* * * *

> "Cast your whole vote, not a strip of paper merely,
> but your whole influence. A minority is powerless
> while it conforms to the majority;...but it is
> irresistible when it clogs by its whole weight."
>
> -- "Civil Disobedience" (1849)
> Henry David Thoreau

A friend of mine at Engineering School at the U. of Ottawa, Subesh -- a Tamil Sri Lankan -- told me his story one day. His whole family had fled Sri Lanka in the mid-1980's, arrived in Canada, and applied for political asylum, saying the father's life was in danger from the Sri Lankan Government for political reasons. A racially-motivated civil war was raging.

The Canadian immigration authorities were dubious of the claim. Ya know, proof, right? How do you prove a claim of political asylum? It's really difficult.

Until one day, at a diplomatic party in Sri Lanka, a Sri Lankan official said right out loud to a Canadian Embassy diplomat that they couldn't wait to get their hands on Subesh's father, in a malevolent tone that was unmistakable.

Subesh's family were granted political asylum forthwith.

* * * *

> "When you go into a house, greet the family. If it is a family that
> listens to you, allow your greeting to stand. But if it is not receptive,
> take back your greeting. If anyone doesn't welcome you or listen to
> what you say, leave that house or city, and shake its dust off your
> feet. I can guarantee this truth: Judgment day will be better for
> Sodom and Gomorrah than for that city."
>
> -- Matthew 10:12-15

The "little" details that make up a truly free and democratic society.

The U.S. Bill of Rights, Constitution, and Declaration of Independence provided the core concepts 240 years ago.

First, the basics: freedom of speech, and with it freedom of the press. Freedom of religion, freedom of conscience, the right to be free from arbitrary arrest and detention, the right to be free from arbitrary undocumented search & seizure, the right to remain silent, the right to a competent lawyer, if you can't afford one.

Now the details: open courts, the right of appeal in court proceedings, mandatory bail, except on showing of risk of flight, jury trials available for all criminal proceeding, a non-corrupt judicial appointment process (requiring all judicial appointments to be lawyers rather than a reward system for local politicians – a disgrace occurring in Canada) sunset laws, sunshine laws, term limits for elected officials, recall elections, whistle-blower protection statutes, binding referenda, and citizen initiatives on the voting ballot, freedom of information laws, and reporter's sources protected from disclosure by law, when in the public interest.

A system of checks and balances between legislative, executive and judicial branches, unlike the top-down pyramidal centralized power structure of Canada with absolute power of the executive branch.

Accountability. Not a Supreme Court that says incompetence is an absolute defense against government, official and professional incompetence (applicable to themselves!). Wholesale violations of the Charter of Rights without punishment. Impunity to gross police incompetence, and d the gamut of misconduct: police malfeasance, misfeasance, and nonfeasance. Routine police perjury in Court proceedings that casts the entire judicial system as a joke.

We are now accessories to U.S, war crimes and other outrages in Iraq and Afghanistan: the 1984 Convention Against Torture (and its federal criminal counterpart, 18 U.S.C. §2340), the Geneva convention, the Child Soldier Convention, and the Charter right for Canadian citizens to leave or enter country without obstruction.

There is no police accountability or transparency -- the Special Investigation Unit (SIU) for investigating civilian deaths or injuries is stacked with pigs (47 out of 54 in Toronto are ex-cops). Out of 3400 SIU investigations, 95 charges were laid, 16 officers were convicted, and 3 received jail time [Darryl T. Davis, Carleton University, Ottawa (2010)].

No punishment, no improvement; government by scandal & public outrage when the occasional word leaks out. Anti-Islamic witch-hunt politics that allowed what happened to Mahar Arar -- his kidnapping by the U.S., his torture in Syria, and his libeling by the Canadian authorities, in "leaks" to cover it up in the press. The TASER killing of an upset, brand new Polish immigrant by the RCMP with no disciplinary action taken against the officers responsible.

I am talking mostly about Canadians in a position of power in the government. That have the power and responsibility to find issues of importance, and deal with them. Maintenance, corrective action, regular inspection, and evolution in a correct direction to insure that flaws or improper behavior of government is found and directed, rather than ignored or covered up.

And worst of all, that they don't care about, and don't change.

* * * *

> "I remember, I remember
> The fir trees dark and high;
> I used to think their slender tops
> Were close against the sky:
> It was a childish ignorance,
> But now 'tis little joy
> To know I'm farther off from Heaven
> Than when I was a boy.
>
> -- "Past and Present" (1843)
> Thomas Hood

But alas, Canada is my home. I have no where else to go. Nowhere that I could feel even close to that relaxed, fundamentally at ease, secure feeling of your home, your " 'hood", your native land, your country.

But all that is gone, now. I hope it comes back, but it is one hope where faith eludes me. I officially renounce my Canadian citizenship, effective the second before my death. I do this as a symbolic rejection of the nation that betrayed one of its own for no reason, so horribly, and without final justice, or satisfactory resolution.

I wish to be cremated, and my ashes scattered to the wind in France somewhere. And if that is not possible, I wish my ashes to be scattered into the Atlantic Ocean within the French territorial waters of the Atlantic Ocean islands -- near the Canadian Province of Newfoundland -- of St. Pierre or Miquelon, in the summer, when the sun is shining bright, and when the sea is calm.

It will be my final act of defiance, rejection, and dismissive contempt of a nation that has shamed itself, and doesn't deserve me in death, as it didn't in life.

Let me say it clearly and loudly. As a whole, collectively. By action and inaction. Canada, you disgrace yourself. Completely and utterly and unequivocally. A complete disgrace.

Deal with it, or not.

> "Representative government...represents money, not people, and therefore has forfeited our allegiance and moral support."
>
> -- "One Life at a Time, Please" (1988)
> Edward Abbey

And so, as I began this chapter, I end it with a happier thought -- a reminder of better times -- another quote from Kahil Gibran's "The Prophet":

> "For in the dew of little things the heart finds its morning and is refreshed."

Prison Diary 87:   "The Pursuit of Happiness", and All That

> "You can see the summit, but you can't reach it,
> It's the last piece of the puzzle, but you just can't make it fit.
> Doctor says you're cured, but you still feel the pain.
> Aspirations in the clouds, but your hopes go down the drain."
>
> -- "No One is to Blame" (1984)
> Howard Jones

> "Dear Abby:
>
> We have neighbors where the parents act like a couple of teenagers. When the husband comes home from work, the wife runs out to the street to meet him, and he carries her into the house piggy-back! If the husband is repairing the roof, she is right up there with him.
>
> If he's under their car, she's there, too. When they sit in church, they hold hands and read out of the same book when hers is right beside her, closed.
>
> What do you think of people like that?
>
> -- St. Petersburg. Florida"

> "Dear St. Petersburg:
>
> I think your neighbors know the secret of real happiness. God bless them."
>
> -- "Dear Abby" column (ca. 1990)
> Abigail Van Buren

Most people are trapped in their belief system. It's like they're trapped in a prison cell. Some are at least lucky enough to be trapped in a golden cage.

It's as if people are all trapped in a box. "Think outside the box" goes the expression (which actually just means, "Be novel and creative in your thinking.")

Think of what pleasures or advantages **might** be outside that box. But, goddammit, if you're trapped in a box, at least make it a box of ice-cream!

*   *   *   *

> "Where there is the tree of knowledge, there is always Paradise: so says the most ancient and the most modern of serpents."

-- "Genealogy of Morals"
Friedrich Nietzsche

The discerning well-read person, American historian, or student of political science will have noted with interest the phrase, "pursuit of happiness" in the U.S. "Declaration of Independence" -- an amazing document by any standard of intellectual, philosophical, political, and legal excellence -- and what's more, written in the Dark Ages of Freedom almost 250 years ago.

The complete statement -- the second line of the Declaration's text -- is as follows:

> "We hold these Truths to be self-evident, that all men
> are created equal, that they are endowed by their
> Creator with certain unalienable Rights, that among
> these are Life, Liberty, and the pursuit of Happiness."

The Declaration of July 4, 1776, and the courageous men -- including Thomas Jefferson, Benjamin Franklin, and Dr. Benjamin Rush -- who wrote and signed the parchment on which it was written, had willingly offered themselves up in sacrifice as rebels, outlaw fugitives from the despotic English "justice" of the time, and consigned themselves to automatic conviction for treason, followed by a merciless sentence, and the carrying out of execution by hanging.

The phrase "pursuit of happiness" was actually first coined in 1759 by Dr. Samuel Johnson.

But enough of history and background -- I'm writing to point out two items of significance.

The first, was the recognition -- expressed as the pursuit (and presumably the possible attainment) of happiness -- by the Fathers of the American Revolution and the War of Independence of the essential need for "freedom of spirit" as necessary to the point of being a God-given Human Right.

The second, and equally if not of greater importance -- maybe trivial or obvious -- is the addition of one word of note -- that they specifically wrote not that they had a right to be happy, but a right to the *pursuit* of happiness.

There is no certainty. There is no assurance. There are no guarantees. Expectation is fruitless -- a nowhere trip. It's all up to you. With maybe some luck or a hand from fate, perhaps... And there is no likelihood *even* of permanence, should you be lucky enough to attain the goal of happiness. They knew that in 1776, and we should realize it today.

Chase it always. Pursue it with all the energy you can muster, and never give up. Grab it when you can. Hold and tight, and cherish it dearly. Then you will appreciate and see the beauty of life. That's the only way I can make sense of my life, and suffering and pain in general.

Play the hand the Life deals you, whatever that hand is. Even if it's only "King high". It's all you've got. Play it for all it's worth. Don't let them win... Just don't let them win, no matter what.

* * * *

> "So afraid to love you, but more afraid to lose,
> Clinging to a past that doesn't let me choose.
> Once there was darkness, a deep and endless night,
> You gave me everything you had, oh, you gave me light.

> And I will remember you, will you remember me?
> Don't let your life pass you by, weep not for the memories."
>
> -- "I Will Remember You" (1995)
> Sarah McLachlan, Seamus
> Egan, Dave Merenda

"Happy, happy. Joy, joy" was a line uttered periodically by the intelligence-deficient Stimpy character of "The Ren and Stimpy Show", a popular adult (stoner?) cartoon from the early 1990's. "Don't worry. Be happy." was the title and main lyric of a song by Jamaican, Bobby McFerrin that rocketed to the top of the U.S. and Canadian charts in 1988.

"Happiness is a warm gun," was the title of a Beatles song from the late 1960's. The lyric was written by the late John Lennon (who was himself gunned down by a mentally disturbed man in 1981 or so). Lennon was quoted in a "Playboy" interview in the 1970's (when men *did* -- and could! -- read "Playboy" for the quality of the articles -- as was widely claimed by men to leery, skeptical women), devoted solely to Lennon's explanations of the story behind, and meaning -- of cryptic or arcane lyrics -- of Beatles songs, the "warm gun" song title simply came from a headline on a U.S. gun magazine cover that Lennon happened to see one day.

* * * *

> "Ain't no angel gonna greet me,
> It's just you and I, my friend.
> My clothes don't fit me no more,
> I walked a thousand miles,
> Just to slip this skin.
>
> The night has fallen, I'm lyin' awake,
> I can feel myself fadin' away.
> So receive me brother with your faithless kiss,
> Or will we leave each other alone like this?
> On the streets of Philadelphia.
>
> -- "Streets of Philadelphia (1994)
> Bruce Springsteen

I was wounded. I was deeply wounded. The armor had been pierced. And I was bleeding badly.

Peace and serenity seemed like distant memories. And torment, mental pain, and suffering were the order of the day. And the incubus ruled the night, so even the comfort and renewal of sleep was no escape.

> "Happiness I cannot feel,
> And love to me is so unreal.
>
> And so, as you hear these words,
> Telling you now, of my fate,

I tell you to enjoy Life,
I wish I could, but it's too late"

-- "Paranoid" (1970)
Black Sabbath
Ozzy Osborne

Prison Diary 88:   Endgame: The "White Rose" Lives!
=============================

> "Now you are burnt-out husks, your spirits haggard, sere,
> always brooding over your wanderings long and hard,
> your hearts never lifting with any joy – you've suffered
> far too much."
>
> -- "The Odyssey" (ca. 850 B.C.)
> Homer

> "After all, Hitler states in an early edition of ["Mein
> Kampf' (1925)] ...'It is unbelievable, to what extent
> one must betray a people in order to rule them.' "
>
> -- "White Rose" Leaflet No. 2 (1942)
> anti-Nazi student group, crushed
> by the Gestapo after Leaflet No. 6

Campanero Yogi Shan. Ahora y siempre! [Comrade Yogi Shan. Now and forever!]

Don't you just love a true story with a happy ending?

At least happy, in that I stood up for my rights, stood up -- and suffered for it -- on a matter of principle, and was **finally** cleared of all criminal charges. I was wounded deeply, but I was alive, and had a lot to be proud of -- in spite of the fact of my mental wounds -- complex PTSD -- and the fact that peace and serenity and hope eluded me.

And so time, a clear conscience, and a head I can still hold up high, are perhaps the only solution -- the healing medicine -- or at least a bandage, for much.

"Life's too short," we hear said about choosing our battles carefully. Our Time is limited in this world. This is obvious. Death is unavoidable, and only the actual exact time of the event is unknown. Death stalks all of us, from the moment of birth, to our last step, our last breath, our last thought. Our last deed.

Is that not truth enough that our focus should be on the way we live, and the way we would want to be seen as having lived? That even though immortality is unachievable, there is a route to symbolic immortality. That though a man will inevitably die, an idea of his may live on. And that as an idea -- a concept, an ideal lived, promoted, taught, and spread -- we are all capable of immortality if we have tried to help our fellow man, or have tried to make the world – or the future -- a better place.

But as for me, I soldier on, as I've always soldiered on. One more wound, one more scar, a little more tired, and the numbness of my emotions reinforced a little more, the weariness of my soul weighing a little heavier.

But alternatively, numbing yourself to emotion as an escape, I discovered that the words of Jesus Christ can and do provide welcome comfort, and genuine emotional relief and respite.

Profound -- yet simplicity itself --, unquestionably sincere, as timely now as they were when spoken 20 centuries ago, and imbued with a pleasant yet inescapable logic that shines as only the obvious Truth can shine. The gentle encouraging words of but a simple man, who saw things with crystalline clarity, was not afraid to voice them, and who seems to have been the son of God:

> "In the world you'll have trouble. But
> cheer up! I have overcome the World."
>
> -- John 16:33

Rock on, Jesus! How's **that** for meaningful and poignant words of comfort for the heavy of heart?

But now comforted and serene, there are the words of actress, Lillian Hellman to put a fire in my belly, to guide and motivate me. She hand-scrawled a note and passed it to her lawyer. They were sitting at a table before the House Un-American Activities Committee – the anti-communist, anti-leftist, red-hunting "HUAC". It was a particularly organized witch-hunt. She had been subpoenaed to appear, and was going to refuse to testify, and name names, which would result in her black-listing and a ruined name and career.

Her scrawled note merely said:

> "I cannot and **will not** cut my
> conscience to fit this year's fashion"
>
> -- Lillian Hellman (1952)
> note scrawled to her lawyer,
> House Un-American Activities
> Committee (HUAC) hearing

* * * *

Final words.

I remember, it was 1994 or so, and I was having an emotional meltdown, the first I'd ever experienced, as the facade of "warrior" crumbled to dust. And I told my mother, in a howl of despair and pain, if there ever was one, that if I ever needed an epitaph, I wanted her to make sure it consisted of these words:

> "He was a good man.
> And they should have
> treated him better."

And then I broke down completely.

"But I'm not crazy, I'm just a little unwell,

I know right now you can't tell,
But stay awhile and maybe then you'll see,
A different side of me.

I'm not crazy, I'm just a little impaired,
I know right now you don't care,
But soon enough, you're gonna think of me,
And how I used to be."

-- "Unwell" (2004)
Matchbox 20

Prison Diary 89:   Epilogue
=======

> "More and better bombs. Where will this lead is difficult to see. We keep saying 'We have no other course'; what we should say is 'We are not bright enough to see any other course.' "
>
> -- David Lilienthal (1950)
> Chairman of the A.E.C.
> in his diary

> "It seems that nobody dares to be themselves. They are afraid of solitude and because of that, everyone is alone in a mass of lonely people... Chile is at least a place where bread is bread and earth is earth."
>
> -- Victor Jara (1968)
> Chilean folksinger,
> on the West, in letter
> to his wife

I am not a prophet. With all my faults, flaws, and imperfections, if I were a prophet, God's in trouble. Has he run out of choices of anyone better? It would mean he has run out of the Good and the Pure.

Rather this is my howl of outrage and my manifesto for peaceful, meaningful, and long-overdue reform.

I tell you all: talk, write, point out injustice when you see it, stop it if you can, make a fuss, make a noise, make a nuisance, make phone calls, email, write a letter to the editor, vote, and vote as a block, demonstrate alone or in groups, spray funny or relevant graffiti, or agitate as you see fit.

And **especially**, unite with others to organize for the fight, **whatever** that might entail: a political force, or resistance.

Sabotage the System, drop out of materialism, don't buy the latest corporation-made electronic toys, anesthetize yourself with cable TV – cut it off – stop buying CDs and renting DVDs, stop going to the movies (read a library non-fiction book instead), or restaurants, stop buying the latest fashion clothes or so much make-up.

Reduce or eliminate meat and fish consumption, if you can (I know it's hard). Stop wasting your money on lottery tickets. Reduce your booze consumption and quit smoking to deprive the government of its taxes. (Re-discover love, parks, and sobriety.) Share a car with a friend or two, instead of each buying one. Use a motorcycle or scooter, instead.

Drop out of the tax system.

Make the economy hurt, and let them know as an organized movement, what and why you're doing it.

Disrupt and obstruct the actions and performance of the State to shaking and quaking its very foundation. Be an official complainer. Be a witness to police misconduct. Infiltrate the enemy's power structure: the military, the police, the government bureaucracy, big corporations.

Infiltrate in large numbers political parties. There are enough students, leftists, liberals union members, the gay, radicals, and fed-up moderates to unite for change instead of running the rat race, or the girl chase.

Be creative: incite into an organized force for a new law depriving the right to vote for federal, provincial, and municipal government employees, as a **direct conflict of interest**. Disenfranchise the bureaucrats and administrators, the military, military sub-contractors, the police and prison personnel, the justice system, the diplomatic service, security guards. **And** their family members over the age of thirty, as well.

Leave teachers and health care workers, garbage collection workers, and the like alone.

Use your imagination and intelligence. Do and continue doing these things, and others until something changes, and then this book will have served its purpose. Real democracy is *our* choice of the future, not theirs.

Start a movement to have the U.S. annex Canada, convert the Provinces to individual States of the U.S., and the rights and freedoms of the Bill of Rights.

Good luck. After thirty years of sole – and ineffective resistance – my fight is over...

> "The Constitution was not adopted as a means of enhancing the efficiency with which government officials conduct their affairs, nor as a blueprint for ensuring sufficient reliance on administrative expertise. Rather, it was meant to provide a bulwark against infringements that might otherwise be justified as necessary expedients of governing."
>
> -- O'Lone v. Shabazz (1987)
> Justice William Brennan
> dissenting opinion

Appendix A: Chronology of Legal Events
==========================

"[L]egal processes are hardly discretionary. Under them there's no provision for moral anguish, moral conviction, moral resistance. No, it's the old story of law hardening, scrubbing itself of human concern, becoming sterile...Law has become its own end...Law has become its own end. So why should I submit to those processes? They certainly have nothing to do with me..."

-- Father Daniel Berrigan (1970)
letter

"Living is easy with eyes closed,
Misunderstanding all you see.
It's getting hard to be someone.
But it all works out.
It doesn't matter much to me."

-- "Strawberry Fields Forever" (1967)
The Beatles
John Lennon, John Winston

Alleged domestic incident, Gatineau, Quebec	Fri. Aug. 17, 2007	ca.3:00 pm
Voluntary Surrender and Arrest at Police Station, Greber Blvd., Gatineau, Quebec	Wed. Aug. 22, 2007	4:00 pm
Warrant-less search of my Ottawa Home by two Gatineau Police officers	Wed. Aug. 22, 2007	7:00 pm
Arrest by Ottawa Police in Ottawa, Ontario	Wed. Aug. 22, 2007	9:00 pm
Warrant-less search of my Ottawa home continued by Ottawa Police Guns & Gangs Unit	Wed. Aug. 22, 2007	9:00 – 11:00 pm
First Court Appearance/"Show Cause" Hearing	Fri. Aug. 24, 2007	10:00 pm
Arrival in OCDC Jail	Fri. Aug. 24, 2007	3:00 pm
Show Cause [First Bail] Hearing	Tue. Oct. 11, 2007	10:00 pm
Trial (Canceled by Crown at "Start" of Trial)	Thurs. & Fri. Feb., 2008	10:00 am
Second Bail Hearing	Mon. March 3, 2008	2:00 pm
Bail Granted by Justice McKinnon	Mon. March 3, 2008	5:00 pm
Release from Custody on Bail	Mon. March 3, 2008	5:30 pm

Post-Traumatic Stress Disorder Diagnosis	Wed. March 5, 2008	2:45 pm
Kicked bloody by Ottawa Police Constable Heaton (Easter)	Fri. March 21, 2008	ca.10:30 am
Ottawa Police (Guns and Gangs Unit) Come to Arrest Me at My Parent's House	Fri. March 28, 2008	11:30 pm
Arrested by DART Unit for Breach of Bail Conditions	Mon. March 31, 2008	8:00 pm
Court Hearing/Stay of Proceeding; All Criminal Charges Stayed	Tues. April 22, 2008	11:00 am
Release from Detention	Tues. April 22, 2008;	noon
Trial Begins on Original Quebec Criminal Charges	Mon. May 25, 2009	10:00 am
Trial Ends; Acquittal by Judge on both Criminal Charges	Tues. May 26, 2009	3:00 pm

"Did they hurt you, son. Did they hurt you and make you mean mad?...

Sometimes they do. Sometimes they do something to you. They hurt you and you get mad, and then you get mean. And then they hurt you again and you get meaner and meaner, till you ain't no boy nor man anymore. Just a walking chunk of mean mad.

Did they hurt you that way, son?"

-- Tom Joad's mother
"The Grapes of Wrath" (1940)

Appendix B: The Canon   [Pun Intended]
                           =========

    Canon (n.)   an official list of the basic literary works that set the standard of quality and information; the body of fundamental works of literature that are the essential benchmarks -- the pillars of knowledge -- that completely survey a given topic.

In celebration of "Freedom to Read" Week, I present the following bibliographies.

This list is no doubt incomplete. But it is what I found and caught my attention and respect, if not fascination. Some are hard to find. But don't let that discourage you.

    [Some are on the Internet (for now!). ILL (Inter-Library Loan) at your public library is an easy service, and great.]

Garbage is easy to find. In books, and people, and societies. True joy is in finding -- and *knowing* you have finally found -- the essential, pivotal Truth.

That is the final victory of all.

> "They drew first blood. Not me. They drew first blood."
>
> -- John Rambo
> "First Blood" (1982)

Political Homicide and Revolutionary Justice

Baumann, Bommi. "How It All Began: The Personal Account of a West German Urban Guerrilla". Vancouver, Canada: Arsenal Pulp Press (2000).

Bell, J. Bowyer. "Assassin!: The Theory and Practice of Political Violence". NY: St. Martin's Press (1979).

Castellucci, John. "The Big Dance". NY: Dodd, Mead & Co. (1986).

Giorgio. "Memoirs of an Italian Terrorist". NY: Avalon Publishing Group (1981).

Guevara, Ché. "Guerrilla Warfare".

Guillen, Abraham. "Philosophy of the Urban Guerrilla". NY: William Morrow (1973).

Luttwak, Edward. "Coup D'État: A Practical Handbook". Middlesex, England: Penguin Books (1968).

Marghella, Carlos. "Mini-manual for the Urban Guerrilla".

Nomad, Max. "Apostles of Revolution". NY: Collier Books (1961).

Richani, Nazih. "Systems of Violence". Albany, NY: State University of New York (2002).

* * * *

Firearms

Allsop, D, Popelinsky, L. _et al_ . "Military Small Arms", London: Brassey's (1997).

Bull, Gerald & Murphy, Charles. "The Paris Guns and Project HARP". Bonn: Mittler (1988).

Chinn, George M. "The Machine Gun", volume 4 (5 volumes in total). Washington, D.C.: Department of Naval Ordnance (1953). Available on the Internet.

Donnelly, John J. "The Handloader's Manual of Cartridge Conversions". South Hackensack, NJ: Stoeger Publishing Company (1987).

"Engineering Design Handbook: Guns - General", DTIC No. AD830 303, U.S Army Materiel Command (1963).

"Engineering Design Handbook: Breech Mechanism Design", DTIC No. AD-A079 666, Alexandria, Virginia: U.S. Army Materiel Development and Readiness Command, (1979).

Ezell, Edward C., "Small Arms of the World", Stackpole Books (1983).

Goldsmith, Dolf. "The Browning Machinegun", Vol. 3. Cobourg, Ontario, Canada: Collector Grade Publications (2008).

Halberstadt, Hans. "Trigger Men". NY: St. Martin's Griffin (2008).

Hatcher, Julian. "Hatcher's Notebook". Harrisburg, Penn.: Military Service Pub. Co. (1947).

Holmes, Bill. "Home Workshop Guns for Defense and Resistance: Volume 1: The Submachine Gun". Boulder, Colorado: Paladin Press (1977).

Musgrave, Daniel. "German Machineguns". MOR Associates (1971).

Nelson, Thomas B. "The World's Submachine Guns", Volume 1. Alexandria, Virginia: T.B.N. Enterprises (1963).

Nelson, Thomas B. and Musgrave, Daniel B. "The World's Machine Pistols and Submachine Guns", Volume 2A. Alexandria, Va.: T.B.N. Enterprises (1980).

Regan, Paul et al. "Weapon: A Visual History of Arms and Armor". NY: DK Publishing (2006).

Ross, John. "Unintended Consequences", St. Paul, Minnesota: Accurate Press (1996).

"Handbook on Weaponry", Rheinmetall Gmbh (1982).

Sherrill, Robert. "The Saturday Night Special". NY: Penguin Books (1975).

Smith W.H.B., and Smith, Joseph E. "The Book of Rifles". Harrisburg, Pennsylvania: Stackpole Books (1963). (Including excellent coverage of the history and development of rifle actions, and German, U.S.S.R., and U.S. rifles.)

Stevens, Blake. "MG-34 – MG-42". Cobourg, Ontario, Canada: Collector Grade Publications (2002).

Stevens, Blake. "Full Circle". Cobourg, Ontario, Canada: Collector Grade Publications (2006).

* * * *

High Explosives and Demolitions

["AD" books available free on-line from dtic.mil]

Carleone, Joseph, ed. "Tactical Missile Warheads". Reston, Virginia: AIAA (1993).

"CIA Field Expedient Methods for Explosives Preparation". Phoenix, Arizona: Desert Publications (1977).

Cook, M.A. "The Science of High Explosives" (1958).

Cooper, Paul. "Explosives Engineering". NY: Wiley-VCH (1997).

Davis, Tenney L. "The Chemistry of Powder and Explosives". Angriff Press (1943).

"Engineering Design Handbook: Elements of Armament Engineering, Part 1, Sources of Energy", DTIC No. AD830 272, Washington, D.C.: U.S. Army Materiel Command (1964).

"Engineering Design Handbook: Principles of Explosive Behavior". DTIC No. AD900 260. Army Materiel Command (1972).

ETI, "The Blaster's Handbook".

"Explosives and Demolitions", FM 5-25, Dept. of the Army (1967).

Fedoroff and Sheffield. "Encyclopedia of Explosives and Related Items", Vol. 1-10, PATR 2700 [Picatinny Arsenal Technical Report] (1983).

Fieser, Louis F. "The Scientific Method". NY: Reinhold (1964). [Includes the history of napalm, by its inventor.]

"Improvised Munitions Black Book", Vol. 3. Cornville, Arizona: Desert Publications (1983).

"Improvised Munitions Handbook", TM 31-210, U.S. Dept. of the Army (1969).

Kohler, Josef. "Explosives". NY: VCH (1993).

Lloyd, Richard. "Conventional Warhead Systems". Reston, Virginia: AIAA (1998).

Prokosch, Eric. "The Technology of Killing". Atlantic Highlands, New Jersey: Zed Books (1995).

Rothman, David B. "Mr. Death: The Life of a CIA Assassination Expert – By His Son" NY: Playboy Press (1982).

SIPRI. "Incendiary Weapons", Cambridge, Mass.: MIT Press (1975).

Stoffel, Joseph. "Explosives and Homemade Bombs". Springfield, Illinois: Charles C. Thomas Ltd. (1972).

Urbanski, T. "The Chemistry and Technology of Explosives", Vol. 1-4, Pergamon Press (1988).

W.P. Walters and J.A. Zukas. "Fundamentals of Shaped Charges" NY: John Wiley (1989).

* * * *

Commonly Accepted Lies, and Damned Lies, Exposed by the *Real* Truth
----

Bamford, James. "A Pretext for War: 9/11, Iraq, and the Abuse of America's Intelligence Agencies". NY: Doubleday (2004).

[The United States and Israel as the "civilized" modern world's Evil Empire.]

Blum, William. "Rogue State: A Guide to the World's Only Superpower". Monroe, Maine: Common Courage Press (2000).

Dallin, Alexander. "Black Box: KAL 007 and the Superpowers". Berkeley, California: University of California Press (1985).

[The last, most thorough, and unbiased U.S. academic book on the early 1980's, somewhat reasonable, downing of the South Korean 747 civilian airliner, flight KAL 007, which had accidentally (twice) violated their military airspace. The danger of a shoot-down was a severe risk, clearly emblazoned in red, on airline navigation maps.

The Americans and the Israelis, respectively, have both similarly shot down non-threatening Iranian (in July 1988 by the U.S.S. Vincennes), and Egyptian (in the late 1970's) civilian airliners, but U.S. President Ronald Reagan used the incident for propaganda purposes to hypocritically lie about, and denounce the Soviets loudly, viciously, and very publicly.]

Haas, Jeffrey. "The Assassination of Fred Hampton: How the FBI and the Chicago Police Murdered a Black Panther". Chicago: Lawrence Hill Books (2009).

Mayer, Jane. "Dark Money: The Hidden History of the Billionaires Behind the Rise of the Radical Right". NY: Doubleday (2016).

Porter, Gareth. "Manufactured Crisis: The Untold Story of the Iran Nuclear Scare". Charlottesville, Virginia: Just World Books (2014).

[The damned lies, covert machinations, and false propaganda of Israel and the U.S. regarding Iranian nuclear intentions.]

Gibson, James. "The Perfect War: Techno-War in Vietnam". NY: Atlantic Monthly Press (1986).

McCoy, Alfred W. "The Politics of Heroin." Chicago, Illinois: Lawrence Hill Books (1991) (original [shorter] edition: "The Politics of Heroin in Southeast Asia", was published in 1972).

Reinarman, Craig and Levine, Harry G., eds. "Crack in America: Demon Drugs and Social Justice". Berkeley, California: University of California Press (1997).

Suskind, Ron. "The One-Percent Doctrine". NY: Simon & Schuster (2006).

Appendix C: "Lies, Damned Lies, and Statistics"
==========================

> "One must live in the middle of contradiction, because if all contradictions were eliminated at once, life would collapse.
>
> There are simply no answers to some of the great pressing questions. You continue to live them out, making your life a worthy expression of leaning into the light."
>
>   -- "Arctic Dreams" (1986)
>   Barry Lopez

According to the "Small Arms Survey" there are an estimated 640 million small arms – everything from handguns to light machine guns – in the world. Civilians own about 400 million of these, generally 12 gauge shotguns, bolt-action hunting rifles, and (lastly) handguns.

Finland, France, Germany, Iraq, Uruguay, and Yemen have 30-40 guns per hundred people.

The U.S., Switzerland, and Israel have much higher rates of civilian ownership with the U.S having 83-99 firearms per hundred people. 1 million Swiss (over a quarter of population) have fully automatic assault rifle in their homes, as part of their formal military service obligation.

There are 700,000 Swedish reserve forces, in a similar system to the Swiss. But they don't get to take their guns home, like the Swiss do.

Only the U.S. has a high rate of gun violence.

Bingo! Conclusion: the high rate of U.S. shootings and gun deaths are related to social cohesion and inequality. Both racial and economic.

Democratic countries with restrictive personal gun possession laws, and heavy penalties are the U.K, Japan, Jamaica, and – borderline – Canada.

The source of civilian ownership of firearms are: legal weapons, stolen – formerly legal – weapons, smuggled weapons from neighboring weapon-rich or lax gun possession law countries (porous borders, lengthy borders, large cross-border trade, traditional smuggling history; bilateral smuggling: bring in dope, take back guns; corrupt customs agents), civil war, and the hold-overs from colonial war.

Estimates of the number of small arms in citizens hands in countries after civil and colonial wars:

       Mozambique   pop. 23 million   1.5 – 6 million
       Angola   pop. 13 million   700,000 minimum
       South Africa   pop. 49 million   4 million legal; 1 million illegal

                                              *   *   *   *

> "[T]he theories of good men who are enthralled by its delusions are made
> the excuse of the wicked who would rather plunder than work; because it
> stops enterprise, promotes laziness, exalts inefficiency, inspires hatred,

> checks production, assures waste and instills into the souls of the unfortunate
> and the weak hopes impossible of fruition whose inevitable blasting will add
> to the bitterness of their lot."
>
> -- "The Inhumanity of Socialism" (1915)
> Edward Adams

In 2001, 17,427 Americans were killed by guns, 43,051 in car accidents, and 16,274 in falls. In 2009, 24,500 people were killed in car accidents. Additionally, medical accidents kill tens of thousands of Americans every year.

Almost 1,100 American children were murdered in 2007.

The gun death statistics I have shown don't include suicides, and accidental shootings. A significant percentage is criminal-on-criminal (mostly gang-related) shootings. Bingo! The rate of U.S. gun deaths, though spectacular and considered newsworthy, is dwarfed by other deaths.

The U.K.'s "Telegraph" newspaper reported in 2008, that in England and Wales -- where handguns and semi-automatic rifles are banned completely, and rifles and shotguns severely restricted -- there were 400 knife crimes per week. In other words, 20,800 per year, not including Scotland.

So much for gun control being senseless and criminal violence control...

But then, the British elites have been stupid, corrupt, undeserving of their status, and homosexually pedophilic for quite some time, ruling a dullardly, lazy, and drunken populace. Australia and Canada are much the same.

* * * *

> "We were kind of products of the sixties...but not adults
> in the sixties. We were all primed with this rebellion-minded
> worldview, but we missed the train."
>
> -- Mike Watt (ca. 1985)
> "Rip It Up"
> interview by Simon Reynolds

According to the U.S. Department of Justice, there were just under 800,000 sworn law enforcement officers in the U.S. in 2004. Fifty-seven were killed that year, almost all by guns. That works out to a rate of 6.6 per 100,000 cops, versus a national average of 5.5 homicides per 100,000 citizens.

In other words, being an urban street cop is hardly such a dangerous job...

It is a job that attracts cowardly white men, who need society's license of approval for their arbitrary and abusive violence. A job that attracts those of mediocre intellect, and the simply dull-witted, and those with a lust for abuse of power without any consequence at all. A job involving abuse of the innocent and guilty alike. A fascist, racist pig job.

But unfortunately, not a dangerous job.

The cutting edge of society's war of oppression against the weak, the poor, immigrants, minorities, and the different and socially marginalized.

Appendix D: Selected Bibliography
                  =================

> "[B]anning books is so utterly hopeless and futile. Ideas don't die because a book is forbidden reading. If [someone] has written the truth, that truth will survive."
>
> -- Gretchen Knief (1939)
> Chief Librarian, Kern County, California, on the banning of Steinbeck's "The Grapes of Wrath"

[You can *always* get copies of these books through "Inter-Library Loans" if your local public library doesn't have a copy. Just ask a librarian on duty about "ILL". The cost is reasonable (sometimes FREE, but usually $2 - $10), and is mostly just for postage.]

Arendt, Hannah. "Men in Dark Times". NY: Harcourt, Brace & World, Inc. (1955).

Bascunan, Rodrigo and Pearce, Christian. "Enter the Babylon System: Unpacking Gun Culture from Samuel Colt to 50 Cent." Toronto: Random House Canada (2007).

The Bible: "God's Word Translation". Holiday, Florida: Green Key (2003).

Borjesson, Kristina, ed. "Into the Buzz-saw: Leading Journalists Expose the Myth of a Free Press". Amherst, NY: Prometheus Press (2002).

Borovoy, Alan, "At the Barricades". Toronto, Ontario: Irwin Law (2013).

Carr, Matthew. "The Infernal Machine: A History of Terrorism". NY: The New Press (2006).

Chambers, Bradford, ed. "Chronicles of Negro Protest". NY: Parents' Magazine Press (1968).

Chase, Kenneth. "Firearms: A Global History to 1700". NY: Cambridge University Press (2003).

"CRC Handbook of Chemistry and Physics", 53rd edition, Weast, Robert, ed. Cleveland, Ohio: Chemical Rubber Co. (1972).

Davidson, James & Lytle, Mark. "After the Fact: The Art of Historical Detection", Vol. 2, 2nd ed. NY: Alfred A. Knopf (1986).

Fellman, Michael. "In the Name of God and Country". New Haven, CT: Yale University Press (2010).

Ferguson, Charles. "Naked to Mine Enemies: The Life of Cardinal Wolsey", Vol. 1 & 2. NY: Time-Life Books (1958).

Gantzel, Klaus and Schwinghammer, Torsten. "Warfare Since the Second World War". London: Transaction Publishers (2000).

Goldacre, Ben. "Bad Science". London: Fourth Estate (2008).

Goodman, Mitchell. "The Movement Towards a New America". NY: Alfred A. Knopf (1970).

Gore, Al. "The Assault on Reason". NY: Penguin Press (2007).

Hoag, David G. "The History of Apollo On-board Guidance, Navigation, and Control". AIAA "Journal of Guidance Control and Dynamics" 6:1 (1982).

Hillsborough, Romulus. "Samurai Sketches: From the Bloody Final Years of the Shogun". San Francisco: Ridgeback Press (2001).

Hughes, Thomas P. "Rescuing Prometheus". NY: Pantheon Books (1998).

King, Charles. "Extreme Politics: Nationalism, Violence, and the End of Eastern Europe". NY: Oxford University Press (2010).

Latimer, Dean and Goldberg, Jeff. "Flowers in the Blood: The Story of Opium". NY: Franklin Watts (1981).

Leonard, Jonathan. "Early Japan". NY: Time-Life Books (1968).

McMahon, Robert J. "The Cold War: A Very Short Introduction". NY: Oxford University Press (2003).

Manchester, William. "The Arms of Krupp". NY: Bantam Books (1964).

Mitchell, James. ed. "Random House Encyclopedia", 3rd edition NY: Random House (1990).

"Modern Steels and Their Properties", 7th edition. Bethlehem, Penn.: Bethlehem Steel Corp. (1972).

Ogawa, Morihiro, ed. "Art of the Samurai: Japanese Arms and Armor, 1156-1868". NY: Metropolitan Museum of Art (2009).

Paglen, Trevor. "Blank Spots on the Map". NY: Penguin Group (2010).

Palmer, R.R. "Twelve Who Ruled: The Year of the Terror in the French Revolution". Princeton: Princeton University Press (1941; reprinted 1989).

Peck, Scott M. "People of the Lie". NY: Simon & Schuster (1983).

Ratti, Oscar and Westbrook, Adele. "Secrets of the Samurai". Boston, Mass.: Charles E. Tuttle (1973); reprinted by Castle Books (1999).

Richardson, Peter. "A Bomb in Every Issue: How the Short, Unruly Life of Ramparts Magazine Changed America". NY: New Press (2009).

Stalker, John. "The Stalker Affair". NY: Penguin (1988).

Street, Arthur and Alexander, William. "Metals in the Service of Man", 10th edition. NY: Penguin Books (1994).

Venetsky, S. "From the Camp Fire to the Plasma". Moscow: Mir Publishers (1989).

Watson, Peter. "A Terrible Beauty: The People and Ideas that Shaped the Modern Mind". London: Orion Books (2000).

Whiteson, Leon. "A Terrible Beauty: An Exploration of the Positive Role of Violence in Life, Culture and Society". Oakville, Canada: Mosaic Press (undated; ca. 2003).

Wills, Gary. "Bomb Power: The Modern Presidency and the National Security State". NY: Penguin Press (2010).

Windholz et al., eds. "The Merck Index", 10th edition. Rahway, NJ: Merck & Co. (1983).

Wright, Mike. "What They Didn't Teach You About the 60s". Novato, California: Presidio Press (2001).

www.ingramcontent.com/pod-product-compliance
Lightning Source LLC
Chambersburg PA
CBHW082318220526
45470CB00008B/2350